Theoretical and Computational Models for Organic Chemistry

NATO ASI Series

Advanced Science Institutes Series

A Series presenting the results of activities sponsored by the NATO Science Committee, which aims at the dissemination of advanced scientific and technological knowledge, with a view to strengthening links between scientific communities.

The Series is published by an international board of publishers in conjunction with the NATO Scientific Affairs Division

A Life Sciences	Plenum Publishing Corporation
B Physics	London and New York
C Mathematical	Kluwer Academic Publishers
and Physical Sciences	Dordrecht, Boston and London
D Behavioural and Social Sciences	
E Applied Sciences	
F Computer and Systems Sciences	Springer-Verlag
G Ecological Sciences	Berlin, Heidelberg, New York, London,
H Cell Biology	Paris and Tokyo
I Global Environmental Change	

NATO-PCO-DATA BASE

The electronic index to the NATO ASI Series provides full bibliographical references (with keywords and/or abstracts) to more than 30000 contributions from international scientists published in all sections of the NATO ASI Series.
Access to the NATO-PCO-DATA BASE is possible in two ways:

– via online FILE 128 (NATO-PCO-DATA BASE) hosted by ESRIN,
Via Galileo Galilei, I-00044 Frascati, Italy.

– via CD-ROM "NATO-PCO-DATA BASE" with user-friendly retrieval software in English, French and German (© WTV GmbH and DATAWARE Technologies Inc. 1989).

The CD-ROM can be ordered through any member of the Board of Publishers or through NATO-PCO, Overijse, Belgium.

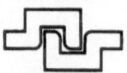

Series C: Mathematical and Physical Sciences - Vol. 339

Theoretical and Computational Models for Organic Chemistry

edited by

Sebastião J. Formosinho
Department of Chemistry,
University of Coimbra, Coimbra, Portugal

Imre G. Csizmadia
Department of Chemistry,
University of Toronto, Toronto, Canada

and

Luís G. Arnaut
Department of Chemistry,
University of Coimbra, Coimbra, Portugal

Springer Science+Business Media, B.V.

Proceedings of the NATO Advanced Study Institute on
Theoretical and Computational Models for Organic Chemistry
Praia de Porto Novo, Portugal
August 26–September 8, 1990

Library of Congress Cataloging-in-Publication Data

Theoretical and computational models for organic chemistry / edited by
    Sebastião J. Formosinho, Imre G. Csizmadia, and Luis G. Arnaut.
        p.   cm. -- (NATO ASI series. Series C, Mathematical and
    physical sciences ; vol. 339)
      Proceedings of the NATO Advanced Study Institute on Theoretical
    and Computational Models for Organic Chemistry, Praia de Porto Novo,
    Portugal, Aug. 26-Sept. 8, 1990.
      "Published in cooperation with NATO Scientific Affairs Division."
    Includes index.
      ISBN 978-94-010-5589-5   ISBN 978-94-011-3584-9 (eBook)
      DOI 10.1007/978-94-011-3584-9
      1. Chemistry, Organic--Data processing--Congresses. 2. Chemistry,
    Organic--Mathematical models--Congresses.   I. Formosinho, Sebastião
    J., 1943-   . II. Csizmadia, I. G. III. Arnaut, Luis G., 1960-   .
    IV. NATO Advanced Study Institute on Theoretical and Computational
    Models for Organic Chemistry (1990 : Praia do Porto Novo, Portugal)
    V. Series: NATO ASI series. Series C, Mathematical and physical
    sciences ; no. 339.
    QD255.5.E4T48 1991
    547'.00285--dc20
                                                        91-18386
                                                          CIP

ISBN 978-94-010-5589-5

Printed on acid-free paper

Table of Contents

PREFACE

The papers in this volume were presented at the NATO Advanced Study Institute held in Porto Novo, Portugal, August 26 - September 8, 1990. The Institute has been able to cover a wide spectrum of the Theoretical and Computational Models for organic molecules and organic reactions, ranging from the *ab initio* to the more empirical approaches, in the tradition established in the previous Institutes at S. Feliu de Guixols (Spain) and Altinoluk (Turkey). The continuity with this work was achieved by inviting half of the lecturers present in those meetings. But other important subjects were also covered at Porto Novo by new lecturers, both from universities and the industry. Molecular Mechanics, Protein Structure and Unidimensional Models were introduced by the first time. The concept of building on the expertise already acquired and available, both in terms of methods and contents, to develop in new directions, was appreciated by participants and lecturers.

The Institute first considered the fundamentals of molecular orbital computations and *ab initio* methods and the construction of Potential Energy Surfaces. These subjects were further explored in several applications related with optimization of equilibrium geometries and transition structures. Practical examples were studied in Tutorial sessions and solved in the computational projects making use of the Gaussian 88 and Gaussian 90 programs.

Empirical models can be complementary to the quantum-mechanical ones in equilibrium geometry optimizations. Thus, the Institute also devoted a considerable amount of time to the presentation of the basis and recent developments of Molecular Mechanics, also explored through computational projects with the MM2 program, and to topological and "template empirical methods" for the design of the protein structures.

Subsequently, the problems of reactivity in small molecules were approached via the "Variational Transition State Theory" and Dynamical Studies. The POLYRATE program was available to the computational laboratory users. To deal with reactivity in larger molecules, both in ground and excited states, or to find trends of reactivity in families of reactions, simpler models have to be employed. A representative sample of such simple methods was presented, ranging from the quantum mechanical Valence Bond Diabatic Methods of Evans, to state correlation diagrams, and to classical models based on the intersection of potential energy curves. Structures-reactivity relationships were also discussed. Emphasis was given to the role of solvent on chemical reactivity and to the dynamical processes occurring in a very short time scale.

This course was centered in the use of computational tools. The participants had direct and unlimited access to a Convex-120 machine with 16 terminals, to 2 clustered VAX Stations with 4 terminals, one IBM PS 2 and Macintoshes. Adequate service was provided both by the tutors and by Convex and Digital personnel. 28 groups opened accounts for the Convex; they used a total of 182 CPU hours and requested a total memory of 6.02×10^6 KBytes. The VAX stations were used virtually 100% during operation time.

We would like to express our gratitude to the NATO Science Division and to JNICT for the award of grants which enabled the meeting to be held, and to CONVEX and DIGITAL for their outstanding support to the computational laboratory.

Organizing Committee

Prof. S.J. Formosinho (Director)

Departamento de Química
Universidade de Coimbra
3049 Coimbra Codex
Portugal

Prof. N. L. Allinger

Department of Chemistry
University of Georgia
Athens, Georgia 30602
United States

Prof. I. G. Csizmadia

Department of Chemistry
University of Toronto
80 St. George St.
Toronto, Ontario M5S 1A1
Canada

Dr. L. G. Arnaut (Co-director)

Departamento de Química
Universidade de Coimbra
3049 Coimbra Codex
Portugal

List of Lecturers

Prof. N. Agmon

Department of Physical Chemistry
The Hebrew University
Jerusalem 91904
Israel

Prof. N. L. Allinger

Department of Chemistry
University of Georgia
Athens, Georgia 30602
United States

Prof. F. Bernardi

Dipartimento di Chimica "G. Ciamician"
Univ. di Bologna
Via Selmi 240126 Bologna
Italy

Prof. J. Bertrán

Dipartament de Química
Universitat Autònoma de Barcelona
08193 Bellaterra (Barcelona)
Spain

Prof. M.J. Calhorda

Centro de Química Estrutural
Instituto Superior Técnico
1096 Lisboa Codex
Portugal

Prof. I. G. Csizmadia

Department of Chemistry
University of Toronto
80 St. George St.
Toronto, Ontario M5S 1A1
Canada

Prof. S.J. Formosinho

Departamento de Química
Universidade de Coimbra
3049 Coimbra Codex
Portugal

Dr. B. C. Garrett

Molecular Science Research Center
Battelle Pacific Northwest Laboratory
Battelle Boulevard
Richland, Washington 99352
United States

Dr. M. R. Lopez

Laboratoire de Chimie Theorique
Université de Nancy I
B.P. 239
54506 Vandoeuvre-Les- Nancy Cedex
France

Dr. G. M. Maggiora

Director of Computational Chemistry
The Upjohn Company
Kalamazoo, Michigan 49001
United States

Prof. P.G. Mezey

Department of Chemistry and Chemical Engineering
University of Saskatchewan
Saskatoon S7N 0N0
Canada

Prof. J. Michl

Center for Structure and Reactivity
Department of Chemistry
University of Texas at Austin
Austin, Tx. 78712-1167
United States

Prof. C. Nicolaides

Theoretical Physical Chemistry Institute
National Hellenic Research Foundation
48, Vassileos Constantinou Av.
Athens 116-35
Greece

Prof. C. Ogretir

Chemistry Department
Anadolu University
26470 Eskisehir
Turkey

Prof. M.A. Robb

Department of Chemistry
King's College
London, Strand, London WC2R 2LS
United Kingdom

Prof. H.B. Schlegel

Department of Chemistry
Wayne State University
Detroit, Michigan 48202
United States

Prof. A.J.C. Varandas

Departamento de Química
Universidade de Coimbra
3049 Coimbra Codex
Portugal

List of Tutors

Dr. Gustavo Arteca

Department of Chemistry
Univ. of Saskatchewan
Saskatoon, Saskatchewan S7N 0W0
Canada

Dr. A. Bottoni

Dipartimento di Chimica
Univ. di Bologna
Via Selmi, 2
40126 Bologna
Italy

Dr. M. Duran

Dep. de Química
Fac. de Ciències
Univ. Autònoma de Barcelona
08193 Bellaterra
Spain

Dr. R. Fausto

Departamento de Química
Universidade de Coimbra
3000 Coimbra
Portugal

Dr. Mike Peterson

Department of Chemistry
Univ. of Saskatchewan
Saskatoon, Saskatchewan S7N 0W0
Canada

Dr. Henry Thomas

Department of Chemistry
University of Georgia
Athens, Georgia 30602
United States

List of Participants

Mr. David Amarilio

N.C.R. "Democritos"
GR-15310, Ag. Paraskevi
Athens P.O. Box 60228
Greece

Dr. Victorya Aviyente

Chemistry Department
Bosphorous University
80815 Bebek
Istambul
Turkey

Dr. Alessandro Bagno

CNR, Meccanismi Reazioni Organiche
Via Marzolo 1
I-35131 Padova
Italy

Prof. Nurettin Balcioglu

Department of Chemistry
Beytepe Campus
Hacettepe University
Ankara
Turkey

Dr. Laura Bonati

Dip. di Chimica Fisica e Elettrochimica
Univ. di Milano
Via Golgi 19
20133 Milan
Italy

Mr. L. Batista Carvalho

Departamento de Química
Universidade de Coimbra
3000 Coimbra
Portugal

Dr. A. Perciras Castro

Dep. de Quimica Fisica
Fac. Quimica
Univ. Santiago
15706 Santiago de Compostela
Spain

Prof. Emine Cebe

Physics Department
Faculty of Science
Uludag University
Bursa
Turkey

Prof. Mustafa Cebe

Chemistry Department
Uludag University
Bursa
Turkey

Dr. Paulo Ribeiro Claro

Departamento de Química
Universidade de Coimbra
3000 Coimbra
Portugal

Mr. Ugo Cosentino

Dip. di Chimica Fisica e Elettrochimica
Univ. di Milano, Via Golgi 19
20133 Milan
Italy

Mrs. M. Céu Costa

DTIQ, LNETI
Estrada das Palmeiras
2745 Queluz
Portugal

Dr. Günay Demirel

Department of Chemistry
Middle East Technical University
06531 Ankara
Turkey

Ms. Özden Özel Evin

Department of Chemistry
Middle East Technical University
06531 Ankara
Turkey

Mr. Adelino Galvão

Centro de Química Estrutural
Instituto Superior Técnico
1096 Lisboa Codex
Portugal

Dr. Elisheva Goldstein

Chemistry
California State Polytechic Univ.
3801 West Temple Av, Pomona CA 91768-4032
United States

Dr. Gülümser Gündogan

Middle East Tecnical University
Department of Chemistry
06531 Ankara
Turkey

Dr.Vera Kolb

Department of Chemistry
University of Wisconsin-Parkside
Kenosha, WI 53141
United States

Mr. Andreas Köster

Theoretische Chemie
University of Hannover
Am Kleinen Felde 30
3000 Hannover 1
Germany

Mr. Michel Loos

Laboratoire de Chimie Théorique
Univ. de Nancy I
54506 Vandoeuvre-Les-Nancy Cedex
France

Mr. J.-F. Marcoccia

Dep. of Chemistry
Univ. of Toronto
80 St. George St.
Toronto, Ontario M5S 1A1
Canada

Ms. M. J. Seixas Melo

Centro de Tecnologia Química e Biológica
R. Quinta Grande, 6
2780 Oeiras
Portugal

Dr. Javier Modrego

Inorganic Chemistry Laboratory
Univ. of Oxford
South Parks Rd.
Oxford, OX1 3QR
United Kingdom

Dr. Giorgio Moro

Dip. di Chimica Fisica e Elettrochimica
Univ. di Milano
Via Golgi 19
20133 Milan
Italy

Mr. George Mousdis

National Hellenic Research Foundation
48 Vassileos Constantinou Ave.
Athens 11635
Greece

Mr. J. J. Soares Neto

Department of Chemistry
Aarhus University
DK 8000 Aarhus C
Denmark

Dr. Massimo Olivucci

Department of Chemistry
King's College London Strand
London WC2R 2LS
United Kingdom

Mr. A. Canelas Pais

Departamento de Química
Universidade de Coimbra
3000 Coimbra
Portugal

Mr. Ian Palmer

Department of Chemistry
King's College London Strand
London WC2R 2LS
United Kingdom

Ms. Ana Paula Paiva

Departamento de Química
Universidade de Lisboa
R. E. Vasconcelos, Ed. C1, 5º piso
1700 Lisboa
Portugal

Prof. N. B. Peynircioglu

Department of Chemistry
Middle East Technical University
06531 Ankara
Turkey

Mr. Joey Pierce

Department of Chemistry
University of Texas at Austin
Austin Tx 78712
United States

Dr. Fernando Pina

Fac. Ciências e Tecnologia
Univ. Nova de Lisboa, Qta. Torre
2825 Monte da Caparica
Portugal

Mr. Ioannis Ragazos

Department of Chemistry
King's College London Strand
London WC2R 2LS
United Kingdom

Dr. M. Mar Reguero

Department of Chemistry
King's College London Strand
London WC2R 2LS
United Kingdom

Ms. M. I. M. Rodriguez

Dep. de Quimica Fisica y Analitica
Univ. Oviedo, Julian Claveria s/n
33006 Oviedo
Spain

Dr. M. Amélia Santos

C.Q.E., Instituto Superior Técnico
Av. Rovisco Pais
1096 Lisboa Codex
Portugal

Dr. Zheng Shi

Department of Chemistry
Dalhousie Univ.
Halifax, Nova Scotia B3H 4J3
Canada

Mr. Alan Smith

Chemical Laboratory
University of Kent at Canterbury
Kent CT2 7NH
United Kingdom

Dr. Canan Ünaleroglu

Department of Chemistry
Beytepe Campus
Hacettepe University
06532 Ankara
Turkey

Dr. G. Suzzi Valli

Dipartimento de Chimica "Ciamician"
Via Selmi 2
40126 Bologna
Italy

Dr. Tereza Varnali

Bogaziçi University
F. E.F Kimya Bebek
80815 Istanbul
Turkey

Dr. A. Venturini

CNR, I. Co. C.E.A.
Via della Chimica 8
40064 Ozzano Emilia (BO)
Italy

Mr. Luís Veiros

C.Q.E. Grupo II, Complexo I
Instituto Superior Técnico
1096 Lisboa Codex
Portugal

Dr. L. Vázquez Vidal

Dep. de Química Física, Fac. Quimica
Univ. Santiago
15706 Santiago de Compostela
Spain

Ms. Wladia Viviani

Laboratoire de Chimie Théorique
Univ. de Nancy I
54506 Vandoeuvre-Les-Nancy Cedex
France

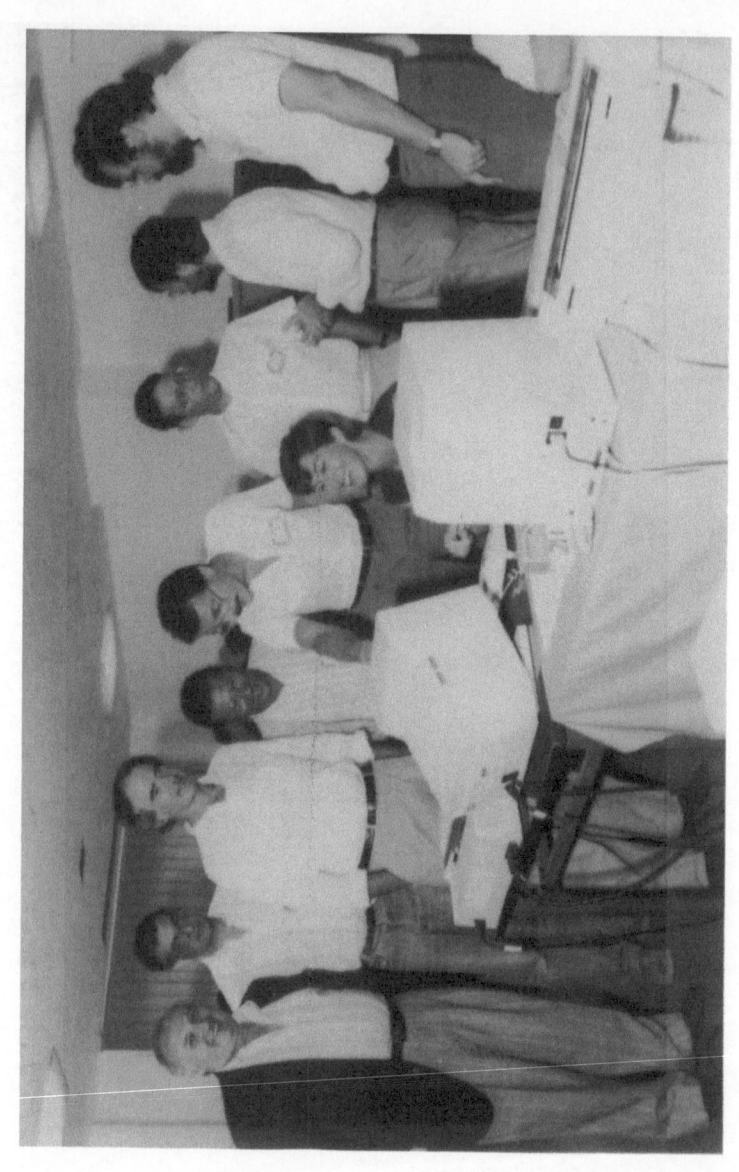

CHEMISTRY AS AN EXACT SCIENCE

I. G. CSIZMADIA
Department of Chemistry
University of Toronto
Toronto, Ontario
Canada M5S 1A1

ABSTRACT. The utility of potential energy surfaces (PES) as a conceptual tool is discussed in the light of historical development that changed chemistry from an empirical science to an exact science.

One does not always notice a dramatic change. For example, on several occasions I flew from North-America to Asia and, by necessity, I crossed the "International Data Line" somewhere between Honolulu and Hong Kong. At one point in the Earth's atmosphere I was in a particular day, let's say in Tuesday and a micrometer further I was in Wednesday. I have lost a day yet I did not notice it! Similary, when every year on December 21 in a given instant we pass over to the astronomical winter, we do not really notice it. Alternatively, when on October 12 in 1492 one of the sailors in the observation basket on Columbus' flag ship cryed out loud: LAND, we have arrived at the New Age, leaving the medieval times behind. Yet in the courts of Europe nobody noticed, at that time, the significance of that event. Analogously in the history of science there are dramatic changes and yet they go unnoticed by the majority of scholars for quite some time.

The 50 year period between 1920 and 1970 may be regardeed as the time slot when chemistry changed from an *empirical* to an *exact* science. The change influenced organic chemistry the most. Needless to say, that this dramatic and history shaping change went unnoticed among many organic chemists.

Although semiempirical theoretical organic chemistry has a relatively longer history, *ab initio* theoretical organic chemistry is only 20 years old. Using the POLYATOM programme system I myself carried out the first *ab initio* Gaussian molecular computation on the first organic molecule (HCOF) in the 1963 / 64 academic year. Although the preliminary results were announced in 1963 in the MIT - SSMTG Quarterly Progress Report the complete work has not been published until 1966. From that point onward molecular Gaussian computations became numerous and by 1970 Professor Pople announced his new computer program system called GAUSSIAN70. Hence a new epoch arrived to the field of Organic Chemistry.

The change in the 50 year period (1920 - 1970) is schematically illustrated in Figure 1.

Many times, a mathematical object have been proved to be invaluable in advancing our chemical understanding even if that mathematical object represented an over-simplification of reality. Take for example the Atomic Orbitals (AO). They are, by every

1

S. J. Formosinho et al. (eds.), Theoretical and Computational Models for Organic Chemistry, 1–3.
© 1991 *Kluwer Academic Publishers.*

2

definition, mathematical objects. They are numerical solutions for the atomic Hartree-Fock equations. This means that they are the result of a mathematical simplification where a one-electron problem (Hartree-Fock equation) is solved instead of a many-electron problem (Schrodinger equation). The atomic orbitals, or single-centered one-electron functions, are nevertheless bona fide, genuine mathematical objects. We may use, from time to time, these mathematical objects to construct multicentred one-electron functions or molecular orbitals (MO) by linear combinations or we may construct from these atomic orbitals valence bond (VB) structures in the form of Slater determinants. Nevertheless, in all these activities we treat atomic orbitals (AO) as mathematical objects. However, beyond these practical utilities, atomic orbitals serve, additional, conceptual purposes.

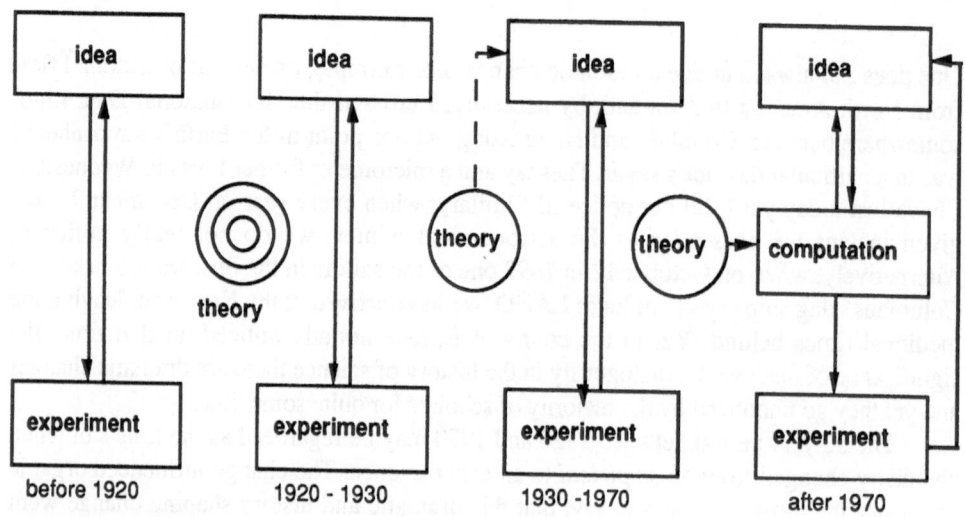

Figure 1. The half century development of chemistry from the state of "Empirical Science" (extreme left) to the state of "Exact Science" (extreme right).

Chemists got a lot of milage out of the concept of atomic orbitals during the past several decades. The concept of orbital overlap in determining bond strength, orbital ionization potentials in identifying various excited states of ions and orbital symmetries in formulating selection rules for electronic excitations are only a few of the most prominent examples. Similarly, molecular orbitals are used in assigning molecular electronic transitions as well as molecular ionization states, not to mention the orbital symmetry rules of Woodward and Hoffmann. All in all, orbitals were more than just mathematical objects in the past several decades. They, in fact, became the conceptual tools of chemists in advancing our understanding in numerous areas of chemistry.

Similarly, we may consider potential energy surfaces as mathematical objects. Potential energy surfaces are artifacts of the Born-Oppenheimer approximation. When one solves the combined electronic and nuclear Schrodinger equation one obtains only energy levels. However, in the fixed nuclear approximation the potential energy surface, as a function of nuclear motion, emerges. So, far all practical purposes the potential energy surface is a mathematical object that is the result of a mathematical simplification of the overall problem. Yet, in the conceptual sense a potential energy surface is more than a mathematical object.

In fact, we might say, without too much exaggeration, that chemistry became an exercise on potential energy surfaces. When, for example, a synthetic organic chemist isolates a certain new compound he proves that a minimum exists for the molecular composition of that molecule he isolates. When the X-ray crystallographic determines its molecular structure he identifies the position of the minimum on that potential energy surface by the molecular geometry. When a vibrational spectroscopist analyses the ir spectrum of that certain new compound he determines the steepness of the potential energy surface around its minimum. If the molecule exists in another isomeric from then we know that the surface must have another minimum. When the thermodynamicist determines the equilibrium constant for the isomerization as the function of temperature then he can make a statement about the relative stabilities of the two isomers. When the pysical organic chemist is studying the isomerization reaction of that certain new compound he is making an exercise on that potential energy surface. Through his kinetic measurements he may determine the height of the barrier that exists between the initial minimum and the minimum associated with the product. In this mechanistic statement he may refer to the path that inter-connects these two minima and passes through one or more transition states (saddle points) of the potential energy surface. All of these are events on a particular electronic ground potential energy surface. Spectroscopists, however, studying the electronic uv-vis spectra of that certain new organic compound are really investigating the transition from its electronic ground potential energy surface to low lying electronic excited state potential energy surfaces. Finally, photochemists are studying reaction paths involving at least two (the ground and excited) potential energy surfaces of that system. All in all, most chemical phenomena is best understood in terms of potential energy surfaces.

Apropriately therefore, the present volume illustrates many of the mathematical and computational features of potential energy surfaces as well as their wide-spread conceptual utility in explaining chemical phenomena.

Computational Bottlenecks in Molecular Orbital Calculations

H. Bernhard Schlegel
Dept. of Chemistry
Wayne State University
Detroit, MI 48202

Michael J. Frisch
Lorentzian, Inc.
127 Washington Ave.
North Haven, CT 06473

ABSTRACT This Chapter examines some of the major steps in molecular orbital calculations, such as energies and gradients at SCF, MCSCF, MPn, CI, CC and QCI levels and SCF second derivatives. The emphasis is on the computational requirements rather than the details of the theory. Specifically cpu times, memory size and disk usage are considered for integral evaluation and transformation, solution of the SCF, CPHF and configuration interaction problems, and calculation of energy derivatives. Conventional, direct and semi-direct algorithms are compared and some guidelines for choosing between them are given.

1. Introduction

Over the past decade, there have been rapid developments in ab initio molecular orbital calculations that have allowed these methods to be applied to larger molecules. With currently available computers, SCF calculations can be carried out for molecules with over 500 basis functions and MP2 calculations for molecules with up to 400 basis functions. These advances come primarily from the use of direct methods, as well as the increase in speed of affordable computers. Direct methods, in turn, have benefitted from improved algorithms for calculating two electron integrals. There have also been advances in both the speed and accuracy of geometry optimizations because of the availablitity energy derivatives, especially for levels of theory that include electron correlation. In this Chapter, we will look at the computation of energies and energy derivatives for SCF, MCSCF, MPn, CI, CC and QCI calculations. The emphasis is on the computational requirements of the various steps in MO calculations and how these costs increase with the size of the molecule. The appropriate equations for the various levels of theory are outlined briefly to show where the computational bottlenecks arise. Details of the theory, however, are left to the text books[1-6] and the original literature[7-44].

S. J. Formosinho et al. (eds.), Theoretical and Computational Models for Organic Chemistry, 5–33.
© 1991 *Kluwer Academic Publishers.*

2. Background

In the non-relativistic approximation for the time independent case, the wavefunction for an n electron system satisfies the Schrödinger equation:

(1) $\hat{H}\Psi = E\,\Psi$

where the Hamiltonian is given by:

(2) $\hat{H} = \sum_{i=1}^{n} [-\frac{1}{2}\nabla_i^2 - \sum_{A}^{nuclei} \frac{Z_A}{r_{iA}}] + \sum_{i<j}^{n} \frac{1}{r_{ij}} + \sum_{A<B}^{nuclei} \frac{Z_A Z_B}{r_{AB}}$

where Z_A is the charge on nucleus A. Except for very simple systems, the Schrödinger equation cannot be solved in closed form; hence various approximations must be made. For an approximate wavefunction, the variational principle gives an upper bound to the energy:

(3) $E_{var} = \langle \Psi|\hat{H}|\Psi \rangle / \langle \Psi|\Psi \rangle > E_{exact}$

Adjustable parameters in the approximate wavefunction Ψ can be varied to minimize E_{var}, yielding an improved energy and wavefunction. The Born-Oppenheimer approximation[7] (clamped nuclei approximation) is used to separate the electronic and the nuclear motions. Further restrictions on the form of the electronic wavefunction and the manner in which the adjustable parameters are determined are governed by level of theory used.

In the Hartree-Fock (HF) model[8], the n electron wavefunction is written as a single Slater determinant[9] of n othonormal spin orbitals, ϕ_i:

(4) $\Psi_0 = \frac{1}{\sqrt{n!}} |\,\phi_1\,\phi_2\cdots\phi_n\,|$

As discussed below, in the Hartree-Fock model each electron sees only the average field of the other electrons. In reality, the electrons must explicitly avoid each other because of their mutual coulombic repulsion; hence their motions are correlated. The difference between the Hartree-Fock energy and the exact, non-relativistic energy is termed electron correlation energy. The Hartree-Fock wavefunction can be improved by taking a linear combination of Slater determinants, yielding a configuration interaction (CI) wavefunction:

(5) $\Psi = a_0\,\Psi_0 + \sum_{ia} a_i^a\,\Psi_i^a + \sum_{\substack{i<j \\ a<b}} a_{ij}^{ab}\,\Psi_{ij}^{ab} + \cdots$

where Ψ_i^a, Ψ_{ij}^{ab}, etc. are Slater determinants in which occupied spin orbital ϕ_i, ϕ_j etc. in the reference determinant Ψ_0 are replaced by unoccupied or virtual spin orbitals ϕ_a, ϕ_b etc. (i.e. determinants that are singly excited, doubly excited, etc.). The amplitudes, a_i^a, a_{ij}^{ab} etc. can be determined variationally, using eq (3) or by perturbation theory, as will be outlined in subsequent sections. If all possible excited configurations are included (full CI), this approach converges (slowly) to the exact wavefunction.

The spin orbitals are constructed from a linear combination of atomic orbitals (LCAO):

(6) $\qquad \phi_p = \sum_\mu C_{\mu p} \chi_\mu \alpha \quad \text{or} \quad \phi_p = \sum_\mu C_{\mu p} \chi_\mu \beta$

where α and β are spin functions and χ_μ are basis functions for the spatial part of the spin orbitals. For simplicity, the formulae presented here will be in terms of spin orbitals; in practice integration over spin reduces the formulae to expressions for each different spin case and involving only spatial orbitals. The basis functions χ_μ are themselves linear combinations of gaussian functions[10] (Slater functions[11] can also be used but the necessary integrals are much more difficult to calculate). Because the computational times depend strongly on the number of basis functions (see Table 1), economic reasons usually place severe restrictions on the number and type of basis functions that can be used.

For any of the levels of theory considered in this Chapter, the total electronic energy corresponding to the Hamiltonian given in eq 2 and a wavefunction such as eq 4 or 5, can be written in terms of integrals over molecular orbitals or directly in terms of integrals over basis functions

(7) $\qquad E = \sum_{pq} P_{pq} h_{pq} + \sum_{pqrs} \Gamma_{pqrs} (pq|rs) + V_{nuc} = \sum_{\mu\nu} P_{\mu\nu} h_{\mu\nu} + \sum_{\mu\nu\lambda\sigma} \Gamma_{\mu\nu\lambda\sigma} (\mu\nu|\lambda\sigma) + V_{nuc}$

where P and Γ are the one and two particle density matrices and V_{nuc} is the nuclear repulsion energy. The one and two electron integrals over basis functions are

(8) $\qquad h_{\mu\nu} = \int \chi_\mu^* \left[-\frac{1}{2}\nabla^2 - \sum_A^{nuclei} \frac{Z_A}{r_A} \right] \chi_\nu \, d\tau \; ; \quad S_{\mu\nu} = \int \chi_\mu^* \chi_\nu \, d\tau$

(9) $\qquad (\mu\nu|\lambda\sigma) = \int \int \chi_\mu^*(1) \chi_\nu(1) \frac{1}{r_{12}} \chi_\lambda^*(2) \chi_\sigma(2) \, d\tau_1 \, d\tau_2$

Since the molecular orbitals are expressed as linear combinations of basis functions, the integrals over molecular orbitals can be obtained by a linear transformation:

(10) $\qquad h_{pq} = \int \phi_p^* \left[-\frac{1}{2}\nabla^2 - \sum_A^{nuclei} \frac{Z_A}{r_A} \right] \phi_q \, d\tau = \sum_{\mu\nu} C_{\mu p}^* C_{\nu q} h_{\mu\nu}$

$$(11) \quad (pq|rs) = \int \int \phi_p^*(1) \, \phi_q(1) \, \frac{1}{r_{12}} \, \phi_r^*(2) \, \phi_s(2) \, d\tau_1 \, d\tau_2 = \sum_{\mu\nu\lambda s} C_{\mu p}^* \, C_{\nu q} \, C_{\lambda r}^* \, C_{\sigma s} \, (\mu\nu|\lambda\sigma)$$

A number of standard conventions are used in this Chapter to simplify the notation. Greek subscripts refer to basis functions, e.g. χ_μ. Indices i, j, k $\cdots$ run over occupied molecular orbitals; a, b, c $\cdots$ run over unoccupied or virtual orbitals; p, q, r $\cdots$ run over both. Unless otherwise stated the sums run over these implied ranges. The computational resources required for the various calulations depend strongly on the number of functions used in the expansion of the wavefunctions. The number of basis functions is N; the number of occupied orbitals is O, and the number of unoccupied or virtual orbitals is V (since some orbitals may be frozen, $O + V \leq N$). The number of first derivatives with respect to the atom positions is N_a (equal to 3 times the number of atoms). The notation $O(n)$ indicates a number of the order of n, e.g. there are formally $O(N^4)$ two electron integrals. Table 1 outlines the formal and the actual size dependence of various electronic structure calculations on N, O, V and N_a. These will be discussed in greater detail as each level of theory is considered in turn.

3. Integrals

All molecular orbital calculations require the evaluation of integrals over the basis functions (for leading references on methods for evaluating these integrals, see ref. 12). The one electron integrals are not time-consuming since there are only $O(N^2)$ of them. On the other hand, the two electron integrals are quite expensive to compute since there are $O(N^4)$ of them. Permutational symmetry reduces the number of integrals by a factor of 8 ($\mu \leftrightarrow \nu$, $\lambda \leftrightarrow \sigma$ and $\mu\nu \leftrightarrow \lambda\sigma$ in $(\mu\nu|\lambda\sigma)$); if available, spatial symmetry can reduce the number of integrals further. One of the major choices to be made in molecular orbitals calculations is how to calculate and store the numerous two electron integrals. There are four possible approaches:

_AO_   The two-electron integrals over the atomic orbitals are generated once and stored externally (on disk). This is the approach used by traditional SCF calculations. Formally this requires $N^4/8$ cpu and disk; in practice the work and disk space grows as $O(N^{3.5})$ because integrals that are near zero can be eliminated before calculation.

_MO_   The atomic orbital integrals are generated once and stored externally, then transformed to the molecular orbital basis. The transformed (MO) integrals are also stored externally. The four index transformation, eq 11, requires N^5 work and N^4 disk storage (see the Integral Transformation section). This is the approach used until recently for all correlated energy methods.

In-Core   The atomic orbital integrals are generated once and stored in canonical order in main memory (i.e., including zeroes). This requires _large_ amounts of memory - $N^4/8$

Table 1: Size-Dependance of Electronic Structure Methods

Method	Formal			Actual	
	CPU	Memory	Disk	CPU	Disk
Conventional SCF	N^4	N^2	N^4	$N^{3.5}$	$N^{3.5}$
In-core SCF	N^4	N^4	----	N^4	N^2
Direct SCF	N^4	N^2	----	$N^{2.7}$	N^2
Conventional MP2	ON^4	N^2	N^4	ON^4	N^4
Direct MP2 Energy	ON^4	OVN	----	O^2N^3	N^2
Semidirect MP2 Energy	ON^4	N^2	VN^2	O^2N^3	VN^2
Conventional MP2 Grad	ON^4	N^2	N^4	ON^4	N^4
Direct MP2 Grad	ON^4	N^3	----	ON^4	N^2
Semi-direct MP2 Grad	ON^4	N^2	N^3	ON^4	N^3
MP3, CISD, QCISD	O^2N^4	N^2	N^4	O^2N^4	N^4
MP4, QCISD(T)	O^3V^4	N^2	N^4	O^3V^4	N^4
Full CI	$((O+V)!/O!V!)^2$				

N = number of basis functions, O = number of occupied orbitals correlated, V = number of vitrual orbitals

words. Keeping the full set of two electron integrals in memory allows the integrals to be processed using matrix operations and eliminates all of the I/O. Consequently this approach is very fast; however, the N^4 memory dependence limits it to small molecules and/or large memory machines.

Direct   The atomic orbital integrals are recomputed as needed. This does not require $O(N^4)$ internal or external storage but does involve additional computational effort. For large molecules, substantial savings are possible which compensate for this additional effort. However, direct methods are the only choice when memory and disk are exhausted and consequently are inevitably used for the largest calculations.

4. Integral Transformation

Calculations that go beyond the Hartree-Fock approximation require the transformation of the one and two electron integrals over the atomic basis to integrals over the molecular orbitals. The one electron integrals are few in number, hence are easy to transform (eq 10, $O(N^3)$ work and $O(N^2)$ storage). The transformation of the two electron integrals is more challenging[13]. A full transformation of the two electron integrals requires $O(N^5)$ work, provided that the indices are transformed one at a time. However, many levels of theory beyond Hartree-Fock do not need explicitly all N^4 two electron integrals in the MO basis; it is sufficient to form (iplqr), where one of the indices runs only over the occupied MO's. This reduces the work to $O(ON^4)$ and the storage to $O(ON^3)$.

There are several choices for the manner in which the transformation is carried out, depending on how the AO integrals are handled and how much memory is available.

Unsorted AO integrals   Because integrals are written to disk in the order that they ar generated and only the non-zero integrals are written, the transformation of the first index cannot be vectorized. The transformation of the remaining indices is vectorizable.

Sorted AO integrals   If the integrals can be sorted into canonical order (with zeros included), then the entire transformation is vectorizable. The cpu time for the sort is small but the I/O demands are heavy unless enough memory and disk are available. If the two electron integrals can be held in core, they can be generated in canonical order.

If there is $O(N^3)$ memory, the partially transformed integrals, (ivl$\lambda\sigma$) for one or more i's can be held in core and no I/O is necessary for the transformation of the remaining indices. If only $O(N^2)$ memory is available, then the half transformed integrals, (ipl$\lambda\sigma$) are accumulated on disk before the last two indices are transformed.

5. Hartree-Fock and MCSCF Energies

The Hartree-Fock approximation obtained by using a single determinantal wavefunction, eq 4, constructed from molecular orbitals given by eq 6, in the expression for the variational energy, eq 3, and minimizing the energy with respect to the molecular orbital coefficients, $C_{\mu i}$.

(12) $E_{HF} = \langle \Psi_0 | \hat{H} | \Psi_0 \rangle / \langle \Psi_0 | \Psi_0 \rangle$; $\dfrac{\partial E_{HF}}{\partial C_{\mu i}} = 0$

This leads to the matrix form of the Hartree-Fock equations, i.e. the Roothaan-Hall equations[14]:

(13) $\displaystyle\sum_v F_{\mu v} C_{vi} = \varepsilon_i \sum_v S_{\mu v} C_{vi}$; $F_{\mu v} = h_{\mu v} + \sum_{\lambda\sigma} [(\mu v | \lambda\sigma) - (\mu\sigma | \lambda v)] P_{\lambda\sigma}$; $P_{\lambda\sigma} = \sum_i C_{\lambda i}^* C_{\sigma i}$

(14) $\displaystyle E_{HF} = \sum_{\mu v} P_{\mu v} h_{\mu v} + \frac{1}{2} \sum_{\mu v \lambda\sigma} [P_{\mu v} P_{\lambda\sigma} - P_{\mu\sigma} P_{\lambda v}] (\mu v | \lambda\sigma) + V_{nuc}$

Given the Fock matrix, $F_{\mu v}$, the MO coefficients can be found from eq 13 by standard diagonalization methods. Since the Fock matrix, depends on the MO coefficients through the density matrix, $P_{\lambda\sigma}$, these equations must be solved iteratively until self consistency is achieved[15] (hence, this is also known as the self consistent field (SCF) method). In effect, each spin orbital is optimized in the average field of all the other orbitals. Hartree-Fock calculations come in a number of flavors, depending on the constraints placed on the molecular orbitals. In restricted Hartree-Fock (RHF) calculations, the alpha (spin up) and the beta (spin down) MO coefficients are the same, whereas in the unrestricted Hartree-Fock (UHF) method, the alpha and beta MO coefficients are allowed to vary independently.

Some systems require a linear combination of determinants for a qualitatively correct description of the electronic wavefunction, e.g. homolytic bond breaking. The Hartree-Fock method can be extended by replacing the single Slater determinant by a linear combination of Slater determinants, eq 5, resulting in a multi-configurational self consistent field (MCSCF) method. The variational energy is minimized with respect to both the MO coefficients, $C_{\mu i}$, and the CI coefficients, a_i, appearing in eq 5. The CI coefficients can be determined in a manner similar to the conventional CI method (see below); the MO coefficients can be determined from a set of Fock-like matrices. As in the Hartree-Fock approach, the equations must be solved iteratively and the most time-consuming part of this process (for small CI expansions) is the construction of the Fock-like matrices. Details of MCSCF methods can be found elsewhere[16].

The performance issues which arise for SCF calculations concern how and when the integrals are to be generated, and whether and in what format they are to be stored. A flow chart of the conventional SCF method is shown in Figure 1a. The two-electron integrals are computed once and stored externally. The wavefunction is then optimized iteratively, with each iteration consisting of forming a Fock matrix from the stored integrals and the density from the current wavefunction, and then diagonalizing the Fock matrix. Diagonalization requires only $O(N^3)$ work and thus is not a rate determining step. However, both the generation of integrals and formation of the Fock matrix formally require $O(N^4)$ arithmetic operations, external storage, and I/O since there are $O(N^4)$ two electron integrals. For larger systems, many small integrals need not be computed, and empirically the calculations grow in cost as about $N^{3.5}$. On most systems, the amount of available external storage is as much of a limitation on the size of conventional SCF calculations as the availability of cpu cycles.

The primary alternative is direct SCF, pioneered by Almlöf and coworkers[17], illustrated in Figure 1b. Here the two-electron integrals are recomputed as needed (once per iteration). This eliminates the need for external storage and the associated I/O. It might seem that direct SCF would be preferred only when disk space is insufficient, but this is not the case. Since the density matrix is available when the integrals are computed, the determination of which integrals can be neglected can take into account small density matrix elements, resulting in many more integrals being discarded. Furthermore, only the change

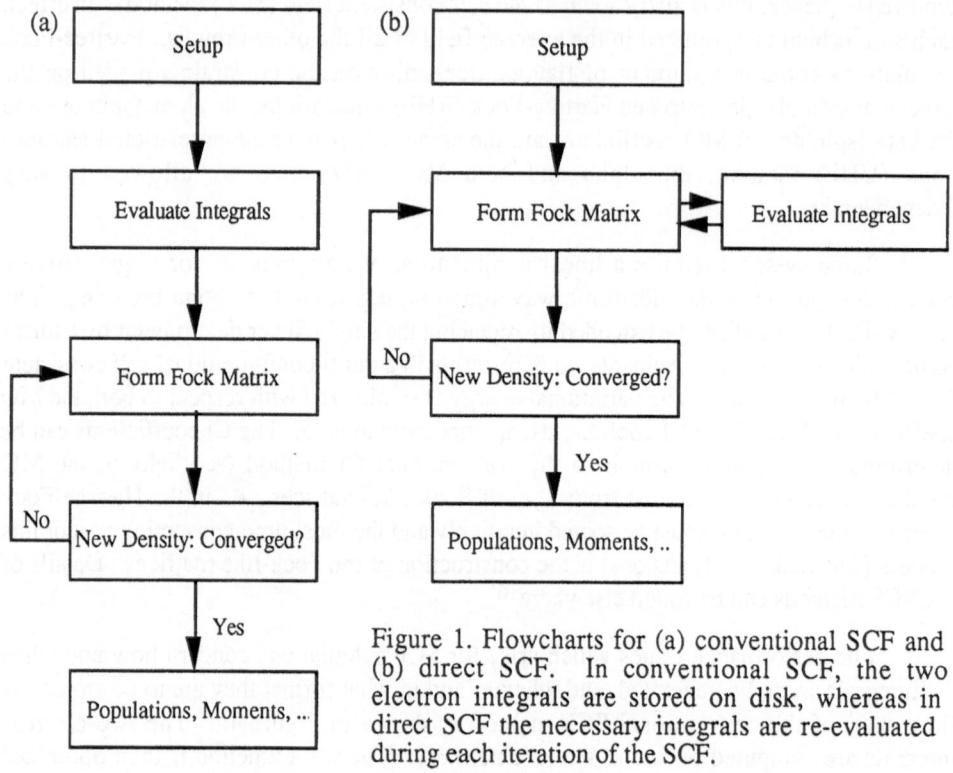

Figure 1. Flowcharts for (a) conventional SCF and (b) direct SCF. In conventional SCF, the two electron integrals are stored on disk, whereas in direct SCF the necessary integrals are re-evaluated during each iteration of the SCF.

in Fock matrix need be computed each iteration, so as convergence is approached and the changes in the density matrix become small, even more integrals can be neglected. The result is that the CPU time required for direct SCF rises as about $N^{2.7}$. Consequently, a point is reached at which it actually consumes less CPU time to recompute the integrals as they are needed every time rather than compute all the integrals and store them externally. Where this crossover occurs depends on the speed of the integral evaluation in direct SCF and varies with the integral algorithms used as well as the type of computer involved. Thus there are several distinct cases where direct SCF is to be preferred over conventional SCF:

1. When storing the integrals exceeds available disk capacity.

2. When the calculation is being run on a workstation with a fast cpu but slow disks, so that I/O wait times are prohibitive.

3. When the calculation is big enough that the use of cutoffs and inherent efficiency in the direct SCF procedure more than compensates for the recomputation of integrals. This crossover point, above which all SCF calculations should be run direct even if sufficient disk space is available, is about 90 basis functions using *Gaussian 90*[18] on

vector machines such as the Cray or Convex, but about 180 basis functions on scalar machines such as the Sun or VAX.

Hence, with typical machine configurations one runs conventional SCF until out of disk space on scalar machines, runs all medium to large jobs direct on vector machines, and has a tradeoff for larger jobs on workstations.

The other alternative strategy for SCF calculations is to store the integrals in main memory rather than externally. This in-core SCF procedure eliminates the external storage and I/O while maximizing the speed of Fock matrix formation. The integrals can be generated and stored in memory in canonical order without any sorting. Formation of the Fock matrix is then a simple matrix operation, which is fast on vector machines. The principle limitation of in-core SCF is the requirement of $N^4/8$ words of main memory for the integrals. This translates to 100 Megabytes of memory for 100 basis functions, which can be found on many machines, 1.6 Gigabytes of memory for 200 basis functions, which is only available on the Cray-2, and 8.1 Gigabytes of memory for 300 basis functions, which cannot currently be obtained on any machine. Since 200-300 basis function direct SCF calculations are quite feasible on departmental machines and super-workstations, this illustrates that memory availability controls the usefulness of in-core SCF and that larger systems will inevitably be done using direct SCF. It is also the case that the pure N^4 size dependance of in-core SCF would cause it to become slower than direct SCF for sufficiently large systems; however, this asymptotic limit has not been reached in practice.

The same considerations apply to the closed-shell restricted, open-shell restricted and open-shell unrestricted SCF procedures. Likewise, both standard iteration procedures such as DIIS and quadratically convergent schemes can be utilized with any approach to handling the two-electron integrals. Only conventional procedures are currently available for MCSCF methods, although the same considerations apply in principle.

6. Hartree-Fock and MCSCF Gradients

Analytical gradients, or first derivatives of the energy with respect to atom positions[19] can improve the efficiency of optimizing equilibrium structures and searching for transition states is an order of magnitude more efficient[20]. The flow chart for a typical geometry optimization is shown in Figure 2. The energy and gradient are evaluated and a quasi-Newton algorithm is used to predict the next estimate of the optimized geometry; these steps are repeated until convergence on the gradient and the displacement are achieved. Second derivatives can be calculated by numerical differentiation of the analytical gradients, and proceeds by a flow chart similar to Figure 2 (see below for analytical second derivatives).

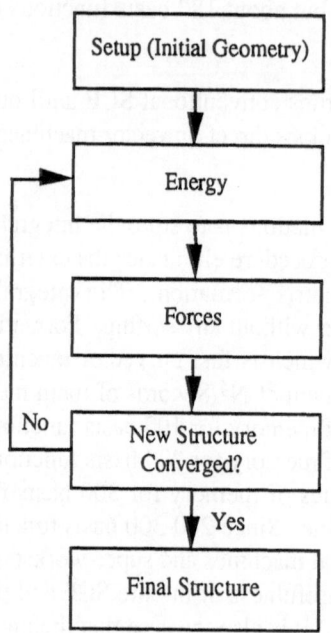

Figure 2. Flowchart for geometry optimization. Starting from an initial guess of the geometry, the energy and forces are calculated. The forces are used to predict a new structure. If the forces and the changes in the geometry are not below a preset threshold, the process is repeated.

The first derivative of the Hartree-Fock energy, eq 14, can be reduced to:

$$(15) \quad E^x_{HF} = \sum_{\mu\nu} P_{\mu\nu} h^x_{\mu\nu} + \frac{1}{2} \sum_{\mu\nu\lambda\sigma} P_{\mu\nu} P_{\lambda\sigma} (\mu\lambda||\nu\sigma)^x + \sum_{\mu\nu} W_{\mu\nu} S^x_{\mu\nu} + V^x_{nuc}$$

$$W_{\mu\nu} = -\sum_{\lambda\sigma} P_{\mu\lambda} F_{\lambda\sigma} P_{\sigma\nu}$$

where the superscript x denotes differentiation with respect to x. The term in W arises from the orthonormality constraint on the MO coefficients (or the idempotency condition for the density matrix, $P S P = P$). Since the wavefunction has been optimized variationally, the derivative of the wavefunction is not needed. The MCSCF gradients can be cast into a similar form, except that the two particle density matrix is not fully separable into the product of one particle density matrices and the construction of W is a bit more involved[21].

The derivatives of SCF and MCSCF energies with respect to nuclear coordinates present no new tradeoffs. In some ways, calculation of derivatives is simpler than the computation of the energy because the integral derivatives need not be stored since they can be contracted immediately with the density to form the contribution to the gradient. Since each two electron integral contributes to at most 12 nuclear coordinate derivatives, the cpu time grows only as the number of integrals $O(N^4)$, not $O(N^4 N_a)$ where N_a is the number of derivatives with respect to nuclear coordinates. MCSCF derivatives are similar, except that the active space two particle density must be back transformed to the AO basis as the integral derivatives are computed. This does not present storage problems because the number of orbitals in the active space is typically very small.

7. Hartree-Fock Second Derivatives

Analytic SCF second derivatives with respect to nuclear coordinates are useful for geometry optimizations and for calculating vibrational spectra. In addition to the force constants, which determine the fundamental frequencies, the dipole derivatives (second derivatives: once with respect to an electric field and once with respect to nuclear coordinates) determine the infra-red intensities and the polarizability derivatives (third derivatives: twice with respect to an electric field and once with respect to nuclear coordinates) determine the Raman intensities. Each of these requires the first derivatives of the MO coefficients along with first and second derivatives of the appropriate integrals.

The second derivatives of the Hartree-Fock energy with respect to the nuclear positions can be written as[22]:

$$(16)\quad E_{HF}^{xy} = \sum_{\mu\nu} P_{\mu\nu}\, h_{\mu\nu}^{xy} + \frac{1}{2} \sum_{\mu\nu\lambda\sigma} P_{\mu\nu}\, P_{\lambda\sigma}\, (\mu\lambda||\nu\sigma)^{xy} + \sum_{\mu\nu} W_{\mu\nu}\, S_{\mu\nu}^{xy} + \sum_{\mu\nu} P_{\mu\nu}^{x}\, F_{\mu\nu}^{y} + \sum_{\mu\nu} W_{\mu\nu}^{x}\, S_{\mu\nu}^{y}$$

$$F_{\mu\nu}^{x} = h_{\mu\nu}^{x} + \sum_{\lambda\sigma} [(\mu\lambda||\nu\sigma)^{x}\, P_{\lambda\sigma} + (\mu\lambda||\nu\sigma)\, P_{\lambda\sigma}^{x}] \; ; \; P_{\mu\nu}^{x} = \sum_{i} [C_{\mu i}^{x*}\, C_{\nu i} + C_{\mu i}^{*}\, C_{\nu i}^{x}]$$

$$W_{\mu\nu}^{x} = -\sum_{\lambda\sigma} [P_{\mu\lambda}^{x}\, F_{\lambda\sigma}\, P_{\sigma\nu} + P_{\mu\lambda}\, F_{\lambda\sigma}^{x}\, P_{\sigma\nu} + P_{\mu\lambda}\, F_{\lambda\sigma}\, P_{\sigma\nu}^{x}]$$

The derivatives of the MO coefficients or the density matrix can be obtained by solving the the corresponding derivatives of the Hartree-Fock equations, i.e. the coupled perturbed Hartree-Fock (CPHF) equations[23]. These can be solved either in the MO basis[23] or in the AO basis[24]. After some manipulation, the CPHF equations in the MO basis can be reduced to:

$$(17)\quad \sum_{bj} [(\varepsilon_i - \varepsilon_a)\, \delta_{ab}\, \delta_{ij} - (ab||ij) - (aj||ib)]\, P_{bj}^{x} = [h_{ai}^{x} - \varepsilon_i\, S_{ai}^{x} + \sum_{j} (aj||ij)^{x} - \sum_{jk} (aj||ik)\, S_{jk}^{x}]$$

for the virtual-occupied block of the density. The occuped-occupied block is given by the orthonormality constraint, $P_{ij}^{x} = -S_{ij}^{x}$, and the virtual-virtual block is zero. The CPHF equations in the AO basis can be written as:

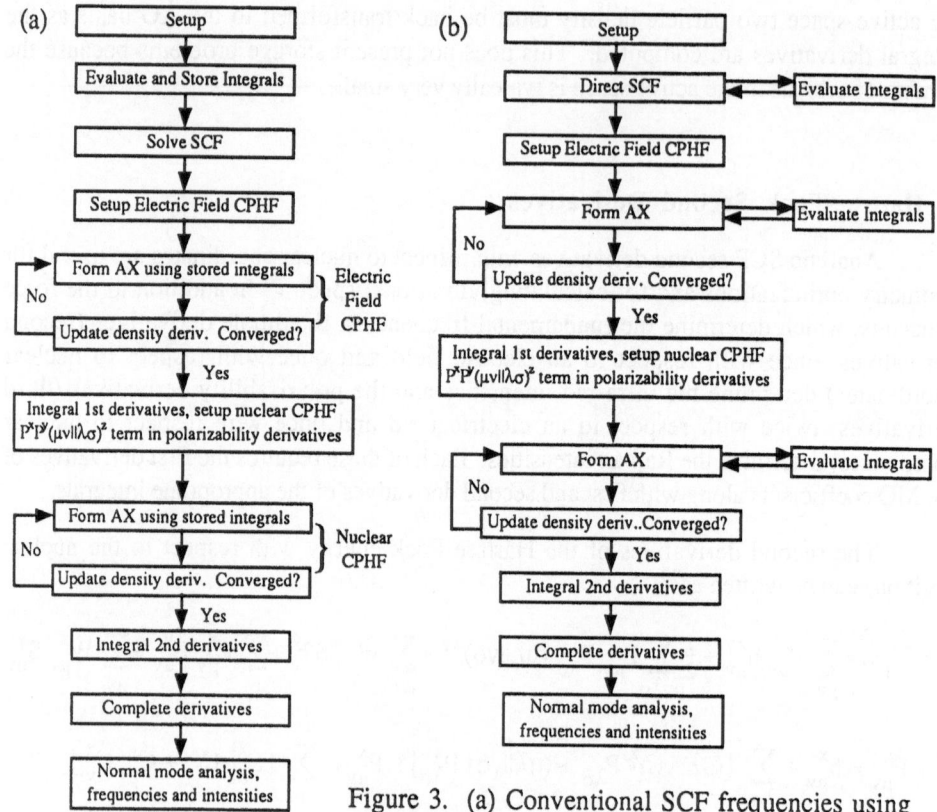

Figure 3. (a) Conventional SCF frequencies using integrals stored on disk. (b) Direct SCF frequencies, avoiding storage of the two electron integrals.

$$(18) \quad \sum_{\lambda\sigma} {}^{ov}[F_{\mu\lambda} S_{\sigma v} - S_{\mu\lambda} F_{\sigma v} - (\mu\sigma\|v\lambda)] {}^{ov}P_{\lambda\sigma}^{x}$$

$$= {}^{ov}h_{\mu v}^{x} + \sum_{\lambda\sigma} {}^{ov}[(\mu\sigma\|v\lambda)^{x} P_{\lambda\sigma} - (\mu\sigma\|v\lambda) {}^{oo}S_{\lambda\sigma}^{x} - F_{\mu\lambda} P_{\lambda\sigma} {}^{ov}S_{\sigma v}^{x}]$$

where superscripts oo and ov denote projection on the occupied-occupied and occupied-virtual spaces; as in the MO basis ${}^{oo}P_{\mu v}^{x} = - {}^{oo}S_{\mu v}^{x}$.

The CPHF equations are linear in the density derivatives and have the form $AX = B$ for each perturbation. In the MO basis A has O^2V^2 elements and X and B have OV elements per perturbation. A straightforward solution of the linear equations by inversion of A would require O^2V^2 memory and O^3V^3 arithmetic operations. Instead, an iterative solution of these equations[22] permits A to be stored on disk and reduces the operation count to $O(N_aO^2V^2)$, where N_a is the number of derivatives (3 times the number of atoms in the case of nuclear coordinates). Thus CPHF has quintic dependance on the size of the system. Similarly, the transformation of the integrals to the molecular orbital basis requires ON^4 operations and O^2V^2 storage. All other components of the second derivatives are no worse than quartic in cost.

Alternatively, the same iterative solution of the CPHF equations in the AO basis forms the products of AX from the integrals over atomic orbitals and the density derivatives without forming or storing A explicitly[24]. The two electron contributions to AX in the CPHF equations are the same as the two electron contributions to the Fock matrices in the HF equations but formed with density derivatives, reflecting the derivative nature of the CPHF equations. Neither integral transformation nor external storage for the transformed integrals is required. The CPHF solution requires $O(N_aN^4)$ arithmetic operations, i.e. still quintic but with a larger constant factor than for solution in the MO basis. This increase in cost is compensated by the elimination of the transformation step and, in practice, solution in the AO basis is comparable in CPU cost to MO solution and hence preferred.

Figure 3a illustrates all the steps in an SCF frequency calculation using externally stored AO integrals. One term in the polarizability derivatives involves two density matrix derivatives with respect to electric field directions and two-electron integral derivatives with respect to nuclear coordinates. Since the integral derivatives must be computed before the nuclear coordinate CPHF (they contribute to the B's in A X = B), it is necessary to solve the electric field CPHF problem first. After the integral derivatives, the nuclear CPHF problem can be solved and the force constants completed with the integral second derivative terms. Finally, a normal mode analysis produces the frequencies and intensities from the Cartesian derivatives. If only the IR intensities are of interest, the nuclear and electric field CPHF problems can be solved at the same time. If only the force constants are required (i.e., if the second derivatives are to be used to assist a geometry optimization) then the electric field CPHF can be omitted altogether.

The SCF and CPHF equations may also be solved using integrals stored in main memory. This has the same advantages of speed and disadvantages of large memory requirements as for in-core SCF.

Finally, the integrals can be recomputed during the SCF and CPHF iterations[25], as shown in Figure 3b. If N_aN^2 memory is available, so that in a given CPHF iteration all perturbations can be handled at once, then the solution using recomputed integrals does not add to the quintic phase of the calculation. Because many Fock matrices are formed at a time, the integral cutoffs in direct CPHF are not as effective as in direct SCF, so that direct CPHF does not become faster than conventional for large systems, but the additional overhead of direct solution is modest (20-30%). Unlike the conventional solution, the extra overhead of separate solution of the electric field CPHF adds significantly (about 20%) to a direct frequency calculation, because the integrals are recomputed while solving for only 3 density derivatives.

8. MP2 Energy

One of the simplest means of including electron correlation energy is by second order perturbation theory:

$$(19)\ (\hat{H}_0 + \lambda\hat{V})\ (\Psi_0 + \lambda\sum_s a_s\,\Psi_s \cdots) = (E^{(0)} + \lambda\,E^{(1)} + \lambda^2\,E^{(2)} + \cdots)\ (\Psi_0 + \lambda\sum_s a_s\,\Psi_s \cdots)$$

$$\hat{H}_0\,\Psi_s = E_s\,\Psi_s\ ;\ \ \hat{V} = \hat{H} - \hat{H}_0$$

$$(20)\quad E^{(0)} + E^{(1)} = \langle\Psi_0|\hat{H}_0 + \hat{V}|\Psi_0\rangle = E_{HF}$$

$$(21)\quad E^{(2)} = \sum_s a_s\,\langle\Psi_0|\hat{V}|\Psi_s\rangle\ ;\ \ a_s = \langle\Psi_s|\hat{V}|\Psi_0\rangle\,/\,(E_0 - E_s)$$

The most convenient choice for $\hat{H}_0$ is the Fock operator since the Hartree-Fock wavefunction is an eigenfunction of the Fock operator, as are the excited configurations derived from it by replacing occupied orbitals by virtual orbitals. This choice of $\hat{H}_0$ yields Møller-Plesset (MP) perturbation theory or many-body perturbation theory (MBPT)[26,27]. Through second order, the energy is:

$$(22)\quad E_{MP2} = E_{HF} + E^{(2)} = E_{HF} + \frac{1}{4}\sum_{ijab} a_{ij}^{ab}\,(ij\|ab)\ ;\ \ a_{ij}^{ab} = -\,(ij\|ab)\,/\,(\varepsilon_a + \varepsilon_b - \varepsilon_i - \varepsilon_j)$$

Note that single excitations do not contribute because of Brillouin's theorem[28] and excitations higher than double do not contribute because $\hat{V}$ is only a two electron operator.

A conventional MP2 energy calculation, illustrated in Figure 4a, consists of an SCF calculation followed by an integral transformation and the formation of the antisymmetrized integrals which are summed into $E^{(2)}$. The only quintic step is the transformation; since only transformed integrals involving two occupied and two virtual orbitals are needed, this has $O(ON^4)$ cost.

The second-order energy can also be computed using direct methods[29], Figure 4b. Following an SCF calculation (which might be either conventional or direct), a transformation is done in which the two-electron integrals are computed as needed. This can be done very efficiently by computing a double-length set of integrals (eliminating the $\mu\nu \leftrightarrow \lambda\sigma$ permutation, yielding $N^4/4$ rather than $N^4/8$ integrals) which permits vectorization of both the integral evaluation and all steps in the transformation. Thus the direct method eliminates the use of external storage while at the same time offering maximal efficiency in the transformation step. Main memory requirements are kept bounded by restricting the number of occupied orbitals in the outer loop which are handled in each batch - that is, transformed integrals $(ij\|ab)$ are formed for a batch of i's and all occupied j, and virtual a,b at a time. A minimum of OVN memory is required and in this limit the integrals are computed O times (equivalent to $2\,O$ canonical integral evaluations).

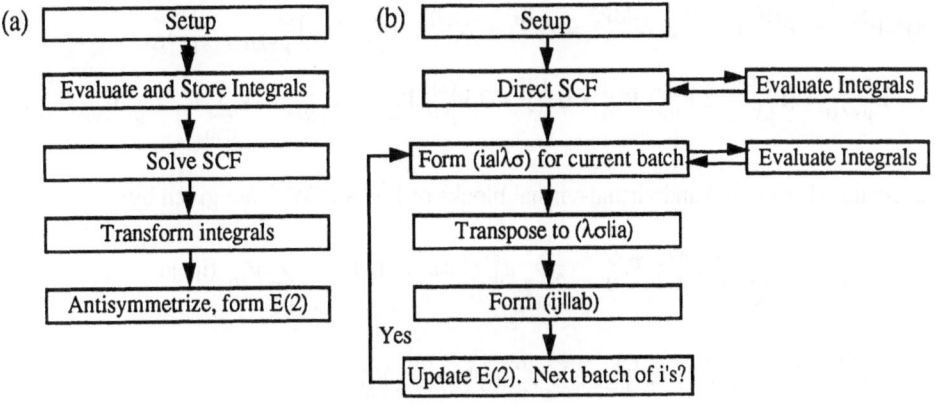

Figure 4. (a) Conventional MP2 energy and (b) direct MP2 energy.

The memory requirements for direct MP2 can be kept quadratic and the number of integral evaluations kept down by using a semi-direct algorithm[30], similar to Figure 4b. This involves storing the half-transformed integrals (ia|λσ) and transposing them using external storage. The amount of external storage used can be held at any fixed amount by restricting the number of occupied orbitals i in a batch. Memory requirements are reduced to $O(N^2)$ and disk requirements are $O(VN^2)$ per occupied orbital in the batch. In the limit of a single pass, the semi-direct MP2 method uses OVN^2 disk, which is less than corresponding conventional algorithms and is considerably more efficient. Thus the combination of conventional SCF and one pass of semi-direct MP2 supersedes traditional MP2 algorithms entirely.

Finally, MP2 can be done with all integrals in memory. A double-length integral list is required, so that twice as much memory is required as for in-core SCF. This is again very CPU-efficient but has even more stringent memory requirements than in-core SCF.

9. MP2 Gradient

First derivatives of the MP2 energy are needed not only for geometry optimization but also for one electron properties such as the dipole moment. After some algebra, the gradient of the MP2 energy can be cast into a form similar to the Hartree-Fock gradient[22,31]:

$$(23) \quad E_{MP2}^x = \sum_{\mu\nu} P_{\mu\nu}^{MP2} h_{\mu\nu}^x + \frac{1}{2} \sum_{\mu\nu\lambda\sigma} \Gamma_{\mu\nu\lambda\sigma}^{MP2} (\mu\nu|\lambda\sigma)^x + \sum_{\mu\nu} W_{\mu\nu}^{MP2} S_{\mu\nu}^x + V_{nuc}^x$$

where P, Γ and W are given by:

(24) $\quad P^{MP2} = P^{HF} + P^{(2)}$; $\quad W^{MP2} = W^{HF} + W^{(2)}$; $\quad \Gamma^{MP2}_{\mu\nu\lambda\sigma} = \Gamma^S_{\mu\nu\lambda\sigma} + \Gamma^{NS}_{\mu\nu\lambda\sigma}$

$$\Gamma^S_{\mu\nu\lambda\sigma} = (P^{HF}_{\mu\nu} + 2\, P^{(2)}_{\mu\nu})\, P^{HF}_{\lambda\sigma} - (P^{HF}_{\mu\sigma} + 2\, P^{(2)}_{\mu\sigma})\, P^{HF}_{\lambda\nu} \; ; \quad \Gamma^{NS}_{\mu\nu\lambda\sigma} = 2 \sum_{ijab} a^{ab}_{ij}\, C^*_{\mu i}\, C_{va}\, C^*_{\lambda j}\, C_{\sigma b}$$

The occupied-occupied and virtual-virtual blocks of $P^{(2)}$ and $W^{(2)}$ are given by:

(25) $\quad P^{(2)}_{ij} = -\dfrac{1}{2}\displaystyle\sum_{kab} a^{ab}_{ik}\, a^{ab}_{jk}$; $\quad W^{(2)}_{ij} = \dfrac{1}{2}\displaystyle\sum_{kab} a^{ab}_{ik}\, (ki\|ab) - \varepsilon_i\, P^{(2)}_{ij} - \displaystyle\sum_{pq} P^{(2)}_{pq}\, (ip\|jq)$

$$P^{(2)}_{ab} = \dfrac{1}{2}\sum_{ijc} a^{ac}_{ij}\, a^{bc}_{ij} \; ; \quad W^{(2)}_{ab} = \dfrac{1}{2}\sum_{ijc} a^{bc}_{ik}\, (ij\|ca) - \varepsilon_a\, P^{(2)}_{ab}$$

The occupied-virtual block requires the solution of a single CPHF-like equation:

(28) $\quad \displaystyle\sum_{bj} \left[(\varepsilon_a - \varepsilon_i)\, \delta_{ab}\, \delta_{ij} + (ab\|ij) + (aj\|ib) \right] P^{(2)}_{ai} = L_{ai}$

$$L_{ai} = \dfrac{1}{2}\sum_{jbc} a^{ab}_{ij}\, (ja\|bc) + \dfrac{1}{2}\sum_{jkb} a^{ab}_{ij}\, (jk\|ib) + \sum_{bc} P^{(2)}_{bc}\, (ic\|ab) + \sum_{jk} P^{(2)}_{bc}\, (ki\|ja)$$

$$W^{(2)}_{ai} = \dfrac{1}{2}\sum_{jkb} a^{ba}_{ik}\, (jk\|ib) - \varepsilon_i\, P^{(2)}_{ai}$$

These equations can be derived by differentiating the expression for E_{MP2} and using the Z-vector method[32] to avoid solving the CPHF equations explicitly for each of the derivatives of the MO coefficients.

In a conventional MP2 gradient calculation, illustrated in Figure 5a, an SCF calculation is followed by an integral transformation producing all MO integrals involving at least one occupied orbital. The Lagrangian is then formed by multiplying integrals and the CPHF equation solved for the generalized density. The non-separable part of two-particle density matrix, Γ^{NS}, is transformed back to the AO basis and sorted into the order in which the integral derivatives will be evaluated. The integral derivatives can then be evaluated and immediately contracted with the sum of the non-separable part two-particle density (read from disk) and the separable part, Γ^S, which is a product of one-particle densities stored in memory. There are two quintic steps: the integral transformation and the two-particle density back transformation, both $O(ON^4)$. External storage is required for the AO integrals, MO integrals, two-particle density and for sorting the partially and fully back transformed two-particle density, each $O(N^4)$.

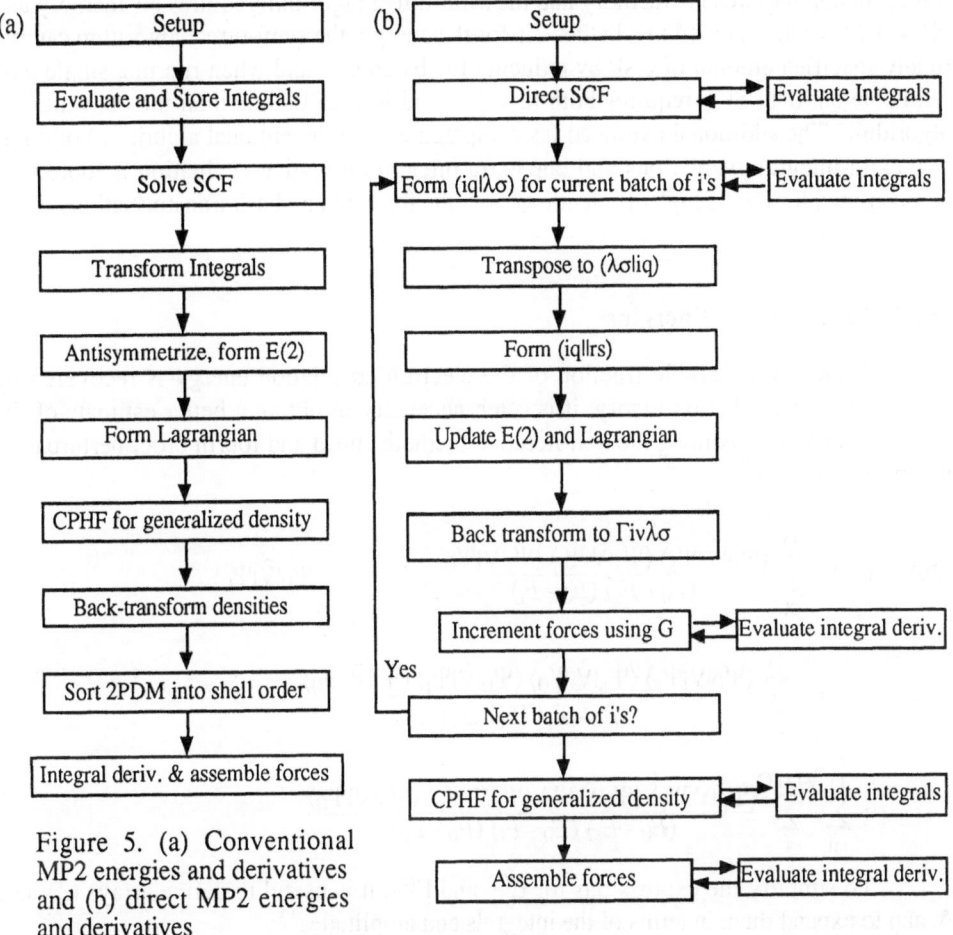

Figure 5. (a) Conventional MP2 energies and derivatives and (b) direct MP2 energies and derivatives

The MP2 gradient can also be computed using direct and semi-direct techniques[30], Figure 5b. Again the SCF is followed by a transformation involving batches of occupied orbitals, i. The transformed integrals are formed as for evaluation of just the MP2 energy, except that integrals (iq‖rs) are formed for all q, r, and s and the current batch of i's. These are used to form $E^{(2)}$ and to add contributions to L_{ai}, $P^{(2)}$ and $W^{(2)}$. Then the two-particle density for the current batch is formed and back transformed. The integral derivatives are evaluated and contracted with the back-transformed two-particle density. Thus one double-length integral evaluation and one integral derivative evaluation are done for each batch. After all batches have been processed and the construction of Lagrangian is finished, the CPHF equation is solved and the computation of $P^{(2)}$, $W^{(2)}$ and Γ^S is completed. Finally, the contribution of the Γ^S and integral derivatives can be computed with an additional integral derivative evaluation. Thus if there are b batches, the complete MP2 gradient requires twice the time for a direct SCF plus (2b+1) integral evaluations and (b+1) integral derivative evaluations. Since the transformed integrals over all q,r,s are required, the fully direct version requires N^3 memory and the semi-direct algorithm requires N^2 memory and N^3 disk for each occupied in a batch. As for the energy, the semi-direct algorithm can run in any specified amount of disk by reducing the batch size and when run in a single pass with conventional SCF requires both less disk and less CPU time than the conventional algorithm. The additional overhead (as compared to the conventional algorithm) of doing two extra integral evaluations and one extra integral derivative evaluation is more than compensated for by the high efficiency of both the forward and back transformations.

10. MP3 and MP4 Energies

Although a sizeable fraction of the electron correlation energy is recovered by second order perturbation theory, it is often necessary to obtain a better estimate of the correlation energy. Among the non-iterative methods, third and fourth order perturbation theory are practical[27]:

$$(29) \quad E^{(3)} = \sum_{st}^{D} \frac{\langle \Psi_0 | \hat{V} | \Psi_s \rangle \langle \Psi_s | \bar{V} | \Psi_t \rangle \langle \Psi_t | \hat{V} | \Psi_0 \rangle}{(E_0 - E_s)(E_0 - E_t)} \; ; \; \bar{V} = \hat{V} - \langle \Psi_0 | \hat{V} | \Psi_0 \rangle$$

$$(30) \quad E^{(4)} = -\sum_{st}^{D} \frac{\langle \Psi_0 | \hat{V} | \Psi_s \rangle \langle \Psi_s | \hat{V} | \Psi_0 \rangle \langle \Psi_0 | \hat{V} | \Psi_t \rangle \langle \Psi_t | \hat{V} | \Psi_0 \rangle}{(E_0 - E_s)(E_0 - E_t)^2}$$

$$+ \sum_{su}^{D} \sum_{t}^{SDTQ} \frac{\langle \Psi_0 | \hat{V} | \Psi_s \rangle \langle \Psi_s | \bar{V} | \Psi_t \rangle \langle \Psi_t | \hat{V} | \Psi_u \rangle \langle \Psi_u | \hat{V} | \Psi_0 \rangle}{(E_0 - E_s)(E_0 - E_t)(E_0 - E_u)}$$

To simplify the expressions for $E^{(3)}$ and $E^{(4)}$, it is useful to define arrays u, v and Δ, and to expand them in terms of the integrals and amplitudes[33]:

$$(31) \quad u_s = \sum_{t}^{D} \langle \Psi_s | \bar{V} | \Psi_t \rangle \langle \Psi_t | \hat{V} | \Psi_0 \rangle / (E_0 - E_t)$$

$$(32) \quad u_i^a = -\sum_{jb} (ja\|ib) \, a_j^b - \frac{1}{2} \sum_{jbc} (ja\|bc) \, a_{ij}^{bc} - \frac{1}{2} \sum_{jkb} (jk\|ib) \, a_{jk}^{ab}$$

$$(33) \quad u_{ij}^{ab} = \sum_{c} [(ab\|cj) \, a_i^c - (i \leftrightarrow j)] - \sum_{k} [(kb\|ij) \, a_k^a - (a \leftrightarrow b)]$$

$$+ \frac{1}{2} \sum_{cd} (ab\|cd) \, a_{ij}^{cd} + \frac{1}{2} \sum_{kl} (kl\|ij) \, a_{kl}^{ab} - \sum_{kc} [(kb\|jc) \, a_{ik}^{ac} - (a \leftrightarrow b) - (i \leftrightarrow j) + (ia \leftrightarrow jb)]$$

(34) $\quad u_{ijk}^{abc} =$

$$\sum_e [(bc||ek)a_{ij}^{ae} - (a \leftrightarrow b) - (a \leftrightarrow c) - (i \leftrightarrow k) - (j \leftrightarrow k) + (ia \leftrightarrow kb) + (ja \leftrightarrow kb) + (ia \leftrightarrow kc) + (ja \leftrightarrow kc)]$$

$$+ \sum_m [(cm||jk) a_{im}^{ab} - (a \leftrightarrow c) - (b \leftrightarrow c) - (i \leftrightarrow j) - (j \leftrightarrow k) + (ia \leftrightarrow jc) + (ja \leftrightarrow kc) + (ib \leftrightarrow jc) + (jb \leftrightarrow kc)]$$

(35) $\quad v_s = \sum_t^Q \sum_u^D \langle \Psi_s | \bar{V} | \Psi_t \rangle \langle \Psi_t | \bar{V} | \Psi_u \rangle \langle \Psi_u | \hat{V} | \Psi_0 \rangle / (E_0 - E_t) (E_0 - E_u)$

(36) $\quad v_i^a = \frac{1}{2} \sum_{jkbc} (jk||bc) [a_i^b a_{jk}^{ca} + a_j^a a_{ik}^{cb} + 2 a_j^b a_{ik}^{ac}]$

(37) $v_{ij}^{ab} = \frac{1}{4} \sum_{jkbc} (kl||cd) [a_{ij}^{cd} a_{kl}^{ba} - 2(a_{ij}^{ac} a_{kl}^{bd} + a_{ij}^{bd} a_{kl}^{ac}) - 2(a_{ik}^{ab} a_{jl}^{cd} + a_{ik}^{cd} a_{jl}^{ab}) + 4(a_{ik}^{ac} a_{jl}^{bd} + a_{ik}^{bd} a_{jl}^{ac})]$

(38) $\quad \Delta_i^a = (\varepsilon_a - \varepsilon_i) \, ; \; \Delta_{ij}^{ab} = (\varepsilon_a + \varepsilon_b - \varepsilon_i - \varepsilon_j) \, ; \; \Delta_{ijk}^{abc} = (\varepsilon_a + \varepsilon_b + \varepsilon_c - \varepsilon_i - \varepsilon_j - \varepsilon_k) \, ; \;$ etc.

where $(p \leftrightarrow q)$ denotes the term obtained by permuting indices p and q. The formation of u_i^a and v_i^a is only $O(N^5)$ work. The computation of u(ab,ij) involves $O(O^2 N^4)$ cpu; the full transformation of the integrals can be avoided be constructing the (ab||cd) contribution directly from the AO integrals. The u_{ijk}^{abc} require $O(O^3 V^4)$ work and are the most costly terms. The v_{ij}^{ab} can be assembled in stages in which the larges fraction of the work is proportional to $O(O^3 V^3)$.

With $a_i^a = 0$ and $a_{ij}^{ab} = - (ij||ab) / \Delta_{ij}^{ab}$ in the above the definitions, the MP3 and MP4 energies become[33]

(39) $\quad E_{MP3} = E_{MP2} + \frac{1}{4} \sum_{ijab} a_{ij}^{ab} u_{ij}^{ab}$

(40) $\quad E_{MP4} = E_{MP3} - E^{(2)} \frac{1}{4} \sum_{ijab} |a_{ij}^{ab}|^2 - \sum_{ia} |u_i^a|^2 / \Delta_i^a - \frac{1}{4} \sum_{ijab} |u_{ij}^{ab}|^2 / \Delta_{ij}^{ab}$

$$- \frac{1}{36} \sum_{ijkabc} |u_{ijk}^{abc}|^2 / \Delta_{ijk}^{abc} + \frac{1}{4} \sum_{ijab} a_{ij}^{ab} v_{ij}^{ab}$$

The five terms in the MP4 energy are the renormalization, singles, doubles, triples and quadruples contributions, respectively.

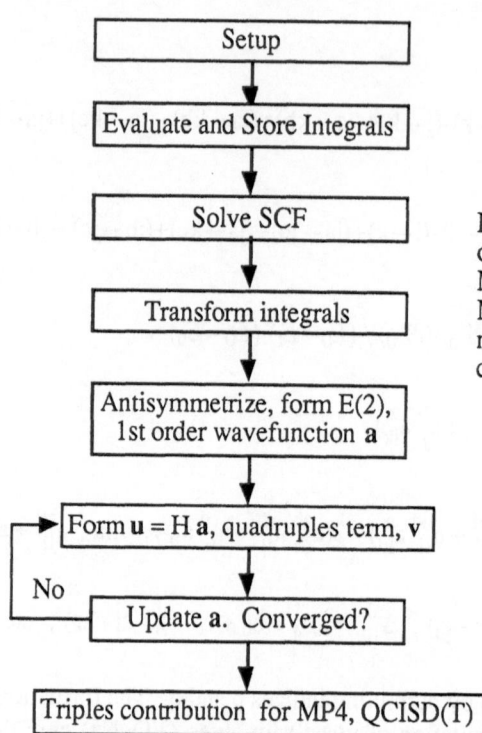

Figure 6. MP3, MP4, CI, CC or QCI energies. For MP3 and MP4, there is no iteration; for MP3, CID and CISD there is no triples or quadruples contribution.

11. CI, CC and QCI energies

The CI energy and wavefunction can be found by substituting the CI expansion of the wavefunction, eq 5, into the expression for the variational energy, eq 3, and solving the corresponding matrix eigenvalue problem.

$$(41) \quad \sum_t H_{st} \, a_t = E_{CI} \, a_s \; ; \quad \Psi_{CI} = \sum_s a_s \, \Psi_s \; ; \quad H_{st} = \langle \Psi_s | \hat{H} | \Psi_t \rangle$$

However, the Hamiltonian matrix H_{st} is too large to be stored explicitly, e.g. $O^4 V^4$ elements if the CI wavefunction is limited to single and double excitations. Furthermore, H_{st} is sparse with only $O(O^2 V^2 N^2)$ non-zero matrix elements. Hence the equations for CI and related methods are solved iteratively, with the sparsity of H_{st} explicitly taken into account in the formation of the product $\sum_t H_{st} \, a_t$.

To simplify the discussion of the CI, CC and QCI methods, it is desireable to define single, double, triple, etc. substitution operators:

$$(42) \quad \hat{T}_1 = \sum_{ia} a_i^a t_i^a \; ; \quad \hat{T}_2 = \frac{1}{4} \sum_{ijab} a_{ij}^{ab} t_{ij}^{ab} \; ; \quad \hat{T}_2 = \frac{1}{36} \sum_{ijkabc} a_{ijk}^{abc} t_{ijk}^{abc} \quad \text{etc.}$$

With these operators, the CI doubles and the CI singles and doubles wavefunctions are:

(43) $\Psi_{CID} = (1 + \hat{T}_2)\, \Psi_0 \,; \quad \Psi_{CISD} = (1 + \hat{T}_1 + \hat{T}_2)\, \Psi_0$

and the CISD equations can be written as:

(44) $\langle \Psi_0|\hat{H}|\hat{T}_2\Psi_0\rangle = E_{corr} \,; \quad \langle \Psi_i^a|\overline{H}|(\hat{T}_1+\hat{T}_2)\Psi_0\rangle = a_i^a E_{corr} \,; \quad \langle \Psi_{ij}^{ab}|\overline{H}|(1+\hat{T}_1+\hat{T}_2)\Psi_0\rangle = a_{ij}^{ab} E_{corr}$

where $\overline{H} = \hat{H} - E_{HF}$. With the definition of the array u from the previous section, these equations can be rewritten[34]:

(45) $E_{corr} = \frac{1}{4}\sum_{ijab} a_{ij}^{ab}\, (ij\|ab) \,; \quad (\Delta_i^a - E_{corr})\, a_i^a + u_i^a = 0 \,; \quad (ab\|ij) + (\Delta_{ij}^{ab} - E_{corr})\, a_{ij}^{ab} + u_{ij}^{ab} = 0$

These equations are solved iteratively for E_{corr} and the amplitudes. The work is dominated by the formation of u at each iteration, $O(O^2N^4)$ cpu.

Although the CI energy is variational (an upper bound to the exact energy), it is not size-consistent (E_{CI} for X and Y at large separation is not the sum of E_{CI} for X and E_{CI} for Y computed individually). This is obviously a severe disadvantage when computing reaction energetics. The MPn energies are size consistent; however, iterative methods can give better estimates of the correlation energy than perturbative methods, especially if the correlation energy is large. Hence, it is desireable to have a CI-like method that is size-consistent. This role is filled by the coupled-cluster (CC) method[35] (for a review see ref 27) and the quadratic configuration interaction (QCI) approach[36]. In the CC method, the wavefunction is written as an exponential of the excitation operators. In CCD and CCSD, the excitations involve $\hat{T}_2$ and $\hat{T}_1 + \hat{T}_2$, respectively:

(46) $\Psi_{CCD} = \exp(\hat{T}_2)\, \Psi_0 \,; \quad \Psi_{CCSD} = \exp(\hat{T}_1 + \hat{T}'_2)\, \Psi_0 \,, \quad \text{where } \hat{T}'_2 = \hat{T}_2 - \frac{1}{2}\hat{T}_1^2$

The equations for CCD and CCSD are:

(47) $\langle \Psi_0|\hat{H}|\hat{T}_2\Psi_0\rangle = E_{corr} \,; \quad \langle \Psi_{ij}^{ab}|\overline{H}|(1+\hat{T}_2+\frac{1}{2}\hat{T}_2^2)\Psi_0\rangle = a_{ij}^{ab} E_{corr}$

(48) $\langle \Psi_0|\hat{H}|\hat{T}_2\Psi_0\rangle = E_{corr} \,; \quad \langle \Psi_i^a|\overline{H}|(\hat{T}_1+\hat{T}_2+\hat{T}_1\hat{T}_2-\frac{1}{3}\hat{T}_1^3)\Psi_0\rangle = a_i^a E_{corr}$

$\langle \Psi_{ij}^{ab}|\overline{H}|(1+\hat{T}_1+\hat{T}_2+\hat{T}_1\hat{T}_2-\frac{1}{3}\hat{T}_1^3+\frac{1}{2}\hat{T}_2^2-\frac{1}{4}\hat{T}_1^4)\Psi_0\rangle = a_{ij}^{ab} E_{corr}$

With the above definitions of u and v, the CCD equations can be reduced to[37]:

(49) $E_{corr} = \frac{1}{4}\sum_{ijab} a_{ij}^{ab}$ (ij‖ab) ; $a_i^a = 0$; (ab‖ij) + $\Delta_{ij}^{ab} a_{ij}^{ab} + u_{ij}^{ab} + v_{ij}^{ab} = 0$

In the QCI approach, only the bare minimum number of excitation operators are added to the CI equations to make them size-consistent. For single and double excitations, this leads to:

(50) $\langle \Psi_0 | \hat{H} | \hat{T}_2 \Psi_0 \rangle = E_{corr}$; $\langle \Psi_i^a | \bar{H} | (\hat{T}_1 + \hat{T}_2 + \hat{T}_1 \hat{T}_2) \Psi_0 \rangle = a_i^a E_{corr}$

$\langle \Psi_{ij}^{ab} | \bar{H} | (1 + \hat{T}_1 + \hat{T}_2 + \frac{1}{2}\hat{T}_2^2) \Psi_0 \rangle = a_{ij}^{ab} E_{corr}$

which can be reduced to:

(51) $E_{corr} = \frac{1}{4}\sum_{ijab} a_{ij}^{ab}$ (ij‖ab) ; $\Delta_i^a a_i^a + u_i^a + v_i^a = 0$; (ab‖ij) + $\Delta_{ij}^{ab} a_{ij}^{ab} + u_{ij}^{ab} + v_{ij}^{ab} = 0$

For each iteration, the worst step is the formation of u and v, each are $O(O^2N^4)$ if only a partial transformation of the integrals has been carried out or $O(O^2V^4)$ for a full transformation (in this case, the $O(O^3V^3)$ step in forming v is the most time consuming). The iterative solution of the CI, CC and QCI equations is illustrated in Figure 6.

For many cases, triple excitations are important as well. CCD, CCSD and QCISD omit the triples contributions to different extents. A simple remedy is to add in the missing contribution in a manner similar to perturbation theory[38]. This adds a single (non-iterative) step of $O(O^3V^4)$ in cpu to an iterative process that is $O(O^2N^4)$ for each step (Fig. 6). An alternative is to add triples in an iterative fashion[39] (at the cost of $O(O^3V^4)$ work and storage per iteration).

It is possible to include all excitations within a fixed basis set size, i.e. full CI. Because the work increases roughly as $((O+V)!/O!V!)^2$ or $(O+V)^n/n!$, where n is the number of electrons correlated, only small systems can be treated - few electrons correlated and/or small active space. However, full CI calculation provide valueable benchmarks for testing other methods for estimating electron correlation energy.

12. MP3, MP4, CI, CC and QCI gradients

The gradients for MP3, MP4 and CI methods[40-42] (Figure 7a) are very similar to MP2 gradients. Similar to the MP2 gradients, the full CPHF can be avoided by the Z vector method[32]. The assembly of L is, however, somewhat more complicated than the MP2 case. In the coupled cluster and QCI approaches, the amplitudes are not calculated in a variational manner. Hence, the derivatives of the amplitudes must be calculated by the

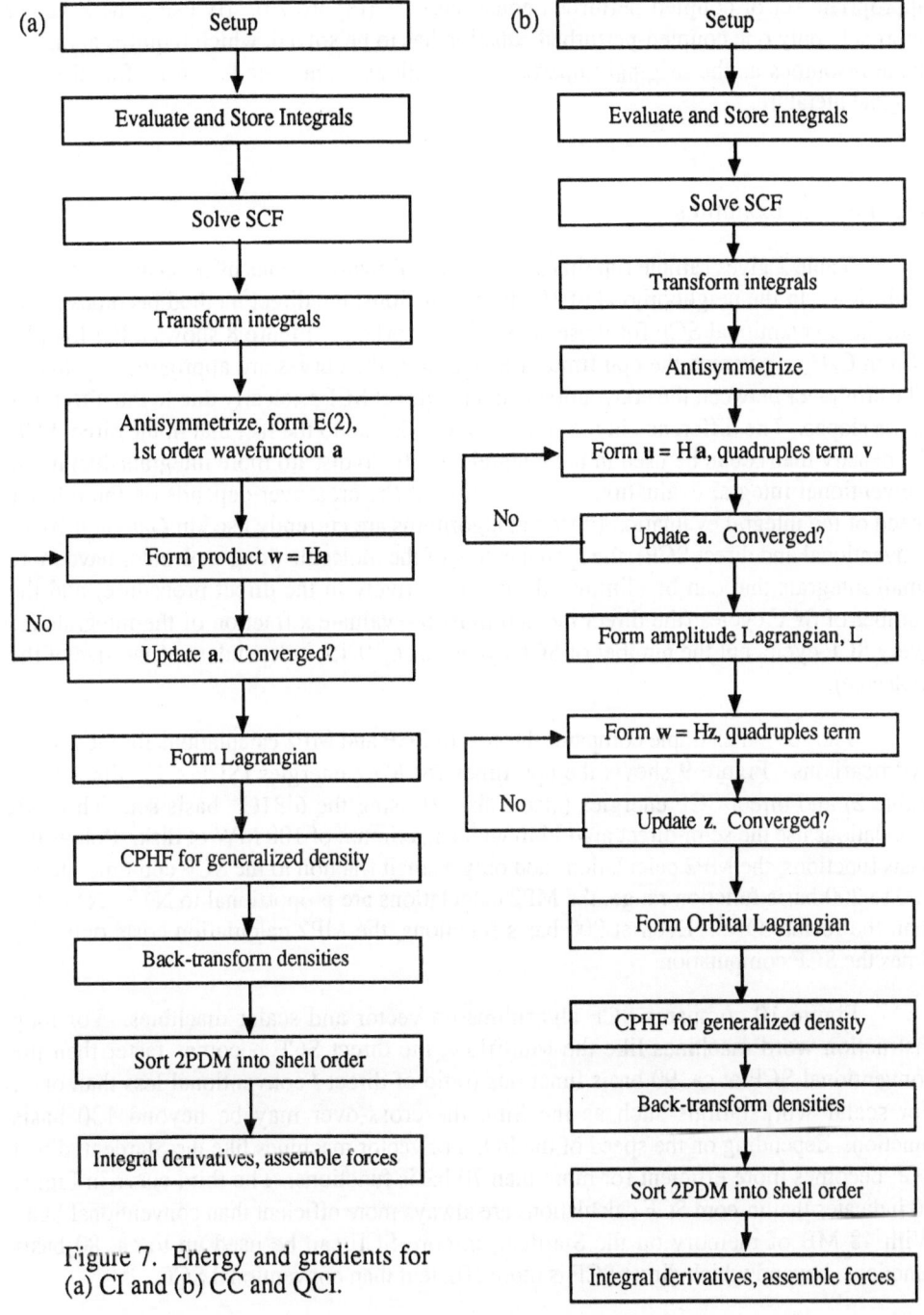

Figure 7. Energy and gradients for (a) CI and (b) CC and QCI.

appropriate set of coupled-perturbed equations[43,44] (Figure 7b). By using the Z vector approach, only one coupled-perturbed equation has to be solved, which requires about the same resources as the original unperturbed equations. The details can be found in the original literature.

13. Timing examples

Table 2 gives sample run times for a series of hydrocarbons using *Gaussian 90* on a Multiflow. In the neighborhood of 100 basis functions, the direct method becomes faster than the conventional SCF for these simple hydrocarbons. Figure 8 shows a log-log plot of n in C_nH_{2n+2} versus the cpu time. Beyond n=3, the curves are approximately linear. The crossover between the conventional and the direct SCF is clearly due to the difference in the slopes. The difference in the slopes, in turn, is due to the fact that in the direct SCF, the density matrix can be used in the integral cut-offs to discard more integrals than in the conventional integral evaluation. The position of the crossover depends on the relative speed of the integral evaluation (different algorithms are currently used in *Gaussian 90* for conventional and direct SCF), the spatial extent of the molecule (long molecules have more small integrals that can be eliminated more effectively in the direct procedure) and the number of SCF cycles (the direct method must re-evaluate a fraction of the integrals for every SCF cycle, but the number of SCF cycles is ca 10-15, independent of the size of the molecule).

The second example compares the cost of SCF and MP2 calculations for the normal hydrocarbons. Figure 9 shows the cpu times for MP2 energies (SCF + E2 times from Table 2) and direct SCF energies (also Table 2) using the 6-31G* basis set. The MP2 calculations use the semi-direct algorithm with a maximum of 100 MW of disk. Below 100 basis functions, the MP2 calculations add only a small fraction to the SCF cpu time. In the 100 to 200 basis function range, the MP2 calculations are proportional to $N^{3.5}$ - N^4 rather than the formal ON^4. Even at 200 basis functions, the MP2 calculation costs only $2^1/_2$ times the SCF computation.

Figure 10 compares SCF algorithms on vector and scalar machines. For long instruction word machines like the Multiflow, the direct SCF becomes faster than the conventional SCF at ca. 90 basis functions (ratio of direct / conventional less than one). For scalar workstations such as the Sun, the cross-over may be beyond 130 basis functions, depending on the speed of the I/O. For vector machines like the Stardent, direct SCF becomes more efficient for more than 70 basis functions. The third curve in Figure 10 indicates that in-core SCF calculations are always more efficient than conventional SCF. With 48 MB of memory on the Stardent, in-core SCF can be used up to ca. 80 basis functions, beyond which direct SCF is more efficient than conventional SCF.

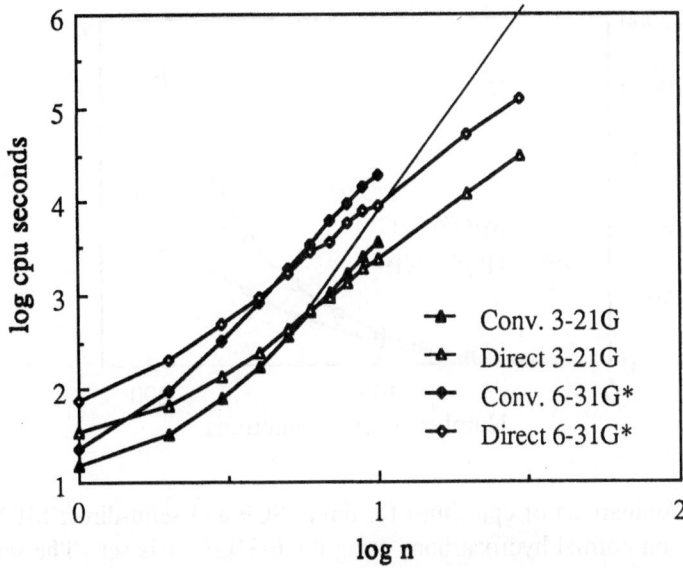

Figure 8. Log-log plot of cpu time vs chain length for normal hydrocarbons, C_nH_{2n+2} (all trans) for conventional and direct SCF using the 3-21G and 6-31G* basis sets.

Table 2: Computational times for SCF calculations on normal hydrocarbons C_nH_{2n+2}

		3-21G			6-31G*		
n	N	Conventional	Direct	N	Conventional	Direct	E2
1	17	16	35	23	23	76	10
2	30	32	65	42	97	208	43
3	43	77	132	61	332	501	103
4	56	174	246	80	860	940	275
5	69	360	431	99	1890	1679	606
6	82	652	685	118	3532	2868	1156
7	95	1048	966	137	6181	3759	2224
8	108	1729	1325	156	9229	5727	3503
9	121	2642	1891	175	13907	8041	7411
10	134	3699	2379	194	18771	9061	12452
20	264		12060	384		52906	
30	394		29910	574		127542	

[1] in seconds on a Multiflow Trace 14/300, E2 calculations semi-direct limited to 100MW disk

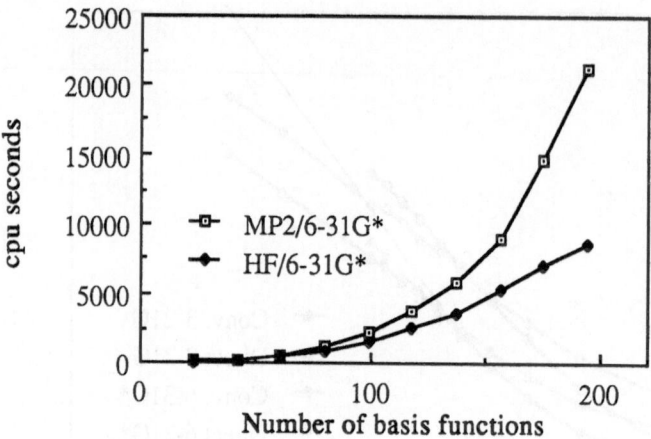

Figure 9. Comparison of cpu times for direct SCF and semi-direct MP2 calculations on normal hydrocarbons using the 6-31G* basis set. The semi-direct MP2 calculation was limited to a maximum of 100 MW of disk.

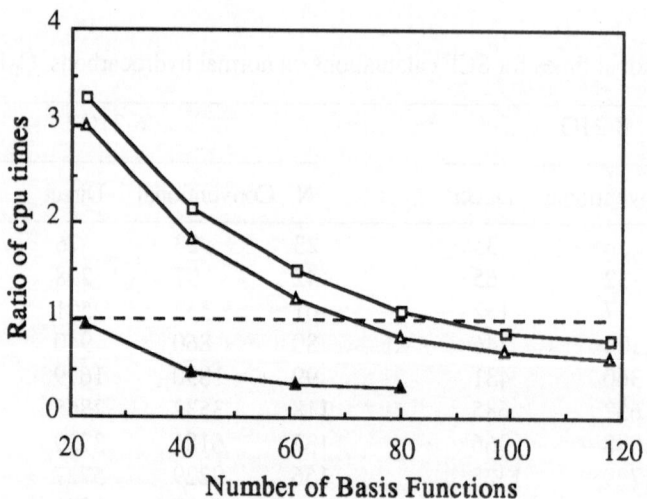

Figure 10. Ratio of cpu times for different SCF algorithms on vector and scalar machines: squares - direct / conventional on a Multiflow Trace 14/300, open triangles - direct / conventional on a Stardent 3020, filled triangles - in-core / conventional on a Stardent using up to 48 MB of memory.

14. References

1. Hehre, W. J., Radom, L., Schleyer, P. vR., and Pople, J. A., *Ab Initio Molecular Orbital Theory,* Wiley-Interscience, New York, 1986.

2. Szabo, A. and Ostlund, N. S., *Modern Quantum Chemistry: Introduction to Advanced Electronic Structure Theory,* Macmillan, New York, 1982.

3. R. McWeeny and Craig, D. P., *Methods of Molecular Quantum Mechanics,* Academic Press, London, 1989.

4. Wilson, S., *Electron Correlation in Molecules,* Clarendon Press, Oxford, 1984.

5. Daudel, R., Leroy, G., Peeters, D. and Sana, M., *Quantum Chemistry,* Wiley-Interscience, Chichester, 1983.

6. Jørgensen, P. and Simons, J., *Second Quantization -Based Methods in Quantum Chemistry,* Academic Press, New York, 1981.

7. Born, M. and Oppenheimer, J. R., *Ann. Physik,* **84**, 457 (1927).

8. Hartree, D. R. *Proc. Cambridge Phil. Soc.,* **24**, 89, 11, 426 (1928); Fock, V. *Z. Phys.,* **61**, 126, (1930); **62**, 795, (1930); **75**, 622, (1932).

9. Slater, J. C., *Phys. Rev.,* **34**, 1293 (1929); **35**, 509 (1930).

10. Boys, S. F., *Proc. Roy. Soc. (London),* **A200** 542 (1950); for a discussion of the utility of gaussians in quantum chemistry see Shavitt, I., *Methods in Computational Physics,* vol 2, Wiley, New York, 1962.

11. Slater, J. C., *Phys. Rev.,* **36**, 57 (19309)

12. Head-Gordon, M. and Pople, J. A., *J. Chem. Phys.,* **89**, 5777 (1988) and references therein.

13. Bauschlicher, C. W. *Theor. Chim. Acta,* **76**, 187 (1989).

14. Roothaan, C. C. J., *Rev. Mod. Phys.,* **23**, 69 (1951); Hall, G. G., *Proc. Roy. Soc. (London),* **A205**, 541 (1951).

15. For a discussion of SCF stability and convergence see Schlegel, H. B.; McDouall, J. J. W. in *Computational Advances in Organic Reactions,* Ögretir, C. Csizmadia, I. G., eds., (Kluwer Academic, the Netherlands), 1990.

16. For a review of MCSCF methods see Werner, H. -J. *Adv. Chem. Phys.,* **69**, 1 (1987); Shepard, R. *Adv. Chem. Phys.,* **69**, 63 (1987); Roos, B. O., *Adv. Chem. Phys.,* **69**, 399 (1987).

17. Almlöf, J.; Korsell, K.; Faegri, K., Jr. *J. Comput. Chem.,* **3**, 385 (1982).

18. Frisch, M. J.; Head-Gordon, M.; Trucks, G.; Foresman, J. B.; Schlegel, H. B.; Raghavachari, K.; Robb, M. A.; Binkley, J. S.; Gonzalez, C.; DeFrees, D. J.; Fox, D. J.; Whiteside, R. A.; Seeger, R.; Melius, C. F.; Baker, J.; Martin, L. R.;

32

Kahn, L. R.; Stewart, J. J. P.; Fluder, E. M.; Topiol, S.; Pople, J. A. *GAUSSIAN 90*, Gaussian, Inc., Pittsburgh PA, 1990.

19. For some reviews of analytical gradients see Pulay, P., *Adv. Chem. Phys.*, **69**, 241 (1987); Jørgensen, P., and Simons, J. (Eds.), *Geometrical Derivatives of Energy Surfaces and Molecular Properties*, Reidel, Dordrecht, 1986; Gaw, J. F., and Handy, N. C., *Annu. Rep. Prog. Chem. Sec. C*, **81**, 291 (1985); Fogarasi, G., and Pulay, P., *Annu. Rev. Phys. Chem.*, **35**, 191 (1984).

20. Schlegel, H. B., *Adv. Chem. Phys.*, **67**, 249 (1987).

21. Kato, S.; Morokuma, K. *Chem. Phys. Lett.*, **65**, 19 (1979); Dupuis, M. *J. Chem. Phys.*, **74**, 5758 (1981); Osamura, Y.; Yamagichi, Y.; Schaefer III, H. F. *J. Chem. Phys.*, **75**, 2919 (1981); **77**, 383 (1982); Schlegel, H. B.; Robb, M. A. *Chem. Phys. Lett.*, **93**, 43 (1982).

22. Pople, J. A.; Krishnan, R.; Schlegel, H. B.; Binkley, J. S. *Int. J. Quantum Chem., Quantum Chem. Symp.*, **13**, 225 (1979).

23. Gerratt, J.; Mills, I. M. *J. Chem. Phys.*, **49**, 1719, 1730 (1968).

24. Osamura, Y.; Yamagichi, Y.; Schaefer III, H. F. *J. Chem. Phys.*, **77**, 383 (1982).

25. Frisch, M. J.; Head-Gordon, M.; Pople, J. A. *Chem. Phys. Lett.*, **141**, 189 (1990).

26. Møller, C.; Plesset, M. S. *Phys. Rev.*, **46**, 618 (1934).

27. For a review of MBPT and coupled cluster methods see Bartlett, R. J. *Annu. Rev. Phys. Chem.*, **32** 359 (1981).

28. Brillouin, L., *Actualities Sci, Ind.*, **71**, 159 (1934).

29. Taylor, P. R. *Int. J. Quantum Chem.*, **31**, 521 (1987); Saebø, S.; Almlöf, J. *Chem. Phys. Lett.*, **154**, 83 (1989); Head-Gordon, M.; Pople, J. A.; Frisch, M. J. *Chem. Phys. Lett.*, **153**, 503 (1989).

30. Frisch, M. J.; Head-Gordon, M.; Pople, J. A. *Chem. Phys. Lett.*, **166**, 275, 281 (1990).

31. Rice, J. E.; Amos, R. D. *Chem. Phys. Lett.*, **122**, 585 (1985); Simandiras, E. D.; Amos, R. D.; Handy, N. C. *Chem. Phys.*, **114**, 9 (1987)

32. Handy, N. C., and Schaefer, H. F., *J. Chem. Phys.*, **81**, 5031 (1984).

33. Krishnan, R.; Pople, J. A. *Int. J. Quantum Chem.*, **14**, 91 (1978); Krishnan, R.; Frisch, M. J.; Pople, J. A. *J. Chem. Phys.*, **72**, 4244 (1980).

34. Pople, J. A.; Seeger, R.; Krishnan, R. *Int. J. Quantum Chem., Quantum Chem. Symp.*, **11**, 149 (1977).

35. Cizek, J. *Adv. Chem. Phys.*, **14**, 35 (1969).

36. Pople, J. A.; Head-Gordon, M.; Raghavachari, K. *J. Chem. Phys.*, **87**, 5968 (1987).

37.   Hurley, A. C. *Electron Correlation in Molecules*, Academic, New York, 1976; Pople, J. A.; Krishnan, R.; Schlegel, H. B.; Binkley, J. S. *Int. J. Quantum Chem.*, **14**, 545 (1978).

38.   Raghavachari, K. *J. Chem. Phys.*, **82**, 4607 (1985); Raghavachari, K.; Trucks, G. W.; Pople, J. A.; Head-Gordon, M. *Chem. Phys. Lett.*, **157**, 479 (1989).

39.   Urban, M.; Noga, J.; Cole, S. J.; Bartlett, R. J. *J. Chem. Phys.*, **83**, 4041 (1985); Noga, J.; Bartlett, R. J. *J. Chem. Phys.*, **86**, 7041 (1987) and references therein.

40.   Fitzgerald, G.; Harrison, R.; Laidig, W. D.; Bartlett, R. J. *J. Chem. Phys.*, **82**, 4375 (1985).

41.   Gauss, J.; Cremer, D. *Chem. Phys. Lett.*, **138**, 131 (1987); **153**, 303 (1988).

42.   Krishnan, R.; Schlegel, H. B.; Pople, J. A. *J. Chem. Phys.*, **72**, 4654 (1980); Brooks, B. R.; Laidig, W. D.; Saxe, P.; Goddard, J. D.; Yamagichi, Y.; Schaefer II, H. F. *J. Chem. Phys.*, **72**, 4652 (1980).

43.   Scheiner, A. C.; Scuseria, G. E.; Rice, J. E.; Lee, T. J.; Schaefer II, H. F. *J. Chem. Phys.*, **87**, 5361 (1987); Salter, E. A.; Trucks, G. W.; Bartlett, R. J. *J. Chem. Phys.*, **90**, 1752 (1989).

44.   Gauss, J.; Cremer, D. *Chem. Phys. Lett.*, **150**, 280 (1988); *ibid*, **163**, 549 (1989).

VARIATIONAL TRANSITION STATE THEORY CALCULATIONS OF CONCERTED HYDROGEN ATOM TUNNELING IN WATER CLUSTERS AND FORMALDEHYDE/WATER CLUSTERS

Bruce C. Garrett
Molecular Science Research Center
Pacific Northwest Laboratory[*]
Richland, WA 99352

Carl F. Melius
Sandia National Laboratories
Livermore, CA 94550

ABSTRACT. The direct participation of water molecules in aqueous phase reaction processes has been postulated to occur via both single-step mechanisms as well as concerted hydrogen atom or proton shifts. In the present work, simple prototypes of concerted hydrogen atom transfer processes are examined for small hydrogen-bonded water clusters – cyclic trimers and tetramers – and hydrogen-bonded clusters of formaldehyde with one and two water molecules. Rate constants for the rearrangement processes are computed using variational transition state theory, accounting for quantum mechanical tunneling effects by semiclassical ground-state adiabatic transmission coefficients. The variational transition state theory calculations directly utilize selected information about the potential energy surface along the minimum energy path as parameters of the reaction path Hamiltonian. The potential energy information is obtained from *ab initio* electronic structure calculations with an empirical bond additivity correction (the BAC-MP4 method). Tunneling is found to be very important for these concerted rearrangement processes – the semiclassical ground-state adiabatic transmission coefficients are estimated to be as high as four orders of magnitude at room temperature. Effects of the size of the cluster (number of water molecules in the cyclic complex) are also dramatic – addition of a water molecule is seen to change the calculated rates by orders of magnitude.

1. Introduction

Hydrogen atom and proton transfer are among the simplest of elementary processes that can occur in aqueous solutions. These processes are important in proton transfers in enzyme and other biochemical processes, chemical reactions occurring in aqueous solutions, and oxidation of organic waste in supercritical aqueous solutions. Large proton transfer rates between oxygen atoms in organic acids have been observed in alcohol and aqueous solutions of acids, and these have been attributed to multiple proton jumps involving direct participation of solvent molecules [1-3]. Using simple models for the potential energy surfaces, the relative energetics of reaction pathways have been used to examine the competition between stepwise mechanisms and concerted processes of proton transfer [4-8]. *Ab Initio* electronic structure methods have also been employed to investigate the mechanisms of these proton transfer processes [9-11]. In some instances

35

S. J. Formosinho et al. (eds.), Theoretical and Computational Models for Organic Chemistry, 35–54.
© 1991 *Kluwer Academic Publishers.*

large measured rates for proton transfer have been attributed to concerted processes [12-14].

In this paper prototypes of concerted hydrogen atom transfer in hydrogen-bonded clusters are examined using modern computational techniques for obtaining reaction energetics and for calculating reaction rates. The effects of the size of the cyclic hydrogen-bonded cluster on calculated rates of hydrogen atom transfer processes are studied for water trimers and tetramers, diagrams 1 and 2 below, and for formaldehyde hydrolysis by one and two water molecules, diagrams 3 and 4 below. We treat the unimolecular process of reacting from the bound hydrogen-bonded cluster. In aqueous phase the breaking up and reformation of these cyclic structures will be facile and calculation of the rates of these processes in condensed phase will be more complicated. Furthermore, other mechanisms may be important, for example, it may be more realistic to treat reaction 4 as a formaldehyde molecule which is solvated and reacting with a water dimer in bulk water rather than a unimolecular rearrangement. In addition, these complexes will be solvated by water molecules that do not directly participate in the hydrogen atom transfer and this will also change the reaction energetics. These effects are beyond the scope of the current paper which represents a first step towards addressing some of these more complicated issues.

Rate constants for these unimolecular processes are calculated using variational transition state theory with semiclassical adiabatic ground-state transmission coefficients [15-25]. These calculations include important quantum mechanical tunneling effects by the small-curvature semiclassical adiabatic ground-state method which includes the effects of curvature of the reaction path. Compared to conventional transition state theory [26] with simple tunneling correction factors based on quadratic expansions of the potential energy surface near the saddle point [27], variational transition state theory with semiclassical adiabatic ground-state transmission coefficients uses a more extended description of the potential along a reaction path and is able to provide more quantitative predictions of reaction rates. Even though variational transition state theory requires more extensive descriptions of the potential energy surface, these calculations can directly utilize potential information obtained from modern electronic structure methods, i.e., the energy and derivatives of the energy along a reaction path, providing an efficient means of estimating reaction rates from first principles calculations [15].

1

2

3

4

In the present work, information about the potential energy surfaces for these systems is obtained by the BAC-MP4 method [28-33]. This method has been very successful for predicting the thermochemistry of molecules and radical species, and has been extended to calculating the potential information along reaction paths needed for the variational transition state theory calculations. In the latter case, the method has been shown to be capable of quantitative predictions for a gas phase chemical reaction [33]. In the present study our interests are in estimates of the order of magnitude of reaction rates, and in studies of qualitative trends such as the effect of cluster size on the magnitude of quantum tunneling. The methods employed here are more than adequate for these types of studies.

The previous studies of concerted hydrogen atom and proton transfer in hydrogen-bonded systems have been limited to studies of reaction pathways for simple model systems [4-8] with simple, reduced dimensionality methods for including quantum tunneling [13,14,34,35]. The applicability of modern computational methods to such systems is exemplified by more recent studies of the energetics of intramolecular hydrogen atom transfer in molecules such as malonaldehyde [36-39] and models of tunneling in malonaldehyde [39,40]. Williams and coworkers [10] have used *ab initio* electronic structure methods to investigate the concerted proton transfer reaction mechanism in addition to carbonyl groups. We extend the studies of Williams *et al.* to include higher levels of electron correlation and follow the entire reaction pathway in order to apply variational transition state theory and to treat the tunneling process.

Variational transition state theory, including semiclassical transmission coefficients, has been extensively review in the literature [15-25], and the use of variational transition state theory with the BAC-MP4 method has also been described in detail [33]. The reader is referred to the previous papers for details of the theory and the computational procedures; a brief outline of the procedure is provided in section 2. Results of the calculations for the concerted hydrogen motion in the four systems $(H_2O)_3$, $(H_2O)_4$, CH_2O-H_2O, and $CH_2O-(H_2O)_2$, are presented and discussed in section 3. Conclusions are presented in section 4.

2. Computational methods

Ab initio electronic structure calculations were performed with the Gaussian 88 set of codes [41]. The first step in the calculations is the identification of the relevant equilibrium geometries for the reactant and product and the saddle point connecting these two stable points. These geometries are optimized at the Hartree-Fock level with a basis set denoted 6-31G* [42]. The minimum energy path connecting the saddle point with the reactant and product is then located by following the paths of steepest descent in both directions from the saddle point in a mass weighted coordinate system such that each degree of freedom has the same reduced mass in the kinetic energy expression. The distance s is defined as the arc length along the reaction path from the saddle point in this mass weighted coordinate systems. For each system studied here, we chose the reduced mass to be the total mass of that system. The first step from the saddle point is taken along the eigenvector of the negative eigenvalue of the Hessian, and subsequent steps are taken in the direction of the negative of the gradient (the first derivative of the energy). In the present calculations, the minimum energy path is obtained using the local quadratic approximation of Page and McIver [43]. This approach requires the first and second derivatives along the reaction path which are also obtained at the Hartree Fock level with the 6-31G* basis. One concern in the calculation of transition state structures at the HF level is the applicability of a single determinate wavefunction in the interaction region. Limited multiconfiguration SCF calculations performed on the transition-state structure of the cyclic water trimer indicate that the electronic structure is dominated by a single configuration [44].

A step size δs of 0.01 a_0 was used in the local quadratic approximation for following the reaction path for all of the systems. The accuracy of the calculated reaction path was tested by comparing computed rate constants (including tunneling effects) with at least one other reaction path determined with another step size. In all cases the rate constants for temperatures above 300 K are numerically converged to better than 50% and those at 200 K are numerically accurate to about a factor of 2. Tests were also made of the importance of the extent of the reaction path in the tunneling calculations and the calculations are converged to better than 50% at all reported temperatures.

In the BAC-MP4 method it is assumed that geometries and frequencies can be obtained to a good approximation from a Hartree-Fock calculation using a moderately sized basis, 6-31G*. The absolute energy for a given geometry is much harder to accurately compute and requires including a large percentage of the electron correlation energy. As a first step in this direction, at the optimized HF geometries and for the points along the minimum energy path, the energy is computed by full fourth-order Møller-Plesset perturbation theory [45] (MP4-SDTQ), including the effects of all single, double, triple, and quadruple excitations in the contracted atomic orbital basis denoted 6-31G** [42] in which the 1s electrons of the oxygen and carbon atoms are frozen. Although the MP4 method does include a large fraction of the electron correlation energy, there are still errors due to basis set and higher order correlation effects, so a correction based upon bond additivity – the BAC – is applied. In all the calculations reported here the minimum energy path and the first and second derivatives along the minimum energy path are obtained at the Hartree-Fock level with the 6-31G* basis. We then distinguish the method for computing the energy along the reaction path as HF, MP4 (which is computed with the 6-31G** basis), and BAC-MP4 when the bond-additivity correction is added to the MP4 results.

The bond additivity correction is typically computed only for atom pairs with interatomic distances within a preset cutoff. For studying chemical reactions in which bonds are made and broken it is necessary to force the BAC to be computed for some pairs of atoms at all distances to insure continuity of potential energy curves. In the present study the BAC is applied to all the hydrogen-bonded interactions, although in the standard application it is not. This introduces approximately an additional 0.8 kcal/mol of binding for each hydrogen-bonded interaction at the equilibrium hydrogen-bonded internuclear distance.

Rate constant calculations were carried out using the POLYRATE program [23]. These calculations directly utilize the energy and the first and second derivatives along the minimum energy path. Rate constants are calculated by the canonical variational theory (CVT) [24] with multidimensional quantum mechanical tunneling effects included by the small-curvature semiclassical adiabatic ground-state (SCSAG) transmission coefficient [46,47]. The small-curvature semiclassical adiabatic ground state (SCSAG) method includes the effect of the reaction-path curvature to induce the tunneling path to 'cut the corner' and shorten the tunneling length. A central quantity in both the CVT and SCSAG calculations is the ground-state adiabatic potential curve

$$V_a^G(s) = V_{MEP}(s) + \sum_{m=1}^{F-1} \varepsilon_{vib,m}^{GT}(n_m=0,s) \tag{1}$$

where $V_{MEP}(s)$ is the value of the potential along the reaction path at s, the sum is over the F-1 bound vibrational modes of the generalized transition state at s (F=3N-6, where N is the number of atoms), and $\varepsilon_{vib,m}^{GT}(n_m,s)$ is the bound energy level for state n_m in mode m at the generalized transition state. It is also useful to define the adiabatic potential relative to the total zero-point energy of the reactants

$$\Delta V_a^G(s) = V_a^G(s) - \sum_{m=1}^{F} \epsilon_m^R(n_m=0) \tag{2}$$

where $\epsilon_m^R(n_m)$ is the reactant energy level for state n_m in mode m and the sum goes over F bound modes (instead of the F-1 modes at the generalized transition state where the reaction coordinate motion is the Fth mode). Choosing the zero of energy such that $V_{MEP}(s)$ is zero at the reactant equilibrium geometry, the maximum of $\Delta V_a^G(s)$ is the adiabatic threshold for reaction. In the current studies we treat all bound degrees of freedom harmonically, and the energy levels are given simply by

$$\epsilon_{vib,m}^{GT}(n_m,s) = (n_m+\tfrac{1}{2})\hbar\omega_m(s) \tag{3}$$

where $\omega_m(s)$ is the harmonic frequency for mode m at the generalized transition state located at s. The harmonic frequencies are obtained from the gradient vector and Hessian matrix using the projection technique of Miller, Handy, and Adams [48].

More accurate calculations of the reaction rates require a careful investigation of the effects of anharmonicity, especially for the the low frequency modes in these loosely bound complexes. As discussed elsewhere [33], effects of anharmonicity are approximately included in the empirical BAC correction. A manifestation of anharmonicity in these hydrogen-bonded complexes is multiple minima in the equilibrium region and multiple transition states that correspond to different orientations of the terminal hydrogens in these cyclic complexes. For example, in the cyclic water trimer, orienting the three hydrogens labeled H2, H4, and H6 in diagram 1 below the plane of the 6-atom ring is slightly higher in energy than to have two hydrogens below and one above the plane. These two minima are similar in energy – they differ by about two kcal/mol – and are separated by a transition state of less than a kcal/mol. Similarly, in the transition state region for the reaction in diagram 1, the two transition structures with all terminal hydrogens below the plane is higher in energy by about two kcal/mol than for two below and one above the plane. The harmonic approximation for these H wagging modes will underestimate the contribution to the partition functions for these modes. However, since these modes are treated the same in the reactant and transition state, the errors will tend to cancel.

Another concern in the dynamical calculations is the accuracy of the approximations used in the calculations of the quantum tunneling corrections. The concerted processes considered here are light hydrogen atom transfers between heavy groups. In analogy with gas phase reactions of light atom transfer between two heavy moieties, these types of systems have regions of the reaction path with high reaction-path curvature. For reactions in which tunneling occurs in regions of high reaction-path curvature, the SCSAG method only gives qualitative estimates of the tunneling factor;

more reliable estimates can be obtained using the large-curvature ground-state (LCG) method [22,49] which has recently been extended to polyatomic systems [50]. However, in its present implementation, it can only be applied with global potential energy surfaces, although there are indications that the method can be adapted to using only limited information about the potential energy surface within the framework of the reaction path Hamiltonian [50].

The transmission coefficients are obtained from Boltzmann averages of the probabilities for tunneling through the ground-state adiabatic barrier. For the unimolecular processes considered here the reactant and product species correspond to local wells in the adiabatic potential separated by the adiabatic barrier. For this case, averaging over a continuum of translational energies is only an approximation since tunneling occurs from discrete energy levels in the bound wells of the adiabatic potential [51,52]. For the systems studied in this paper, the barriers are high and the adiabatic potentials will support many states. Thus, the continuum limit will yield a reasonable approximation. Quantitative accuracy in calculations of reaction rates for polyatomic systems presents a challenge because of limited accuracy in the computed energies and the approximations in the dynamical calculations of the rates. However, the methods used here will provide good estimates of the orders of magnitude for these reactions and will be very useful for predicting relative rates in homologous series of reactions.

3. Results and Discussion

3.1. STRUCTURE AND ENERGETICS OF WATER CLUSTERS

The heat of formation of water at 0 K using the BAC-MP4 method is taken to be -57.1 kcal/mol, the same as the experimental value [53]. The accuracy of the BAC-MP4 method for predicting the structure and binding in hydrogen-bonded clusters is tested by comparing the calculated values with experiment [54,55] and with previous calculations [56,57]. The calculated geometry is shown in diagram $\underline{5}$ (bond lengths are in units of Å).

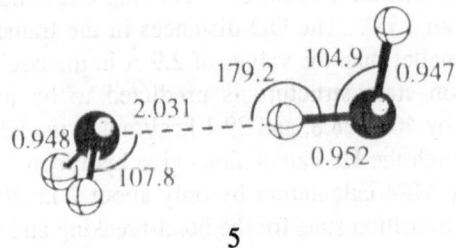

$\underline{5}$

The bond angles and lengths are in reasonable agreement with previously calculated results. The computed hydrogen-bond energy (neglecting zero-point energy) in the dimer is 5.5, 6.7, and 7.4 kcal/mol computed by the HF, MP4, and BAC-MP4 levels of theory. The heat of formation of the dimer from two water molecules at 0 K is computed to be

-5.2 kcal/mol at the BAC-MP4 level. This is about 2 kcal/mol lower in energy than the experimental value and is consistent with the reliability of the BAC-MP4 method.

The lowest energy equilibrium cyclic configuration of the trimer is shown in diagram 6. This agrees with previous *ab initio* electronic structure calculations [58,59]. The (OH)₃ ring is nearly planar – the largest distortion out of plane is about 9° – and two of the terminal H atoms are below the plane with one above. The hydrogen bonds are not quite symmetric – bond lengths range from 1.99 to 2.01 Å – and the OHO bond angle varies from 150° to 152°. Relative to three separated water molecules (at their equilibrium geometries), the cyclic trimer is calculated to be bound by -17.1, -21.5, and -23.6 kcal/mol at the HF, MP4, and BAC-MP4 levels of theory (these energies represent the electronic energies without zero-point energies), and the heat of formation relative to three water molecules is predicted by the BAC-MP4 method to be -17.7 kcal/mol at 0 K.

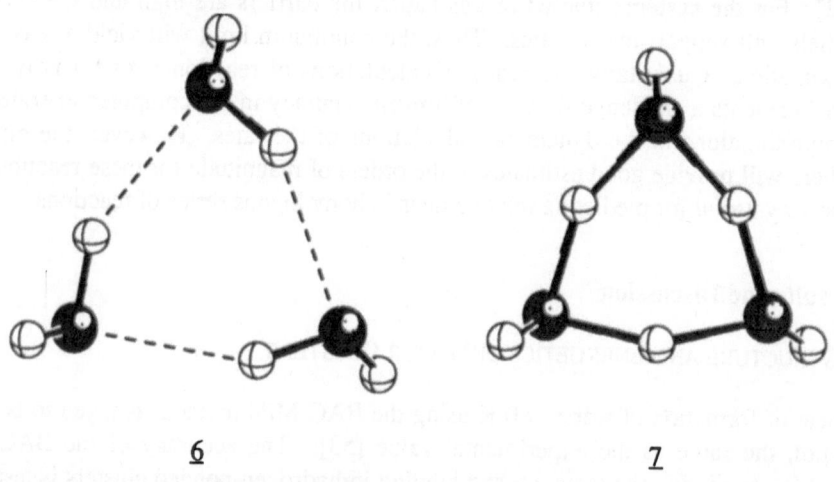

6   7

The rearrangement of the hydrogens as indicated in diagram 1 proceeds via the transition state structure shown in diagram 7. The transition state is also nearly planar – its largest out-of-plane distortion is about 10°. The ring OH bonds are all 1.21 Å and the OHO bond angles are all 153°. The OO distances in the transition state structure are about 2.3 Å which is smaller than the values of 2.9 Å in the equilibrium geometry. The energy of the transition state structure is predicted to be higher than that of the equilibrium structure 6 by 46.6, 28.8, and 29.1 kcal/mol at the HF, MP4, and BAC-MP4 levels of theory. Although the HF calculations give hydrogen-bond energies that differ from the higher quality MP4 calculation by only about 2 kcal/mol for each hydrogen bond, the energy of the transition state for the bond-breaking and bond-making process is nearly 18 kcal/mol higher at the HF level than the MP4 level.

Frequencies calculated as a function of location along the reaction path are presented in figure 1. The highest frequency modes corresponds the the OH stretch for the terminal hydrogens. These frequencies change only slightly as a function of the reaction coordinate. The next (lower) set of frequencies correlate with the 'in-plane' OH stretch in the equilibrium configurations. At the saddle point the OH bond is stretched relative to the equilibrium geometries and this frequency is greatly reduced.

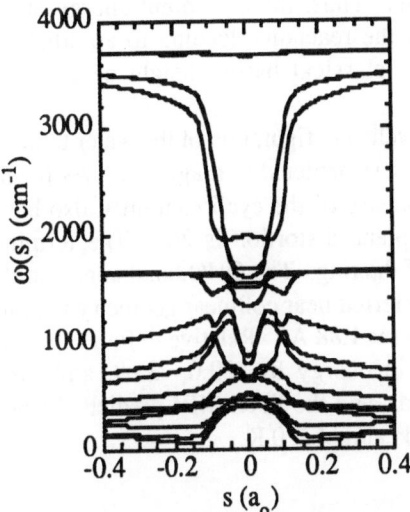

Figure 1. Harmonic frequencies for bound modes orthogonal to the reaction path as a function of the reaction coordinate for the concerted hydrogen atom transfer in the water trimer (see diagram 1).

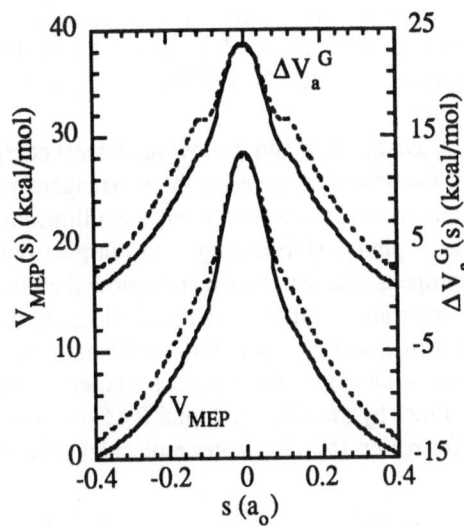

Figure 2. Potential energy along the minimum energy path and relative ground-state adiabatic potential energy [eq. (2)] as a function of reaction coordinate for the concerted hydrogen atom transfer in the water trimer. The zero of energy for the potential along the minimum energy path is the at the equilibrium geometry of the reactants. Solid lines are calculated using the BAC-MP4 calculations and the dashed curves are computed using the MP4 calculations.

Rapid changes in the frequencies occur near $s=\pm0.1$ a_0 which is the region of the reaction path with the highest reaction path curvature. Near the saddle point, the reaction coordinate motion corresponds mostly to concerted motion of the three hydrogen atoms in the ring and this motion occurs along nearly straight lines in cartesian coordinates. However, as is evident from diagrams 6 and 7 the oxygen atoms must also move away from each other in proceeding along the reaction coordinate from the saddle point to the equilibrium geometry. Near $s=\pm0.1$ a_0 the reaction coordinate motion incorporates more oxygen atom motion and the coupling of the oxygen and hydrogen motions along the reaction path leads to more reaction-path curvature.

The energy on the minimum energy path (neglecting zero-point energy effects) $V_{MEP}(s)$ and the ground-state adiabatic potential curve relative to the reactant zero point energy $\Delta V_a^G(s)$ are plotted in figure 2. The BAC-MP4 (solid curves) and MP4 energies (dashed curves) agree surprisingly well, indicating that the bond additivity correction nearly cancels out between reactants and points along the reaction path.   Besides the large drop in the three 'in plane' OH stretches in going from the reactants to the saddle point, as indicated above, there is one extra bound vibrational mode in reactants that becomes the unbound reaction coordinate motion at the saddle point. This is the mode which contributes to the zero-point energy of the reactants, but has no contribution to the adiabatic potential along the reaction coordinate. Thus, the zero-point energy at the saddle point is nearly 5 kcal/mol lower than at the reactants, leading to an adiabatic threshold of only 24 kcal/mol compared to the classical barrier height of about 28 kcal/mol.

Diagrams $\underline{8}$ and $\underline{9}$ show the lowest energy cyclic configuration of the water tetramer and the transition state for the rearrangement process depicted in diagram $\underline{2}$. As for the cyclic trimer, the lowest energy equilibrium geometry of the cyclic tetramer also has a nearly planar $(OH)_4$ ring – the largest out-of-plane distortion is 20°. The terminal hydrogens alternate above and below the plane of the ring. The OHO bond angles in the tetramer are about 167°, which is closer to the preferred near collinear geometry than for the trimer, and the hydrogen-bond lengths are all near 1.88 Å. Relative to four separated water molecules, the cyclic tetramer is lower in energy by -29.0, -35.9, and -39.2 kcal/mol by the HF, MP4, and BAC-MP4 methods, and the BAC-MP4 predicts the heat of formation from the water molecules to be 30.4 kcal/mol at 0 K.

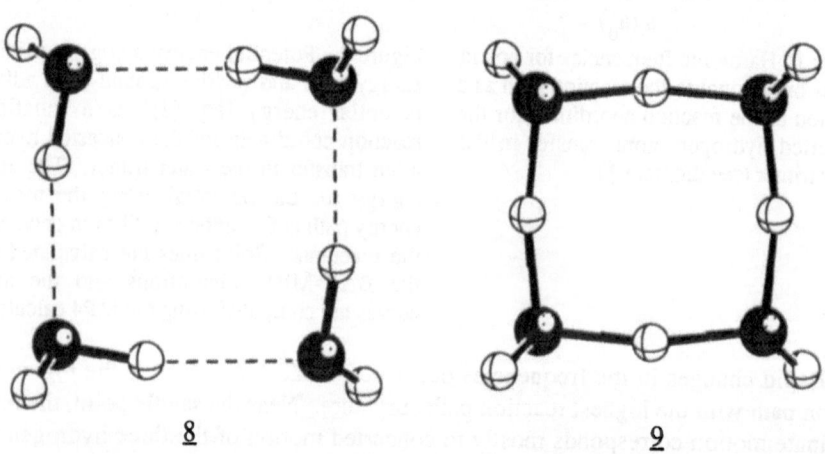

$\underline{8}$ $\underline{9}$

The transition state structure has a planar ring – its largest distortion is only 2°. The OHO bond angles are 168°, very closer to those for the equilibrium structure, and all the OH bond lengths in the ring are 1.20 Å. Similar to the trimer structures, the OO distances change from 2.8 Å in the equilibrium structure to 2.4 Å in the transition state structure. Relative to the reactant complex, the transition state is higher in energy by 45.2, 25.0, and 25.1 kcal/mol at the HF, MP4, and BAC-MP4 methods. As was the case for the water trimer, the HF method is much higher in energy than the MP4 method, but the MP4 and BAC-MP4 methods agree very well.

Frequencies and potential energy curves for the concerted hydrogen atom transfer process in the tetramer (diagram $\underline{2}$) are presented in figures 3 and 4. The variations of the frequencies with s are very similar to those for the trimer; for the OH bonds in the ring, the stretching frequencies exhibit large changes in going from the reactant to the transition state. The potential curve $V_{MEP}(s)$ for the tetramer is very similar in shape to that for the trimer (shown as a dashed curve in figure 4) but the barrier height is lower in energy by about 4 kcal/mol. The changes in zero-point energies in going from the reactant to the saddle point are greater in the tetramer than the trimer. The difference in adiabatic thresholds is nearly 8 kcal/mol. The adiabatic potential for the tetramer also shows more structure and is a little broader than that for the trimer.

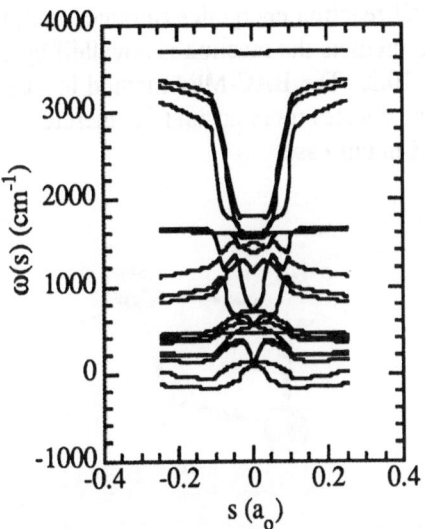

Figure 3. Same as figure 1 except for the concerted hydrogen atom transfer in the water tetramer (see diagram 2).

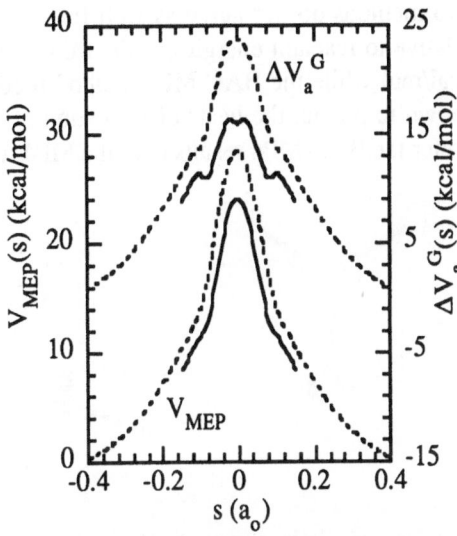

Figure 4. Comparison of potential energy along the minimum energy path and relative ground-state adiabatic potential for the concerted hydrogen atom transfer in the water trimer (dashed curves) and water tetramer (solid curves), using BAC-MP4 calculations.

3.2 STRUCTURES AND ENERGETICS OF FORMALDEHYDE-WATER CLUSTERS

Diagrams 10, 11, and 12 show the reactant, transition state, and product structures for formaldehyde hydrolysis by one water molecule (see diagram 3). The reactant complex is nearly planar for the COHO forming a 'ring' – the torsion angle for these four atoms is only 11°. The OH hydrogen bond has a length of 2.1 Å and the distance for the oxygen in the water molecule to the carbon atom is 3.2 Å. The dimer is bound relative to separated formaldehyde and water by 5.2, 6.6, and 7.3 kcal/mol by the HF, MP4 and BAC-MP4 methods. For the products, diagram 12, the CO bonds lengths are 1.4 Å with OH bond lengths of 0.95 Å. Relative to separated formaldehyde and water, the diol is more bound by 14.6, 14.2, and 17.5 kcal/mol at the HF, MP4, and BAC-MP4 levels of theory.

The transition state shown in diagram 11 is much nearer to planar than the reactant with an out-of-plane torsion of less than 0.1°. Compared to the reactants, the OH hydrogen bond has shortened to 1.3 Å, the distance from the carbon to the water oxygen is only 1.6 Å, and the 'active' OH bond in water has lengthened from 0.95 Å to 1.15 Å. The energy of the transition state relative to the reactant dimer is 53.5, 42.3, and 41.0 kcal/mol by the HF, MP4, and BAC-MP4 methods. Once again the HF method gives an energy which is much larger than either the MP4 or BAC-MP4 methods. As was the case for the water clusters, the MP4 and BAC-MP4 methods agree well for the relative energetics going from the hydrogen-bonded species to the transition state. However, the two methods do not agree as well for the overall reaction energetics (product energies relative to reactant energies) – the MP4 method predicts the reaction is downhill by 7.7 kcal/mol while the BAC-MP4 method predicts 10.2. The BAC-MP4 method has been shown to predict the heats of formation of bound species more accurately, therefore we prefer the BAC-MP4 results over the MP4 results in this case.

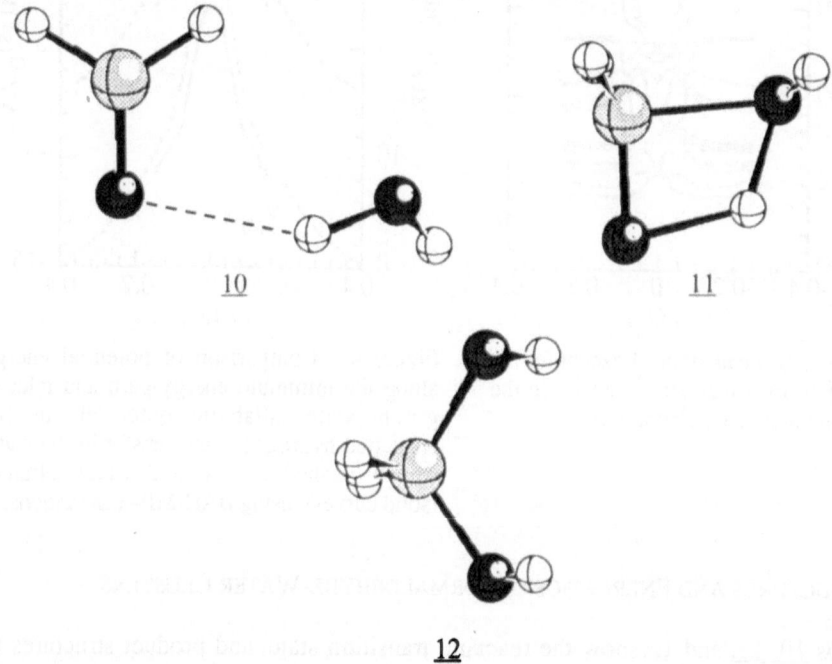

10

11

12

Plots of frequency and potential energies versus reaction coordinate are shown in figures 5 and 6. As for concerted hydrogen atom transfer in water clusters, some of the harmonic frequencies exhibit dramatic changes in going from the reactant to the saddle point. The high frequency mode which changes the most corresponds to the OH bond of the water that is being broken and will become the new OH bond in the diol. Unlike the water clusters which indicate a lower zero-point energy at the saddle point compared to the hydrogen-bonded reactant, for formaldehyde hydrolysis the adiabatic threshold and classical barrier heights agree to within 0.3 kcal/mol. Obviously the marked decrease in one high frequency mode is offset by the more gradual increase in many of the low frequency modes. The classical potential $V_{MEP}(s)$ and the relative adiabatic potential $\Delta V_a^G(s)$ are nearly identical in height and shape for this process.

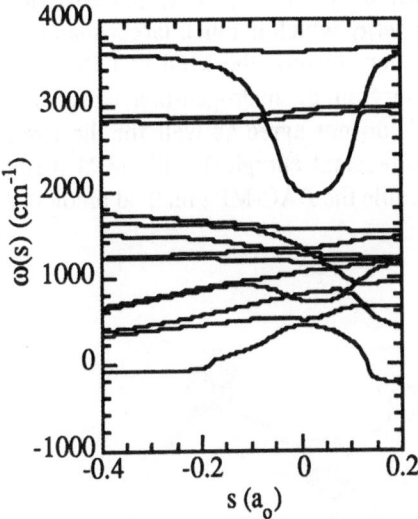

Figure 5. Same as figure 1 except for formaldehyde hydrolysis by a single water molecule (see diagram 3).

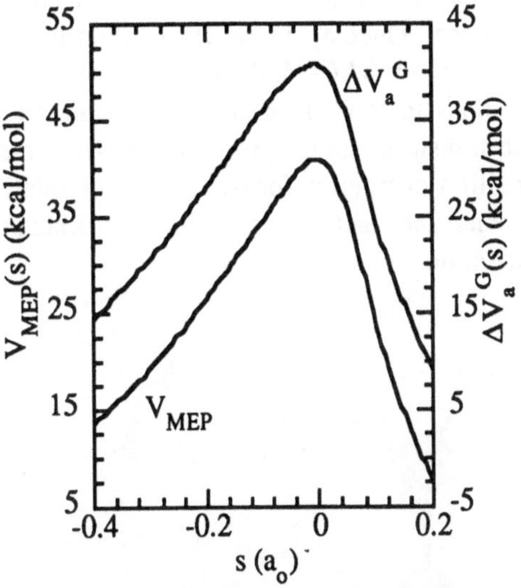

Figure 6. Potential energy along the minimum energy path and relative ground-state adiabatic potential curves as a function of reaction coordinate for formaldehyde hydrolysis, by one water, using BAC-MP4 calculations.

Diagrams 13, 14, and 15 show the reactant, transition state, and product structures for formaldehyde hydrolysis by two water molecules (see diagram 4). In the reactant complex, the OHOHO forming the ring is nearly planar – the largest out-of-plane angle is only 9° – but the carbon atom is distorted from planar by 22°. The OH hydrogen bond between the carbonyl oxygen and water has a length of 2.1 Å, the water-water OH hydrogen-bond length is 2.0 Å, and the distance for the oxygen in water to the carbon atom is 2.8 Å. The dimer is bound relative to separated formaldehyde and 2 water molecules by 13.5, 16.4, and 16.0 kcal/mol by the HF, MP4 and BAC-MP4 methods.

For the product, diagram 15, the CO bonds lengths are 1.4 Å with OH bond lengths of 0.95 Å. The OH hydrogen-bond lengths are 2.0 Å for the distance from the water O to alcohol H, and 2.1 Å for the distance from the alcohol O to the water H. The OH groups that help form the ring are nearly planar – the largest torsion angle is only 5° – but the two CO bonds are at angles of 21° and 38° to this plane. Relative to separated formaldehyde and two water molecules, the diol-water complex is more bound by 18.6, 20.7 and 23.7 kcal/mol at the HF, MP4, and BAC-MP4 levels of theory.

For the transition state shown in diagram 14, the atoms OHOHO that form the ring with C is much nearer to planar than the reactant – the largest out-of-plane torsion is about 5° – but the carbon atom is still out of the plane – the largest out-of-plane torsion is about 30°. All OH bonds in the ring are intermediate between a covalent OH bond length of 0.95Å and a hydrogen-bond OH bond length of nearly 2.0 Å. The distance from the carbon atom to oxygen atom in the water molecule is 1.6 Å and the carbonyl CO bond length has lengthened from 1.2 Å to 1.3 Å. The energy of the transition state relative to the reactant dimer is 40.1, 24.2, and 20.3 kcal/mol by the HF, MP4, and BAC-MP4 methods. Once again the HF method gives an energy which is much larger than either the MP4 or BAC-MP4 methods. As for the water clusters, the MP4 and BAC-MP4 methods agree well for the relative energetics going from the hydrogen-bonded species to the transition state. However, the two methods do not agree as well for the overall reaction energetics (product energies relative to reactant energies) – the MP4 method predicts the reaction is downhill by 4.3 kcal/mol while the BAC-MP4 method predicts 7.7 kcal/mol.

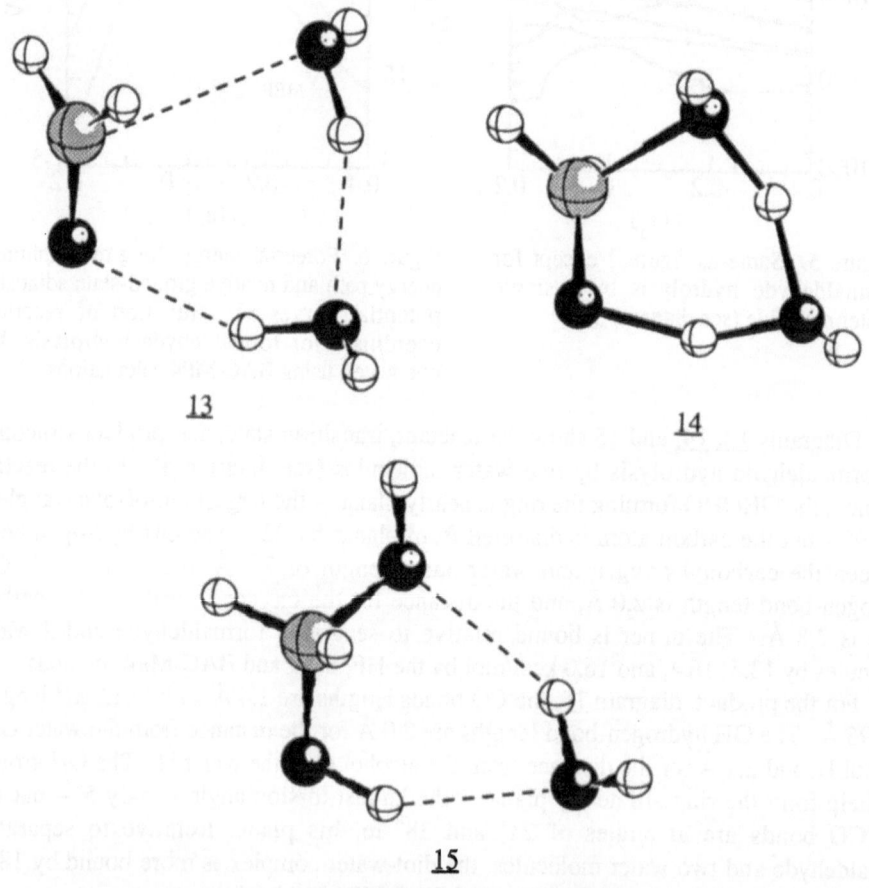

13

14

15

The frequencies as a function of reaction coordinate presented in figure 7 show the same general behavior seen for the formaldehyde-water cluster. As in the previous case, the zero-point energy of the reactant is within 0.3 kcal/mol of the zero-point energy of the reactants. Therefore, the relative adiabatic potential at the saddle point is nearly equal to the classical barrier height. As seen in figure 8, the maximum of the adiabatic potential occurs about 0.06 a_0 before the barrier and is about 1.3 kcal/mol higher than the value at the saddle point. Compared to the formaldehyde-single-water reaction, the classical barrier for the reaction with two water molecules is lowered by over 20 kcal/mol and the adiabatic threshold is lowered by 17.5 kcal/mol. The shapes of the two classical barriers are very similar for the two systems, but the adiabatic barrier for the system with two waters is slightly broader than that for the single water system.

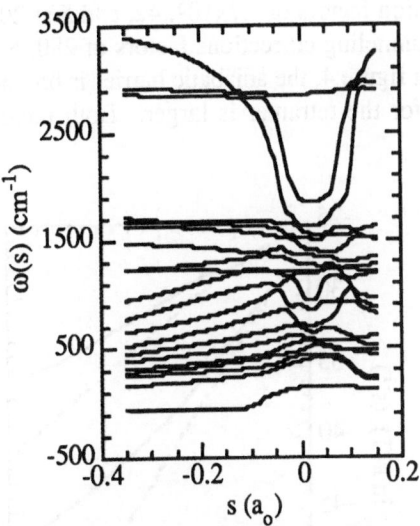

Figure 7. Same as figure 1 except for formaldehyde hydrolysis by two water molecules (see diagram 4).

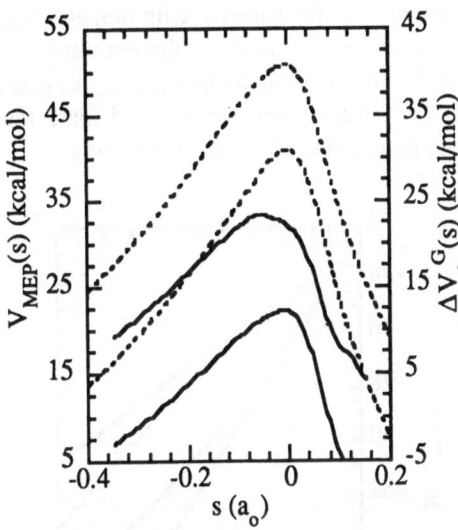

Figure 8. Comparison of potential energy along the minimum energy path and relative ground-state adiabatic potential as a function of reaction coordinate for formaldehyde hydrolysis by one water molecule (dashed curves) and two water molecules (solid curves), by BAC-MP4 calculations. For each set, the top curve is the the adiabatic potential to lower curve is the potential energy.

3.3 REACTION RATES

In this paper, the direct participation of water molecules in the concerted proton transfer process has been addressed, but the large solvating effect of surrounding water molecules on the stabilization of the transition state has not been considered. For the reactions studied here, the barriers are quite high, ranging from ~16 kcal/mol to ~40 kcal/mol. As a consequence, the resulting gas-phase rate constants are quite low and would not be measurable in the temperature range from room temperature up to about 600 K. However, it is interesting to see how these rates vary with the number of participating water molecules in the concerted proton transfer and how the number of participating water molecules effects the importance of tunneling. Rates for the hydrogen atom transfer processes in water clusters (diagrams 1 and 2) are shown in figure 9 and those for the two formaldehyde hydrolysis reactions (diagrams 3 and 4) are compared in figure 10.

The rate constants for the concerted hydrogen atom transfer in cyclic water clusters are seen to change dramatically with the number of waters in the clusters. The CVT/SCSAG rate constants for the tetramer are enhanced by factors of 4.5×10^4, 7500, and 4 relative to the trimer for temperatures of 200, 300, and 400 K. Tunneling is very important for these reactions, as can be seen by comparing the CVT results (which neglects tunneling) with the CVT/SCSAG results in figure 9. Tunneling is more important for the trimer – with tunneling correction factors of 1.7×10^5, 42, and 5 at 200, 300, and 400 K – than for the tetramer – with tunneling corrections factors of 280, 5.4, and 2.1 for the same temperatures. As is seen in figure 4, the adiabatic barrier is broader for the tetramer and the reduced mass factor for the tetramer is larger. Both factors contribute to the diminished tunneling.

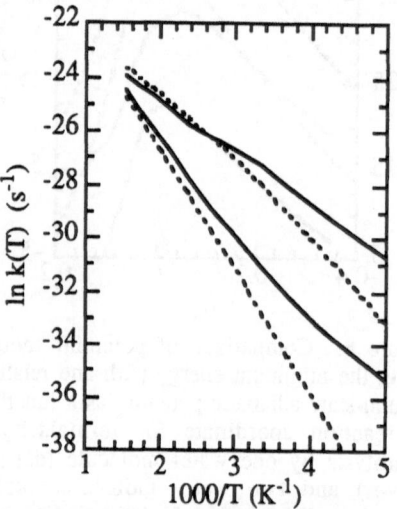

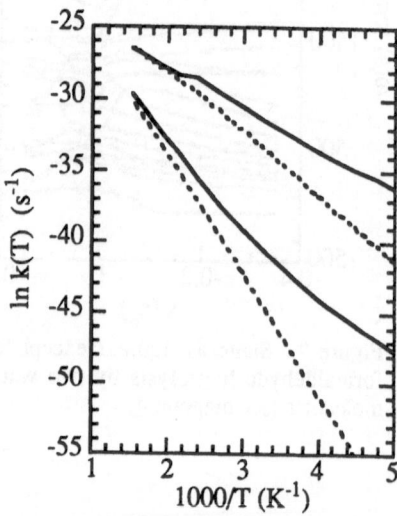

Figure 9. Rate constants calculated by the CVT (dashed curves) and CVT/SCSAG (solid curves) methods for concerted hydrogen atom transfer in cyclic water trimers (lower set of curves) and tetramers (upper set of curves).

Figure 10. Same as figure 9 except for formaldehyde hydrolysis by one water molecule (lower set of curves) and by two water molecules (upper set of curves).

The changes in the rate constants with addition of water molecules are even larger for formaldehyde hydrolysis. Compared to the reaction with one water molecule, the rates of the reaction with two water molecules are enhanced by 3.4×10^{11}, 2.1×10^9, and 2.2×10^7 at 200, 300, and 400 K. Tunneling is also important for formaldehyde hydrolysis: the tunneling correction factors are 6.1×10^{11}, 1.6×10^4, and 77 with one water, and 12, 2.5, and 1.6 for two waters at 200, 300, and 400 K. As for the concerted hydrogen atom transfer in the water clusters, the broader adiabatic barrier and heavier reduced mass diminish the tunneling in the larger cluster.

In an aqueous environment, the surrounding medium of additional water molecules will further stabilized the transition state structure of the proton transfer reaction through hydrogen bonding. This solvation stabilization is greater for the transition state than for the reactants due to the greater ionic character of the transition state. Using an equation of state with critical parameters scaled from changes in the atomic charge of each hydrogen [60], we estimate that the free energy of activation for the reaction in diagram 2 is lowered by 18.3 kcal/mol. This large lowering of the activation barrier will significantly flatten the adiabatic potential along the reaction coordinate, thereby decreasing the effect of tunneling significantly. On the other hand, the overall rate constant will be greatly increased due to the lower adiabatic activation energy. Thus, we expect the effect of increasing the polar strength of the solvent will be to decrease the tunneling contribution to the rate constant, but increase the proton transfer rate. More reliable estimates of the rates of these processes in aqueous solutions will also require calculations of the equilibrium populations of these cyclic complexes in solution. One expects the population of the complexes to decrease with increase in the number of water molecules in the complex; therefore, even though the rates of reaction are larger for the larger complexes, they may be less important in the overall mechanism. The calculations of the rates presented here will be of more direct relevance to these clusters in aprotic, yet polarizable, solvents. We are pursuing further research on the effects of the solvating medium on the concerted proton transfer reaction within the variational transition state theory formalism.

4. Conclusions

Rate constants have been computed over a temperature range from 200 to 600 K for the concerted hydrogen atom transfer in cyclic water trimer and tetramer, and for formaldehyde hydrolysis in clusters with one and two water molecules. The rates are computed by variational transition state theory with semiclassical adiabatic ground-state transmission coefficients using limited information about the potential energy surface along the reaction path. This type of information about the potential is obtained directly from *ab initio* electronic structure calculations of the energy and its first and second derivative with respect to coordinates. In the present calculations, *ab initio* information is empirically adjusted by the BAC-MP4 method to yield more reliable predictions of the reaction energetics.

Dramatic effects of the number of participating water molecules on the reaction rates are predicted. At 300 K, enhancements of nearly four orders of magnitude and over nine overs of magnitude are obtained for the addition of a water molecule to the cyclic water trimer and the formaldehyde water dimer, respectively. Tunneling is also predicted to be important for these systems but the effect diminishes as additional water molecules participate actively in the proton transfer.

52

References

* Pacific Northwest Laboratory is operated for the U. S. Department of Energy by Battelle Memorial Institute under contract DE-AC06-76RLO 1830.

[1] Grunwald, E.; Jumper, C. F.; Meiboom, S.; *J. Am. Chem. Soc.* **1963**, *85*, 522; Grunwald, E.; Meiboom, S.; *J. Am. Chem. Soc.* **1963**, *85*, 2047; Grunwald, E.; Jumper, C. F.; *J. Am. Chem. Soc.* **1963**, *85*, 2051; Luz, Z.; Meiboom, S.; *J. Am. Chem. Soc.* **1963**, *85*, 3923.

[2] Vinogradov, S. N.; Linnell, R. H.; *Hydrogen Bonding*; Van Nostrand Reinhold Co., New York, 1971; p. 217.

[3] Grunwald, E.; Eustace, D. in *Proton Transfer Reactions*, Caldin, E.; Gold, V; Ed.: Chapman and Hall, London, 1975; p. 103

[4] Bell, R. P.; Millington, J. P.; Park, J. M.; *Proc. Roy Soc. A* **1968**, *303*, 1; Bell, R. P.; Evans, P. G.; *Proc. Roy Soc. A* **1966**, *291*, 297.

[5] More O'Ferral, R. A.; *J. Chem. Soc. B* **1970**, 274.

[6] Hine, J.; *J. Am. Chem. Soc.* **1972**, *94*, 5766.

[7] Critchlow, J. E.; *J. Chem. Soc. Faraday Trans.***1972**, *68*, 1774.

[8] Hibbert, F.; *Adv. Phys. Organic Chem.* **1986**, *22*, 113, and references therein.

[9] Gandour, R. D.; Maggiora, G. M.; Schowen, R. L.; *J. Am. Chem. Soc.* **1974**, *96*, 6967.

[10] Williams, I. H.; Maggiora, G. M.; Schowen, R. L.; *J. Am. Chem. Soc.* **1980**, *102*, 7831; Williams, I. H.; Spangler, D.; Femec, D. A.; Maggiora, G. M.; Schowen, R. L.; *J. Am. Chem. Soc.* **1983**, *105*, 31; Williams, I. H.; *J. Am. Chem. Soc.* **1987**, *109*, 6299.

[11] Kong, Y. S.; Jhon, M. S.; Löwdin, P.; *I. J. Quant. Chem.* **1987**, *14*, 189.

[12] See for example, Hibbert, F. in *Comprehensive Chemical Kinetics*, Bamford, C. H.; Tipper, C. F. H., Ed.; Elsevier: New York, 1977; p. 97, and references therein.

[13] See for example, Crooks, J. E. in *Comprehensive Chemical Kinetics*, Bamford, C. H.; Tipper, C. F. H., Ed.; Elsevier: New York, 1977; p. 197, and references therein.

[14] Bell, R. P.; Critchlow, J. E.; *Proc. Roy. Soc. A* **1971**, *325*, 35.

[15] Isaacson, A. D.; Truhlar, D. G.; Rai, S. N.; Steckler, R.; Hancock, G. C.; Garrett, B. C.; Redmon, M. J. *Computer Phys. Comm.* **1987**, *47*, 91.

[16] Truhlar, D. G.; Garrett, B. C. *Acct. Chem. Res.* **1980**, *13*, 440.

[17] Walker, R. B.; Light, J. C. *Ann. Rev. Phys. Chem.* **1980**, *31*, 401.

[18] Pechukas, P. *Ann. Rev. Phys. Chem.* **1981**, *32*, 159.

[19] Truhlar, D. G.; Isaacson, A. D.; Skodje, R. T.; Garrett, B. C. *J. Phys. Chem.* **1982**, *86*, 2252; *J. Phys. Chem.* **1983**, *87*, 4554.

[20] Truhlar, D. G.; Hase, W. L.; Hynes, J. T. *J. Phys. Chem.* **1983**, *87*, 2664.

[21] Truhlar, D. G.; Garrett, B. C. *Ann. Rev. Phys. Chem.* **1984**, *35*, 159.

[22] Truhlar, D. G.; Isaacson, A. D.; Garrett, B. C. *The Theory of Chemical Reaction Rates*; Baer, M., Ed.; CRC Press: Boca Raton, FL, 1985; Vol. 4, p 65.

[23] For Version 1.1 see Ref. 9 ; For Version 1.5 see Isaacson, A. D.; Truhlar, D. G.; Rai, S. N.; Hancock, G. C.; Lauderdale, J. G.; Truong, T. N.; Joseph, T.; Garrett, B. C.; Steckler, R. *POLYRATE Program manual*, 1988.

[24] Garrett, B. C.; Truhlar, D. G. *J. Phys. Chem.* **1979**, *83*, 1079; *J. Phys. Chem.* **1983**, *87*, 4553.

[25] Garrett, B. C.; Truhlar, D. G.; Grev, R. S.; Magnuson, A. W. *J. Phys. Chem.* **1980**, *84*, 1730; *J. Phys. Chem.* **1983**, *87*, 4553.

[26] Glasstone, S.; Laidler, K. J.; Eyring, H.; *The Theory of Rate Processes*; McGraw-Hill: New York, 1941.

[27] Wigner, E. Z.; *Phys. Chem. B* **1932**, *19*, 203.

[28] Melius, C. F.; Binkley, J. S.; Paper WSS/CI 83-16, 1983 Fall Meeting of the Western States Section of the Combustion Institute, University of California, Los Angeles, Oct. 17-18, 1983.

[29] Melius, C. F.; Binkley, J. S. *The Chemistry of Combustion Processes* (ACS Symp. Ser. No. 249); Sloane, T. M., Ed.; American Chemical Society: Washington, D.C., 1984; p 103.

[30] Melius, C. F.; Binkley, J. S.; *Symp. (Int.) Comb. [Proc.]* **1985**, *20*, 575.

[31] Perry, R. A.; Melius, C. F.; *Symp. (Int.) Comb. [Proc.]* **1985**, *20*, 639.

[32] Ho, P.; Coltrin, M. E.; Binkley, J. S.; Melius, C. F.; *J. Phys. Chem.* **1985**, *89*, 4647; *J. Phys. Chem.* **1986**, *90*, 3399.

[33] Garrett, B. C.; Koszykowski, M. L.; Melius, C. F.; Page, M.; *J. Phys. Chem.* **1990**, *94*, 7096.

[34] Bell, R. P.; *The Proton in Chemistry, 2nd Ed.*; Chapman and Hall, London, 1973.

[35] Lewis, E. S. in *Proton Transfer Reactions*, Caldin, E.; Gold, V; Eds: Chapman and Hall, London, 1975; p. 317

[36] Del Bene, J. E.; Kochenour, W. L.; *J. Am. Chem. Soc.* **1976**, *98*, 2041.

[37] Karlstrom, G.; Wennerstrom, H.; Jonsson, B.; Forsen, S.; Almlöf, J.; Roos, B.; *J. Am. Chem. Soc.* **1975**, *97*, 4188; Karlstrom, G.; Jonsson, B.; Roos, B. Wennerstrom, H.; *J. Am. Chem. Soc.* **1976**, *98*, 6851.

[38] Bouma, W. J.; Vincent, M. A.; Radom, L.; *Int J. Quantum Chem.* **1978**, *14*, 767.

[39] Bicerano, J.; Schaefer, H. F., III; Miller, W. H.; *J. Am. Chem. Soc.* **1983**, *105*, 2550.

[40] Baughcum, S. L.; Smith, Z.; Wilson, E. B.; Duerst, R. W.; *J. Am. Chem. Soc.* **1984**, *106*, 2260.

[41] *Gaussian 88*, Frisch, M. J.; Head-Gordon, M.; Schlegel, H. B.; Raghavachari, K.; Binkley, J. S.; Gonzalez, C.; Defrees, D. J.; Fox, D. J.; Whiteside, R. A.; Seeger, R.; Melius, C. F.; Baker, J.; Martin, R.; Kahn, L. R.; Stewart, J. J. P.; Fluder, E. M.; Topiol, S.; Pople, J. A.; Gaussian Inc., Pittsburgh, PA, **1988**.

[42] Krishnan, R.; Binkley, J. S.; Seeger, R.; Pople, J. A.; *J. Chem. Phys.* **1980**, *72*, 650.

[43] Page, M.; McIver, J. W., Jr.; *J. Chem. Phys.* **1988**, *88*, 922.

[44] Kendall, R. A.; private communication.

[45] Pople, J. A.; Binkley, J. S.; Seeger, R. *Int. J. Quantum Chem. Symp.* **1976**, *10*, 1; Pople, J. A.; Krishnan, R.; Schlegel, H. B.; Binkley, J. S.; *Int. J. Quantum Chem. Symp.* **1978**, *14*, 545.

[46] Skodje, R. T.; Truhlar, D. G.; Garrett, B. C.; *J. Chem. Phys.* **1982**, *77*, 5955.

[47] Skodje, R. T.; Truhlar, D. G.; Garrett, B. C.; *J. Phys. Chem.* **1981**, *85*, 3019.

[48] Miller, W. H.; Handy, N. C.; Adams, J. E.; *J. Chem. Phys.* **1980**, *72*, 99.

[49] Garrett, B. C.; Truhlar, D. G.; Wagner, A. F.; Dunning, T. H., Jr.; *J. Chem. Phys.* **1983**, *78*, 4400.

[50] Garrett, B. C.; Joseph, T.; Truong, T. N.; Truhlar, D. G.; *Chem. Phys.* **1989**, *136*, 271.

[51] Lauderdale, J. G.; Truhlar, D. G.; *J. Am. Chem. Soc.* **1985**, *107*, 4590; *Surf. Sci.* **1985**, *164*, 558; *J. Chem. Phys.* **1986**, *84*, 1843.

[52] Rice, B. M.; Garrett, B. C.; Koszykowski, M. L.; Foiles, S. M.; Daw, M. S.; *J. Chem. Phys.* **1990**, *92*, 775.

[53] Chase, M. W., Jr.; Davies, C. A.; Downey, J. R., Jr.; Frurip, D. J.; McDonald, R. A.; Syverud, A. N.; "JANAF Thermochemical Tables, 1985 Supplement" *J. Physical Chemical Reference Data* **1985**, *14*.

[54] Dyke, T. R.; Mack, K. M.; Muenter, J. S.; *J. Chem. Phys.* **1977**, *66*, 498.

[55] Curtiss, L. A.; Frurip, D. J.; Blander, H.; *J. Chem. Phys.* **1979**, *71*, 2703.

[56] Smith, B. J.; Swanton, D. J.; Pople, J. A.; Schaefer, H. F. III; Radom, L.; *J. Chem. Phys.* **1990**, *92*, 1240.

[57] Curtiss, L. A.; M. Blander, M.; *Chem. Rev.* **1988**, *88*, 827.

[58] Popkie, H.; Kistenmacher, H.; Clementi, E.; *J. Chem. Phys.* **1973**, *59*, 1325; Clementi, E.; Kołos, W.; Lie, G. C.;, Ranghino, G.; *Int. J. Quantum Chem.* **1980**, *17*, 377; Niesar, U.; Corongiu, G.; Huang, M.-J.; Dupuis, M.; Clementi, E.; *Int. J. Quantum Chem. Symp.* **1989**, *23*, 421.

[59] Chałasiński, G.; Szcześniak, M. M.; Cieplak, P.; Scheiner, S.; preprint.

[60] Melius, C. F.; Bergan, N.; Shepherd, J. E.; *Symp. (Int.) Comb. [Proc.]* **1990**, *23*, 0000.

Double Many-Body Expansion Potential Energy Surface for $O_4(^3A)$, Dynamics of the $O(^3P) + O_3(^1A_1)$ Reaction, and Second Virial Coefficients of Molecular Oxygen

A.J.C. Varandas and A.A.C.C. Pais
Departamento de Química, Universidade de Coimbra
3049 Coimbra Codex, Portugal

ABSTRACT. We report a six-dimensional potential energy surface for the $O_4(^3A)$ system using the double many-body expansion (DMBE) method. The four-body energy terms of this surface have been calibrated from available dispersion coefficients for the $O_2(\tilde{X}\,^3\Sigma_g^-)$ – $O_2(\tilde{X}\,^3\Sigma_g^-)$ interaction, second virial coefficients of molecular oxygen and experimental activation energy data for the $O(^3P) + O_3(^1A_1) \rightarrow 2O_2(\tilde{X}\,^3\Sigma_g^-)$ reaction. A dynamics study of the title reaction has been also carried out, yielding results for the corresponding thermal rate coefficients as a function of temperature in good agreement with experiment. The main features of the relevant triplet-state O_4 surface are also analyzed.

1 Introduction

The reaction between atomic oxygen and ozone to give molecular oxygen,

$$O(^3P) + O_3(^1A_1) \rightarrow 2O_2(\tilde{X}\,^3\Sigma_g^-), \tag{1}$$

has been the subject of several experimental studies[1]-[4], being of crucial importance for a complete understanding of the chemical and photochemical processes of the earth's stratosphere and mesosphere. It is also an important sink for "odd" oxygen in those regions of the atmosphere[1]. A recent evaluation of kinetic and photochemical data for atmospheric chemistry that includes the title reaction is [5], while a survey of the rate data on this reaction up to 1973 may be found in [2].

   In C_s symmetry, the ground state reactants $O(^3P) + O_3(^1A_1)$ produce one surface of $^3A'$ symmetry and two surfaces of $^3A''$ symmetry. The first arises from the $^1A'$ surface of $O_3(^1A_1)$ and the $^3A'$ component of $O(^3P)$, while the other two arise from the $^1A'$ component of $O_3(^1A_1)$ and the two $^3A''$ components of $O(^3P)$. In turn, the products $O_2(\tilde{X}\,^3\Sigma_g^-) + O_2(\tilde{X}\,^3\Sigma_g^-)$ in non-planar and planar geometries have $^{1,3,5}A''$ and $^{1,3,5}A'$ symmetries[6]. Only the $^3A''$ ($^3A'$) electronic species of the term manifold of O_4 correlates adiabatically both with the $O(^3P) + O_3(^1A_1)$ reactants and the $O_2(\tilde{X}\,^3\Sigma_g^-) + O_2(\tilde{X}\,^3\Sigma_g^-)$ products. It is this $O_4(^3A)$ potential energy surface that we assume to be of relevance here.

   The interaction between two oxygen molecules in their ground electronic state determines the second virial coefficient of molecular oxygen (in addition to other thermophysical properties) and may be described by knowing the relevant potential

S. J. Formosinho et al. (eds.), Theoretical and Computational Models for Organic Chemistry, 55–78.
© 1991 *Kluwer Academic Publishers.*

energy surfaces of the tetra-atomic molecule, in particular the lowest-triplet state potential energy surface of O_4. Thus, information on the $O_2(\tilde{X}\,^3\Sigma_g^-) - O_2(\tilde{X}\,^3\Sigma_g^-)$ interaction can be obtained from measure ments[7,8] of those coefficients and their temperature dependence. A number of theoretical studies using effective one-dimensional isotropic potential energy curves[9] and di-Lennard-Jones potential functions[10] to mimic the experimental second-virial coefficient data of gaseous oxygen as a function of temperature have also been published.

Ab initio studies on tetra-oxygen have focused on its lowest closed-shell electronic state, which is presumed to correspond to a structure having four single bonds between oxygen atoms[11]-[14]. Although this singlet-state potential energy surface may, by the above mentioned Wigner-Witmer spin-spatial correlation rules, dissociate to two $O_2(\tilde{X}\,^3\Sigma_g^-)$ product molecules, it fails, by violation of spin-conservation, to correlate with the reactants $O(^3P) + O_3(^1A_1)$. Thus, it will not be discussed any further in the present work.

The importance of the title system on its own, the wealth of available independent kinetic and thermophysical measurements, and the scarcity of fully six-dimensional potential energy surfaces for tetra-atomic systems have prompted us to carry out a consistent study of the $O(^3P) + O_3(^1A_1)$ reaction dynamics and second virial coefficients of molecular oxygen using a single-valued potential energy surface for the lowest-triplet state of O_4 based on the double many-body expansion (DMBE)[15]-[19] method. We note that, except for a formal application[20] to H_4, the present study represents the first practical application of this method to a tetra-atomic system. Indeed, although the DMBE method has been successfully applied to many triatomic systems, including both chemically stable molecules such as O_3[20]-[22] and HO_2[23]-[25], and weakly bound van der Waals complexes such as HeH_2[26,23], $HeLi_2$[23], H_3[27,20], and FH_2[28] (these latter two systems being also well known prototypes in reactive scattering and chemical kinetics), it remains to be seen how easily and efficiently it can be used to obtain the potential energy surfaces of larger polyatomic systems. In particular, an important question to be answered for systems with more than three atoms is whether the $(n \geq 4)$-body terms can be modelled by relatively simple functional forms. Previous experience on the many-body expansion[29] method suggests that the main features of the potential energy surface are contained in the one-body (which, for the present case of a single-valued surface, can be taken as the reference energy and hence not be explicitly considered), two-body, and three-body terms with the main role of the four-body terms being to reproduce the correct relative energies of stable and metastable structures. As it will be shown in the present work, these findings seem to be true also for the DMBE method. In particular it is shown that relatively simple functional forms for the four-body energy terms suffice to reproduce the main features of both the valence and long range regions of the $O_4(^3A)$ potential energy surface.

The structure of this paper is as follows. Section 2 provides a brief review of the DMBE method while an application to the lowest-triplet potential energy surface of O_4 is described in section 3. Section 4 summarizes the approach used to calculate the second virial coefficients of molecular oxygen, and section 5 reports the results of a classical trajectory study for the title reaction. All calculations have employed the fully six-dimensional O_4 DMBE potential energy surface from the present work. Some conclusions are gathered in section 6.

2 The double many-body expansion method

A systematic strategy to obtain the potential energy functions of small polyatomic systems consists of writing the potential energy as a double cluster expansion usually referred to as DMBE, the acronym of double many-body expansion[17]. Accordingly, the electronic potential energy is partitioned into an extended-Hartree-Fock-type energy (which includes the nondynamical correlation of the valence electrons or nearly degenerate orbitals, and is denoted EHF) and the dynamical correlation energy (which arises from the true dynamic correlation of the electrons, and is abbreviated dc), with both these energy contributions being developed as a many-body expansion[29]. Although the DMBE method has primarily been designed to provide a functional representation of accurate *ab initio* potential surface calculations or experimental information, as it is the case in the present work, it may also be used as a semiempirical predictive tool[30]. Since a detailed account of this method has been given elsewhere[17,19], only those aspects which are relevant for the present work are reviewed in this section.

For a four-atom single-valued potential energy surface the DMBE energy development assumes the form

$$V = V_{EHF}(\mathbf{R}) + V_{dc}(\mathbf{R}) \qquad (2)$$

where

$$V_x = \sum_{i=1}^{6} V_{x,i}^{(2)}(R_i) + \sum_{i<j<k} V_{x,ijk}^{(3)}(R_i, R_j, R_k) + V_x^{(4)}(\mathbf{R}) \qquad x = EHF, dc \qquad (3)$$

with the second summation comprising all sets of three distances defining a triatomic fragment. In addition, x specifies the energy contribution, $\mathbf{R} = (R_1, R_2, \ldots, R_6)$ represents a collective variable, and the superscripts indicate the n-body order of the contribution.

A convenient form to represent the EHF energy is (see, *e.g.*, [17])

$$V_{EHF}^{(n)} = P^{(n)}(\mathbf{R}^n) T^{(n)}(\mathbf{R}^n) \qquad (4)$$

where $\mathbf{R}^n \subset \mathbf{R}$, $P^{(n)}$ denotes a n-body multinomial in the $\mathbf{R}^n$ interparticle co-ordinates, and $T^{(n)}$ is a n-dimensional range-determining factor. The parameters contained in both $P^{(n)}$ and $T^{(n)}$ are determined, as usual, from available *ab initio* and experimental data.

The dynamical correlation energy is generally approximated semiempirically from the relevant long-range dispersion interactions suitably damped to account for orbital overlap and electron exchange effects. Accordingly, the two-body dynamical correlation is approximated by[15,17]

$$V_{dc,i}^{(2)} = \sum_n C_n \chi_n(R_i) R_i^{-n} \quad n = 6,8,\dots \tag{5}$$

where C_n are the long-range dispersion energy coefficients for the i-th atom-atom interaction, and χ_n are dispersion damping functions defined by[31,32]

$$\chi_n = \left[1 - \exp\left(-A_n x - B_n x^2\right)\right]^n \tag{6}$$

where $x = R/\rho$, with ρ being a suitable scaling distance, and[15]

$$A_n = \alpha_0 n^{\alpha_1} \tag{7}$$
$$B_n = \beta_0 \exp(-\beta_1 n) \tag{8}$$

with $\alpha_0 = 25.9528$, $\alpha_1 = 1.1868$, $\beta_0 = 15.7381$, and $\beta_1 = 0.09729$ being universal parameters for all isotropic interactions.

Similarly, the three-body dynamical correlation is semiempirically approximated by the form[15,17]

$$V_{dc,ijk}^{(3)} = \sum_{i=1}^{3} \sum_n C_n \chi_n(R_i) \left\{1 - \frac{1}{2}\left[g_n(R_j)h_n(R_k) + g_n(R_k)h_n(R_j)\right]\right\} R_i^{-n} \tag{9}$$

where C_n and χ_n ($n = 6,8,\dots$) have the meaning assigned above, $j = i+1(\bmod\,3)$, $k = i+2(\bmod\,3)$, and the g_n and h_n are auxiliary functions containing parameters to be deduced from atom-diatom dispersion coefficients as a function of the diatomic bond distance.

For interactions having an important long range electrostatic contribution (the same applying to systems for which the induction energy plays an important role), it is convenient to treat the electrostatic energy separately from the rest of the EHF-type energy. At the two-body level this interaction may occur only when both atoms are in non-S-states and have therefore permanent electric moments. In this case the two-body electrostatic interaction energy may be approximated from the corresponding asymptotic inverse power series expansion suitably damped to account for charge-overlap effects. The result is an expression similar to eqn (5) in which the C_n are long-range electrostatic coefficients and χ_n are electrostatic damping functions, with n running over the appropriate powers of R ($n \geq 5$).

Because the long-range electrostatic interaction energy for the $^3\Sigma_g^-$ ground electronic state of O_2 is zero, the leading long range electrostatic contribution that needs to be considered occurs at the three-body level for the $O(^3P) - O_2(\tilde{X}\,^3\Sigma_g^-)$ interactions. A general form to represent this three-body electrostatic energy contribution is[23,17]

$$V_{elec,ijk}^{(3)} = \frac{1}{2}\sum_{i=1}^{3} \sum_n \left[C_n G_n(R_j)h_n(R_k) + C_n' G_n(R_k)h_n(R_j)\right] \chi_n(R_i) R_i^{-n} \tag{10}$$

where C_n is now the long range electrostatic coefficient for the atom-diatom inter-action involving one of the atoms of the i-th bond with the j-th diatomic, and C'_n has a similar meaning but referring to the interaction of the remaining atom of the i-th bond with the k-th diatomic; $n = 4, 5$.

One could attempt an approach similar to that described above for the lower-order ($n < 4$) terms of the electrostatic and dynamical correlation energy. However, a complete description of all asymptotic channels is far from trivial, and is not attempted here. Instead, we suggest a simple procedure, which has been found useful to impose those features of the potential energy surface that are required to describe the kinetic rate and second virial coefficient data available for the title system.

3 Application to the O_4 potential energy surface

We first examine the three-body and four-body long range electrostatic energy con-tributions. At the three-body level such electrostatic energy is associated with the interaction between the quadrupole moments of atomic and molecular oxygen in their ground electronic states. If both species were rigid then this interaction would assume the form[33]

$$V^{(3)}_{elec}(r, \omega) = C_5(\omega) r^{-5} \tag{11}$$

where ω is a collective variable of all angular coordinates, and r is the atom-diatom centre of mass distance; for the coordinates used in this work, see Figure 1. The angular-dependent long-range electrostatic coefficient is in turn defined by[33]

$$
\begin{aligned}
C_5(\omega) = \ & \frac{3}{16} \Theta_a \Theta_b \left[1 - 5 \cos^2 \theta_a - 5 \cos^2 \theta_b - 15 \cos^2 \theta_a \cos^2 \theta_b \right. \\
& \left. + (4 \cos \theta_a \cos \theta_b + \sin \theta_a \sin \theta_b \cos \phi_{ab})^2 \right],
\end{aligned}
\tag{12}
$$

where Θ_a and Θ_b are the quadrupole moments of the A and B interacting species, and θ_a, θ_b, ϕ_a, and ϕ_b are the usual angular coordinates ($\phi_{ab} = \phi_a - \phi_b$); in this discussion we have assumed that A represents the O atom and B the O_2 molecule.

However, eqn (11) treats the O-atom formally as a diatomic molecule the quadrup-ole moment of which is attached to the diatomic axis, and hence cannot reorient as the atom moves around the O_2 molecule. As a result the electrostatic energy so calculated vanishes when averaged with equal weight over all orientations of one of the interacting quadrupoles. This is illustrated in Figure 2(a), which shows contours for the angular dependence of the long range (undamped) electrostatic energy as the O-atom quadrupole (keeping the angles θ_a and ϕ_{ab} fixed at $\theta_a = \pi/2$ and $\phi_{ab} = 0$) moves around that for the O_2 molecule. As shown by the appearance of solid (which correspond to negative values of the interaction energies) and dashed (corresponding to positive energies) contours, the interaction energy vanishes when averaged with equal weight over all orientations.

Because electrons move much faster than nuclei, a more realistic classical model consists of letting the electronic distribution of the O-atom instantaneously adjust

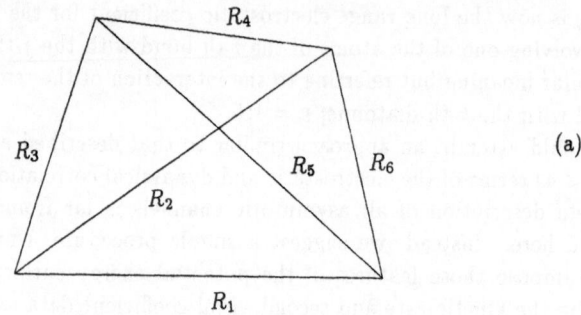

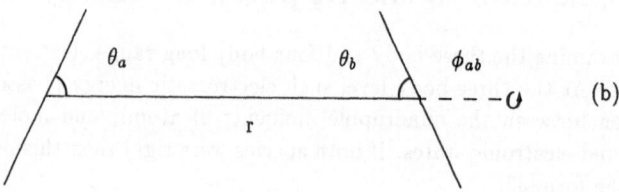

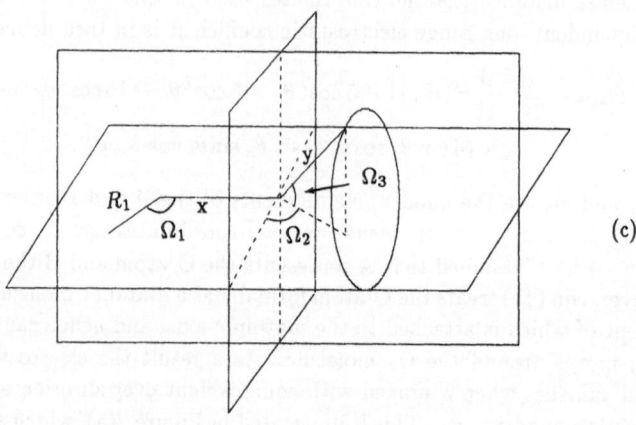

Figure 1. Coordinates used in this work to define: (a) the interatomic distances; (b) the intermolecular $O_2 - O_2$ interaction; (c) the plots of Figures 4 and 5.

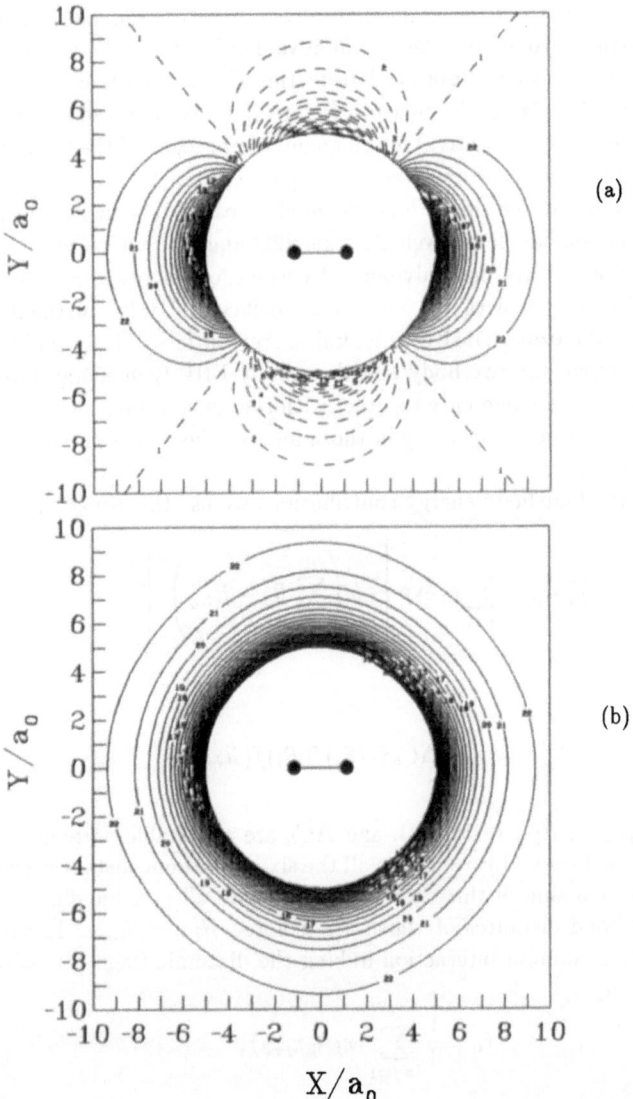

Figure 2. Energy contour plots for the angular dependence of the un-damped electrostatic energy: (a) with θ_a and ϕ_{ab} fixed at $\theta_a = \phi_{ab} = 0$; (b) using the classical-optimized-quadrupole model. The solid lines indicate negative energies, starting at $-4.4 \times 10^{-4}\, E_h$. The dashed lines indicate positive energies, starting at $0\, E_h$. Contours are equally spaced by $2 \times 10^{-5}\, E_h$. Only the larger values of r $(r \geq 5\, a_0)$ are shown. O_2 is at its equilibrium geometry.

to the O_2 molecule, whatever the angle of approach, so as to produce the lowest energy. This is equivalent to finding the optimum value of θ_a for each value of θ_b. The undamped electrostatic energy obtained from this classical-optimized-quadrupole model[18]-[22] (see also [34,35]) is shown in Figure 2(b). We observe that the interaction energy is negative (attractive) for all angles of approach of the O-atom to the O_2 molecule. A more detailed analysis of this problem together with the relevant formulas for defining the optimized long range electrostatic coefficients in (10) for the $O - O_2$ interaction has been given elsewhere[22] and will not be repeated here.

The two-body and three-body dynamical correlation energy contributions are given by eqns (5) and (9) with the numerical coefficients[1] taken without change from our work on the ozone DMBE potential energy surface[22]. In addition, the forms used to represent the two-body and three-body EHF-type energy terms were taken unchanged from this reference together with the specific numerical values used to represent the coefficients appearing in those forms. The reader is referred to the original paper for details.

To represent the four-body energy contributions we use the forms

$$V_{EHF}^{(4)} = \sum_{i=1}^{2} a_i \exp\left[-b\left(\sum_{j=1}^{6} R_j - R_{\Sigma,j} \right)^2 \right] \tag{13}$$

and

$$V_{elec}^{(4)} = \sum_{i=1}^{3} \Delta C_5 \tilde{\chi}_5(\tilde{x}_i) f(R_i) f(R_{i+3}) \tilde{r}_i^{-5} \tag{14}$$

where a_i $(i = 1,2)$, b, $R_{\Sigma,j}$ $(j = 1,2)$, and ΔC_5 are adjustable parameters. Note that the summation in eqn (13) runs over all the six interatomic distances while that in eqn (14) involves a sum of three terms associated to all possible diatom-diatom interactions, the bond distances of which are denoted R_i and R_{i+3}. In turn, $\tilde{r}_i$ is an averaged diatom-diatom interaction linking the diatomic fragments with bond distances R_i and R_{i+3},

$$\tilde{r}_i = \frac{1}{4} \sum_{i \neq j=1}^{3} (r_j + r_{j+3}), \tag{15}$$

and the damping function $\tilde{\chi}_5$ is given by

$$\tilde{\chi}_5 = \left[1 - \exp\left(-\tilde{A}\tilde{x}_i - \tilde{B}\tilde{x}_i^2 \right) \right]^5 \tag{16}$$

where

$$\tilde{x}_i = \frac{\tilde{r}_i}{\tilde{\rho}} \tag{17}$$

---

[1] Unless mentioned otherwise, all quantities reported in this work are in atomic units: a_0 (Bohr)= a.u. of length $= 0.529177 \text{Å} = 5.29177 \times 10^{-11}\ m$; E_h (Hartree)= a.u. of energy $= 27.211652\ eV = 627.509\ kcal\ mol^{-1} = 4.3598\ aJ = 315773.22\ K.$

and $\tilde{\rho} = 17.763\, a_0$ is a scaling distance corresponding approximately to that of an isotropic $O_2 - O_2$ interaction. We have therefore assumed the same form and numerical values (i.e., $\tilde{A} = A_5$ and $\tilde{B} = B_5$) as for the case of two-body interactions; see (6). Moreover, $f(R)$ is a gaussian-type form that terminates the effect of $V_{elec}^{(4)}$ when the bond distances R_i and R_{i+3} deviate from the equilibrium geometry of ground-state O_2, i.e., when the system fragmentates or two atoms tend to collapse into the corresponding united-atom. It has the explicit form

$$f(R_i) = \exp\left[-b'\left(R_i - R_m\right)^2\right] \qquad (18)$$

where R_m is the equilibrium distance for a ground-state O_2 molecule.

Except for b', which was arbitrarily fixed at $b' = 1$, and $R_{\Sigma,1}$, which was fixed by a procedure to be described below, all other parameters in eqns (13) and (14) have been numerically defined by an approximate least-squares fitting procedure such as to correct the excessive attractiveness of the potential surface when the DMBE is truncated at three-body terms. It is on this calibration procedure that we focus next.

The constant ΔC_5 in eqn (14) is chosen from the requirement that the long-range part of the spherically-averaged component of the complete DMBE potential energy surface reproduces that obtained from theoretical calculations. Thus it partly cancels the quadrupole-quadrupole interactions arising at the three-body level such as to bring the unweighted spherically average of the DMBE potential energy surface for the $O_2 - O_2$ interaction into agreement with the theoretical long-range potential curve in which there is no r^{-5} term. Note, however, that the electrostatic long-range interaction for a specific relative orientation of two O_2 molecules does not necessarily vanish. Furthermore, $V_{elec}^{(4)}$ corrects in an approximate way for the dispersion terms varying as R^{-n} ($n = 6, 8, 10$), since these energy contributions have also been considered when calculating ΔC_5 from the difference between the long-range spherically averaged $O_2 - O_2$ interaction potential and the theoretical curve of Battaglia et al.[9] at $r = 30\, a_0$. We observe that this intermolecular separation is sufficiently large to assume that the interaction energy is given by its long range component.

Figure 3(a) shows a contour plot of the O_4 DMBE potential energy surface considering only two-body and three-body terms for an oxygen atom moving in the plane of an ozone molecule fixed at its equilibrium geometry with the origin located at one of the terminal atoms. The point marked × in this Figure represents a local minimum of this potential energy surface, which has been found to be at a geometry defined by the collective variable (in a_0) $\mathbf{R}^\circ = (2.4648, 2.4648, 4.0936, 4.1553, 4.1553, 4.9895)$ with a well depth of $-0.013958\, E_h$ relatively to the $O+O_3$ dissociation limit. The geometry of this minimum has been calculated by numerical optimization techniques using the fully six-dimensional potential energy surface. It corresponds to a kite-type structure of O_4, which is unlikely to have physical reality. Indeed, experimental measurements suggest an activation energy for the title reaction of ca. $4.5\, kcal\, mol^{-1}$ (see, e.g., [2] and [5]) while the results of exploratory classical trajectory calculations on this DMBE potential energy surface yielded thermal rate coefficients more than

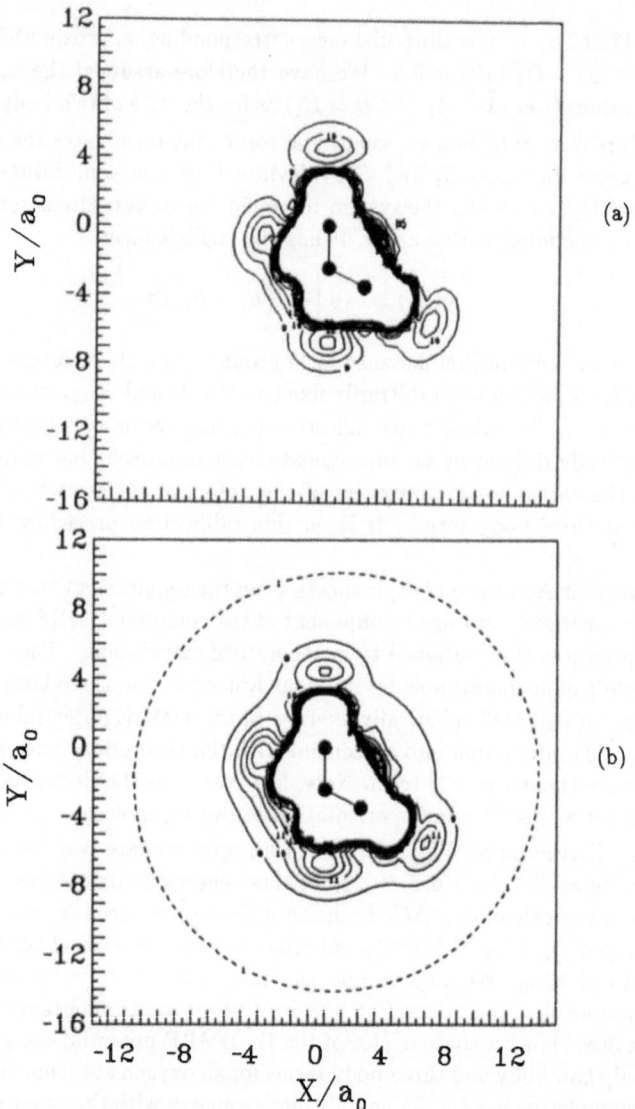

Figure 3. Energy contour plots for an oxygen atom moving around a non-optimized ozone molecule. (a) Considering only two-body and three-body energy terms in the DMBE. (b) Using the complete O_4 potential energy surface. For both plots, the contours are equally spaced by $0.02\,E_h$, starting at $-0.015\,E_h$. The dashed line in plot (b) indicates the energy contour corresponding to the $O + O_3$ dissociation limit, $i.e.$, to the zero of energy. The $\times$ in plot (a) represents a minimum; see the text.

a_1	a_2	$R_{\Sigma,1}$	$R_{\Sigma,2}$
0.02900	0.00143	22.32	35.55
b	b'	ΔC_5	R_m
0.02840	1.00	5.334	2.2818

Table 1. Numerical values, in atomic units, of the parameters used to define the four-body energy contributions.

one order of magnitude larger than experiment. In addition, a calculation of the second virial coefficients for gaseous oxygen over the temperature range $90 \leq T \leq 400\,\mathrm{K}$ led to values far too negative in comparison with experiment, and failed to reproduce the Boyle temperature that, according to ref. [8], should be near 400 K.

Based on these findings, the parameters a_1 and $R_{\Sigma,1}$ in the first term of eqn (13) have been determined such that $V_{EHF}^{(4)}$ ensures the existence of a potential barrier for the title reaction. Of these, $R_{\Sigma,1}$ has been chosen from the condition $R_{\Sigma,1} = \sum_{i=1}^{6} R_i^o$. In turn, a_1 has been chosen by trial-and-error such as to make the threshold energy of the calculated excitation function similar to the experimental activation energy for the $O(^3P) + O_3(^1A_1)$ reaction; see section 5. Finally, the optimum values of b, a_2, and $R_{\Sigma,2}$ were determined from a least-squares fitting procedure to the available information on the second virial coefficients of gaseous oxygen as it is described in section 4. Table 1 summarizes the final numerical values of all parameters.

Figures 3(b) to 5(a,b) illustrate important topographical features of the complete O_4 DMBE potential energy surface. It is seen from the contour plot of Figure 3(b) that there is a potential barrier for the in-plane attack of an oxygen atom to an equilibrium ozone molecule starting at a center of mass distance of about $10\,a_0$ [cf. Figure 3(a)]. Moreover, as can be seen from Figures 4(a) to 5(b), this barrier exists for all angles of approach. Note especially, from Figure 3(b), that any approach involving three collinear oxygen atoms is very unfavourable. This result stems largely from the fact that the three-body EHF contributions are highly repulsive for these geometries, which can be explained by the high energy barrier encountered for bending the ozone molecule until linearity[22]. In fact, the most favourable in-plane directions for an O-atom to attack an equilibrium O_3 molecule are predicted to be for angles $\Omega_2 \sim 45°$ and 135°. A detailed graphical analysis for the out-of-plane approach of O to O_3 has shown no significantly different features. Thus, except for Figures 5(a,b), which show an O-atom approaching the terminal atom of O_3 at angles $\Omega_3 = 45°$ and $\Omega_3 = 90°$, we omit the details of that analysis. Finally, we observe from Figures 4(b) and 5(a) that the saddle point for the title reaction occurs at a non-planar geometry. Indeed, using a numerical procedure, we have obtained for its geometry $R_1 = 2.42\,a_0$, $x = 2.51\,a_0$, $y = 4.21\,a_0$, $\Omega_1 = 117.56°$, $\Omega_2 = 111.86°$, and $\Omega_3 = 34.84°$), with the coordinates being defined in Figure 1(c).

66

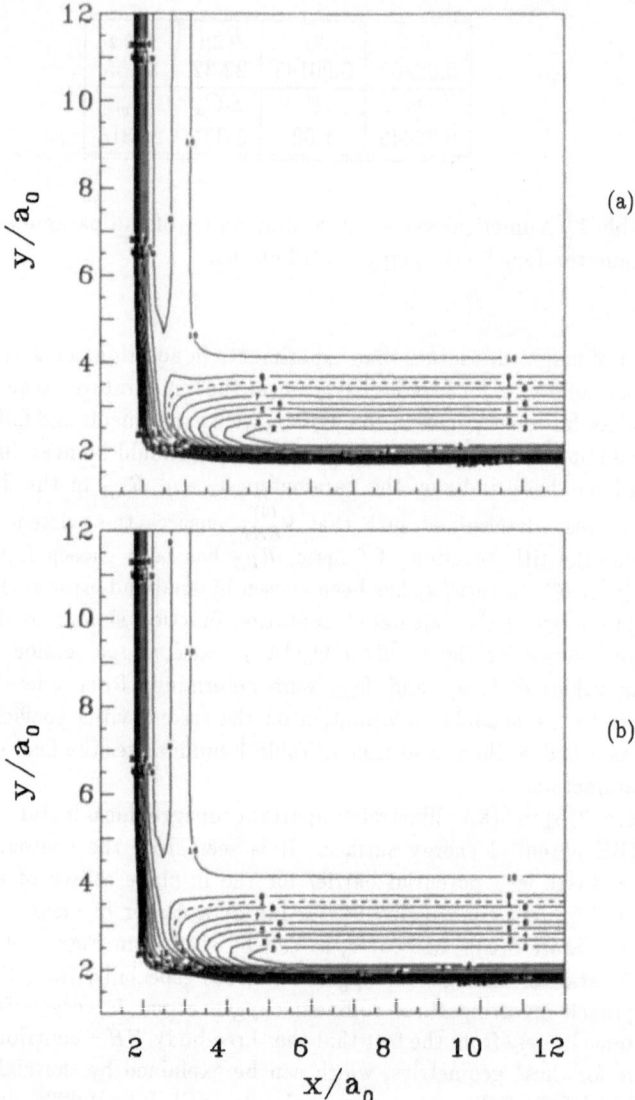

Figure 4. Energy contour plots for the in-plane ($\Omega_3 = 0°$) attack of O to O_3: (a) $\Omega_2 = 45°$; (b) $\Omega_2 = 135°$. Contours are equally spaced by $0.02\,E_h$, starting at $-0.15\,E_h$. The dashed line indicates the energy contour corresponding to the $O + O_3$ dissociation limit.

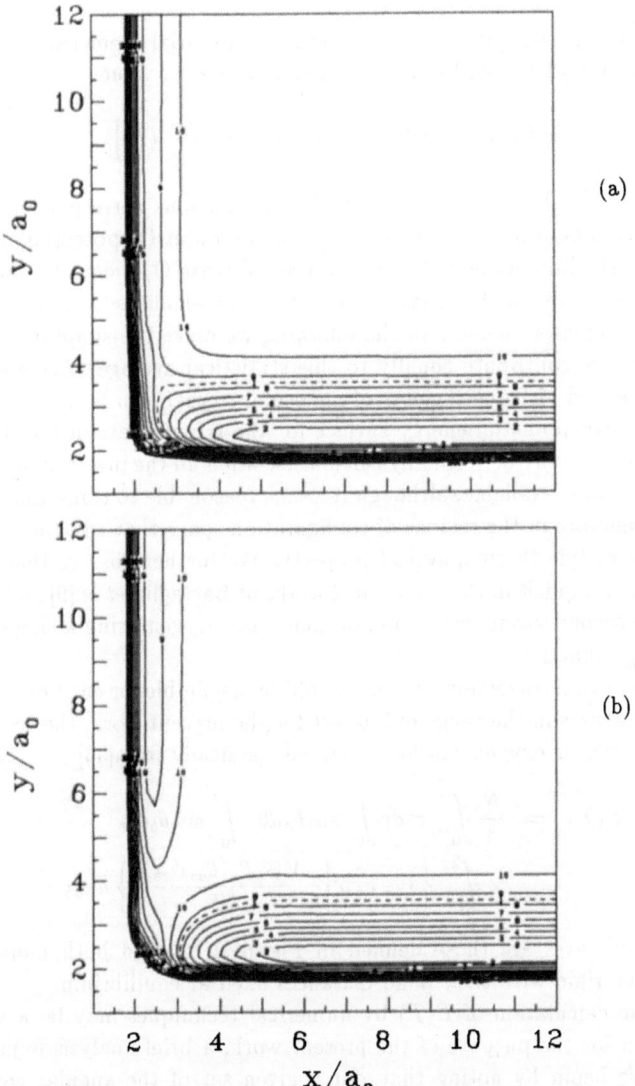

Figure 5. Energy contour plot for the out-of-plane attack of O to O_3:
(a) $\Omega_3 = 45°$; (b) $\Omega_3 = 90°$. In both plots Ω_2 was fixed at $\Omega_2 = 90°$.
Contours as in Figure 4.

4 Second virial coefficients of molecular oxygen

Taking into account the spin symmetries, the second virial coefficient of molecular oxygen should be represented by the statistically averaged value

$$B(T) = \frac{1}{9} \left[B^{(1)}(T) + 3B^{(3)}(T) + 5B^{(5)}(T) \right] \qquad (19)$$

where $B^{(1)}(T)$, $B^{(3)}(T)$, and $B^{(5)}(T)$ stand in an obvious correspondence for the contributions coming from the singlet, triplet, and quintet O_4 potential energy surfaces on which the interaction between two ground state O_2 molecules may evolve; for a related discussion on the virial coefficients of the alkali metal vapours, see [36] and [37], and references therein. In the following we make the simplifying assumption that all states contribute equally to this statistical average. Accordingly, the experimental second virial coefficients of molecular oxygen can be fitted using only the current DMBE potential energy surface for the triplet state of O_4. Of course, this assumption can only be tested by calculation when all the involved potential energy surfaces become available, although it seems reasonable to think that all states are nearly degenerate in the regions of configuration space that contribute most for the calculation of this thermophysical property. We further observe that a similar approximation is implicit in the theoretical works of Battaglia et al.[9] and Wells[10] who fitted the second virial coefficients of molecular oxygen using a single effective potential energy function.

Ignoring quantum corrections, which should be negligible for entities as heavy as O_2 and temperatures in the range of interest for the present work, the second virial coefficients of gaseous oxygen can be estimated classically from[33]

$$\begin{aligned}
B(T) &= \frac{N}{4} \int_0^\infty r^2 dr \int_0^\pi \sin\theta_a d\theta_a \int_0^\pi \sin\theta_b d\theta_b \\
&\times \int_0^{2\pi} \left\{ 1 - \exp\left[-\frac{V(r,\theta_a,\theta_b,\phi_{ab})}{kT} \right] \right\} d\phi_{ab}
\end{aligned} \qquad (20)$$

where the coordinates are those defined in Figure 1(b), and both molecules are considered to be rigid with their bond distances fixed at equilibrium.

Because the calculation of $B(T)$ by numerical techniques may be a very time consuming step for the purpose of the present work, a brief analysis is justified at this point. We begin by noting that, for a given set of the angular coordinates $(\theta_1, \theta_2, \phi_{12})$, the integrand has a single root (besides the trivial one for $r = 0$) at $r = \sigma(\theta_1, \theta_2, \phi_{12})$, the distance for which the potential energy vanishes. This is, of course, invariant with T. Thus, we have chosen to use a modified version of a generalization[38] of our original procedure for atomic gases[37]; see also [39]. It consists of splitting, for each set of angles $(\theta_1, \theta_2, \phi_{12})$, the integration in r into two parts, one up to $\sigma(\theta_1, \theta_2, \phi_{12})$, and the other from this root up to infinity. However, each evaluation of $\sigma(\theta_1, \theta_2, \phi_{12})$ using standard numerical techniques involves many evaluations of an elaborate function. To reduce this computational labour, we have chosen an hybrid approach that consists of assuming a common $\bar{\sigma}$ value for all sets of angles appearing in the numerical integration over the angular coordinates. It has been taken as $\bar{\sigma} = 7.5\,a_0$, a value that is close to the distance of zero-energy for the most popular $O_2 - O_2$ isotropic interaction potential energy curves.

The integrations in r, for each set of the angular variables, have then been evaluated by using, respectively, a Gauss-Legendre and a Gauss-Laguerre quadratures. We have found sufficient for the present purposes to use panels of 16 pivots for each of these parts of the integration in r. For the angular part, we have also used a Gauss-Legendre quadrature with 16 pivots for each of the angles. The overall cost of this numerical integration procedure would amount therefore to $2 \times 16^4 = 131072$ evaluations of the O_4 potential energy function, which is performed only once for any prespecified number of temperatures. Note though, that because of symmetry reasons, all pivots need to cover only one fourth of the molecular configuration space, which reduces the amount of function evaluations by the same factor. A comparison with a similar calculation based on 32 pivots for each coordinate gives us confidence to have assured an accuracy of $ca.$ $0.2\,cm^3$ over the whole range of temperatures, which is completely satisfactory for the present purposes.

For the least-squares fitting procedure, the second virial coefficients have been calculated at $T = 90, 150, 200, 250, 300,$ and $400\,K$, and then compared with the experimental values[8]. The first derivatives of eqn (20) with respect to the least-squares parameters, also used in this least-squares fitting procedure, have been calculated numerically.

Figure 6 compares the experimental values for the second virial coefficients of molecular oxygen with those calculated from the present O_4 DMBE potential energy surface. Also included for comparison in this figure are those of Wells[10] calculated using a di-Lennard-Jones potential and of Battaglia $et\ al.$[9] using an effective spherically averaged curve. Note that these potentials have been specially designed to fit the second-virial coefficients data (in addition to viscosities in the case of the di-Lennard-Jones function), and hence give no hint about the form of the complete O_4 potential energy surface. It is seen that the present results are in fairly good overall agreement with experiment, even in the moderately high temperature range ($T > 300\,K$) for which the relative error increases. It should also be noted that the form of the four-body energy terms allows one to correct, separately, the attractive and repulsive parts of the $O_2 - O_2$ interaction potential irrespective of the relative orientation of the two oxygen molecules. The agreement between the calculated second virial coefficients and the experimental values could, therefore, be improved by introducing anisotropic terms. However, this would necessarily lead to a more elaborate potential form with important time-consuming consequences for applications in molecular dynamics.

Finally, we compare in Figure 7 the unweighted spherically averaged $O_2 - O_2$ interaction potential obtained from the O_4 DMBE potential energy surface of the present work with that obtained from the quasi-isotropic di-Lennard-Jones function of Wells[10]. Also included for comparison in this figure is the isotropic curve of Battaglia $et\ al.$[9]. It is seen that the second-virial coefficients calculated from our fully six-dimensional O_4 DMBE potential energy surface agree reasonably well with

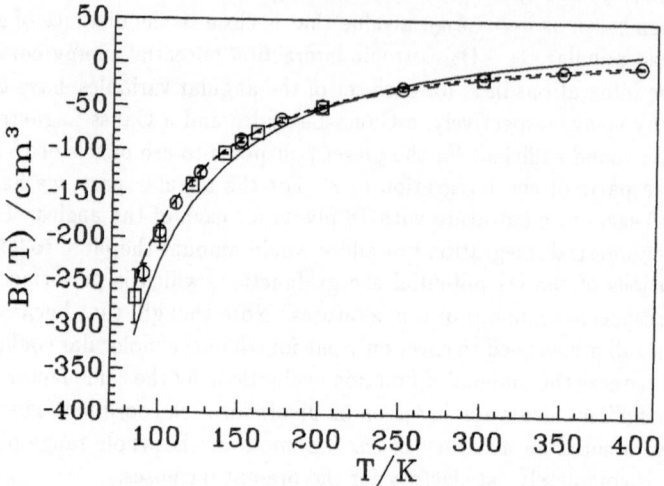

Figure 6. Second virial coefficients for molecular oxygen. Experimental: open circles, [8]; open squares, [7]. Calculated: small-dash line, [9]; large-dash line, [10]; solid line, this work.

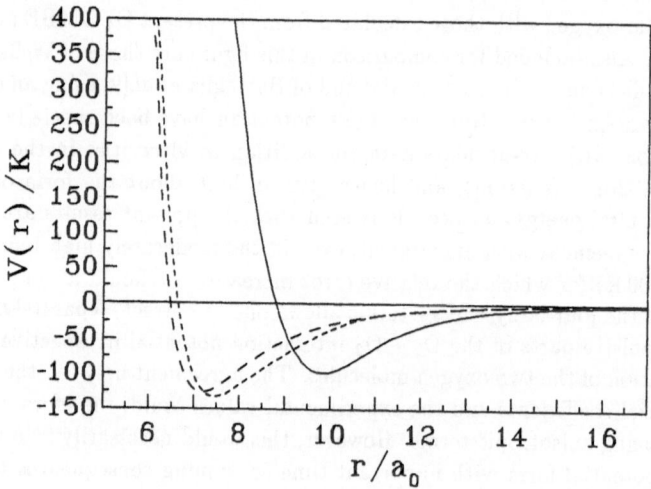

Figure 7. Spherically averaged potentials, in K, for the $O_2 - O_2$ interaction: small-dash line, [9]; large-dash line, [10]; solid line, this work.

experiment. Thus, the difference in position and well depth of our unweighted spherically averaged curve for the $O_2 - O_2$ interaction when compared with the effective spherical potentials of Wells and Battaglia *et al.* may partly be attributed to the anisotropy of this interaction. Indeed, as observed by Smith and Tindell[40], effective spherical potentials obtained by taking an unweighted average over all mutual orientations of the two interacting molecules are, except for the case of very small anisotropies, considerably different from the effective spherically averaged curves obtained by direct inversion of the thermophysical measurements.

5 Classical trajectory calculations for the $O + O_3$ reaction

Although the classical and quasi-classical trajectory methods are well documented in the literature for atom-diatom reactions[41]-[45], the number of applications of these methods to atom+triatom and diatom+diatom collisions is considerably smaller. An example of the latter is the study of $Cl + O_3 \rightarrow ClO + O_2$ using a many-body expansion potential energy surface[46]. However, the problem of establishing the Hamilton equations for a four-atom system is conceptually similar regardless the dimensionality of the problem. Thus, we focus here only on the strategy adopted for setting the initial conditions of the trajectories and on the treatment of the energy partitioning in the products.

For the classical trajectory study of the title reaction we have used the MERC-URY[47] program suitably modified to accommodate the O_4 DMBE potential energy surface of the present work, and to recognize the in principle available reaction channels. Calculations have been carried out for atom-triatom relative energies of 5, 6, 7, 10 and 15 $kcal\,mol^{-1}$, having in mind that the activation energy has to be placed near 4.5 $kcal\,mol^{-1}$ in accordance with most experimental results. For all calculations, the ozone molecule was kept in its ground vibrational state and a fixed normal mode energy sampling used. The initial normal mode coordinates and their time derivatives were obtained using the standard procedure[47] through

$$Q_i = A_i \cos(2\pi\eta_i) \qquad (21)$$
$$\dot{Q}_i = -\omega_i A_i \sin(2\pi\eta_i) \qquad (22)$$

where A_i denotes the amplitude of normal mode i, and η_i is a random number chosen to attribute a random phase $2\pi\eta_i$ to this normal mode.

For each collisional energy, the rotational energy about each principal axis has been considered to be $RT/2$, and the rotational temperature taken as 300 K, which corresponds to an intermediate value in the temperature range $200 - 400$ K for which the thermal rate coefficients were determined. All other initial conditions have been set using the standard procedures[47].

Batches of 100 trajectories per collisional energy were run to determine the maximum value of the impact parameter (b_{max}) that leads to reaction. Batches of 1000 trajectories were then run for each translational energy. Table 2 summarizes the

E (kcal mol^{-1})	$b_{max}(\mathring{A})$	N_r
5	2	40
6	2.5	41
7	2.5	64
10	3	107
15	3	174

Table 2. Collision energy, E, maximum value of impact parameter that leads to reaction, b_{max}, and number of reactive collisions, N_r.

values of b_{max} used, and the number of reactive collisions obtained. The results further indicate that reaction occurs through an equally probable attack of the O-atom to the terminal atoms of the ozone molecule.

Figure 8 illustrates the opacity function obtained for the collisional energy $E = 15\,kcal\,mol^{-1}$, which is typical of a reaction occurring with a potential energy barrier[41]. Figure 8 shows a distance *versus* time plot for a typical reactive trajectory. It is seen that most of the energy is released as vibration in the product O_2 molecule carrying that reactant oxygen atom. Indeed, a study of the partitioning of energy release between the product molecules has typically shown the values 45:17:17 and 2:2:17 (vibrational:rotational:translational, in %), respectively for the molecule carrying the reactant oxygen atom and for the remaining O_2 molecule. A similar result has been obtained for the $Cl + O_3$ reaction by Farantos and Murrell[46]. It may be rationalized by analogy with atom-diatom exothermic reactions, for which the energy barrier occurs early in the reactant channel, providing one makes the assumption that the remaining O_2 fragment in O_3 acts as a rigid block.

Figure 10 displays the calculated excitation function. It is seen that the reactive cross section increases with the collisional energy for the collisional energies of interest in this work. The calculated excitation function was then used to calculate the thermalized rate coefficients $k(T)$, after being adjusted to the form[49]

$$\sigma^r(E) = C(E - E_0)^n \exp[-m(E - E_0)] \quad E \geq E_0$$
$$\sigma^r(E) = 0 \qquad E < E_0. \tag{23}$$

where C, n, m and E_0 are least-squares parameters. All these quantities, including the threshold energy E_0, have been determined by chosing the least-squares weights such that the fit passed through the calculated error bars. The optimum values have been found to be $C = 1.7266$, $n = 1.0204$, $m = 0.0089$, and $E_0 = 4.1$; the units are such that with the energy in $kcal\,mol^{-1}$ the cross section comes in a_0^2.

From the total reactive cross section and by assuming a Maxwell-Boltzmann distribution over the translational energy one obtains[41]

$$k(T) = f_e(T) \left(\frac{8}{\pi\mu}\right)^{1/2} (kT)^{-3/2} \int_0^\infty E \exp(-E/kT)\,\sigma^r(E)dE \tag{24}$$

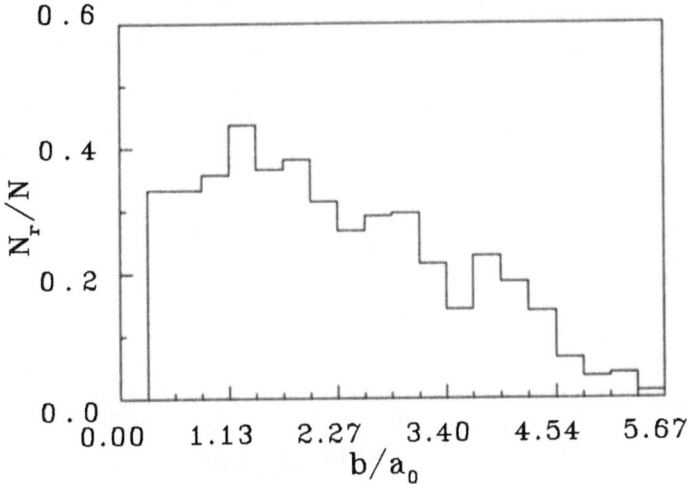

Figure 8. Probability of reaction as a function of the impact parameter for a collisional energy of $E = 15\,kcal\,mol^{-1}$.

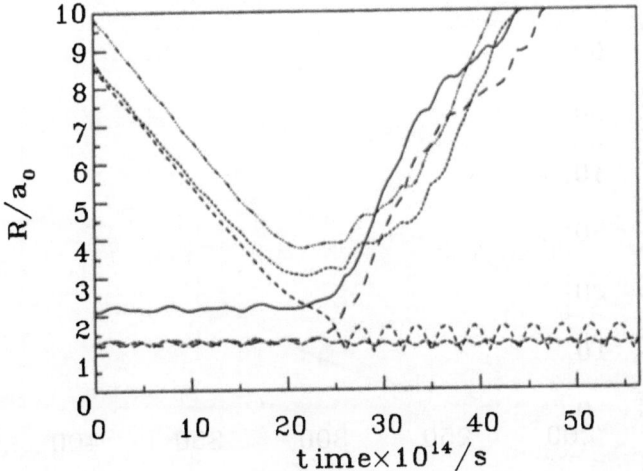

Figure 9. Distance *versus* time plot for a typical reactive trajectory.

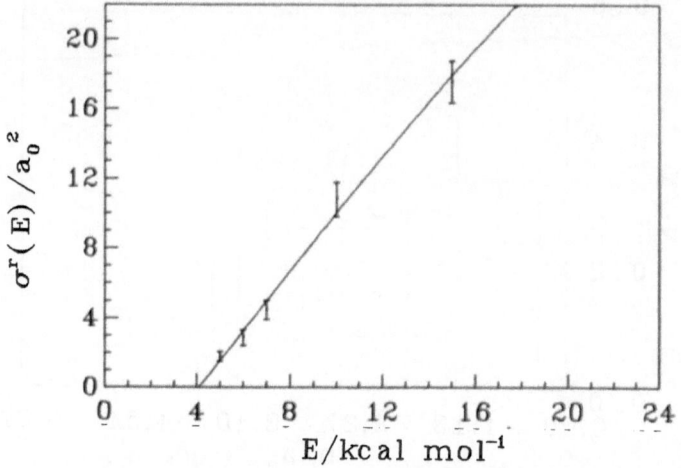

Figure 10. Excitation function for reaction 1. The solid line corresponds to eqn (23). The error bars in the calculated points are for a 68% confidence interval.

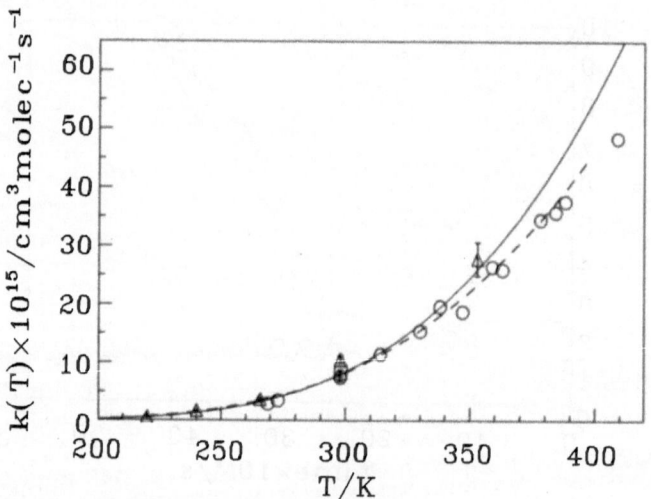

Figure 11. Thermal rate coefficients for the $O(^3P) + O_3(^1A_1) \rightarrow 2O_2(\tilde{X}\,^3\Sigma_g^-)$ reaction. Experimental: open circles, [1]; open triangles, [2]; open square, [3]; dashed line, [5]. Calculated: solid line, this work eqn (26). Also shown are the experimental error bars.

where the factor that accounts for the electronic degeneneracies in the title reaction has been assumed to be

$$f_e(T) = \frac{3}{5 + 3\exp(-228/T) + \exp(-326/T)} \tag{25}$$

which goes asymptotically to $\frac{1}{3}$ at high temperatures and to $\frac{3}{5}$ at low temperatures; all other symbols have their usual meaning. The factor 3 is the numerator accounts for the fact that the potential energy surface of relevance for the intermediate complex is a triplet while the denominator is due to the reactants being $O(^3P)$ and $O_3(^1A_1)$. Note that $O(^3P)$ consists of five states in the lowest $^3P_2$ level, three in the $^3P_1$ level which is $158\,cm^{-1}$ higher in energy, and one in the $^3P_0$ state another $68\,cm^{-1}$ above that[48], while $O_3(^1A_1)$ is singlet and non-degenerate. Using eqn (23), the thermal rate coefficient $k(T)$ can be obtained in closed analytical form yielding[49]

$$\begin{aligned} k(T) &= f_e(T)C\left(\frac{8RT}{\pi\mu}\right)^{1/2}\frac{(RT)^n\exp\left(-E_0/RT\right)}{(1+mRT)^{n+2}} \\ &\times [\Gamma(n+2) + \Gamma(n+1)(1+mRT)E_0/RT] \end{aligned} \tag{26}$$

Figure 11 compares, as a function of temperature, the results of $k(T)$ from the present work with some of the available experimental determinations. The agreement of the different experimental values with each other and with our calculations is shown to be good. The agreement with the experimental measurements of Davis *et al.* is particularly noteworthy since it extends over the entire range of temperatures covered by experiment. Finally, we observe that numerical calculations of $k(T)$ based on other representations for the excitation function do not modify this good agreement.

6 Conclusions

We have been stressing in previous papers the importance of having a reliable description of the long-range forces for reaction dynamics calculations, and we have shown that the DMBE method offers a general strategy to this problem in the case of triatomic systems. Here, we have reported a successful application of this method to a four-atom system, namely to the the lowest triplet state of O_4. To calibrate the four-body energy terms of this O_4 DMBE potential energy surface we have used available dispersion coefficients for the $O_2(\tilde{X}\,^3\Sigma_g^-)$ — $O_2(\tilde{X}\,^3\Sigma_g^-)$ interaction, second virial coefficients of molecular oxygen, and experimental activation energy data for the $O(^3P) + (^1A_1) \rightarrow 2O_2(\tilde{X}\,^3\Sigma_g^-)$ reaction. A detailed dynamics study of this reaction has been also carried out. In particular, the thermal rate coefficients calculated as a function of temperature have been shown to be in good agreement with experiment over the complete range of temperatures covered by the experimental measurements. In addition to this realistic behaviour for the kinetics of the $O(^3P) + O_3(^1A_1)$ reaction, the second-virial coefficients of gaseous oxygen calculated using the present O_4 potential energy surface have been shown to be in reasonable agreement with the best reported measurements. This is most encouraging in that it shows that information on such distinct properties can be satisfactorily reproduced using the same fully six-dimensional O_4 DMBE potential energy surface. To our knowledge, such credit can hardly be given to potential energy surfaces obtained from other existing semiempirical methods.

Acknowledgements

The authors express their gratitude to Professor J.N. Murrell (University of Sussex) for helpful discussions. They are also grateful to the "Instituto Nacional de Investigação Científica" (INIC) for financial support, and to the "Fundação para o Desenvolvimento dos Meios Nacionais de Cálculo Científico – FCCN" for the allocation of time on a CONVEX C220 computer.

References

[1] J.L. McCrumb and F. Kaufman, *J. Chem. Phys.*, 1972, **57**, 1270.

[2] D.D. Davis, W.Wong and J. Lephardt, *Chem. Phys. Lett.*, 1973, **22**, 273.

[3] G.A. West, Jr., R.E. Weston and G.W. Flynn, *Chem. Phys. Lett.*, 1978, **56**, 429.

[4] S.K. Chekin, Y.M. Gershenzon, A.V. Konoplyov and V.B. Rozenshtein, *Chem. Phys. Lett.*, 1979, **68**, 386.

[5] R. Atkinson, D.L. Baulch, R.A. Cox, Jr. R.F. Hampson and J.A. Troe, *J. Phys. Chem. Ref. Data*, 1989, **18**, 881.

[6] G.L. Zarur and Y.-N. Chiu, *J. Chem. Phys.*, 1972, **56**, 3278.

[7] L.A. Weber, *J. Res. natn. Bur. Stand.*, 1970, **74A**, 93.

[8] J.H. Dymond and E.B. Smith, *The virial coefficients of pure gases and mixtures. A critical compilation* (Clarendon Press, Oxford, 1980).

[9] F. Battaglia, F.A. Gianturco, P. Casavecchia, F. Pirani and F. Vecchiocattivi, *Faraday Discuss. Chem. Soc.*, 1982, **73**, 257.

[10] B.H. Wells, *Faraday Discuss. Chem. Soc.*, 1982, **73**, 306.

[11] V. Adamantides, D. Neisius and G. Verhaegen, *Chem. Phys.*, 1980, **48**, 215.

[12] V. Adamantides, *Chem. Phys.*, 1980, **48**, 221.

[13] W.L. Feng and O. Novaro, *Int. J. Quant. Chem.*, 1984, **26**, 521.

[14] E.T. Seidl and H.F. Schaefer, *J. Chem. Phys.*, 1988, **88**, 7043.

[15] A.J.C. Varandas, *J. Mol. Struct. Theochem*, 1985, **120**, 401.

[16] A.J.C. Varandas, in *Structure and Dynamics of Weakly Bound Molecular Complexes*, ed. A. Weber (D. Reidl Publishing Company, 1987), p. 357.

[17] A.J.C. Varandas, *Adv. Chem. Phys.*, 1988, **74**, 255.

[18] A.J.C. Varandas, *J. Mol. Struct. Theochem*, 1988, **166**, 59.

[19] A.J.C. Varandas, to be published.

[20] A.J.C. Varandas, *Int. J. Quant. Chem.*, 1987, **32**, 563.

[21] A.A.C.C. Pais and A.J.C. Varandas, *J. Mol. Struct. Theochem*, 1988, **166**, 335.

[22] A.J.C. Varandas and A.A.C.C. Pais, *Mol. Phys.*, 1988, **65**, 843.

[23] A.J.C. Varandas and J. Brandão, *Mol. Phys.*, 1986, **57**, 387.

[24] A.J.C. Varandas, J. Brandão and L.A.M. Quintales, *J. Phys. Chem.*, 1988, **92**, 3732.

[25] M.R. Pastrana, L.A.M. Quintales, J. Brandão and A.J.C. Varandas, *J. Phys. Chem.*, 1990, **94**, 8073.

[26] A.J.C. Varandas, *Mol. Phys.*, 1984, **53**, 1303.

[27] A.J.C. Varandas, F.B. Brown, C.A. Mead, D.G. Truhlar and N.C. Blais, *J. Chem. Phys.*, 1987, **86**, 6258.

[28] G.C. Lynch, R. Steckler, D.W. Schwenke, A.J.C. Varandas, D.G. Truhlar and B.C. Garrett, *J. Chem. Phys.*, in press.

[29] J.N. Murrell, S. Carter, S.C. Farantos, P. Huxley and A.J.C. Varandas, *Molecular Potential Energy Functions* (Wiley, Chichester, 1984).

[30] A.J.C. Varandas, *J. Chem. Phys.*, 1989, **90**, 4379.

[31] A.J.C. Varandas and J. Brandão, *Mol. Phys.*, 1982, **45**, 857.

[32] A.J.C. Varandas, *Mol. Phys.*, 1987, **60**, 527.

[33] J.O. Hirschfelder, R.F. Curtiss and R.B. Bird, *Molecular Theory of Gases and Liquids* (Wiley, New York, 1954, second printing, 1964).

[34] J.N. Murrell, N.M.R. Hassani and B. Hudson, *Mol. Phys.*, 1987, **60**, 1343.

[35] M.M. Graff and A.F. Wagner, *J. Chem. Phys.*, 1990, **92**, 2423.

[36] V.M.F. Morais and A.J.C. Varandas, *Chem. Phys. Lett.*, 1985, **113**, 192.

[37] C.A. Nieto de Castro, J.M.N.A. Fareleira, P.M. Matias, M.L.V. Ramires, A.A.C. Canelas Pais and A.J.C. Varandas, *Ber. Bunsenges. Phys. Chem.*, 1990, **94**, 53.

[38] A.J.C. Varandas and M.C.A. Gomes, unpublished work.

[39] J.D. Silva, J. Brandão and A.J.C. Varandas, *J. Chem. Soc. Faraday Trans. 2*, 1989, **85**, 1851.

[40] E.B. Smith and A.R. Tindell, *Faraday Discuss. Chem. Soc.*, 1982, **73**, 221.

[41] M. Karplus, R.N. Porter and R.D. Sharma, *J. Chem. Phys.*, 1965, **43**, 3259.

[42] D.L. Bunker, *Meth. Comp. Physics*, 1971, **10**, 287.

[43] E.E. Nikitin and L. Zulicke, *Lecture Notes in Chemistry, Theory of Elementary Processes*, (Springer-Verlag, Heidelberg, 1978), Vol. 8.

[44] D.G. Truhlar and J.T. Muckerman, in *Atom-Molecule collision theory*, ed. R.B. Bernstein (Plenum Press, New York, 1979), p. 505.

[45] R.N. Porter and L.M. Raff, in *Modern Theoretical Chemistry, Dynamics of Molecular Collisions*, ed. W.H. Miller (Plenum Press, New York, 1976), vol. II, p. 1.

[46] S.C. Farantos and J.N. Murrell, *Int. J. Quant. Chem.*, 1978, **14**, 659.

[47] W.L. Hase. Mercury: a general Monte-Carlo classical trajectory computer program, QCPE#453.

[48] A.A. Westenberg and N. de Haas, *J. Chem. Phys.*, 1967, **47**, 4241.

[49] R.L. LeRoy, *J. Chem. Phys.*, 1969, **73**, 4338.

THE SELF-CONSISTENT REACTION FIELD MODEL FOR MOLECULAR COMPUTATIONS IN SOLUTION

Jean-Louis RIVAIL, Daniel RINALDI and Manuel F. RUIZ-LOPEZ
Laboratoire de Chimie théorique, Université de Nancy 1
B.P. 239
54506 Vandœuvre lès Nancy Cedex
France

ABSTRACT. The effect of a macroscopic phase (solvent or adsorbent) interacting with a molecule can be taken into account in quantum chemical computations by means of a *self-consistent reaction field* approach. The macroscopic phase is represented by a continuum with macroscopic dielectric properties. The molecule undergoes an electric potential arising from the polarization by the molecular charge distribution of the macroscopic phase.

In the present approach, the charge distribution of the molecule is expanded in multipoles about an origin, and the electric potential as well as its derivatives of increasing order at this point are related to the moments through *reaction field factors* whose values depend on the geometry of the boundary of the continuum and on its dielectric properties only.

This leads to efficient Hartree-Fock computations and to the evaluation of the electrostatic and induction contributions to the free energy of solvation. The intramolecular correlation effects can be taken into account through a Møller-Plesset perturbation scheme and the dispersion can also be evaluated.

1.0 Introduction

A complete treatment of a small sample of molecular liquid from first principles is still beyond the reach of computational chemistry. In order to better understand what happens in a liquid or a solution at the molecular levels, one is forced to adopt a simplified approach. One may perform a full statistical treatment of the sample, and then the representation of the molecule and the intermolecular forces have to be rather schematic. One may describe this approach as *a true liquid of model molecules.* Conversely, one may wish to analyze in greater detail the structural modifications of a molecule - the solute or a particular molecule of a pure liquid - when it is placed in a liquid environment. One is then led to adopt a quantum chemical treatment of the molecule and to replace its actual surroundings by a simple medium, usually a *continuum*, having the averaged properties of the liquid. One may speak of a *model liquid of true molecules.* The Self-Consistent Reaction Field method belongs to this approach.

S. J. Formosinho et al. (eds.), Theoretical and Computational Models for Organic Chemistry, 79–92.

2.0 The reaction field

Let us imagine a solute molecule (M) surrounded by an arbitrary large number of solvent molecules (S) in a liquid phase. Our purpose is to consider the solute as a quantum system perturbed by its surroundings. In order to analyze this perturbation, let us consider the system in an averaged configuration and let us imagine that the solute is removed without any modification either in the solvent or in the solute structure. We shall examine the energetics of the process later but let us first consider the structure of the liquid. The volume previously occupied by the solute is now empty: we shall call it the solute's cavity. In this cavity, the averaged electric potential is non-zero. It usually varies from one point to another giving rise to a non-uniform electric field which is called the *reaction field*.

The origin of this field is easy to visualize. Let us consider the electric field created by the charge distribution of the solute in its vicinity. This field has two main effects :

(i) it distords the charge distribution of each solvent molecule creating an *induced electric moment* whatever be the nature of these molecules
(ii) it generates an *angular correlation* between the solvent and solute molecules, especially in the case of dipolar species, but also if the charge distribution of the solute is represented by higher moments (or if the solute is charged).

The overall effect is an electric polarization of the solvent around the solute which is at the origin of the electric field in the cavity. The reaction field corresponds to the statistical average of this field.

3.0 The energetics of the solvation process

Let us consider the bulk solvent and the solute molecule separately. The solvation process may be decomposed in a series of steps :

(i) Creation of the cavity in the solvent. The corresponding free energy variation ΔA_C is sometimes called *cavitation energy*
(ii) Polarization of the solvent around the cavity : ΔA_P
(iii) Induced polarization of the solute by the reaction field : ΔA_I
(iv) Electrostatic interaction between the polarized solute and the reaction field : ΔA_E
(v) Contribution of the dispersion and repulsion forces between the solute and the neighbouring solvent molecules : ΔA_D
(vi) Finally, the translational, rotational and vibrational partition functions of the solute : ΔA_T

The free energy of solvation ΔA_S is the sum of all these contributions

$$\Delta A_S = \Delta A_C + \Delta A_P + \Delta A_I + \Delta A_E + \Delta A_D + \Delta A_T \tag{1}$$

so that each individual term ought to be evaluated. Nevertheless, in some special processes, some contributions can be considered as constant during the process. In particular one may assume that if the cavity does not vary much during a process such as the search of the equilibrium geometry of a molecule or a complex, the cavitation and the dispersion-repulsion terms should not vary much. Besides, in such a process, the volume variation may be considered as negligible so that the Gibbs free energy variation reduces to the free energy variation limited to the polarization ΔA_P, induced ΔA_I and electrostatic ΔA_F terms.

4.0 The Continuum Model of the solvent

The reaction field can be computed by a statistical simulation of the solution. Nevertheless, this procedure is still very costly and limited to a somewhat simplified picture of the molecules. In particular it is still not very easy to take the induced polarization into account.

Another approach consists in assimilating the solvent to a macroscopic continuum which is characterized by some well-defined temperature dependent macroscopic quantities such as the dielectric permittivity at a given frequency ω : $\varepsilon(\omega)$, or the surface tension η. In principle, the latter quantity would be used to evaluate the cavitation free energy as soon as the geometry of the cavity is defined [1] Nevertheless, the validity of the continuum model for such a determination is questionable since at the molecular level an averaged macroscopic quantity is somewhat meaningless.

A macroscopic treatment of solute-solvent electrostatic interactions is more justified, mainly because of the long range nature of electrostatic forces. As soon as one has been able to define the surface which separates the molecule from the continuum, one is able to use classical electrostatics to analyze the polarization of the solvent ΔA_P, and of the averaged electrostatic potential in the cavity, which permits us to define the perturbation that the solute undergoes under the influence of the solvent.

5.0 The reaction field factors

Let us consider a molecule placed in a cavity surrounded by a dielectric continuum (fig. 1). The relative dielectric permittivity of the continuum is assumed to be ε and in the cavity it is taken as equal to the permittivity of a vacuum. In the following we shall assume that the charge distribution of the solute is represented by a single center multipole expansion. An equivalent distributed multipole [2,3] representation may be used without further difficulty. We shall use the spherical tensors formalism [4,5] for the multipoles in which the $2\ell + 1$ components of the multipole of rank ℓ at the origin are defined from unnormalized spherical harmonics C_ℓ^m [6] by the equation:

$$M_\ell^m = \sum_i e_i \, r_i^\ell \, C_\ell^m(\theta_i, \chi_i) \tag{2}$$

e_i representing a classical charge whose position is defined by the radius vector $\vec{r}_i(r_i, \theta_i, \chi_i)$.

The reaction of the macroscopic medium generates an electrostatic potential at this origin, which we denote by R_0^0 and if R_ℓ^m denotes the m component of the ℓ derivative of this potential at the origin, the electrostatic energy of interaction of the molecule with the reaction field may be written as:

$$\mathcal{H}'_E = - R_\ell^m M_\ell^m \tag{3}$$

in which the repeated indices stand for a sum on all their possible values: ℓ from 0 to ∞ and m from $-\ell$ to ℓ.

In the linear response approximation, the reaction field components R_ℓ^m which are due to the molecular charge distribution may be written

$$R_\ell^m = f_{\ell\ell'}^{mm'} M_{\ell'}^{m'} \tag{4}$$

in which the *reaction field factors* $f_{\ell\ell'}^{mm'}$ only depend on the dielectric properties of the medium and on the shape of the cavity. The determination of these factors is a key problem of the model. It can be achieved by verifying the continuity conditions at the boundary.

If $V_R(P)$ denotes the value of the reaction potential at a point P whose polar coordinates with respect to the reference frame at the origin are r, θ, χ, this quantity may be obtained by performing a Taylor expansion for V_R around the origin leading to [4]

$$V_R(P) = r^\ell R_\ell^m C_\ell^{m*}(\theta, \chi) \tag{5}$$

This potential is supperposed on the potential $V_M(P)$ created by the molecular charge distribution which is expressed by the usual expression [4]

$$V_M(P) = r^{-(l+1)} C_\ell^{m*} M_\ell^m \tag{6}$$

in which the repeated indices ℓ and m indicate a sum over all the possible values of ℓ (from 0 to ∞) and m.

In the cavity, the electrostatic potential V_C is the sum of these two contributions:

$$V_C = V_M + V_R \tag{7}$$

Outside the cavity, in the continuum, the electrostatic potential V_s has to satisfy the Laplace equation

$$\Delta V_s = 0 \tag{8}$$

and vanish at infinity. Its general form is therefore [7]

$$V_s = \gamma_\ell^m r^{-(l+1)} C_\ell^m(\theta, \chi) \tag{9}$$

The continuity of the electric field implies that at each point S of the surface, one has, for the tangential component of the field:

$$[\vec{\nabla} V_C(S)]_t = [\vec{\nabla} V_s(S)]_t \tag{10}$$

and for the normal component of the field

$$\varepsilon_C [\vec{\nabla} V_C(S)]_n = \varepsilon_s [\vec{\nabla} V_s(S)]_n \tag{11}$$

if ε_C and ε_s are the dielectric constant of the medium in the cavity and of the solvent, respectively. One usually takes $\varepsilon_C = \varepsilon_0$, the dielectric permittivity of the vacuum, and the only parameter is the relative permittivity of the continuum

$$\varepsilon = \frac{\varepsilon_s}{\varepsilon_0} \tag{12}$$

Conditions (10) and (11) offer us the possibility of determining the unknown coefficients γ_ℓ^m of equation (9) numerically and, above all, the reaction field factors in equation (4) in any case. In some special case of a spherical, spheroidal [8] or ellipsoidal cavity [9] , these equations can be solved analytically. For instance, for a spherical cavity of radius a, the reaction field factors reduce to scalars

$$f_{\ell\ell'}^{mm'} = \frac{(l+1)(\varepsilon-1)}{\varepsilon(l+1)+l} \frac{1}{a^{2l+1}} \tag{13}$$

This expression illustrates the dependence of each factor (for a given ℓ) on two quantities only, the relative electric permittivity ε and the shape of the cavity, represented here by its radius.

Knowledge of the reaction field factors allows us to obtain the electrostatic energy of interaction of the molecule with the continuum

$$\mathcal{H}'_E = - M_\ell^m f_{\ell\ell'}^{mm'} M_{\ell'}^{m'} \tag{14}$$

In addition, an important theorem [10] states that the variation of the polarization free energy, ΔA_P is related to $\mathcal{H}'_E$ by the simple equation

$$\Delta A_P = -\frac{1}{2} \mathcal{H}'_E \tag{15}$$

so that if one assumes that the solute molecules are rigid and that the only contribution to the free energy variation comes from the internal energy term, the electrostatic contribution to the free energy of solvation of these molecules is

$$\Delta A_S = -\frac{1}{2} M_\ell^m f_{\ell\ell'}^{mm'} M_{\ell'}^{m'} \tag{16}$$

6.0 The cavity

The key feature of the model is now the shape of the cavity, which has to be defined on physical criteria.

The volume of the cavity can be regarded as the fraction of the total volume of solution which is devoted to one solute molecule. Therefore this volume is taken as the partial molecular volume of the solute.

The geometry is more difficult to define. If one states that the cavity should include the largest fraction of the molecule *i.e.* all the nuclei and as much as possible of the electron cloud, one is then led to limiting the cavity by an electronic density surface. Figure 1 represents such a cavity for the molecule of 1-amino-2-nitroethylene. It is clear that this surface bears some similarity with the surface bounded by the van der Waals spheres, with some differences, nevertheless, which appear to be sensitive to the electronic charge of the atoms.

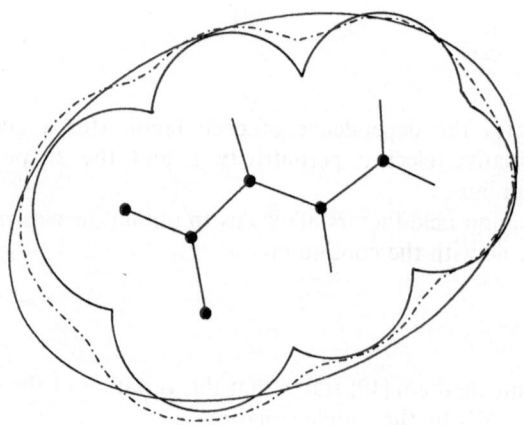

Figure 1. Cross section of the isodensity surface (dashed line), ellipsoidal cavity and Van der Waals surface for the Z isomer of 1-amino-2-nitroethylene.

7.0 The Self-Consistent Reaction Field equations

The equilibrium of the system composed of the solute and the continuum implies that free energy is minimum. The main terms which are assumed to vary during the solvation process being the electrostatic and the induction terms, the geometry and the electron distribution of the solute will be modified in order to minimize the sum of their contributions to the free energy of the system.

Within the framework of the Born-Oppenheimer approximation, this minimization can be reached in two steps: by varying the electronic wavefunction for a given geometry and by varying the geometry until the minimum is reached.

If one further assumes that the only entropy variation arising from a modification of the solute's electronic structure appears in the solvation free energy term the electronic wavefunction at a given geometry can be determined by minimizing the quantity

$$< \Psi | \mathcal{H}_o | \Psi > \; - \; \frac{1}{2} < \Psi | \mathbf{M}_\ell^m | \Psi > f_{\ell\ell'}^{mm'} < \Psi | \mathbf{M}_{\ell'}^{m'} | \Psi > \tag{17}$$

where Ψ is the normalized multielectron wavefunction, $\mathcal{H}_o$ the hamiltonian of the free solute molecule and $\mathbf{M}_\ell^m$ a multiple moment quantum chemical operator. One recognizes in the second term of expression (17) the quantum mechanical equivalent of the free energy of interaction (16).

If one assumes that each orbital ψ_i entering the wavefunction Ψ is a linear combination of atomic orbitals ϕ_μ, which can be written:

$$| \psi_i > \; = \; \sum_\mu c_{\mu i} | \mu > \tag{18}$$

one is able by the usual variational approach to determine the coefficients of combination (18) as being the eigenvalues of the self-consistent reaction field matrix whose (μv) element has for expression [11]:

$$F_{\mu v} \; = \; F_{\mu v}^o \; + \; < \Psi | \mathbf{M}_\ell^m | \Psi > f_{\ell\ell'}^{mm'} < \mu | \mathbf{M}_{\ell'}^{m'} | v > \tag{19}$$

where F^o is the usual Hartree-Fock operator of a free molecule [12]. Therefore, when the reaction field factors are determined, the computation of the electronic wavefunction is not substantially different from a usual SCF computation.

8.0 Optimization of the geometry

The equilibrium structure of the solute is reached when the quantity

$$A' \; = \; < \mathcal{H}_o > \; - \; \frac{1}{2} < \mathbf{M}_\ell^m > f_{\ell\ell'}^{nm'} < \mathbf{M}_{\ell'}^{m'} > \tag{20}$$

is minimum.

Efficient optimization procedures require the computation of the derivatives of the quantity to be minimized with respect to a coordinate k. If one denotes by $\partial_k X$ the partial derivative of X with respect to k, one can write :

$$\partial_k A' \; = \; \partial_k < \mathcal{H}_o > \; - \; < \mathbf{M}_\ell^m > f_{\ell\ell'}^{mm'} \partial_k < \mathbf{M}_{\ell'}^{m'} > \; - \; \frac{1}{2} < \mathbf{M}_\ell^m > \partial_k f_{\ell\ell'}^{mm'} < \mathbf{M}_{\ell'}^{m'} > \tag{21}$$

The first term of this sum is the derivative of the molecular energy which is computed by standard procedures [13]. The derivatives of the moments occurring in the second term are easy to compute analytically. The only difficulty comes from the derivatives of the reaction field factors. The variation of these factors is the consequence of the modification of the shape of the cavity with the molecular geometry. In the general case these derivatives have to be computed numerically. Nevertheless, in the case of an ellipsoidal cavity these derivatives may be computed analytically leading to an efficient optimization algorithm [14].

9.0 Intramolecular electron correlation effects

The previous derivation is limited at the SCF level. The correction of these results by correlation effects will modify the molecular energy. By modifying the molecular electron distribution, it is expected to modify the solute-solvent interaction free energy and therefore the equilibrium geometry of the solute.

A MCSCF approach to this problem has been proposed by other authors [15]. We shall develop a perturbational method here [16] which consists in an extension of the usual Møller-Plesset scheme.

Let us write the hamiltonian operator of the solvated molecule as

$$\mathcal{H} = \mathcal{H}_o - M_\ell^m f_{\ell\ell'}^{mm'} < \Psi \mid M_{\ell'}^{m'} \mid \Psi > \qquad (22)$$

where Ψ denotes the exact (correlated) wavefunction. The Hartree-Fock operator in turn may be written as

$$F = F^o - M_\ell^m f_{\ell\ell'}^{mm'} < \Psi_o \mid M_{\ell'}^{m'} \mid \Psi_o > \qquad (23)$$

where Ψ_o denotes the electronic wavefunction at the Hartree-Fock level.

The hamiltonian may be written as a perturbed expression of the Hartree-Fock operator :

$$\mathcal{H} = F + \left(\mathcal{H}_0 - F^0\right) + M_\ell^m f_{\ell\ell'}^{mm'} \left(< \Psi_0 \mid M_{\ell'}^{m'} \mid \Psi_0 > - < \Psi \mid M_{\ell'}^{m'} \mid \Psi > \right) \qquad (24)$$

The perturbation comprises two parts :

- The standard Møller-Plesset perturbation

$$\mathbb{C} = \mathcal{H}_0 - F^0 \qquad (25)$$

- The perturbation due to the solute-solvent interaction which has the particularity of containing the correlated wavefunction Ψ.

This last term generates a non linear perturbation. The general treatment of such a case has been considered already [17]. It consists in expanding the wavefunction Ψ in terms of the multielectronic eigenfunctions of the unperturbed operator, to successive orders of perturbation and replacing Ψ by this expansion in (23).

If $C_J^{(i)}$ denotes the coefficient of the eigenstate $|J>$ in the corrections of Ψ to order i, and if $\mathcal{H}^{(i)}$ denotes the perturbation operator to the same order, one obtains, with $|0> \equiv |\Psi_0>$

$$\mathcal{H}^{(1)} = \mathbb{C}$$

$$\mathcal{H}^{(2)} = -2\Sigma'_I C_I^{(1)} \mathbf{M}_\ell^m f_{\ell\ell'}^{mm'} <0|\mathbf{M}_{\ell'}^{m'}|I>$$

$$\mathcal{H}^{(3)} = -2\Sigma'_J C_J^{(2)} \mathbf{M}_\ell^m f_{\ell\ell'}^{mm'} <0|\mathbf{M}_{\ell'}^{m'}|J> - \Sigma'_I \Sigma'_P C_I^{(1)} C_P^{(1)} \mathbf{M}_\ell^m f_{\ell\ell'}^{mm'} <I|\mathbf{M}_{\ell'}^{m'}|P>$$

$$\text{(26)}$$

...

In these equations the symbol Σ'_L stands for the sum over all the $|L>$ states except $|0>$. The calculation follows as usual, by comparing coefficients of the Schrödinger equation to successive orders.

9.1 FIRST ORDER

The first order results are the same as the usual Møller-Plesset treatment except that the Hamiltonian contains the solvent-solute interaction energy. The first order energy is $<0|\mathcal{H}_o|0>$. Similarly the first order contribution to the free energy of interaction is

$$\Delta A_S^{(1)} = -\frac{1}{2} <0|\mathbf{M}_\ell^m|0> f_{\ell\ell'}^{mm'} <0|\mathbf{M}_{\ell'}^{m'}|0> \tag{27}$$

so that to the first order, the electrostatic contribution to the free energy of solvation is identical to the result obtained after the Hartree-Fock computation

$$\Delta A_S = <0|\mathcal{H}_0|0> - \frac{1}{2} <0|\mathbf{M}_\ell^m|0> f_{\ell\ell'}^{mm'} <0|\mathbf{M}_{\ell'}^{m'}|0> \tag{28}$$

The first order correction to the wavefunction is also identical to the first order Møller-Plesset (MP1) perturbation

$$C_I^{(1)} = \frac{<I|\mathbb{C}|0>}{E^{(0)} - E_I} \tag{29}$$

in which the states $|I>$ reduce to double excitations of Ψ_0, which we shall denote by $|D>$.

9.2 SECOND ORDER

Owing to the fact that the wavefunction perturbed to first order only contains double excitations $|D>$ and that the electric moment operators occurring in $\mathscr{H}^{(2)}$ are one electron operators, the energy correction reduces to the Møller-Plesset (MP2) correction

$$E^{(2)} = \Sigma'_D \, C_D^{(1)} <0|\mathbb{C}|D> \tag{30}$$

Similarly the correction to the wavefunction is defined by

$$C_J^{(2)} = \Sigma'_D \, C_D^{(1)} \frac{<J|\mathbb{C}|D>}{E^{(0)} - E_J} \tag{31}$$

It allows single, double, triple and quadruple excitations for $|J>$ so that the second order correction to the free energy of interaction is

$$
\begin{aligned}
\Delta A_S^{(2)} = \; & 2\Sigma'_S \, C_S^{(2)} <0|\mathbf{M}_\ell^m|S> f_{\ell\ell'}^{mm'} <0|\mathbf{M}_{\ell'}^{m'}|0> \\
& + \Sigma'_D \Sigma'_{D'} \, C_D^{(1)} \, C_{D'}^{(1)} <D|\mathbf{M}_\ell^m|D'> f_{\ell\ell'}^{mm'} <0|\mathbf{M}_{\ell'}^{m'}|0>
\end{aligned}
\tag{32}
$$

where $|S>$ stands for a single excitation, $|D>$ and $|D'>$ for a pair of double excitations differing by one orbital.

9.3 HIGHER ORDERS

The same reasoning may be extended to higher order without excesive difficulty [16]. It is therefore possible to extend the usual Møller-Plesset treatment of electron correlation to the SCRF model.

10.0 Intermolecular electron correlation. Energy of dispersion

The many body aspect of dispersion forces makes the computation of their role on a solvated molecule far more difficult than the intramolecular effects. Nevertheless the SCRF model can be adapted successfully to the evaluation of dispersion. The treatment is a generalization of Linder's theory [18] of van der Waals interactions in condensed media using reaction field techniques.

One starts from the general expression of the free energy

$$\Delta A_S = -\frac{1}{2} \, R_\ell^m(t) \, M_\ell^m(t) \tag{33}$$

in which R_ℓ^m and M_ℓ^m stand for statistical averages of the corresponding quantities. The time dependent moments $M_\ell^m(t)$ and the related reaction field component $R_\ell^m(t)$ are then decomposed into a permanent part, which gives rise to the electrostatic free energy considered above, and the time dependent fluctuations of these quantities which are responsible for the dispersion contribution.

Moments $\hat{M}_\ell^m(\omega)$ and reaction field $\hat{R}_\ell^m(\omega)$ depending on frequency are introduced by means of Fourier transforms of (33) and we still have the relationship :

$$\hat{R}_\ell^m(\omega) = f_{\ell\ell'}^{mm'}(\omega)\,\hat{M}_\ell^m(\omega) \tag{34}$$

in which the frequency dependent reaction field factor is obtained by considering the frequency dependence of the dielectric permittivity of the medium.

The calculation follows the treatment given by Linder in the dipolar case and has been published previously for a spherical cavity [19]. It gives for the dispersion contribution to the free energy of dispersion :

$$\Delta A_D = -\frac{\hbar}{\pi^2} \int_0^\infty \int_0^\infty d\omega_i\, d\omega_s\, \frac{1}{\omega_i + \omega_s}\, f''^{mm'}_{\ell\ell'}(\omega_s)\, \aleph''^{mm'}_{\ell\ell'}(\omega_i) \tag{35}$$

In this equation indices i and s refer to the solute and the solvent respectively. $f''^{mm'}_{\ell\ell'}(\omega_s)$ stands for the imaginary part of the reaction field factor at frequency ω_s and $\aleph''^{mm'}_{\ell\ell'}(\omega_i)$ for the imaginary part of the corresponding component of the multipole polarizability.

The frequency dependence of the reaction field factor is calculated by using a frequency dependent dielectric permittivity $\varepsilon(\omega)$, which is an experimental quantity related to the solvent. Computation of the frequency dependent multipole polarizabilities is feasible, in principle, by perturbation techniques. Nevertheless this procedure is tedious and one generally prefers some variation-perturbation scheme [20]. In addition, such a computation is still limited to small systems and can scarcely be extended economically to molecules of chemical interest. Hence a further simplification has been proposed. It consists in assuming that the quantities $f''^{mm'}_{\ell\ell'}$ and $\aleph''^{mm'}_{\ell\ell'}(\omega_i)$ are strongly peaked only at the effective frequencies :

$$\omega_i^0 = \frac{\Delta E_i}{\hbar}$$

$$\omega_s^0 = \frac{\Delta E_s}{\hbar} \tag{36}$$

One obtains the general expression :

$$\Delta A_D = -\frac{1}{8}\frac{\Delta E_i\,\Delta E_s}{\Delta E_i + \Delta E_s} \sum_{\ell\ell'}\sum_{mm'} f_{\ell\ell'}^{mm'}(\infty)\,\aleph_{\ell\ell'}^{mm'}(0) \tag{37}$$

in which $\aleph_{\overline{\ell\ell'}}^{\pi\pi'}(0)$ is a component of the static polarizability tensor and $f_{\overline{\ell\ell'}}^{\pi\pi'}(\infty)$ is the corresponding reaction field factor computed by using the "infinite frequency dielectric permittivity" ε_∞ (usually the square of the extrapolated refractive index) instead of ε.

This assumption is in general rather poor and introduces two more parameters into the model, ΔE_i and ΔE_r which are often related to the ionization potentials. Its only advantage is to make the evaluation of ΔA_D feasible provided that one is able to compute the multipole polarizabilities. We proposed [21] a convenient extension of the Kirkwood-Pople-Schofield variational method [22] which leads to reasonable values of ΔA_D, and enables us to derive SCRF Hartree-Fock equations including the electrostatic and dispersion interactions [23].

11.0 Conclusion

In this paper we have developed the main features of a Self-Consistent Reaction Field Model of solvation based on the use of generalized reaction field factors which enable us to relate the perturbation caused by the solvent on the solute to the multipole moments of the solute.

This formalism has proved to be very efficient for computing the electronic wavefunctions of solvated molecules and equilibrium molecular geometries. It also permits the computation of the electrostatic and induction contributions to the free energy of solvation and can be extended to the computation of intramolecular correlation effects as well as intermolecular correlation *i.e.* the dispersion contribution to the free energy of solvation.

This model can be applied to the study of the properties of a solvated molecule, complex or transition state. Its extension to a full treatment of a reaction path requires the computation of the free energy variation and in addition to the terms computed by this model there is another important contribution: the cavitation term. This last term obviously deserves a careful theoretical analysis in order to devise reliable computational techniques. Nevertheless, some empirical approaches of this quantity make possible an estimate of its variations along the reaction path. Therefore, the way to reactivity studies in the liquid state seems to be open.

This approach is not limited to the study of solutions. It has been extended to the case of a molecule interacting with the surface of a solid [24]. In this case, the problem is even simpler provided that the solid is a dielectric since the cavity does not need to be created and one only has to introduce a repulsive potential between the molecule and the solid. It can be extended to the quantum chemical study of a fragment of macromolecule in electrostatic interaction with the rest of the system. This is particularly interesting to study the chemical processes in the active site of a biological macromolecule like an enzyme [25]. Therefore, several fields of chemistry are already concerned with this methodology.

12.0 References

[1] Haliciloglu T. and Sinagoglu O. (1969), *Ann. N.Y. Acad. Sci.*, **158**, 308

[2] Claverie P. (1978), *Intermolecular Interactions : from Diatomics to Biopolymers*, Pullman B. Ed. (John Wiley, Chichester), p. 69

[3] Stone A.J. (1978), *Chem. Phys. Letters*, **83**, 233; Stone A.J. (1985), *Mol. Phys.*,**56**, 1065

[4] Gray C.G. (1976), *Can J. of Phys.*, **54**, 505

[5] Stone A.J. and Tough R.J.A. (1984), *Chem. Phys. Letters*, **110**, 123

[6] Brink D.M. and Satchier G.R. (1962), *Angular Momentum*, Oxford University Press, London

[7] Kirkwood J.G. (1934), *J. Chem. Phys.*, **2**, 351

[8] Harrison S.W., Nolte N.J. and Beveridge D. (1976), *J. Phys. Chem.*, **80**, 2580

[9] Rivail J.L., Terryn B. (1982), *J. Chim. Phys.*, **79**, 1; Rinaldi D., Ruiz-Lopez M.F. and Rivail J.L. (1983), *J. Chem. Phys.*, **78**, 834

[10] Claverie P. (1982), *Quantum Theory of Chemical Reactions*, Vol. III, Daudel R. et al. Ed., (Reidel, Dordrecht) pp. 151-175

[11] Rivail J.L. and Rinaldi D. (1976), *Chem. Phys.*, **18**, 233

[12] One notices that the perturbation has been multiplied by -2 when passing from (17) to (19). The factor 2 appears in the variational process from the quadratic character of the perturbation and the negative sign from the charge of the electrons in the multipole moments.

[13] See for instance : Pulay P. (1987), *Advan. Chem. Phys.*, **69**, 241; Gaw J.F. and Handy N.C. (1985), *Annu. Rep. Prog. Chem. Sec. C*, **81**, 291

[14] Rinaldi D. et al., to be published

[15] Mikkelsen K.V., Ågren H, Jensen H.J.Aa. and Helgaker T. (1988), *J. Chem. Phys.*, **89**, 3086 [16] Rivail J.L. (1990), *C.R. Acad. Sci. Paris*, **311 II**, 307

[17] Surján P.R. and Ángyán J. (1983), *Phys. Rev. A*, **28**, 45

[18] Linder B. (1967), *Advan. Chem. Phys.*, **12**, 225

[19] Costa Cabral B., Rinaldi D. and Rivail J.L. (1984), *C.R. Acad. Sci. Paris*, **298 II**, 675

[20] Bishop D. M. and Cartier A. (1983), *J. Comp. Chem.*, **4**, 170

[21] Rivail J.L. and Rinaldi D. (1976), *C.R. Acad. Sci. Paris*, **283 B**, 111

[22] Rivail J.L. and Cartier A. (1978), *Mol. Phys.*, **36**, 1085

[23] Rinaldi D., Costa Cabral B. and Rivail J.L. (1986), *Chem. Phys. Letters*, **125**, 495

[24] Hoggan P. (1990), *J. Chim. Phys.*, **87**, 1013

[25] Tapia O. (1982), *Molecular Interactions*, Vol.3, Ratajczak II. and Orville-Thomas W.J. Ed. (John Wiley, Chichester), p. 47

NEW SYMMETRY THEOREMS AND SIMILARITY RULES FOR TRANSITION STRUCTURES

PAUL G. MEZEY

Department of Chemistry and

Department of Mathematics,

University of Saskatchewan,

Saskatoon, Canada, S7N 0W0

ABSTRACT. Transition structures of chemical reactions and conformational changes may be viewed as special nuclear configurations. For the computational analysis of molecular processes it is advantageous to regard the family of all possible configurations as a formal, multidimensional space, the nuclear configuration space. This approach leads to intuitive interpretations of many familiar chemical concepts, however, it also has many counterintuitive properties which often lead to misconceptions in computational theoretical chemistry. Several of these problems will be reviewed, in particular those which have importance in the study of transition structures.

After clarifying the relevant concepts and terminology used in reaction topology, some of the fundamental properties of transition structures are reviewed within a general, configuration space approach. Special attention is given to some recently proven symmetry theorems on the presence of critical points in various domains of the configuration space. Based on these results, new symmetry theorems are presented with applications to transition structures. The new theorems may aid the development of computational algorithms devised for the theoretical determination, analysis, and characterization of transition structure nuclear configurations.

1. Introduction

In this chapter some implications of a family of new symmetry theorems will be described, with particular emphasis on transition structure problems. The results have both theoretical and computational consequences for conformational analysis, shape analysis, and reaction path studies in organic chemistry. We shall use the concepts, terminology and notations reviewed in earlier works [1-3]. The reader may find

93

S. J. Formosinho et al. (eds.), Theoretical and Computational Models for Organic Chemistry, 93–110.
© 1991 *Kluwer Academic Publishers.*

additional background information on potential surface and nuclear configuration problems in references [4-17] (and references quoted therein), on reaction topology and molecular topology in [18-23], and on molecular shape analysis in [24-26]. The general, mathematical background to various aspects of topology can be found in references [27-37]. Here only a brief introduction will be given, in order to set the framework for the discussion of the symmetry and shape relations, emphasising the intuitive, pictorial aspects of the molecular topology program. In addition, some of the more problematic, somewhat counterintuitive concepts of the configuration space approach, relevant to transition structure studies, will be reviewed.

Molecular shape changes during conformational processes and chemical reactions depend on the associated nuclear displacements and on the electronic state. One may choose a suitable set of $3N-6$ internal coordinates of a system of N nuclei, for example, internuclear distances and formal bond angles to represent nuclear arrangements. These coordinates define a multidimensional nuclear configuration space, where each point of this space corresponds to a formal nuclear configuration. Note that in organic chemistry the word "configuration" is sometimes used with a meaning limited to chirality problems, however, in the more general terminology we use here, nuclear configuration means an arbitrary relative arrangement of the nuclei. A configuration change can be modelled by a displacement within the nuclear configuration space. A general distance function $d=d(K,K')$ may be defined for any two nuclear configurations K and K', turning this space into a metric space, motivating the notation M for the configuration space. Note that such a general distance function can be defined for all molecular systems, but only if one refrains from using the familiar bond length, bond angle, and torsion angle internal coordinates when defining the configuration space [1]. The distance $d(K,K')$ can be interpreted as a measure of dissimilarity between the two nuclear configurations K and K'. The nuclear configuration space M represents the family of all possible geometrical arrangements of a given set of atoms of a fixed overall stoichiometry. Within nuclear configuration space M the distance $d(K,K')$ between any two nuclear configurations K and K' can be calculated using analytical formulas given in ref. [1].

The molecular energy also depends on the nuclear configuration K, and can be expressed as a function of the internal coordinates. This function may be regarded as a multidimensional potential energy hypersurface $E(K)$ over the configuration space M. It is advantageous to study molecular shape problems and symmetry conditions in the context of potential energy hypersurfaces and nuclear configuration spaces.

2. Relaxed Cross Sections of Nuclear Configuration Spaces

When studying confomational changes and chemical reactions, it is often unnecessary to consider all the degrees of freedom of a polyatomic molecular system. The major changes often occur in only a few of the internal coordinates, and based on chemical experience it is sometimes possible to identify these coordinates before the actual conformational study. These selected coordinates are regarded as the essential variables, and the remaining coordinates are either fixed or optimized (energy-optimized) for each choice of the essential variables. In the former case one obtains a rigid cross section of the nuclear configuration space, and in the latter case a formal, "relaxed" cross section is obtained.

Following the terminology of ref.[3], we shall refer to the essential variables as the *active coordinates*, and to the remaining ones as the *passive coordinates*.

The relaxed cross section is dependent on the electronic state: relaxation (energy optimization) of the passive coordinates is determined by the actual energy function, that is, by the energy hypersurface of the given electronic state. For example, a typical passive coordinate is a C-H bond length that may remain nearly constant in some chemical reaction. However, this C-H bond length may have slightly different optimum values in the lowest singlet and lowest triplet states of the molecule. Hence, the relaxed cross sections in the two electronic states are different.

In many applications, approximate relaxed cross sections are generated by the following simple method: for selected, fixed values of the active coordinates the energy is minimized as a function of the passive internal coordinates. For example, if one studies the conformational problem of a C-C single bond rotation in ethane, then for selected values of the bond rotation angle (the active coordinate), the energy is minimized in terms of the C-C and C-H bond lengths and the C-C-H and H-C-H bond angles (the passive coordinates). However, this technique does not always lead to a subset of the configuration space that is truly relaxed according to the usual condition applied to analogous macroscopic objects. For a relaxed macroscopic surface, all the net forces act tangentially to the surface, whereas for the approximate relaxed crossection obtained by the above technique, this is not necessarily so. The optimized passive coordinates are not necessarily constant throughout the cross section, hence the energy gradient (force) may have components orthogonal to this cross section.

For example, take the two-dimensional model surface

$$z = f(x,y) = x + y + (x - y)^2 . \tag{1}$$

This surface is generated by the parabola $z = w^2$, of variable $w = x - y$, where the parabola is raised at a constant rate as it is translated along the line

$$y = x . \tag{2}$$

Consider now variable x as the active coordinate, and for each x value minimize f as a function of passive coordinate y. By partial differentiation, one obtains

$$y = x - 0.5 \tag{3}$$

for the locus of the points fulfilling the above condition. However, this cross section is not truly relaxed, since at points of this line the energy gradient does not lie within the line. If it were a rubber string, it would move spontaneously, until all forces locally orthogonal to the string would vanish. For example, at the point $(1,0.5)$ of the line (3), the gradient vector is $(2,0)$, clearly not parallel with the direction vector $(1,1)$ of the line (3). Consequently, in this and similar cases the cross section is not fully relaxed.

By contrast, the line (2), $y = x$ of the above example is a fully relaxed cross section of the energy function (1), since there the gradient is the constant vector $(1,1)$, parallel with line (2) at all points.

We may conclude that it is a somewhat misleading practice in computational conformational analysis and reaction surface studies to use the term "relaxed cross section" for the cross sections obtained with the usual method. The significance of this distinction has been pointed out in recent studies [2,3] where truly relaxed cross sections have special importance. Here, also, we shall need truly relaxed cross sections, since they have special properties which can be exploited in symmetry and shape analysis.

The extent of deviation of an actual cross section from a relaxad one is also of importance in obtaining upper and lower bounds on the number of critical points within various regions of the configuration space. As it has been pointed out [38], many of the essential features of potential energy hypersurfaces, for example, valley floors and mountain ridges, are often approximately aligned with the "chemically motivated" internal coordinates, such as bond torsion angles. The valley floors and mountain ridges are among the possible relaxed cross sections of the surface. Since for truly relaxed cross sections the gradient is tangential to the cross section, the search for critical points can be formulated by searching for intersections of relaxed cross sections. This observation has been exploited in estimating the number of minimum points and saddle points of potential energy hypersurfaces, and various formulas have been derived for the number of critical points [38]. In fact, one does not require fully relaxed cross sections for the mere detection of critical points, and in the same study [38] a criterion has been given for what approximate alignment of the actual cross sections with the fully relaxed ones is required for the validity of the enumeration formulas [38] for critical points.

More recently, several studies [39-41] have applied some of the formulas derived in [38]. By renaming some quantities as κ, one of the formulas of [38] has been referred to as the κ-rule [39-41]. In studying some of the actual examples, much attention has been paid to the following question: which domains of the configuration space (and which coordinate systems) obey the rule? The potential surfaces in these domains have been called "ideal" [39-41]. In this context, it is useful to point out that the validity criterion of the original paper has already been proposed for a test [38] that can decide the applicability of the formulas, that is, in the terminology of the more recent applications, which surfaces (in fact, which representations) are ideal. In the more recent applications [39-41], the validity criterion of the original work [38] has not been exploited.

3. Catchment Regions and Symmetry Domains of Nuclear Configuration Spaces

The critical points $K(\lambda,i)$ of potential energy hypersurfaces $E(K)$ are of special importance. At each critical point $K(\lambda,i)$, the energy gradient vanishes, that is, by regarding the energy as a formal vertical dimension, the tangent hyperplane of $E(K)$ at $K(\lambda,i)$ is horizontal. At a critical point $K(\lambda,i)$, the Hessian matrix of second derivatives of the energy has precisely λ negative eigenvalues. The sign of eigenvalues can be used to characterize critical points. For a minimum the index λ is zero, whereas for a simple saddle point $\lambda=1$. In the notation $K(\lambda,i)$, the integer i is a serial index.

For an energy hypersurface $E(K)$ of a specified electronic state, the catchment regions $C(\lambda,i)$ are defined [18] by the formal, vibrationless relaxations of various nuclear configurations. Vibrationless relaxations can be described by formal steepest descent paths on the potential surface if mass weighted coordinates are used. A catchment region $C(\lambda,i)$ is the collection of all formal nuclear configurations K from where a steepest descent on the surface $E(K)$ leads to a given critical point $K(\lambda,i)$. For a continuous and differentiable potential energy hypersurface $E(K)$, the catchment regions $C(\lambda,i)$ generate a complete partitioning of the nuclear configuration space M. The catchment regions are analogous with basins and watersheds of geographical terrains.

The catchment regions can be associated with chemical concepts [1,18]. If the critical point $K(\lambda,i)$ has no negative canonical curvatures, $\lambda=0$, that is, if $K(\lambda,i)$ is a minimum, then the catchment region $C(\lambda,i)$ represents a stable chemical species.

If the critical point $K(\lambda,i)$ has a single negative canonical curvature, that is, if the local Hessian matrix has precisely one negative eigenvalue, $\lambda=1$, then $K(\lambda,i)$ is called a simple saddle point and the corresponding catchment region $C(\lambda,i)$ represents a transition structure (a "transition state", that is not a state at all).

If the critical point $K(\lambda,i)$ has more than one negative canonical curvatures $(\lambda>1)$, then its catchment region $C(\lambda,i)$ represents an unstable family of formal configurations (a formal "species") of little direct chemical importance.

In the following we shall assume that all critical points are isolated, and such degenerate cases like a horizontal line of infinitely many minimum points along a valley bottom [1] do not occur.

We shall study the relations among critical points, catchment regions and three-dimensional point symmetry and shape of molecules. The three-dimensional point symmetry group of a nuclear configuration K will be denoted by $g(K)$. We shall exploit some earlier results, among which the catchment region point symmetry theorem is of the most direct relevance: the three-dimensional nuclear configuration corresponding to the critical point $K(\lambda,i)$ must have the highest point symmetry within its catchment region $C(\lambda,i)$ [see ref. 1, p367, and ref. 3]. Stated differently, the point symmetry group $g(K)$ of any nuclear configuration K from a catchment region $C(\lambda,i)$ is a subgroup of the point symmetry group $g(K(\lambda,i))$ of the critical point $K(\lambda,i)$ of $C(\lambda,i)$:

$$g(K) \subset g(K(\lambda,i)) \tag{4}$$

if

$$K \in C(\lambda,i). \tag{5}$$

Note that each group is regarded as one of its own subgroups.

This theorem has been proven using an earlier result of Pechukas: along a steepest descent path the point symmetry group (as well as the framework group) of nuclear configurations may change only at a critical point, where it must have all those point symmetry elements (framework group elements, resp.) that are present at non-critical points of the path [5].

The following corollary [3] of the catchment region point symmetry theorem is of importance to our present problems: if one considers steepest ascent paths and the formal catchment regions $C^-(\lambda,i)$ of the inverted potential energy hypersurface $-E(K)$, then the theorem applies for all configurations K of $C^-(\lambda,i)$. The critical points $K(\lambda,i)$ of $E(K)$ are evidently the same as the critical points of $-E(K)$; however, their catchment regions, $C(\lambda,i)$ and $C^-(\lambda,i)$, are not in general the same. Consequently, this corollary of the theorem is providing a useful additional tool for symmetry analysis. Note that in the notation $C^-(\lambda,i)$, the index λ of the corresponding critical point $K(\lambda,i)$ refers to the original potential energy hypersurface $E(K)$, and not to the inverted potential $-E(K)$.

If the same point is regarded as a critical point of the inverted hypersurface $-E(K)$, for which the index is denoted by λ^-, then for any nondegenerate critical point $K(\lambda,i)$ the following index relation applies:

$$\lambda^- = 3N - 6 - \lambda . \tag{6}$$

For any k-fold degenerate critical point $K(\lambda,i)$ the following, modified index relation applies:

$$\lambda^- = 3N - 6 - k - \lambda . \tag{7}$$

The point symmetry groups also provide a basis for a partitioning of the nuclear configuration space M into domains. If $g_i = g_i(K)$ is the point symmetry group of nuclear configuration K, then G_i is the subset of all points K of M having the same point symmetry group g_i. This set G_i is not necessarily connected, since several, separate regions of M may have nuclear configurations of the same point symmetry. In such cases, we use a second index j in order to distinguish the various maximum connected components G_{ij} of G_i.

One interesting consequence of the above theorems on the point symmetry domain partitioning of M into subsets G_{ij} is the following "non-crossing rule":

Symmetry boundary non-crossing rule:

No steepest descent (relaxation) path of the potential energy hypersurface E(K) *of any electronic state can cross the boundary of any point symmetry domain* G_{ij} *of a nuclear configuration space* M.

This rule has been pointed out in ref. [3] (for an illustration see Figure 4 in ref. [3]). A steepest descent (relaxation) path may lead to and terminate at a boundary point of G_{ij}, however, actual crossing of the boundary cannot happen. The above symmetry boundary non-crossing rule is equally valid for the inverted potential energy hypersurfaces -E(K).

Whereas the catchment region partitioning of the nuclear configuration space M is based on the individual potential energy hypersurface of a specified electronic state, by contrast, the point symmetry domain partitioning is based on geometric properties. The symmetry relations of critical points (that are the lowest energy points within each catchment region) provide interrelations between energy and geometry [1,3].

An alternative, energy-based partitioning has also been proposed for the nuclear configuration space M: a partitioning based on the eigenvalues of the local Hessian matrices along the chosen potential energy hypersurface ([42], see also [1]). One may consider the index λ of generalized Hessian matrices along the potential energy hypersurface as the basis for partitioning, or one may take the local Hessians in subspaces that are orthogonal to the gradient at each non-critical point [42]. The resulting $D_{\mu j}$ domains of the partitioning are characterized by having precisely μ negative eigenvalues for the local Hessian matrices for every nuclear configuration within the given domain. By analogy with the partitionings discussed above, the second index refers to the j-th maximum connected component of the set containing all nuclear configurations of index μ. The interrelations of these two partitionings, the first one based on the eigenvalues of local Hessian matrices, and the second one on point symmetry, lead to additional relations, which can be exploited in a computational search for critical points [43].

In ref. [20] a general technique has been described in detail for defining neighbor relations for domains (and, in general, for subsets of various dimensions) of a partitioning of the nuclear configuration space M. This technique has been applied for the special case of catchment region neighbor relations [20]. The same neighbor relations are applicable for the partitionings into domains based on the eigenvalues (signature) of local Hessian matrices [42], and on the point symmetry of nuclear configurations [1,3]. The neighbor relation for point symmetry domains G_{ij}, and $G_{i'j'}$ is defined as

$$N(G_{ij}, G_{i'j'}) = \begin{cases} 1 & \text{if } (\text{clos}[G_{ij}] \cap G_{i'j'}) \cup (G_{ij} \cap \text{clos}[G_{i'j'}]) \neq \emptyset \\ \\ 0 & \text{otherwise,} \end{cases} \tag{8}$$

where clos, $\cap$, $\cup$, and $\emptyset$ are the usual set theoretical symbols for closure, intersection, union, and the empty set, respectively. Informally, the closure of a set is the collection of all points of the set and all of its boundary points.

Based on the above neighbor relation, and by direct analogy with the model of ref.[20], we define the *point symmetry graph* g(M,sym) of the nuclear configuration space M of the given stoichiometric family of chemical species by the following relations:

$$V(g(M,sym)) = \{G_{ij}\}, \tag{9}$$

and

$$E(g(M,sym)) = \{ (G_{ij}, G_{i'j'}) : N(G_{ij}, G_{i'j'}) \neq 0 \} . \tag{10}$$

The vertex set V(g(M,sym)) and edge set E(g(M,sym)) define the graph g(M,sym). This graph provides a concise representation of the mutual arrangements of point symmetry domains G_{ij} within the nuclear configuration space M.

By replacing the point symmetry domains G_{ij}, and $G_{i'j'}$ of equations (8)-(10) with the $D_{\mu j}$ and $D_{\mu'j'}$ domains of the eigenvalue sign distribution of local Hessian matrices within the nuclear configuration space M, and by replacing the symbol sym with the symbol hess, one obtains the analogous neighbor relation

$$N(D_{\mu j}, D_{\mu' j'}) = \begin{cases} 1 & \text{if } (clos[D_{\mu j}] \cap D_{\mu' j'}) \cup (D_{\mu j} \cap clos[D_{\mu' j'}]) \neq \emptyset \\ 0 & \text{otherwise.} \end{cases} \tag{11}$$

The corresponding graph g(M,hess), characterizing the local curvatures of the potential energy hypersurface, is defined by its vertex and edge sets as follows:

$$V(g(M,hess)) = \{D_{\mu j}\}, \tag{12}$$

and

$$E(g(M,hess)) = \{ (D_{\mu j}, D_{\mu' j'}) : N(D_{\mu j}, D_{\mu' j'}) \neq 0 \} . \tag{13}$$

Whereas the point symmetry graphs g(M,sym) are characteristic of the nuclear configuration space and are independent of the electronic state, by contrast, the graphs g(M,hess) are dependent on the actual potential energy hypersurface, that is, on the electronic state. (If required, a label v of the electronic state may be specified as a third entry in the parentheses, g(M,hess,v)). The graphs g(M,hess), based on local curvature properties, provide information on the stability of steepest descent paths in various regions of the given potential energy hypersurface [42].

In many cases, it is not necessary and in most cases, it is not practical to consider point symmetry or potential surface curvature relations throughout the entire nuclear configuration space. In such a case one may restrict the analysis and the corresponding neighbor relations to a suitable subset S of the nuclear configuration space M, leading to graphs g(S,sym) and g(S,hess). In particular, the subset S may be chosen as a relaxed cross section of M, as defined by a specified potential energy hypersurface. This is a practical approach, since the point symmetry theorems are applicable, essentially in identical form, to relaxed cross sections, that leads to a reduction in the dimensions to be considered in a computational study. If point symmetry and curvature relations for a specified chemical species are of interest, then the graphs $g(C(\lambda,i),sym)$ and $g(C(\lambda,i),hess)$ of the corresponding catchment region $C(\lambda,i)$ provide the required information.

4. Symmetry Constraints on Transition Structures from Global Symmetry Theorems

We shall give a proof of the following result:

Theorem:

If $B(\lambda,i)$ *denotes the boundary of catchment region* $C(\lambda,i)$, *and if* $K(\lambda',i')$ *is a critical point on the boundary* $B(\lambda,i)$, *then the catchment region* $C(\lambda,i)$ *must contain a point* K *that has a point symmetry group* $g(K)$ *that is a subgroup of the point symmetry group* $g(K(\lambda',i'))$ *of the critical point* $K(\lambda',i')$:

If

$$K(\lambda',i') \in B(\lambda,i) \,, \tag{14}$$

then there must exist some

$$K \in C(\lambda,i). \tag{15}$$

such that

$$g(K) \subset g(K(\lambda',i')) \,. \tag{16}$$

Note that nuclear configuration K is not necessarily unique.

Proof:

In order to prove this result, we shall show that the catchment regions $C(\lambda,i)$ and $C^-(\lambda',i')$ overlap, and we shall apply the catchment region point symmetry theorem to $C^-(\lambda',i')$ and to a point K from the overlapping region.

First we deal with the special case when the critical point $K(\lambda,i)$ is a maximum. In this case the catchment region is its own boundary, and one may take

$$K(\lambda,i) = C(\lambda,i) = B(\lambda,i) = K(\lambda',i') = K \,, \tag{17}$$

where for simplicity we use the same notation for points and for sets composed from a single point. Clearly,

$$g(K) \subset g(K(\lambda',i')) = g(K) ,\qquad(18)$$

hence the theorem is trivially fulfilled for the special case.

Let us consider now the more common cases of $K(\lambda,i)$ being a critical point different from a maximum (that is, $\lambda \leq 3N-6$, unless $K(\lambda,i)$ is a degenerate maximum).

It is clear that $C^-(\lambda',i')$ cannot be a single point catchment region, since there cannot exist a minimum of $E(K)$ on the boundary $B(\lambda,i)$ of any catchment region $C(\lambda,i)$. Furthermore, $C^-(\lambda',i')$ cannot contain points exclusively from the boundary $B(\lambda,i)$, since any infinitesimal neighborhood of $K(\lambda',i')$ must cut into $C(\lambda,i)$, from where a steepest descent path leads away from $K(\lambda',i')$. Hence, for the inverted surface $-E(K)$, the corresponding path, when reversed, is a steepest descent path that leads toward $K(\lambda',i')$. Consequently, there must exist some steepest ascent path of $E(K)$ that leads from an interior point K' of $C(\lambda,i)$ to $K(\lambda',i')$ on the boundary $B(\lambda,i)$. We conclude that this point K' must lie within both catchment regions $C(\lambda,i)$ and $C^-(\lambda',i')$, hence their intersection is not empty,

$$C(\lambda,i) \cap C^-(\lambda',i') \neq \varnothing .\qquad(19)$$

One may take any point K (in fact, one may take the very point K') from the intersection,

$$K \in C(\lambda,i) \cap C^-(\lambda',i') .\qquad(20)$$

The catchment region point symmetry theorem applies to $C^-(\lambda',i')$ of $-E(K)$ and to K, that is one of the points of $C^-(\lambda',i')$, consequently,

$$g(K) \subset g(K(\lambda',i'))$$

must hold, that proves the theorem.

An interesting result can be obtained as a simple consequence of the above theorem:

Corollary:

If the point symmetry group g *is a subgroup of the point symmetry group* g(K') *of each configuration* K' *of a catchment region* C(λ,i), *then* g *is also a subgroup of the point symmetry group* g(K(λ',i')) *of every critical point* K(λ',i') *on the boundary* B(λ,i). *That is, if*

$$g \subset g(K), \quad for\ every \quad K \in C(\lambda,i), \tag{21}$$

then

$$g \subset g(K(\lambda',i')) \tag{22}$$

also holds for any critical point on the catchment region boundary,

$$K(\lambda',i') \in B(\lambda,i) . \tag{23}$$

Proof:

This result follows immediately from the previous theorem, since any point K that fulfills the theorem for the given critical point K(λ',i') (and we know that there must exist such point K) must have a point symmetry group g(K) that contains g as a subgroup. Consequently,

$$g \subset g(K(\lambda',i')),$$

as stated.

Note that if C(λ,i) is the catchment region of a minimum K(0,i), then the critical point K(λ',i') on the boundary B(0,i) can almost always be chosen as a saddle point K(1,i') of a transition structure. Exceptional cases, such as the catchment region C(0,1) on a nearly spherical planet with a single hilltop and a single minimum point and no saddle point, are not expected to be of significance for molecular potential energy hypersurfaces.

The above theorem and corollary represent a constraint on the point symmetry groups of transition structures of reactions with a common reactant, represented by a catchment region $C(\lambda,i) = C(0,i)$. A common subgroup g (not in general the trivial group) exists for the point symmetry groups $g(K(\lambda',i'))$ of all the transition stuctures $K(\lambda',i')$ on the boundary $B(0,i)$ of $C(0,i)$. This fact may be regarded as a geometrical similarity constraint on the transition structure geometries.

5. Domains of Common Three-Dimensional Shape Properties for the Partitioning of Nuclear Configuration Spaces

The shape group method (SGM), reviewed in ref.[2], has been proposed for the analysis of three-dimensional shape properties of formal molecular bodies. For example, by choosing the electronic charge isodensity contours $G(a)$ (of various density values a) as the physical property P for shape representation, and by taking the family of Betti numbers b^k as the topological tool for shape description [2], the similarity of the geometrical shapes of two molecules, A and B, is transformed into an equivalence of their topological shape, expressed as

$$A \ (P,W) \ B .\tag{24}$$

The equivalence classes of the (P,W) topological shape equivalence correspond to the actual shape types $\tau_{(P,W)}$, denoted simply by τ for a fixed choice of the (P,W) pair.

With reference to a given electronic state, the nuclear configuration space M can be partitioned according to the τ_i shape types into subsets denoted by T_i. Within each subset T_i all nuclear configurations K correspond to formal molecular bodies (e.g., electron density distributions) characterized by the shape type τ_i. The maximum connected components of the set T_i are denoted by T_{ij}. By replacing G_{ij} and $G_{i'j'}$ of equations (8)-(10) by T_{ij} and $T_{i'j'}$, and the notation sym by τ, a new neighbor relation and the corresponding shape domain graph $g(M,\tau)$ of the nuclear configuration space M are obtained. The neighbor relation is defined as

$$N(T_{ij}, T_{i'j'}) = \begin{cases} 1 & \text{if } (clos[T_{ij}] \cap T_{i'j'}) \cup (T_{ij} \cap clos[T_{i'j'}]) \neq \emptyset \\ 0 & \text{otherwise.} \end{cases}\tag{25}$$

The corresponding graph $g(M,\tau)$, characterizing the distribution of shape domains within the nuclear configuration space M, is defined by its vertex set and edge set as follows:

$$V(g(M,\tau)) = \{ T_{ij} \}, \tag{26}$$

$$E(g(M,\tau)) = \{ (T_{ij}, T_{i'j'}) : N(T_{ij}, T_{i'j'}) \neq 0 \} . \tag{27}$$

As for the other partitionings mentioned above, these neighbor relations and graphs can be restricted to various subsets S, such as relaxed cross sections, and in particular, to individual catchment regions $C(\lambda,i)$ of the nuclear configuration space M. This approach leads to the local shape domain graphs $g(S,\tau)$ and $g(C(\lambda,i),\tau)$, respectively.

One may analyse the detailed variations of contributions of various nuclear configurations to each shape type τ_i as a function of some continuous parameters, for example, as function of the contour density value a and reference curvature parameter b of isodensity contours $G(a)$. This is equivalent to the analysis of the parameter dependence (for example, (a,b)-dependence) of the T_{ij} subsets within the configuration space M, and in particular, in relaxed cross sections or within each catchment region $C(\lambda,i)$ [44]. These changes can be monitored within the dynamic shape space D, obtained as the product space of the nuclear configuration space M and the space of the actual continuous parameters. This approach has been described in some detail in ref. [44], and various applications can be found in refs. [45-47].

Acknowledgment
The research leading to the developments reviewed above has been supported by both strategic and operating research grants from the Natural Sciences and Engineering Research Council of Canada, and in part by the Computational Chemistry Unit of the Upjohn Laboratories, Kalamazoo, Mich., USA.

108

References

[1]   Mezey, P.G. *Potential Energy Hypersurfaces*;   Elsevier: Amsterdam, 1987.

[2]   Mezey, P.G. Three-Dimensional Topological Aspects of Molecular Similarity. In *Concepts and Applications of Molecular Similarity*, Johnson, M.A., Maggiora, G.M., Eds.;   Wiley: New York, 1990; p 321-368.

[3]   Mezey, P.G. *J. Am. Chem. Soc.* **1990**, *112*,   3791.

[4]   Fukui, K. *J. Phys. Chem.* **1970**, *74*,   4161.

[5]   Pechukas, P. *J. Chem. Phys.* **1976**, *64*, 1516.

[6]   Tachibana, A.; Fukui, K. *Theor. Chim. Acta* **1979**, *51*, 189.

[7]   Leroy, G.; Sana, M.; Burke, L.A.; Nguyen, M.-T. In *Quantum Theory of Chemical Reactions*,   Daudel, R., Pullman, A., Salem, L., Veillard, A., Eds.; Reidel: Dordrecht, 1979.

[8]   Miller, W.H.; Henry, N.C.; Adams, J.E. *J. Chem. Phys.* **1980**, *72*, 99.

[9]   Fukui, K. *Acc. Chem. Res.* **1981**, *14*, 363.

[10]   Truhlar, D., Ed. *Potential Energy Surfaces and Dynamics Calculations*; Plenum: New York, 1981.

[11]   Pulay, P. In *The Force Concept in Chemistry* ;   Deb, B.M., Ed.; Van Nostrand - Reinhold: Toronto, 1981.

[12]   Oie, T.; Maggiora, G. M.; Christoffersen, R. E.; Duchamp, D. J. *Internat. J. Quantum Chem., Quant. Biol. Symp.* **1981**, *8*, 1.

[13]   Maruani, J.; Serre, J., Eds. *Symmetries and Properties of Non-Rigid Molecules*; Elsevier: Amsterdam, 1983.

[14]   Schlegel, H.B. *Theor. Chim. Acta* **1984**, *66*, 333.

[15]   Hall, G.G. *Theor. Chim. Acta* **1985**, *67*, 439.

[16]   Bernardi, F.; Olivucci, M.; Robb, M.A.; Tonachini, G. *J. Am. Chem. Soc.* **1986**, *108*, 1408.

[17]   Murrell, J.N.; Carter, S.; Farantos, S.C.; Huxley, P.; Varandas, A.J.C. *Molecular Potential Energy Functions*; Wiley: New York, 1984.

[18]   Mezey, P.G. *Theor. Chim. Acta* **1981**, *58*, 309.

[19]   Mezey, P.G. *Theor. Chim. Acta* **1982**, *62*, 133.

[20]   Mezey, P.G. *Theor. Chim. Acta* **1982**, *60*, 409.

[21]   Mezey, P.G. *J. Chem. Phys.* **1983**, *78*, 6182.

[22]   Mezey, P.G. *Theor. Chim. Acta* **1983**, *63*, 9.

[23]   Mezey, P.G. *Int. J. Quant. Chem.* **1984**, *26*, 983.

[24]   Mezey, P.G. *Int. J. Quant. Chem. Symp.* **1986**, *12*, 113.

[25]   Mezey, P.G. *J. Comput. Chem.* **1987**, *8*, 462.

[26]   Mezey, P.G. *J. Math. Chem.* **1988**, *2*, 325.

[27]   Gamelin, T.W.; Greene, R.E. *Introduction to Topology;* Saunders College Publishing: New York, 1963.

[28]   Simmons, G.F. *Introduction to Topology and Modern Analysis;* McGraw-Hill: New York, 1963.

[29]   Munkres, J. *Elementary Differential Topology;* Annals of Math. Studies, **54**, Princeton Univ. Press: Princeton, 1963.

[30]   Bishop, R.L.; Crittenden, R.J. *Geometry of Manifolds;* Academic Press: New York, 1964.

[31]   Spivak, M. *Calculus on Manifolds;* Benjamin: Don Mills, Ont., 1965.

[32]   Spanier, E.H. *Algebraic Topology;* McGraw-Hill: New York, 1966.

[33]   Greenberg, M. *Lectures on Algebraic Topology;* Benjamin: New York, 1967.

[34]   Hu, S.-T. *Elements of General Topology;* Holden-Day: San Francisco, 1969.

[35]   Vick, J. *Homology Theory;* Academic Press: New York, 1973.

[36]   Guillemin, V.; Pollack, A. *Differential Topology;* Prentice Hall: Englewood Cliffs, 1974.

[37]   Singer, I.M.; Thorpe, J.A. *Lecture Notes on Elementary Topology and Geometry;* Springer-Verlag: New York, 1976.

[38]   Mezey, P.G. *Chem. Phys. Letters* **1981**, *82*, 100; *ibid* **1981**, *86*, 562.

[39]   Csizmadia, I.G. *J. Mol. Struct.,Theochem* **1986**, *138*, 1.

[40]   Ángyán, J.G.; Daudel, R.; Kucsman, Á.; Csizmadia, I.G. *Chem. Phys. Letters* **1987**, *136*, 1.

[41]   Csizmadia, I.G. In *New Theoretical Concepts for Understanding Organic Reactions*; Bertran, J., Csizmadia, I.G., Eds.; Reidel: Dordrecht, 1989.

[42]   Mezey, P.G. *Theor. Chim. Acta* **1980**, *54*, 95.

[43]   Mezey, P.G., to be published.

[44] Mezey, P.G. *J. Math. Chem.* **1988**, *2*, 299.

[45] Arteca, G.A.; Mezey, P.G. *J. Phys. Chem.* **1989**, *93*, 4746.

[46] Arteca, G.A.; Mezey, P.G. *Int. J. Quant. Chem. Symp.* **1989**, *23*, 305.

[47] Arteca, G.A.; Heal, G.A.; Mezey, P.G. *Theor. Chim. Acta* **1990**, *76*, 377.

A TOPOLOGICAL ANALYSIS OF MACROMOLECULAR FOLDING PATTERNS

Gustavo A. ARTECA and Paul G. MEZEY

Department of Chemistry and Department of Mathematics,
University of Saskatchewan, Saskatoon, Saskatchewan,
Canada S7N0W0

ABSTRACT. In this chapter we discuss the application of topological techniques, namely knot-theoretical methods, for the characterization of some aspects of the shape of large biological macromolecules, in particular, proteins. The essential shape features we describe are those conveyed by a simplified molecular backbone, as determined by the sequence of a-carbon atoms. The methodology allows one to recognize the occurrence of some motifs in the supersecondary structure of proteins, and to compare a sequence of structurally related species in the search of common shape characteristics. We discuss a characterization of the shape changes in the protein fold induced by conformational rearrangements (e.g., during folding-unfolding processes). The proposed procedure, based on computing a family of knots and graphs derived from the original macromolecular space curve, allows one to describe quantitatively the extent of the essential modification of the backbone's fold introduced by a conformational rearrangement.

1. Introduction

The characterization of molecular shape and its changes is of current interest in many fields of applied science. Although there is a number of methods available to tackle this problem, most of the techniques applicable to small molecules become cumbersome and unpractical for analyzing macromolecules.

However, some biomolecules present special features which allow alternative approaches to shape analysis. This is the case, for instance, of proteins, where one finds a hierarchical structural organization [1]. If one disregards the *details* of the atomic arrangements, then the essential features of a protein supersecondary structure are well represented by its a-carbon backbone. This macromolecular backbone provides a model for the characterization of molecular shape at a level of abstraction

S. J. Formosinho et al. (eds.), Theoretical and Computational Models for Organic Chemistry, 111–124.
© 1991 *Kluwer Academic Publishers.*

comparable to that of formal molecular surfaces for small molecules.

Visual inspection on a computer screen is a widespread approach to recognizing the occurrence of systematic structural patterns in proteins [1,2]. However, there is an unavoidable bias in this rather subjective approach to assess molecular shape. On the other hand, numerical or algebraic descriptors provide an alternative description, which is more objective, since the computation of the shape descriptors can be done, in principle, in an automated way.

A number of these descriptors have been proposed in the literature for the characterization of macromolecules [3-13]. Some of them, such as the winding, writhing, twisting, and linking numbers, describe rather global features of the backbone [3,4]. A number of shape descriptors based on the analysis of local features have also been developed. Methods based on graph theory and knot theory belong to this class. Among other applications we can mention the classification of protein structural motifs [5-9], their search and recognition within large databases [10-12], and the description of chirality in globular proteins [13].

If one is interested in the study of changes in the folding pattern of biomacromolecules, it is necessary to have local descriptors. The structural changes that occur during the *folding-unfolding process* are due to dynamical fluctuations in the molecular geometry, which may be viewed as a superposition of local effects in various small atomic neighborhoods. A successful technique to follow these changes in the structural motifs must take into consideration the local changes in the backbone structure. Similar considerations are valid when one seeks to compare homologous proteins, searching for common structural features. This problem is relevant to drug design, since the presence of some common features may indicate that two different molecules catalyze similar biological processes. As it is known, the cavity generated by some sections of the skeleton of an enzyme (a protein) mimic *transition structures* for some reactions. The recognition of features which resemble transition structures, and their persistence under conformational changes, is thus related to the problems mentioned above.

In this work we discuss a method for characterizing the shape of the tertiary structure of proteins which possess the above characteristics. The procedure associates a series of knots to the projections of the space curve corresponding to the protein backbone [14]. Following this approach, the shape features can be reduced to a discrete number of knot symbols and polynomials related to the overcrossing andknotting patterns of curves derived from the molecular space curve. If this procedure is repeated for all possible projections of the backbone to planes tangent to a sphere enclosing the space curve, then one has an intrinsic, direction-independent description. The problem of characterizing a molecular space curve is thus transformed into the characterization of the distribution of shape descriptors on the surface of a sphere. The essential simplification is based on the following observation: the shape descriptors are invariant within some domains on the sphere, hence one has to deal with only a finite number of descriptors. These descriptors can be computed algorithmically, and complement the information provided by visual analysis.

In the next sections we present a brief discussion of the derivation of knots from the molecular space curves representing the protein backbone, using projections to a sphere. The knots are characterized by topological invariants. As in ref. 14, we use the Jones polynomials [15]. The occurrence of basic structural patterns can be recognized in terms of the knots; we discuss briefly some of the results derived in ref. 16. Finally, we comment on the application of this procedure to study conformational motions in proteins, and to recognize the occurrence of essential changes in the shape or folding patterns.

2. Derivation of knots from molecular space curves

If one disregards small scale structural features, the protein backbone can be described by the sequence of C_a atoms of the aminoacid residues. In turn, this sequence can be represented as a parametric space curve $r(t)$, the *molecular space curve*, expressed as

$$r(t) = x(t)\, i + y(t)\, j + z(t)\, k , \quad 0 \le t \le 1, \tag{1}$$

where i, j, and k indicate the three unit vectors of an orthogonal Cartesian framework taken as a reference. If one assumes linear bonds, then the line (1) is a sequence of straight-line segments. In the case of protein backbones, function (1) is single valued, bounded, and continuous. In most cases $r(t)$ will be an open curve (i.e., it will not be a loop). Note that (1) is an oriented curve; for proteins, $r(0)$ and $r(1)$ are the N-terminal and C-terminal ends, respectively.

In ref. 14 we discussed a number of techniques to characterize the essential shape features of the curve (1). All these approaches, both graph- and knot-theoretical, were based on the occurrence of crossings in planar projections of (1), when viewed along some preferential direction. These "crossings" are the results of sections of the space curve passing over one another when viewed from some direction in space. We refer to these as overcrossings, thus reserving the word crossing for an actual crossing in the projected curve. In a degenerate projection two or more overcrossings may be projected to the same point of the plane. In a regular projection all projected overcrossings are separated. A projection can be "regularized" by a small tilt in the viewing direction.

When the two end points $r(0)$ and $r(1)$ of the mathematical curve $r(t)$ are *formally* joined, we obtain an object which is topologically a loop, possibly a knot [17,18]. The use of modern knot theory in chemical applications has an extensive literature [14,15,19,20]. In this work, we use the conventions and notations of ref. 20d for the knots, and the procedure discussed in ref. 14 to derive them from molecular space curves. In what follows, we shall assume that the coordinates specifying the protein backbones are available, for example, in the format of the Protein Data Bank (PDB) of X-ray structures.

A knot-theoretical description allows one to describe some topological features which remain invariant for various placements and deformations of the backbone (excluding breaking or rejoining it). Since the curve is oriented, one can characterize it by the handedness [17] of its overcrossings. Moreover, based on these overcrossings, it is possible to assign polynomials to each knot. These polynomials are topological invariants. There are a number of polynomials which characterize the knots [17,18]. The Jones polynomials are used here [15]; they are easy to compute and they distinguish a large number of different knots. A practical approach to the computation of the Jones polynomials is given in ref. 20d. (See ref. 14 for a table of Jones polynomials for some simple and commonly occurring knots).

We have discussed in ref. 14 the rules to derive a knot from the space curve; the essential steps are as follows: 1. Consider a projection of the molecular space curve to a generic viewing plane. We shall assume that all crossings of the resulting planar curve are nondegenerate (*vide supra*). 2. Attach straight-line segments to $r(0)$ and $r(1)$; these segments will be perpendicular to the viewing plane and pointing away from the viewer. These segments will be long enough so that they reach beyond the most distant point of the original curve $r(t)$. 3. Join the ends of these extended line segments by another straight-line segment, parallel to the viewing plane.

The operations above produce a closed curve from the original space curve $r(t)$. This closed curve is a either simple loop or a knot K_0 characterized by a polynomial $V(K_0)$. In the more detailed and local analysis that follows we will disregard all crossings produced by the closing of the loop, since they introduce information not present in the original curve.

The above operations on the protein backbone produce a loop in most cases. The simple loop (called the "unknot") is the trivial knot, and it has $V(K_0)=1$ as Jones polynomial. Nevertheless, one can derive nontrivial knots when constructing a family of loops from the original curve, by introducing a sequence of formal switches in the original overcrossing pattern.

Here we discuss briefly how the switches are introduced in the case of a projection along a given viewing direction. Suppose that the planar curve obtained after the projection has n crossings, $n \geq 1$, resulting from n regular overcrossings in the original loop K_0. Since the loop K_0 is oriented, each crossing can be characterized by the numbers $C_j = +1$ or -1 (crossing types), representing right- and left-handed crossings, respectively [17]. This information can be collected in the form of a vector

$$C = (C_1, C_2, \ldots C_n) ,\qquad (2)$$

which provides a simple characterization of some of the essential shape features. One can associate a family of possible knots with the same 2D projection by modifying some or all n of the C_j crossing types. This formal transformation gives rise to a set of polynomials, determined from K_0, which provide a characterization to the original loop.

Consider the n-dimensional *switching vector*

$$v = (v_1 , v_2 , \ldots v_n) ,\qquad (3)$$

whose elements $v_n = 1$, or -1, will identify whether or not a formal modification of the initial crossing pattern takes place. Thus, one can derive a new vector of crossing types C_v from the original vector C, as follows

$$C_v = (v_1C_1, v_2C_2, \ldots v_nC_n) \qquad (4)$$

The family of all the possible n-dimensional vectors v of form (3) will generate all possible knots (and links) compatible with the given 2D projection. Let us denote by $\{K_b\}$ the corresponding family of knots (and links) obtained, and by $\{V_{Kb}(t)\}$ the corresponding family of Jones polynomials, with t the polynomial variable (not to be confused with the parameter of the space curve $r(t)$). For simplicity, we consider only the subset of single switches, which is given by the vectors: $v_0=(1,1,1,...1)$, $v_1=(-1,1,1,...1)$, $v_2=(1,-1,1,...1)$, ..., $v_n=(1,1,1, ... -1)$.

The Jones polynomials of the knots $\{K_n\}$, obtained by the switches specified by vectors $\{v_n\}$, are in general different from the polynomial of the actual, original knot K_0. Consequently, they provide a more detailed characterization of the projection. We shall use the complete set of knots $\{K_n\}$ as a shape descriptor, following a formal vector notation (*knot vector* K):

$$K = (K_0, K_1, K_2, ... K_n) \ . \qquad (5)$$

In Figure 1 we present an example, which illustrates the results obtained by our procedure. The example chosen is one view of a small protein, the pancreatic trypsin inhibitor [21,22]. The image in the left-hand side of Fig. 1 represents schematically the protein backbone, as it appears in the crystal structure. The protein has only 58 aminoacid residues. In this projection, the protein shows 12 overcrossings; a simple analysis [14] gives the following shape descriptor,

$$K = (3_1,0_1,0_1,0_1,0_1,3_1{}^*,0_1,0_1,0_1,4_1,3_1{}^*,0_1) \ . \qquad (6)$$

The symbols in (6) are a short-hand notation for the knots found [20d]. The notations $0_1,3_1,3_1{}^*$, and 4_1 identify the unknot, left-handed trefoil knot, right-handed trefoil knot, and the figure-eight knot, respectively. The corresponding Jones polynomials are as follows [14]:

$V_{0_1}(t) = 1,\quad V_{3_1}(t) = -t^4 + t^3 + t,\quad V_{3_1*}(t) = -t^{-4} + t^{-3} + t^{-1},$

$V_{4_1}(t) = t^2 - t + 1 - t^{-1} + t^{-2} .$

As Fig. 1 reveals, some of the crossings occur just by a very small margin. We refer to these as *marginal overcrossings*. These crossings would not appear if the view were slightly tilted, or if one takes into account the uncertainty in the atomic positions due to the inaccuracies in the X-ray resolution of a structure. We can estimate the essential shape features appearing in a given view, by comparing the knot vector $\mathbf{K}$ with another one, $\mathbf{K}^\dagger$, obtained by neglecting all marginal overcrossings. As a cut-off to decide whether an overcrossing is marginal, one could use the X-ray resolution. The right-hand side of Fig. 1 shows how this procedure applies to the pancreatic trypsin inhibitor. The original space curve $\mathbf{r}(t)$ exhibits 12 overcrossings, while the simplified curve $\mathbf{r}^\dagger(t)$ has only 8, with the following shape characterization:

$$\mathbf{K}^\dagger = (3_1, 0_1, 0_1, 0_1, 0_1, 4_1, 3_1{}^*, 0_1) . \tag{7}$$

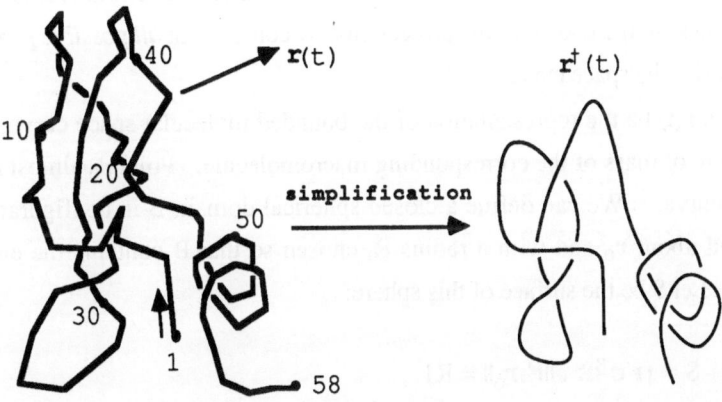

Figure 1. One view of the backbone structure of the pancreatic trypsin inhibitor (left-hand side), and its simplification by disregarding the "marginal crossings". The space curve $\mathbf{r}(t)$ is characterized by the knot vector $\mathbf{K}$, and the simplified curve $\mathbf{r}^\dagger(t)$ by vector $\mathbf{K}^\dagger$.

As it has been discussed in ref. 16, it is possible to recognize the occurrence of some structural motifs from the sequence of single-switch knots. For example, the triplet $0_1, 4_1, 3_1^*$ is characteristic of the overcrossing pattern of two consecutive turns of a right-handed α helix. Analogously, the triplet $0_1, 3_1^*, 0_1$ is characteristic of the crossing between an α helix and a β strand. Thus we can recognize from the knot description the occurrence of a very short α-helix in the terminal aminoacids of the trypsin inhibitor.

The above description refers to a single projection, the choice of which is in principle arbitrary. The description by means of a single view will not be very informative. For the comparison of a family of related proteins one can choose a number of projections; the three Cartesian views defined by the axes of inertia are a natural choice for homologous polypeptides. However, this choice will perform poorly in the case of studying a folding rearrangement, since the changes in the axes of inertia can be significant.

3. Characterization of foldings by projecting to a sphere

In this section we discuss an alternative approach that circumvents the above arbitrariness in the choice of the projections, by considering *all possible projections* to characterize the space curve.

Let $r(t)$ be the representation of the bounded molecular space curve, and r_0 be the centre of mass of the corresponding macromolecule. Point r_0 almost never falls on the curve. We can define a closed spherical domain B in configuration space, centered about r_0 and with a radius R, chosen so that B contains the entire space curve. Let S be the surface of this sphere:

$$S = \{ r' \in {}^3\mathbb{R} : \| r' - r_0 \| = R \} , \tag{8}$$

where $\| r - r_0 \| \leq R$, for all r on the space curve. The radius R will be taken as the smallest value that satisfies this condition.

The points on the sphere can be used to generate all possible viewing directions, and to each such point we can associate a family of knots. In order to characterize the curve **r**(t) one can us the sphere S (the *reference sphere*) as follows.   Consider an arbitrary point **r'** on S as a viewing point for the space curve **r**(t). From this viewing point a projection is defined, as one to a plane perpendicular to the r_0-r vector (tangent plane to the sphere S at **r'**). This projection can be characterized by graph-theoretical or knot-theoretical methods as described above. Let us denote by s(**r'**) the "shape type" of the curve as viewed from **r'**, using some shape descriptor (say, the knot vector **K**).   This shape analysis can be applied to every point **r'** on the reference sphere S.

The shape, as defined by the shape descriptor **K**, may be invariant to most small changes of the viewing point **r'**.   As a matter of fact, the sphere S will have only a finite number of domains with a distinct shape type. All the points within each domain lead to projections with the same shape characterization in terms of vector **K**. We shall call these domains the *shape domains* of the reference sphere.

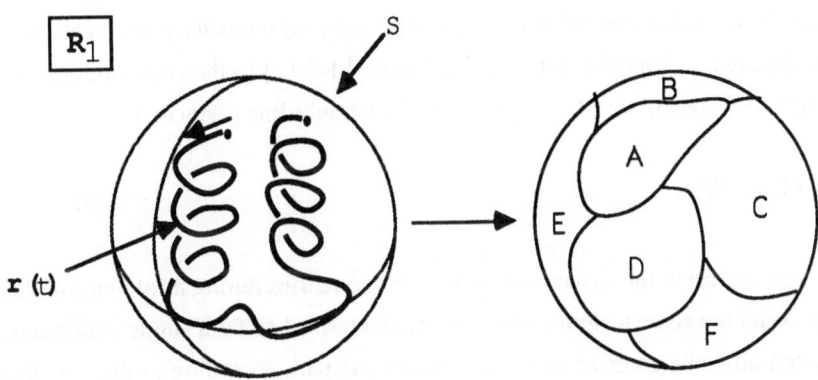

Figure 2. Schematic representation of the characterization of the folding pattern of a space curve **r**(t), associated with a configuration R_1. The left-hand-side drawing represents the sphere S enclosing the space curve **r**. To the right we indicate how the sphere can be subdivided into regions according to the shape classification of the space curve.   Each of the letters indicates a different shape type of the shape descriptor (e.g., a new knot vector **K**).

This approach is schematically represented in Figure 2 for an arbitrary space curve, with a geometry indicated by $\mathbf{R}_1$. On the left-hand side we have a bounded space curve, enclosed by the sphere S. On the right, one finds the resulting subdivision of the spherical surface into shape domains. Each of the regions is characterized by a different knot description of the curve $\mathbf{r}(t)$. A different letter identifies a different vector $\mathbf{K}$.

The distribution of shape domains on the sphere S enclosing the molecular curve provides a detailed description of the shape of the curve. This approach avoids the choice of an arbitrary projection since all the possible projection directions are taken into account. From now on, *the reference sphere S, with its subdivision into shape domains, will be our shape descriptor.* Then, one can apply then to S itself some of the methods proposed in the literature for the shape characterization of molecular surfaces [23].

The procedure just described can be applied to configurations $\mathbf{R}(p)$, found along a parametrized conformational or reaction path. Let $p(p)$ be such a path; it can be seen as a continuous assignment from the unit interval $I=[0,1]$ to the configuration space $3NR$ (for a backbone made of a sequence of N-1 straight-line segments):

$$p : I \to 3NR \ . \tag{9}$$

Path p can represent the dynamical change of geometries during a folding-unfolding process. With the same technique we have characterized so far a single configuration R, one can now characterize entire sections of the path according to the new shape descriptor [24]. To this purpose, one must follow the changes in shape description of the subdivision pattern of the reference sphere S along the given reaction path $p(p)$.

Figure 3 shows an illustrative example of the type of behavior expected along a reaction path. The left-hand side figure identifies the reference sphere S discussed in Fig. 2, corresponding to a configuration $\mathbf{R}_1$.

Along the reaction path a number of changes can be expected in the shape description. For example, configuration $\mathbf{R}_2$ in Fig. 3 corresponds to a situation in which we can still see the same shape domains as for $\mathbf{R}_1$, but their areas on S have changed. By constrast, one can reach a configuration such as R_3 where not only the domain areas change, but also new shape domains (new K types) appear. This latter change represents a more essential modification in the shape features, since it may correspond to the occurrence of a new motif or to the loss of another. A shape change along the reaction path where the reference sphere exhibits a different pattern of shape domains can be seen as a formal structural transition, defined in terms of shape.

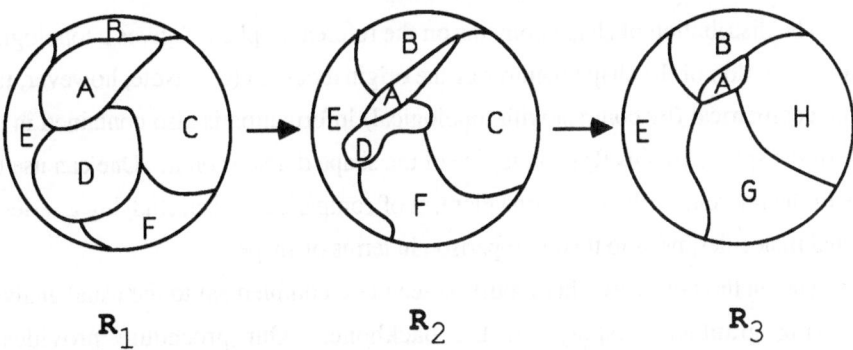

$$R_1 \qquad\qquad R_2 \qquad\qquad R_3$$

Figure 3. Schematic representation of the characterization of the structures found along a conformational folding path of a space curve $r(t)$. The figure indicates the changes in the distribution of shape domains over the surface of the observational sphere S, when the configuration changes from R_1 to R_3. Notice that for some configuration one may find the same type of shape description but different sizes of the shape regions over the sphere (cf. R_1 and R_2). On the other hand, the configurational change may lead to the occurrence of topologically different shape descriptions (cf. R_1 and R_3).

A practical implementation of the above procedure can be obtained by replacing the sphere S by a quasi-regular convex polyhedron with a large number of vertices. The buckminsterfullerene (60 vertices) is a good initial candidate. The implementation of this procedure to analyze the dynamical changes in small enzymes is under development [25].

4. Final comments

The procedure discussed above provides an alternative method for the study of the occurrence of certain features of the supersecondary structure of proteins, and to follow their changes with the configurational rearrangements. The same technique can be applied to the study of other macromolecules, even if they do not have a hierarchical structure or basic structural motifs.

The distribution of shape domains on the reference sphere S gives a topological characterization of the shape features of the original curve $r(t)$. Note, however, that some geometrical (by contrast with topological) information is also contained in the size of the sphere (radius R) and the size of the shape domains on it. One can use the size of the reference sphere as a first element of comparison, when studying a series of related molecules, prior to their comparison in terms of shape.

The method discussed here must be seen as a complement to the usual analysis by using graphical displays of the backbone. Our procedure provides a characterization which can be displayed jointly with the actual 3D structure on a computer screen, thus helping to assess the molecular shape in a less subjective manner.

Acknowledgments

We would like to acknowledge fruitful discussions with G. Maggiora on this and related subjects. One of us (GAA) would like to thank helpful discussions with O. Tapia and O. Nilsson, while statying at the Department of Physical Chemistry (Uppsala, Sweden). This work has been supported by both operating and strategic grants from the Natural Sciences and Engineering Research Council (NSERC) of Canada.

References

[1]   (a) Richardson, J.S. *Adv. Protein Chem.* **1981**, *34*, 167.   (b) Richardson, J.S. *Methods in Enzymol.* **1985**, *115*, 359

[2]   (a) Carson, M.; Bugg, C.E. *J. Mol. Graph.* **1986**, *4*, 121.   (b) Carson, M., *J. Mol. Graph.* **1987**, 5, 103.   (c) Lesk, A.M.; Hardman, K.D. *Science* 1982, *216*, 539.   (d) Lesk, A.M.; Hardman, K.D. *Methods in Enzymol.* **1985**, *115*, 381.   (e) Dearden, T. *J. Comput. Chem.* **1989**, *10*, 529.   (f) Jaenicke, R. *Prog. Biophys. Molec. Biol.* **1987**, *49*, 117.   (g) Kikuchi, T.; Némethy, G.; Scheraga, H.A. *J. Comput. Chem.* **1986**, *7*, 67.

[3]   Delbrück, M.   *Proc. Symp. Appl. Math.* **1962**, *14*, 55.

[4]   (a) Fuller, F.B.   *Proc. Symp. Appl. Math.* **1962**, *14*, 64.   (b) Fuller, F.B. *Proc. Natl. Acad. Sci. USA* **1971**, *68*, 815.

[5]   Le Bret, M.   *Biopolymers* **1979**, *18*, 1709.

[6]   De Santis, P.; Morosetti, S.; Palleschi, A.   *Biopolymers* **1983**, *22*, 37.

[7]   Hao, M.-H.; Olson, W.K.   *Biopolymers* **1989**, *28*, 873.

[8]   Mitchell E.M.; Artymiuk, P.J.; Rice, D.W.; Willett, P. *J. Mol. Biol.* **1990**, *212*, 151.

[9]   Liebman, M.N.; Venanzi, C.A.; Weinstein, H.   *Biopolymers* **1985**, *24*, 1721.

[10] Rawlings, C.J.; Taylor, W.R.; Nyakairu, J., Fox, J.; Sternberg, M.J.E.   *J. Mol. Graph.* **1985**, *3*, 151.

[11] Richards, F.M.; Kundot, C.E.   *Protein Struct. Funct. Genet.* **1988**, *3*, 71.

[12] Abagyan, R.A.; Maiorov, V.N.   *J. Biomol. Struct. Dynam.* **1988**, *5*, 1267.

[13] Maggiora, G.M.; Mezey, P.G.; Mao, B.; Chou, K.C. *Biopolymers* **1990**, *30*, 211.

[14] Arteca, G.A.; Mezey, P.G.   *J. Mol. Graph.* **1990**, *8*, 66.

[15] (a) Jones, V.F.R.   *Bull. Am. Math. Soc. (NS)* **1985**, *12*, 103.   (b) Freyd, P.; Yetter, D.; Hoste, J.; Lickorish, W.B.R.; Millett, K.; Ocneanu, A. *Bull. Am. Math. Soc. (NS)* **1985**, *12*, 239.

[16] Arteca, G.A.; Tapia, O.; Mezey, P.G. *J. Mol. Graph.*, submitted.

[17] See, for example: Crowell, R.H.; Fox, R.H. *Introduction to Knot Theory.* Springer-Verlag, Berlin, 1977.

[18] (a) Dowker C.H.; Thistlethwaite M., *Comp. Rend. Acad. Sci. (Canada)* 1982, *VI 2*, 129; *Topology and Its Applicat.* **1982**, 16, 19. (b) Thistlethwaite M., *London Math. Soc. Lecture Notes* **1985**, *93*, 1.

[19] Walba, D.M. Stereochemical Topology, in: King, R.B. (Ed.), *Chemical Applications of Topology and Graph Theory.* Elsevier, Amsterdam, 1983.

[20] (a) Walba, D.M. *Tetrahedron* **1985**, *41*, 3161. (b) Wasserman, S.A.; Cozzarelli, N.R. *Science* **1986**, *240*, 110. (c) Connolly, M.L; Kuntz, I.D.; Crippen, G.M. *Biopolymers* **1980**, *19*, 1167. (d) Mezey, P.G. *J. Am. Chem. Soc.* **1986**, *108*, 3976. (e) Millett, K.C. *J. Comp. Chem.* **1987**, *8*, 536. (f) Simon, J., *J. Comput. Chem.* **1987**, *9*, 718. (g) Sumners, D.W. *J. Math. Chem.* **1987**, *1*, 1.

[21] Levitt, M.; Warshel, A., *Nature* **1975**, *293*, 693.

[22] Cantor, C.R.; Schimmel, P.R. *Biophysical Chemistry, Part I: The Conformation of Biological Macromolecules*, Freeman, San Francisco, 1980.

[23] (a) Mezey, P.G. *Int. J. Quantum Chem QBS* **1986**, *12*, 113. (b) Mezey, P.G. *J. Comput. Chem.* **1987**, *8*, 462. (c) Mezey, P.G. *J. Math. Chem.* **1988**, *2*, 299. (d) Arteca, G.A.; Mezey, P.G. *J. Comput. Chem.* **1988**, *9*, 554.

[24] (a) Arteca, G.A.; Mezey, P.G. *Int. J. Quantum Chem. QBS* **1988**, *14*, 113. (b) Arteca, G.A.; Mezey, P.G. *J. Phys. Chem.* **1989**, *93*, 4746

[25] Arteca, G.A.; Tapia, O.; Nilsson, O.; Mezey, P.G., in preparation.

MOLECULAR MECHANICS

N. L. ALLINGER
Department of Chemistry
School of Chemical Sciences
University of Georgia
Athens, Georgia 30602
USA

ABSTRACT. Molecular mechanics is a computational scheme whereby one can calculate the structure of a molecule using a force field which is developed from data derived either from experiment, or from *ab initio* calculations.For those classes of compounds for which good data are available, the calculations give molecular structures and energies which are competitive with experiment, and they can be reliably used to make predictions. Since the calculation can be done one or two orders of magnitude more rapidly than experiment, they are pratically useful. The principle limitation of the method is when insufficient data exist for the force field to be fully developed for a particular class of compounds. Usually approximate data are available, and therefore an approximate force field can be developed, and the calculations can still be carried out, although with limited accuracy.

1. History

In 1930 there appeared in the *Physical Review* a paper on vibrational spectroscopy by D. H. Andrews [1]. It was mentioned in passing that if we better understood more of the details in this latter field, it would be possible to calculate structures and many properties of molecules, as we now do in the subject we call Molecular Mechanics. It was, of course, not possible at that time to actually carry out these calculations in any useful way, but the general principles behind the method were already becoming clear.

Sometime later, in 1946, there appeared several papers, especially those by Westheimer, where in such calculations were actually carried out [2]. The calculations, while rather straightforward if one has a computer, were lengthy and tedious in the days of a desk calculator. However, Westheimer showed in studies of the rotational barriers of hindered biphenyls that one actually could calculate the activation energies, and obtain results in the satisfactory agreement with experiment. The potential power of the method, but not its usefulness, was therefore established that long ago.

For molecular mechanics calculations to be of practical use, however, an electronic computer was required. These were developed fairly rapidly beginning in the decade of the 1950′s, and in 1961, Hendrickson published the first of a series of papers in which he described calculations on a number of medium ring compounds [3]. The conformations of these molecules had previously been the subject of considerable speculation, but Hendrickson was able to make some predictions as to which conformations should be

S. J. Formosinho et al. (eds.), Theoretical and Computational Models for Organic Chemistry, 125–135.
© 1991 *Kluwer Academic Publishers.*

preferred, and in cases where the predictions could be tested, there was general agreement with experiment, and an indication that the method was indeed going to be powerful and useful.

Wiberg, in 1965, published a description of a computer program that would carry out these calculations in a very general way for any kind of molecule (in principle) [4], and for a great many real molecules in practice. At this point the tools needed for molecular mechanics were all available. It was then necessary to refine the details of molecular mechanics so that the method gave results of experimental accuracy, and at a speed that made them useful to the average chemist. The subject of molecular mechanics was reviewed in a book in detail in 1982 [5]. For earlier reviews, see for example Engler, Altona, Dunitz, Allinger, Ermer, Niketic, Warshel, White, and Osawa [6].

2. Principles

The fundamentals principles behind the molecular mechanics method can perhaps be best understood by first considering the proper quantum mechanical approach to the problem of molecular structure [7], and then see how the calculations may be simplified through the use of molecular mechanics.

It is usual in attacking molecular problems with quantum mechanics to first introduce the Born-Oppenheimer approximation, which says that the motions of the electrons and nuclei may be considered separately, because of the large difference in the masses of the particles. The calculation typically begins by assuming some geometry for the nuclei. This may be from standard bond lengths and angles, or from some other method. A starting wave function is then assumed, usually based on some simple approximation such as an extended Huckel calculation, for example. What we want to do is to find the geometry of the molecule. This is the position on the potential surface where the energy is a minimum. We find the energy by first optimizing the electronic structure (the wave function) for our starting nuclear configuration. When we have a good wave function for the nuclear configuration, there will ordinarily be net forces acting on most or all of the atoms. That is to say, we are not at the energy minimum. Accordingly, we move the nuclei in the directions indicated by the forces, and there are various more or less sophisticated methods to indicate just how far we might move the nuclei. Having moved the nuclei, we then have to reoptimize our wave function. When this is done, we again check to see if there are significant net forces acting. If there are, we continue to move the nuclei, and then optimize the wave function and calculate the net forces until we are satisfied that we are close enough to the energy minimum so that we have to an adequate geometry.

The advantage of this method is that one needs to know very little at the outset. One needs to somehow obtain a crude starting geometry for the molecule, one needs to know things like the value of Plank´s constant, and the mass of the electron, but the problem is solved essentially "from the beginning". The disadvantage of solving the problem this way is that the calculations are quite time consuming. Accordingly, as long as one wants to deal only with quite small molecules, or with limited precision, one can obtain a great deal of information in this way. As computers get faster, and programs get more efficient, the situation will continue to improve. Currently, one can get useful answers by this method for problems which involve a molecule containing maybe five or six atoms, and with somewhat less accuracy maybe up to about ten. However, for larger molecules of general interest, the *ab initio* approach is at present too time consuming to be of general use.

How can we simplify the *ab initio* calculation, so as to make it applicable to large molecules, but fast enough to be of practical use? This is a question chemists asked about thirty years ago, and at that time there were two apparent possibilities. The first possibility is the semi-empirical method, where one omits most of the 3- and 4-centered electronic integrals, which are so time consuming to calculate because of their vast numbers. One then introduces empirical approximations in order to make up for this omission. This semi-empirical method has been developed and refined over the years, and is useful. It is certainly fast, compared to the *ab initio* method, but the accuracy is often less than one would desire. The method works fine for some kinds of problems, but it´s not adequate for other kinds of problems, and trying to decide if the problem at hand falls into the first or second group is not always easy.

The other approach to the problem is what is now called "molecular mechanics". Here, instead of calculating the electronic structure of the molecule in detail from first principles, one simply assumes that the electrons contribute to the forces acting on the nuclei. One does the calculations directly on the forces, and not on the electronic structure of the molecules. The electrons are then considered only implicitly, not explicitly. The calculations are enormously simplified when carried out in this way, however, they cannot begin from first principles. Rather, one has to have quite a lot of information about the basic features of structure for some simple molecules related to the molecule at hand. These basic features can then be translated into the forces in more complicated molecules. So the procedure employed is formally similar to what is done with *ab initio* methods. That is, one needs a starting geometry, one calculates the forces on the nuclei, and then one moves the nuclei in response to these forces. The simplification comes in that one never has to worry about the electronic structure of the molecule. These calculations are sufficiently fast that they may be used to study very large molecule (including proteins, with up to several thousand atoms) in a practical way at the current time. The accuracy of the calculations depends critically on the "force field". That is, one needs to have good data on small molecules, which can be transferred to problems involving larger molecules. Such data are currently available from experiments, and also in a few cases but increasingly from *ab initio* calculations. As the *ab initio* calculations are carried out with higher accuracy on more and more small molecules, they yield the information necessary for the molecular mechanics calculations on still larger molecules, and this kind of "boot strapping" will doubtlessly become increasingly important as time goes on.

3. Force Fields

In general we have a starting structure, some approximation to the molecular structure we are interested in, and we have to learn enough about the potential surface in the vicinity of that structure so as to be able to locate the energy minimum. In other cases, we might want to find much more of the potential surface, several minima, and perhaps saddlepoints connecting them. It might be thought that since we are working with a molecular system, which is quite small, quantum effects would be important, and that the calculations would have to be done using quantum mechanics. This proves not to be true, for the most part. The general philosophy here is to use classical mechanics if you can, and quantum mechanics only when you must. The reason being that the quantum mechanical calculation is very much more complicated, and should be avoided if the classical calculation will give adequate results. It turns out that one can determine as much of the potential surface as needed, and one can even locate the lowest vibrational levels accurately enough using classical mechanics. The molecule can be treated as though it were an assemblage of weights connected by springs, which undergo vibrational motion. This approximation is good enough for most of we want to do.

Molecules are usually studied in terms of their internal coordinates (that is, bond lengths, bond angles, and torsion angles). However, for mathematical reasons it is usually convenient to carry out the calculations in the cartesian coordinates. It is straightforward if laborious to go back and forth between cartesian and internal coordinates. Such transformations were exceedingly difficult before the advent of computers, but now, standard computer programs are available that will carry this transformation out automatically, very quickly, and with little assistance from the user.

If we think about the potential surface for a molecule, which is multi-dimensional, and involves the locations of all of the atomic nuclei, there is somewhere on the surface one or more energy minima, and we want to find those, as they will correspond to the stable conformational structures of the molecule. The potential energy surface, while not known to us at the outset, can be expressed as a Taylor's series expansion. If we are near to one of the energy minima, the practical problem in general is to find that energy minimum, as that will correspond to the structure we are interested in. A Taylor series expansion of the molecule in internal coordinates gives equations for stretching and bending of a given bond or bond angle which look as follows:

$$E_s = k_s \, (l - l_0)^2 + k_{s1} \, (l - l_0)^3 + k_{s2} \, (l - l_0)^4 + ... \text{ (stretching)} \qquad (1)$$

$$E_\theta = k_\theta \, (\theta - \theta_0)^2 + k_{\theta 1} \, (\theta - \theta_0)^3 + ... \text{ (bending)} \qquad (2)$$

Where E is the energy of a particular distortion, l_0 and θ_0 are the natural bond length and angle (the value that would be taken up in the absence of other forces) and the k's are coefficients from the Taylor's expansion.

3.1. Objectives

Chemistry has always been an experimental science. Why do we wish to carry out molecular mechanics calculations when we might determine the same quantities experimentally? There are actually several reasons.

It has been said that if you cannot calculate something exactly, you probably do not understand it very well. That would seem to be true here. Hence one reason we want to be able to carry out these calculations is to make sure that we do understand the molecule at hand, in terms of our model.

A second reason is a very practical one. In order to carry out molecular mechanics calculations, one must have all of the necessary force field parameters. In many cases these will not be available, and if it is necessary to determine them, that may be a major effort. However, if they are available, then the calculation can be done quickly, reliably, and accurately. In that case, rather than a complicated experiment which may take weeks or months, we can do a calculation, perhaps in a matter of hours, and get the same information. The practical advantage of saving time can be very large.

3.2. Methods

Since the time of the earlier work discussed in the introduction to this chapter, a long sequence of force fields has been presented in the literature by various authors. In the 1960's and early 70's, these differed from one another in a very marked ways. The reason for the difference was basically that different people were interested in different classes of compounds, and they adjusted their parameters to fit the things in which they were interested. The result was usually that the force field did a good job on whatever was of major interest, but a poorer job on other kinds of compounds. As time has gone on, force fields have become more general , and they have also converged upon one another to a large extent.

A force field is a series of equations which give the energy in terms of the distortion of a molecule, when bonds are stretched, or angles are bent, or torsion angles are twisted. To be useful, such a force field must be transferable to other molecules in a general way. We have no guarantee from first principles that force fields will be transferable. Chemists assumed long ago that they would be, at least to some approximation, because the physical and chemical properties of classes of compounds depend primarily on the class of compound; i.e. these properties are transferable, so the force field that describes the property should be transferable. However, chemists also know that this is only a first approximation. That is to say, ketones, for example, as a class, have certain properties. However, the range of reactivity of different members of the class may vary by several orders of magnitude. So the properties are not exactly transferable in a simple way. How transferable is the force field? Actually, we don't know the answer to the latter question. It is transferable, if complicated enough, as far as we have gone with it to this time.

In principle, the force field contains terms not only for stretching of bonds, and bending of bonds, but also for simultaneously stretching and bending bonds and angles (a stretch-bend interaction), or for stretching two bonds at the same time (a stretch-stretch interaction), and so on. Thus the number of possible terms in a force field is very much larger than elementary considerations might indicate. Considering only the stretching and bending, and not their interactions, is clearly a first approximation. Considering only the quadratic term for bending or stretching is clearly an approximation. In principle, one should use an infinite series of terms. Is the truncation of an infinite series to just one term adequate? Well, experience shows that it is adequate for solving many problems with sufficient accuracy, but it certainly leads to noticeable errors in many other cases. The equations for stretching and bending were given earlier. Primitive force fields truncate each expression after the first term. Better force fields now in use many consider two or three terms. As far as determining structures of molecules, two or three terms are adequate for all cases that have been studied to date.

The torsional motion of a molecule can be represented by a quadratic term similar to that used for bond stretching or bending as a first approximation. However, rotational barriers are often quite low in molecules, and hence it is better to represent this motion with periodic (cosine) type term. By adjusting the height of the curve, one can reproduce not only the correct potential surface at the energy minimum, but also elsewhere, up to and including through the saddlepoint. The torsional term for a given bond is commonly written as a Fourier series.

$$E_\omega = V_1/2 \, (1 - \cos \omega) + V_2/2 \, (1 - \cos 2\omega) + V_3/3 \, (1 - \cos 3\omega) + ... \quad (3)$$

For ethane type molecules, the V_3 term is the important term. For ethylene types, the V_2 term is similarly the important term. The V_1 term is kind of a catch-all, to absorb errors that come other approximations, but it may also have contributions from more fundamental interactions. For example, in 1,2-dichloroethane, there is an electrostatic interaction between the two C-Cl dipoles. To the extent that this is not taken into account exactly with the electrostatic part of the calculation, it can be accounted for using the V_1 term.

Electrostatics is unimportant in saturated hydrocarbons, and in several other kinds of molecules. That is to say, whether one includes electrostatics as carefully and completely as possible, or whether one simply leaves it out, the results are, for practical purposes, indistinguishable. The electrostatic interactions in hydrocarbons are sizable, and one must therefore adjust the other parameters in the calculations to reproduce the experimental facts for simple molecules. But after this is done, the effect of the electrostatics essentially cancels out.

With more complicated and more polar structures, such as 1,2-dichloroethane as a simple example, or such as a peptide as a more complicated example, electrostatics can become not only important, but of major importance in describing some of the properties of the molecule. There is at present no general agreement as to just how electrostatics is best treated. Often Coulomb's law is used, and point charges are placed at atomic nuclei, and allowed to interact. Another approximation is to place point dipoles in bonds, and allow those to interact via dipole-dipole interactions. At long distances, the interactions between four point charges or two point dipoles give the same result. However, at short distances, this is only approximately true. Which method gives the better result in general is uncertain, but the two methods seem to give similar results on the whole. The point charge approach has the advantage that the calculation is faster.

Better approximations would appear to include not only monopoles, and dipoles, but higher multipoles in the electrostatic calculation. Of course, this would greatly complicate the calculation, and it has not been demonstrated that this refinement has any practical general use. One of the reasons why it has been difficult to decide how to handle the electrostatics intramolecularly comes from the fact that polar molecules are ordinarily studied in polar solvents, so that solvation is also an important aspect of the problem, and it is usually mixed in with the electrostatics in such a way that it is difficult to decide how much of the calculational error is due to the improper treatment of the intramolecular electrostatics, and how much of it is due to the improper treatment of solvation. Further complicating the problem of a polar molecule in a polar solvent is the fact that hydrogen bonding intramolecularly, as well as between the solvent and the solute, generally is occuring. Hydrogen bonding is to large extent an electrostatic phenomenon, but that is not the whole story. And how to treat the remainder of the hydrogen bond is also not completely clear at this time. Electrostatics, solvation, and hydrogen bonding are tangled together in such a way that one might make an error in one of these, and compensate for it in another, and obtain good results in a particular case, but in a fundamentally poor way. This area of investigation is quite active at present.

3.3. VAN DER WAALS´ INTERACTIONS

All atoms and molecules, including those with closed shells of electrons, show interactions with one another which are usually referred to as van der Waals´ interactions. These are interactions which are a result of electron correlation between the interacting particles. The simplest example would be the interaction of two helium atoms. While the helium atom is neutral overall, at any given instant the nucleus of the atom is someplace, and the electrons are not spherically distributed about it, but have definite if unknown locations. When a second helium atom comes up near the first one, the electrons repel one another, but attract the nuclei, not only of their own atom, but of the other atom as well. Hence the electronic motions of the electrons on one atom correlate with those of the electrons on the second atom, in such a way as to minimize the electrostatic energy. This leads to a small net attractive force between the two helium atoms. Indeed, helium condenses to a liquid only because of this van der Waals´ interaction. This interaction is sometimes referred to as a "dispersion force", or an "induced dipole - induced dipole" interaction. When *ab initio* calculations are carried out to the Hartree-Fock level, this correlation effect is not included. Accordingly, Hartree-Fock calculations do not reproduce interactions of the van der Waals´ type. Hartree-Fock calculations can be improved by including electron correlation, and usually either a configuration interaction calculation is used for the purpose. Molecular mechanics simply includes a direct pairwise van der Waals´ interaction between all atoms which are not bound together, or to a common atom. The omission of the interaction in the latter two cases is based on the fact that the appropriate van der Waals´ interaction is absorbed into the bond stretching or bond angle bending deformation.

Van der Waals´ interactions are always treated as two body interactions. This has been an adequate approximation so far, but many-bodied interactions may be needed at a later date as force fields become more refined.

The van der Waals´ attraction between two first row atoms is pretty small, of the order of 0.02 kcal / mol. The interactions become an order of magnitude larger as we go down in the periodic table, because there are more electrons around the atom, and correlation becomes more important. But their real importance comes when one is considering not just two atoms, but whole molecules. In a typical molecule with say 50 atoms, when it interacts with another molecule with say 50 atoms, the total number of interactions on an atom-atom basis is $(50)^2$ or 2500, so that even at 0.02 kcal / mol, this adds up to be a significant amount of energy.

If one has good van der Waals´ and electrostatic properties for a molecule, and its structure, then one should be able to calculate the crystal packing accurately, and also the heat of sublimation of the crystal. One can also calculate, with more difficulty, the density and heat of vaporization of a liquid. This has been done, and it is a necessary (but not sufficient) condition that good results can be obtained in such calculations to be sure that the van der Waals´ functions are reasonable.

Van der Waals´ functions are generally written with two terms (but again these are leading terms of infinite series). One term is attractive, and is proportional to r^{-6}. This is a relatively long range term. The short range term is repulsive, and is sometimes written as proportional to r^{-12}, although r^{-9} is more accurate. Theoretically, this term should in fact be proportional to ae^{-r}. All of these approximations are found in various force fields in the literature. The r^{-12} is probably the most commonly found, although the poorest, approximation. As in other cases, it is used, not because it is a poor approximation, but because the computations are faster with that approximation, and the approximation seems adequate for many purposes.

3.4. CONJUGATED SYSTEMS

Delocalized electronic systems present special problems in molecular mechanics. From what has been described previously, one clearly needs to know properties on a bond basis in order to carry out molecular mechanics calculations. That is, for each bond one must know the stretching force constant, the value for l_0, and other items. If we talk about an x-y bond, for example, these numbers are usually constants and present no particular problem, as long as their values are known. However, in delocalized system, there are a whole spectrum of bonds which vary in length, and in force constant and other properties, between X and Y. Consider naphthalene as an example. The bonds are somewhat benzenoid, but not exactly. Some are longer, some are shorter. How do we describe this with molecular mechanics? There is the brute force method, where one simply gives the different atoms in naphthalene, for example, different atom type numbers, and then gives different bond properties to the different bonds. This would be OK for this molecule, but if one wants to consider a whole series of molecules, where bond lengths vary incrementally over a large range, it's not a very practical way to go. One way to treat such a system is to separate the pi parts and sigma parts, and then do a pi-system calculation to determine pi bond orders. It is assumed (and seems to be approximately true) that the stretching force constant and l_0 (and other properties) of a bond are related to the pi bond order (usually taken to be linearly related). Hence what one does in practice is a pi-system calculation on the conjugated part of the molecule, from which bond orders are obtained. These are translated into the necessary parameters that are needed for the molecular mechanics calculation. From that point on, the conjugated system is just an ordinary molecule, and the calculations are carried out in the usual way.

Non-planar systems are a little more complicated. In a non-planar molecule, the pi-system and the sigma-system are not orthogonal, and hence the sigma/pi separation assumption cannot be validly made. Accordingly, what is done is to initially carry out a "planarization" of the non-planar pi-system. That is, the direction cosine terms are removed from the geometry expression, and the molecule is thereby flattened into a planar conformation. The pi-system calculation can then be carried out in a valid way. When the appropriate parameters are then applied to this non-planar system, there are out-of-plane bending forces and torsional forces that twist the molecule back to its non-planar conformation. Hence one gets the correct structure for the molecule anyway, as one would hope. But the pi-system calculation is done on a (hypothetical) planar system where the sigma / pi separation is valid. The calculations seem to work well in practice.

4. Results

The previous discussion outlines how a force field, and the corresponding multi-dimensional energy surface, may be constructed for a molecule. In the long range, what we want to do is collect a set of parameters from which we can construct a force field for any kind of a molecule, or molecular fragment. This method is best for studying energy minima, that is, stable molecules. It can be used equally well for things like free radicals, which are stable in the sense that they lie at energy minima on the potential surface. Whether or not they are stable in the sense that one can collect them and put them in a bottle is not relevant.

One can study transition states such as rotational barriers, quite easily. However, transition states for chemical reactions where bonds are being broken are another matter, and they are beyond the scope of the general method described here. Indeed, the method can be extended with some additional assumptions and approximations to deal with such cases, but these are usually better studied by *ab initio* methods. In some cases, where one is studying a bond breaking in a very large and complicated molecule, for example, it may be advantageous to study the actual local system where the bond is broken in terms of an *ab initio* calculation, and to study the remainder of the molecule with a molecular mechanics calculation, and superimpose the two.

To study the wide variety of molecular types that exist, or might exist, would require an extremely large parameter set, and these parameters are largely unavailable at present.

Methods exist for estimating these parameters, so that one can carry out molecular mechanics calculations on all sorts of molecules with somewhat limited accuracy, and on a smaller variety of structural types with increasing accuracy.

When one studies structure by molecular mechanics, one minimizes the energy. Of course, one learns something about the energy of the system in the process. Hence things like conformational equilibria, and other thermodynamics properties, may be estimated approximately from crude force fields, or more accurately from better force fields. While the studies of thermodynamic properties so far reported are rather limited, indications are that with a good force field, one can obtain numbers that are of experimentally useful accuracy.

The use of molecular mechanics is therefore pretty straightforward, and the method is applicable to a wide variety of problems in structural chemistry. It is being so used today by large numbers of chemists interested in structural problems. It should perhaps be added in closing that one additional advantage of molecular mechanics is that it may be used to uncover parts of structural chemistry where our present understanding is incomplete. When one develops a good force field, one does calculations, and one expects to get results that agree well with experiment. If one uses a good force field, and one gets results that do not agree well with experiment, then there is some kind of "effect" occuring in the systems at hand which are not adequately understood, and not adequately included in the force field. There are quite a few such "effects" known to organic chemists already, and, of course, these must be taken into account if one is to have a good force field. It is anticipated that additional such "effects" are likely to be uncovered as studies became increasingly accurate.

References

[1]   Andrews, D. H. *Phys. Rev.*, **1930**, *36* , 544.

[2]   Westheimer, F. H.; Mayer, J. E. *J. Chem. Phys* . **1946**, *14* , 733.

[3]   Hendrickson, J. B. *J. Am. Chem. Soc* . **1961**, *83*, 4537.

[4]   Wiberg, K. B. *J. Am. Chem. Soc* . **1965**, *87*, 1070.

[5]   Burkert, U.; Allinger, N. L. *Molecular Mechanics*, American Chemical Society, Washington, DC, **1982**.

[6]   (a) Engler, E. M.; Andose, J. D.; Schleyer, P. v. R. *J. Am. Chem. Soc.* **1973**, *95*, 8005. (b) Altona, C. L.; Faber, D. H. *Top. Curr. Chem.* **1974**, *45*, 1. (c) Dunitz, J. D.; Burgi, H. B. *MTP Int. Rev. of Science, Series 2: Chemical Crystallography* Robertson, J. M. R., Ed.; Butterworths: London, 1975; p.81. (d) Allinger, N. L. *Adv. Phys. Org. Chem.* **1976**, *13*, 1. (e) Ermer, O. *Struct. Bonding (Berlin)* **1976**, *27*, 161. (f) Niketic, S. R.; Rasmussen, K. *The Consistent Force Field*: Springer

Verlag: Berlin, 1977. (g) Warshel, A. *In Modern Theoretical Chemistry* ; Segal, G. Ed.; Plenum: New York, 1978; Vol. 7, p. 133. (h) White, D. N. J. *Molecular Structure by Diffraction Methods*; Chemical Society: London, 1978; Vol. 6, p. 38. (i) Osawa, E.; Musso, H. *Top. Stereochem.* **1982**, *13* , 117.

[7]   Hehre, W. J.; Radom, L.; Schleyer, P. v. R.; Pople, J. A. *Ab Initio Molecular Orbital Theory*, J. Wiley and Sons, **1986**.

PREDICTING THE THREE-DIMENSIONAL STRUCTURE OF PROTEINS BY HOMOLOGY-BASED MODEL BUILDING

Gerald M. Maggiora
S. Lakshmi Narasimhan
Cindy A. Granatir
James R. Blinn
Joseph B. Moon
Upjohn Laboratories
The Upjohn Company
Kalamazoo, MI 49001

ABSTRACT. *A priori* prediction of the 3-D structure of proteins remains an important unsolved problem. Of the theoretical methods employed today, only homology-based model building produces protein structures of sufficient accuracy to be of use in structure-function and related studies. The present work provides a discussion of the features of protein structure which bear on homology-based model building in addition to a detailed description of the homology-based approach, with examples drawn from our own work.

1. Introduction

One of the greatest challenges facing theorists today is the prediction of the 3-D structure of a protein starting from its amino acid sequence. Due to the size and complexity of proteins, theoretical methods currently used to investigate small organic molecules are not directly applicable in the study of protein structure. Current approaches to protein-structure prediction can be broadly classified into *a priori* and *heuristic* methods. A number of reviews [1-5] and books [6,7] which address 3-D-structure prediction and theoretical aspects of protein folding are available and should be consulted for additional details.

A priori methods seek to predict the 3-D structure of a protein without the use of information derived from known protein structures. Reliable predictions using these methods are not yet a reality, but efforts towards this goal show promise, and the methods developed are useful in supplementing heuristic methods of structure prediction. *A priori* methods rely on the use of potential-energy functions to model the intra- and inter-molecular interactions in the protein-solvent system at the atomic level (see Table 1 in [5]). An optimization method, typically energy minimization, is then used to locate a structure that resembles the native one. A serious impediment to the successful application of such methods is the multiple-minimum problem [8], which arises from the large number of degrees of freedom and resultant numerous local minima. Since there is no algorithm available at present that can determine the "global minimum", energy minimization will in general produce only local minima, making

S. J. Formosinho et al. (eds.), Theoretical and Computational Models for Organic Chemistry, 137–158.
© 1991 *Kluwer Academic Publishers*.

energy-based structure prediction impractical. More recent methods utilize "simplified" potential-energy functions and employ Monte Carlo and/or simulated annealing procedures [9-11] to overcome the multiple-minimum problem. However, additional work is needed to bring the reliability and robustness of these methods to a satisfactory level.

In contrast, heuristic approaches make use of information on protein structure such as that available in the Brookhaven Protein Data Bank (PDB). This information is often supplemented by energy-based procedures. Currently, the most reliable and hence most popular heuristic approach is *homology-based model building* [12,13], which seeks to predict the structure of a protein *("the target")* from the known structure of a related protein *("the template")*. Implicit in this approach is the notion that protein structure possesses inherent regularities which limit the variety of 3-D structures that can be formed from related sequences of amino acids.

Ideally, any model of a protein should meet the following criteria:

1. It should be reliable.

2. It should be obtained relatively quickly compared to x-ray crystallography and multi-dimensional NMR.

3. It should be consistent with existing experimental data on the protein in question, and it should be consistent with general features of protein structure.

4. It should be detailed enough to provide useful insights into structure-function relationships.

To date, more than 14,000 protein sequences have been determined. This number contrasts sharply with the approximately 450 3-D structures that have been determined by x-ray crystallographic and NMR methods. Thus, development of reliable 3-D models by homology-based model building can provide useful structural information that will benefit a variety of structure-function studies. For example, site-specific mutagenesis experiments rely heavily on 3-D structure information, both for design and for the interpretion of results [14]. Information on 3-D structure can also significantly reduce the effort needed to design molecules which specifically influence a protein's function (see *e.g.* [15]).

In Section 2, a review of the salient structural features of proteins is presented, followed in Section 3 by a discussion of homology-based model building. An example of the model-building process, as applied to Domain I of CD4 (CD4-I) - the T-cell surface protein [16] which binds to Human Immunodeficiency Virus (HIV), is presented to illustrate the various steps of the process. Section 4 discusses a number of issues which bear on the future of homology-based model building.

2. Overview of Protein Structure

One of the most notable features of protein structure is its hierarchical character, which is illustrated in Figure 1. Thus, the study of protein structure can best be accomplished by examining the manner in which smaller structural units pack together to form larger, more complex ones, ultimately resulting in the unique, functional, 3-D structure of a protein. The simplest level in the hierarchy is the amino acid sequence of the protein, which describes the atomic composition and covalent connectivity of the constituent atoms and is referred to as *primary structure*. The peptide backbone may be grouped into regions with distinct, regular folding patterns and regions without such regularity. There are three main motifs observed in these "regular" regions: α-helix, β-strand and turns. Regions that do not conform to those motifs are classified as coils. Together the

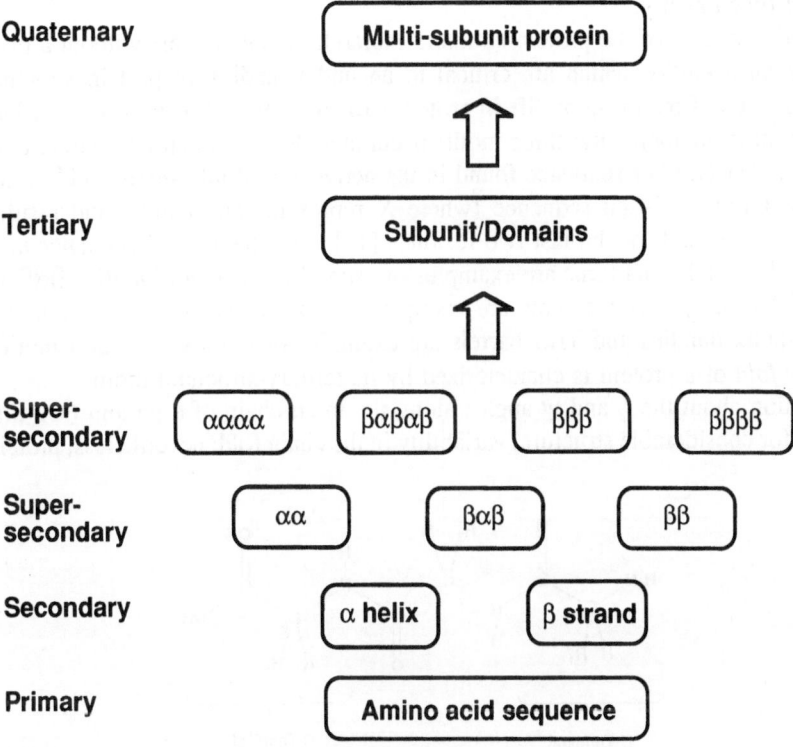

Figure 1. Diagram of the protein-structure hierarchy beginning at the bottom with the most fundamental element, primary structure, which is made up of individual amino acid residues linked together by peptide bonds (see Figure 2) to form polypeptide chains of appropriate sequence. Moving up the diagram each level corresponds to greater structural complexity than the preceeding one; at the top of the diagram the highest level of structure, quaternary structure, is reached. See text for further discussion.

helices, strands, and turns constitute the *secondary structures* observed in proteins, and form the smallest "building blocks" of 3-D structure. As the structures of more proteins became available in atomic detail, recurring patterns of association of secondary-structural elements emerged, leading to the next level in the structural hierarchy, *supersecondary structure*. Supersecondary-structural elements are characterized by the relative positions of their secondary-structural elements, as dictated by packing and energetic considerations. The next higher level in the hierarchy is that of *tertiary structure*, which describes the complete 3-D structure of an individual protein molecule or subunit, and which may contain one or more *structural domains*. Although the concept of a domain is difficult to define rigorously (see *e.g.* [5]), it can operationally be thought of as a compact globular region of a protein, which arises from all or a portion of a single polypeptide chain and potentially is capable of independent folding. Lastly, *quaternary structure* is the highest order structure formed when individual subunits assemble into the multi-subunit structures required by some proteins for full functionality.

A description of the protein-structure hierarchy is incomplete without a discussion of *structural motifs*, which are critical to an understanding of protein structure [17]. Identification of recurring motifs in protein structures has refined our knowledge of the protein-structure hierarchy: these motifs occur at all levels from primary to tertiary. The Phe-Asp-Thr-Gly-Ser sequence found in the active site of all aspartic acid proteinases, and the Gly-Gly-X-Leu sequence (where X represents any amino acid residue) that predicts a β-strand for the last two residues [17], are examples of *sequence motifs*; α-helices, β-strands, and turns are examples of *secondary-structural motifs*; βαβ and βxβ units, β-hairpins, and Greek keys are examples of *supersecondary-structural motifs*; and four-α-helix bundles and TIM barrels are examples of tertiary-structural motifs. The *tertiary fold* of a protein is characterized by its tertiary-structural motif.

Rotation about the ϕ and ψ angles along the main chain of a protein (see Figure 2) allows for considerable structural variability in the chain fold; nevertheless, proteins tend

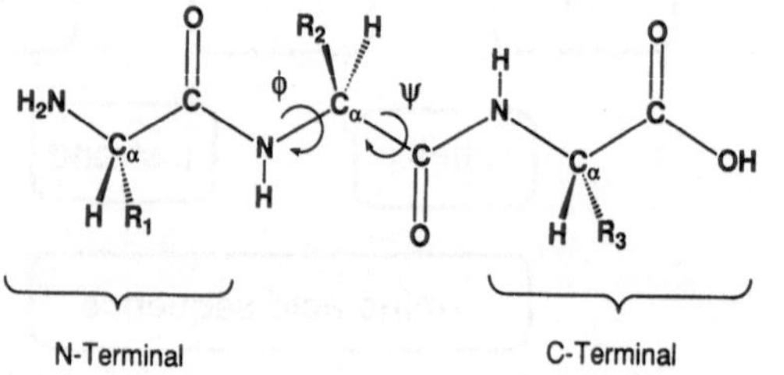

Figure 2. An example of the polypeptide backbone of a protein showing the N- (*amino*) and C- (*carboxyl*) residues and the ϕ and ψ angles which define the geometry of the chain. The sidechains, labelled R_1, R_2, and R_3, determine the nature of the individual amino acids making up the polypeptide chain. By convention the direction of the chain runs from its N-terminal residue (R_1) to its C-terminal residue (R_3).

to adopt a relatively small number of tertiary-fold motifs. The Ramachandran diagram depicted in Figure 3 [18] shows that the possible φ-ψ conformations lie within limited regions of the diagram which are determined by steric interactions among main chain atoms: the distinct asymmetry of the diagram is due to the L-stereochemistry about the α-carbon of the amino acids. The clusters of dots in Figure 3 indicate that the distribution of φ-ψ angles in real proteins, which include sidechain interactions and bond length and valence angle variations not considered by Ramachandran, adhere closely to the "allowed" regions of the Ramachandran diagram [19]. However, such conformational restrictions do not completely account for the limited number of chain

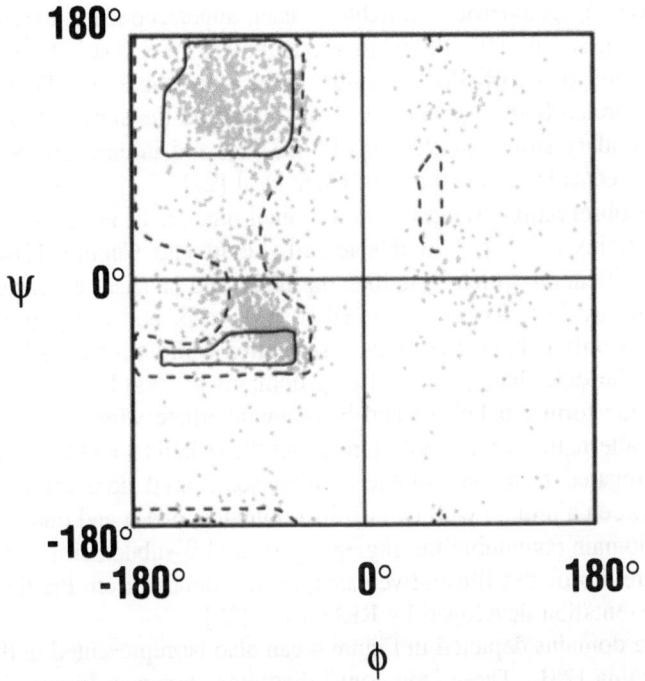

Figure 3. Ramachandran diagram depicting the regions of "close contact" between backbone atoms. The regions designated by the irregular shaped areas outlined by the solid lines correspond to normal contact distances, while those outlined by dashed lines correspond to lower-limit contact distances [18]. The outlined regions in the upper-left quadrant correspond primarily to β-sheet structures, those in the lower-left quadrant to right-handed helical structures. The region in the upper-right quadrant corresponds to left-handed helical structures, which essentially are not observed in proteins made up of the usual L-amino acids. The light-grey dots represent the φ-ψ angle values of amino acid residues of 13 proteins taken from the work of Levitt [19].

folds observed (*vide infra*).

From a more "global" perspective, all water-soluble, globular proteins share two important characteristics: they are close-packed structures [20], in that they do not possess significant "holes" in their interiors, and they have a laminar structure in which a predominantly *hydrophobic core* is surrounded by a predominantly *hydrophilic shell* which constitutes the outer surface of the protein and is in contact with the aqueous solvent environment. Although these characteristics play a role in determining the nature of a protein's tertiary fold and the sequences of amino acids which permit such a fold to exist, other factors have also been shown to be of importance [1,21,22].

Due to the small number of secondary-structural types found in proteins, only a limited number of simple supersecondary-structural motifs such as $\alpha\alpha$, $\beta\beta$, $\beta\alpha\beta$,..., are possible. Moreover, as discussed by Chothia and co-workers in a number of papers [23,24], the packing geometries available to each supersecondary-structural motif are also quite limited; specifically, the relative orientations of two α-helices, of an α-helix and a β-sheet, and of two β-sheets exhibit limited distributions. These distributions have been interpreted both in geometric terms, based on the surface topography of the interacting secondary-structural elements [20,23-26], and in energetic terms, based on the molecular mechanics calculations of Chou et al [27].

The above observations suggest that a limited number of tertiary-fold motifs exist in globular proteins (*cf.* [1]). In this regard, Levitt and Chothia [28] developed a systematic classification which describes the helix and/or sheet character of globular protein domains by five classes: all-α, all-β, α/β, $\alpha+\beta$, and everything else. All-α domains possess only α-helical secondary structures while all-β domains possess only β-strands. On the other hand, α/β and $\alpha+\beta$ domains possess both types of secondary structures. In the former α-helices and β-strands alternate with one another, while in the latter such alternation occurs only rarely and the α-helical and β-strand regions are somewhat segregated from one another. With some $\alpha+\beta$ domains it is difficult to distinguish between a protein with two distinct domains, one α and one β, and a protein with a single domain containing the segregated α- and β-subdomains required for $\alpha+\beta$ proteins. Figure 4 provides illustrative examples of domains from the first four classes using the representation developed by Richardson [29].

Each of the domains depicted in Figure 4 can also be represented in the notation of Levitt and Chothia [28]. These "topology" diagrams shown in Figure 5 illustrate the location of secondary-structural elements along the main chain as well as their relationship to one another. Based on analyses of many proteins, Chothia [20] has suggested three rules for the *chain topologies* of globular protein domains:

1. Consecutive elements of secondary structure tend to be in contact in the tertiary fold of the domain.

2. Connections in $\beta x\beta$ units are generally right-handed.

3. Segments connecting secondary-structural elements do not cross each other or form knots in the chain.

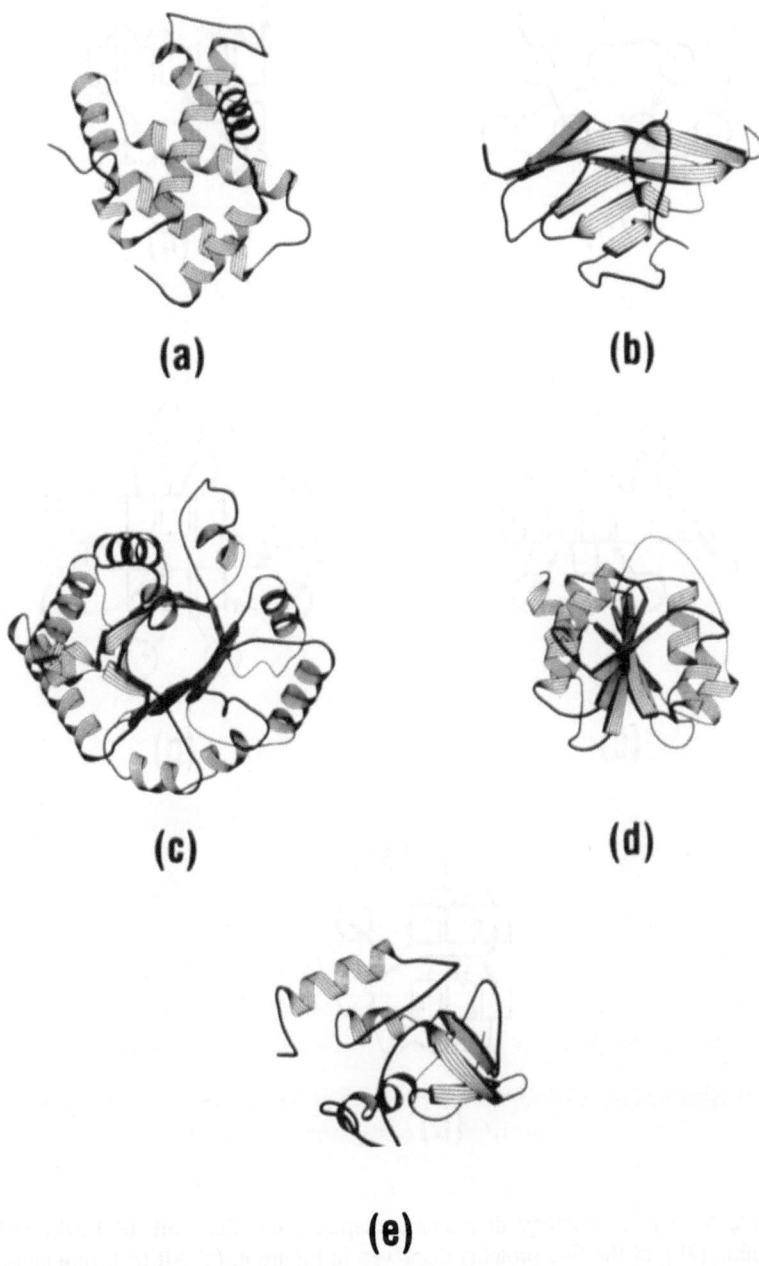

(a)

(b)

(c)

(d)

(e)

Figure 4. "Richardson diagrams" [29] of five proteins illustrating four classes of tertiary-fold motifs typically found in globular proteins. (a) All-α: hemoglobin, β subunit. (b) All-β: Immunoglobulin variable domain. (c) α/β: triose phosphate isomerase. (d) α/β: alcohol dehydrogenase, domain 2. (e) α+β: Staphylococcal nuclease.

144

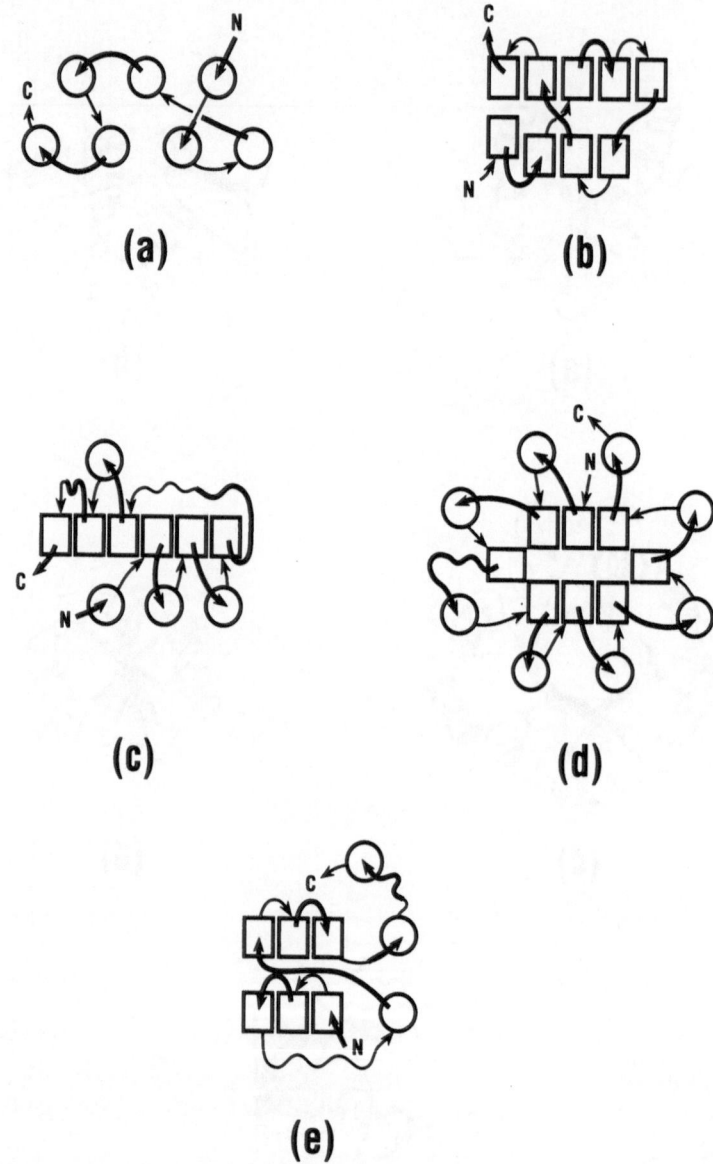

Figure 5. "Chain topology diagrams", adapted from the work of Levitt and Chothia [28], of the five proteins displayed in Figure 4. (a) All-α: hemoglobin, β subunit. (b) All-β: Immunoglobulin variable domain. (c) α/β: triose phosphate isomerase. (d) α/β: alcohol dehydrogenase, domain 2. (e) α+β: Staphylococcal nuclease. The squares represent β-strands and the circles α-helices. The darker lines connecting the secondary-structural elements represent the backbone chain lying above the plane of the page, while the lighter lines represent the backbone chain lying below the plane of the page. The discrepancy between the staphylococcal nuclease topology in this figure and in the ribbon diagram of Figure 4 is due to the use of different methods of secondary-structural assignment.

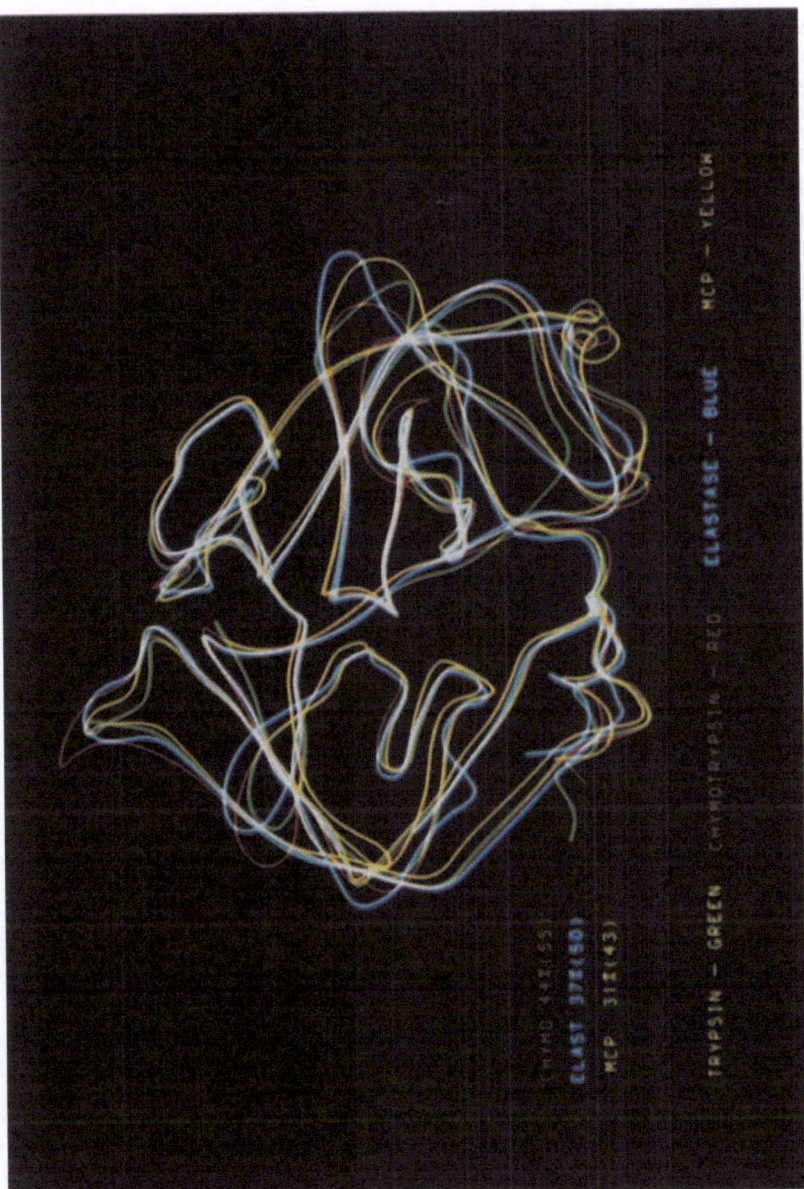

Figure 6. Thin-ribbon overlay of four serine proteinase structures obtained from the PDB: trypsin (green), chymotrypsin (red), elastase (blue), and mast-cell proteinase (yellow); the degree of sequence homology relative to trypsin is also given in the figure: the numbers in parentheses correspond the amount of sequence homology with respect to conserved residues, while the other numbers correspond to sequence homology with respect to identical residues.

146

The above discussion shows quite clearly that despite the potential for great structural diversity in proteins, such diversity has not been observed.

In summary, protein structures are formed from α-helices and β-strands which pack together in a limited number of ways to form higher-level structures. These structures constitute the hydrophobic core of proteins and exclude any turns or loops. In contrast, the hydrophilic outer shell of a protein contains turns, loops (*cf.* [30]), and the hydrophilic portions of α-helices and β-strands.

The limited number of possible tertiary-folds also explains why the molecular architecture of proteins within a family remains somewhat impervious to amino acid substitution. In fact, in cases where several structures within the same family have been determined experimentally, the overall tertiary-structure motif is conserved, demonstrating that *structural homology persists even though sequence homology may have undergone significant changes* [12,31,32]. Figure 6 illustrates this point for four proteins in the serine proteinase family, *viz.* trypsin, chymotrypsin, elastase and mast cell

```
TRP  IVGGYTCGANTVPYQVSLNS.....GYHFCGGSLINSQWVVSAAHCYKS.
CHT  IVNGEEAVPGSWPWQVSLQDKT...GFHFCGGSLINENWVVTAAHCGVT.
ELA  VVGGTEAQRNSWPSQISLQYRSGSSWAHTCGGTLIRQNWVMTAAHCVDRE
MCP  IIGGVESIPHSRPYMAHLDIVTEKGLRVICGGFLISRQFVLTAAHCKG..

TRP  .....GIQVRLGQDNINV.VEGNQQFISASKSIVHPSYNSNTL...NNDIM
CHT  ....TSDVVVAGEFDQGSSSE.KIQKIKIIAKVFKNSKYNSLTI...NNDIT
ELA  ....LTFRVVVGEHNINQ.NNGTEQYVGVQKIVUHPYWNTDDVAAGYDIA
MCP  ....REITVILGAHDVRK.RESTQQKIKVEKQIIHESYNSVPN...LHDIM

TRP  LIKLKSAASLNSRVASISLP.T...SCASAGTQCLISGWGNTKSSGTSYPD
CHT  LLKLSTAASFSQTVSAVCLP.SASDDFAAGTTCVTTGWGLTRY..ANTPD
ELA  LLRLAQSVTLNSYVQLGVLPRA.GTILANNSPCYITGWGLTRTN.GQLAQ
MCP  LLKLEKKVELTPAVNVVPLP.SPSDFIHPGAMCWAAGWGKTGVR.DPTSY

TRP  VLKCLKAPILSNSSCKS..AYPGQITSNMFCAGYLQGGKDSCQGDSGGPV
CHT  RLQQASLPLLSNTNCKK..YWGTKIKDAMICAGA..SGVSSCMGDSGGPL
ELA  TLQQAYLPTVDYAICSSSSYWGSTVKNSMVCAGGDG.VRSGCQGDSGGPL
MCP  TLREVELRIMDEKACVD...YRYYEYKFQVCVGSPTTLRAAFMGDSGGPL

TRP  VCS......GKLQGIVSWGSG..CAQKNKPGVYTKVCNYVSWIKQTIASN
CHT  VCKKN..GAWTLVGIVSWGSS.TCSTS.TPGVYARVTALVNWVQQTLAAN
ELA  HCLVN..GQYAVHGVTSFVSRLGCNVTRKPTVFTRVSAYISWINNVIASN
MCP  LCA......GVAHGIVSYGHP....DAKPPAIFTRVSTYVPTINAVIN..
```

Figure 7. Alignment of the sequences of the serine proteinases shown in Figure 6. TRP, CHT, ELA, and MCP correspond to trypsin, chymotrypsin, elastase, and mast-cell proteinase, respectively. The shaded boxes designate identical or highly conserved residues, and the bold horizontal lines represent structurally-conserved regions of these proteins derived from the structural overlay presented in Figure 6.

proteinase. The overlay indicates that there is significant similarity among the structures in the *"core-framework"* even though these proteins possess a range of sequence homology relative to trypsin: from 44% (55%) for chymotrypsin, to 37% (50%) for elastase, and only 31% (42%) for mast cell protease, where the numbers in parentheses represent the percent homology allowing for conservative amino acids substitutions. Figure 7 presents an alignment of the sequences of these proteins in which the boxes indicate conserved residues and the horizontal bold lines indicate the structurally-conserved regions present in all four proteins. As is seen from the figure, the highly-conserved residues tend to fall into structurally-conserved regions, while non-conserved residues tend to fall into structurally-variable regions.

As noted above, the core-framework of a protein contains "packed" secondary-structural elements and is free of loops. Since proteins within a given family possess the same tertiary-fold motif, mutations of residues in the core-framework tend to preserve the general features of the tertiary fold [20,23,24] as well as the physico-chemical characteristics of the mutated amino acids [33]. Moreover, as the core-framework is highly hydrophobic, conservation of hydrophobic residues is expected and is indeed observed to be the case [33,34]. Figure 7 also demonstrates this point, where the hydrophobic residues located within the structurally-conserved regions (denoted by bold horizontal bars) are seen to be well conserved.

3. Overview of Homology-Based Model Building

As noted in the Introduction, *a priori* structure prediction at present is not sufficiently robust for use as a general method, and the best current approach to the prediction of the 3-D structure of proteins is homology-based model building. The structural properties of proteins such as their hierarchical character and the limited number of structural motifs at each level of the hierarchy plus the growing number of high-resolution structures now available make homology-based model building quite feasible and the method of choice today. However, an important assumption implicit in all successful applications of this type is that the essential structural "scaffolding" of the target and template proteins not differ significantly. If such a condition can be reasonably assumed, then homology-based model building usually can be applied in a procedure similar to that illustrated in Figure 8. In this section various aspects of the model-building process will be illustrated with examples taken from our own work on CD4-I (*vide supra*).

3.1. TEMPLATE SELECTION

The first step in homology-based model building involves the choice of an appropriate template protein related to the target protein. Two important issues in this choice are the degree of homology between the target and template proteins and the accuracy of the structure of the template protein; both of these factors will influence the accuracy of the predicted structure. If the target protein is within the same family as one or more

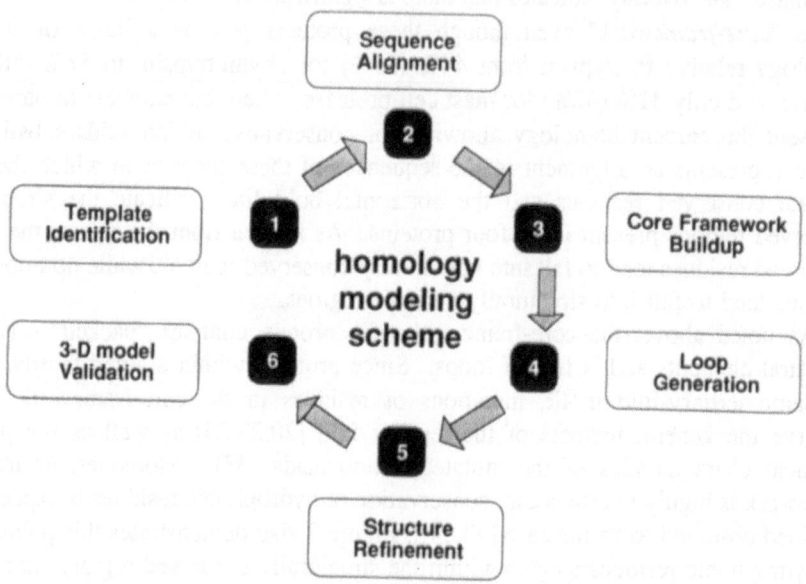

Figure 8. Scheme for homology-based model building of protein structures. See text for further details.

proteins of known structure (*e.g.* the serine protease family discussed earlier) and has greater than 25% sequence homology to the template (with respect to conservative replacement of residues), it is generally possible to obtain a satisfactory model of the target protein: the greater the sequence homology the greater the chance of obtaining an acceptable model, given that template structures of suitable accuracy are available. In the case of CD4-I comparisons with immunoglobulin variable domains exhibit about 33% sequence homology among similar residues. In cases where a closely-related protein is not available and/or where sequence homology is less than 25%, choice of an appropriate template is difficult, and homology-based model building may not even be feasible. Such cases must be dealt with individually, and require ancillary data such as is obtained in photoaffinity labelling and site-specific mutagenesis experiments.

For example, in the case of the polymerase domain of the reverse transcriptase from HIV-1 (HIV-RT), the only suitable template protein whose structure is known is *E. coli* DNA polymerase I (Pol-I) [35], although the sequence of homology between these two proteins is considerably less than 25%. However, the presumed functional similarities between the two proteins and the fact that affinity-labelling chemical modification experiments [36] have identified key corresponding lysine residues provide important data supporting the feasibility of modeling HIV-RT on the basis of the Pol-I structure. An additional difficulty in modeling HIV-RT from Pol-I is the low resolution of the Pol-I structure. Currently, only the α-carbon (see Figure 2) coordinates of Pol-I are available, and thus any attempt to build an accurate structure of HIV-RT will be further limited by the accuracy with which the full 3-D structure of Pol-I can be reconstructed from its α-carbon coordinates alone (*cf.* [37]).

3.2. ALIGNMENT OF TARGET AND TEMPLATE SEQUENCES

The most crucial step in the homology-based model-building process is properly aligning the target and template sequences. Alignment determines the segments of the target sequence which correspond to the core-framework and the loops, and thus are critical to the generation of a correct 3-D structure. The difficulty of obtaining a proper alignment is directly proportional to the amount of sequence homology between the template and target sequences; Figure 9 summarizes various issues related to template-target sequence alignment.

When sequence homology exceeds 50%, sequence alignment is relatively straightforward and can be accomplished by any one of a number of automated procedures [38]. As the sequence homology drops below 50% the process of sequence alignment becomes considerably more difficult, and auxillary information is generally required to obtain a reliable alignment. The problem is one of identifying appropriate *"anchor points"*, which are residues or groups of residues present in both the template and target proteins that possess structural or functional significance. Anchor points provide a guide for insuring correct structural correspondence between the two proteins. Identification of anchor points can be accomplished in a number of ways; one is to align the template and target proteins with other proteins in the same or closely related families: the anchor points are the residues that are conserved within the family. Structural alignment of potential template proteins [31], identification of cysteine residues in disulfide bonds, site-directed mutagenesis and affinity-labeling chemical-

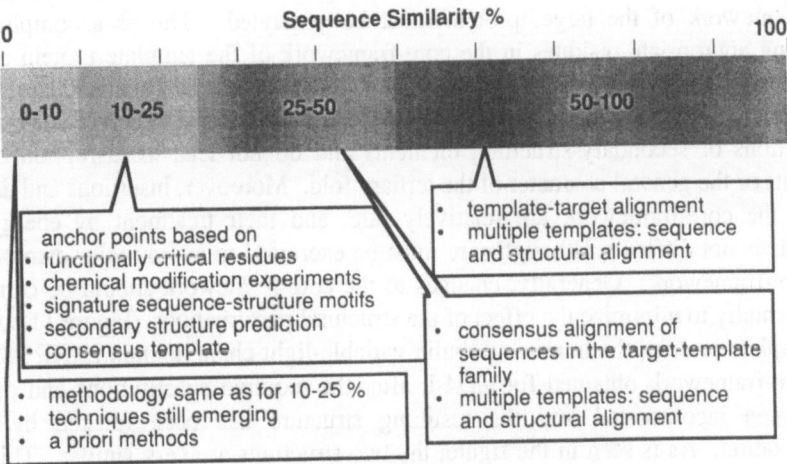

Figure 9. Summary of the key issues affecting sequence alignment as a function of the amount of sequence homology. See text for further details.

modification experiments on the template and target proteins also help in identifying appropriate anchor points, but such data may not always be available. Another useful criterion that can be applied to target-template alignment is that insertions, deletions, and non-conservative mutations should be confined to turn and loop regions, while conserved hydrophobic residues should be confined to the core-framework.

When sequence homology falls below 25% one is entering what Doolittle [39] has called the "twilight zone". In this region sequence alignment is error prone, and a variety of tools in addition to those shown in Figure 9 need to be employed to identify a sufficient number of anchor points to insure proper structural correspondence between the sequences. These tools include spectroscopic estimation of the amount of secondary structure [40-42], predicted location of secondary-structure elements along the main chain [6,7,43-45], and determination of sequence-structure motifs [45]. Finally, when sequence homology drops below 10% *a priori* methods are generally required, except in rare cases where it is possible to align the target and template sequences.

While it is generally possible in cases of low sequence homology to obtain a target-template sequence alignment, it may be difficult to obtain an unambiguous alignment. In order to avoid the consequences of an improper alignment, it is advisable to construct several models based on a number of reasonable alignments. Models produced in this way can then be evaluated and compared in the validation step (see Section 3.6).

The necessity for developing several models based on alternate alignments to the same template was encountered in our work on CD4-I. Despite sequence homology of ~ 33% and well-established anchor points, a 30-residue segment in the middle of the target sequence could not be unambiguously aligned with the template. Thus, several alignments, illustrated in Figure 10, were considered in our work.

3.3. GENERATION OF CORE-FRAMEWORK

Once a working alignment of the target and template sequences has been obtained, the core-framework of the target protein must be generated. This is accomplished by replacing appropriate residues in the core-framework of the template protein and then taking care of any required insertions or deletions. As noted earlier [23,24], simple mutations in the core-framework tend only to perturb the relative positions and orientations of secondary-structural elements and do not lead to disruption of these elements or the general character of the tertiary fold. Moreover, insertions and deletions within the core-framework are relatively rare, and their treatment by energy-based methods is not difficult, although care must be exercised to avoid major disruptions of the core-framework. Generally, changes to the core-framework should be carried out incrementally to minimize the effect of the structural perturbations. Figure 11 compares the template protein, the immunoglobulin variable-light chain domain 2RHE [46], with the core-framework obtained for CD4-I, after the required substitutions and deletions have been incorporated and the resulting structure has been relaxed by energy minimization. As is seen in the figure, the two structures are very similar. This result is consistent with a key premise of homology-based model building that the core-framework of the target protein not differ significantly from that of its template (*vide*

```
2RHE    ESULTQPPSASGTPGQRUTISCTGSATDIG..SNSU

CD4#1   ......KKUULGKKGDTUELTCTASQKKS....IQF
CD4#2   ......KKUULGKKGDTUELTCTASQK.......KS
CD4#3   ......KKUULGKKGDTUELTCTASQKKSIQFHWKN


2RHE    IUYQQUPGKAPKLLIY...UNDLLPSGU....SDRFSASKSGT

CD4#1   HUKNSNQIKILGNQGS...FLTKGPSKLNDRADSRRSLUDQG.
CD4#2   IQFHUKNSNQIKILGNQGSFLTKGPSKLNDRADSRRSLUDQG.
CD4#3   SNQIKILGNQGSFLTK...GPSKLNDRA....DSRRSLUDQG.


2RHE    SASLAISGLESEDEADYYCARUNDS..UDEPGFGGTKLTUL.

CD4#1   NFPLIIKNLKIEDSDTYICEUEDQKEEUQLLUFG.LTANSDT.
CD4#2   NFPLIIKNLKIEDSDTYICEUEDQKEEUQLLUFG.LTANSDT.
CD4#3   NFPLIIKNLKIEDSDTYICEUEDQK..EEUQLLUFGLTANSDT
```

Figure 10. Three possible alignments of CD4-I with the target 2RHE immunoglobulin variable domain. The shaded boxes indicate identical or highly-conserved residues. Note the significant regions of low sequence homology, which make unique alignment difficult or, at best, subject to considerable uncertainty. This is especially true of the alignment in the second panel.

supra).

In dealing with substitutions or insertions it is necessary to determine the conformation of the amino acid sidechains. For substitutions, the appropriate sidechains are generally replaced keeping one or more χ angles fixed at their template values and using the most probable values observed for proteins in the PDB for the other sidechain angles [47-51]. Energy minimization can then be used to relieve any "strain" produced by unfavorable van der Waals contacts (*cf.* [52]). In the case of insertions all χ angle values must be supplied.

3.4. LOOP GENERATION

Another critical step in homology-based model-building is the generation of loop structures. The procedures used for this step can be roughly divided into two categories, *knowledge-based* and *energy-based*. Knowledge-based procedures [12,53-55] are fast and are capable of producing accurate results for short loops (4-6 residues) and for medium-sized loops which are members of recurring structural classes [54]. Such procedures rely on known structures in the PDB, and are currently restricted by the small size and limited structural diversity of the PDB. Short loops, which have an

152

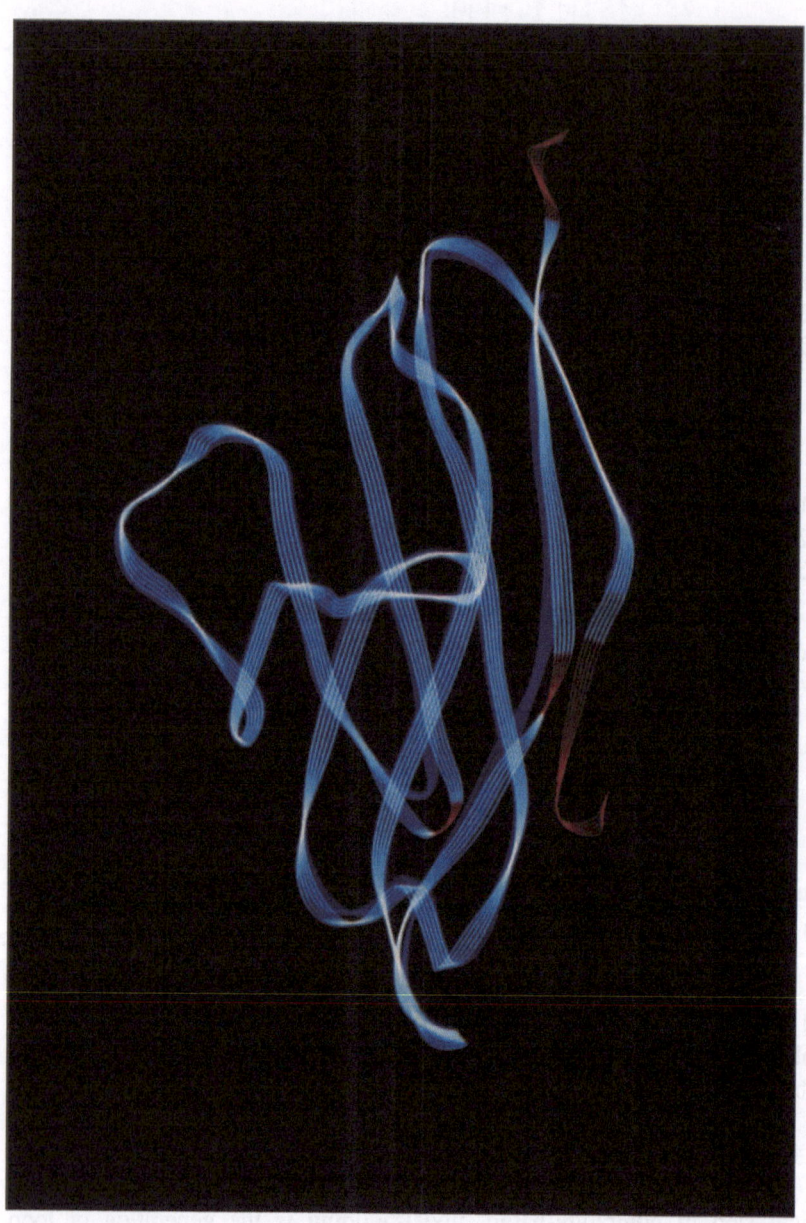

Figure 11. An example of a core-framework constructed for CD4-I after all required substitutions and deletions, compared to the 2RHE template, had been accounted for and energy minimization had been carried out to relieve any severe contacts. The dark blue ribbon represents the original 2RHE template structure; the light blue ribbon represents the CD4-I core framework model; and the red areas represent regions of the template where substitutions and deletions were required.

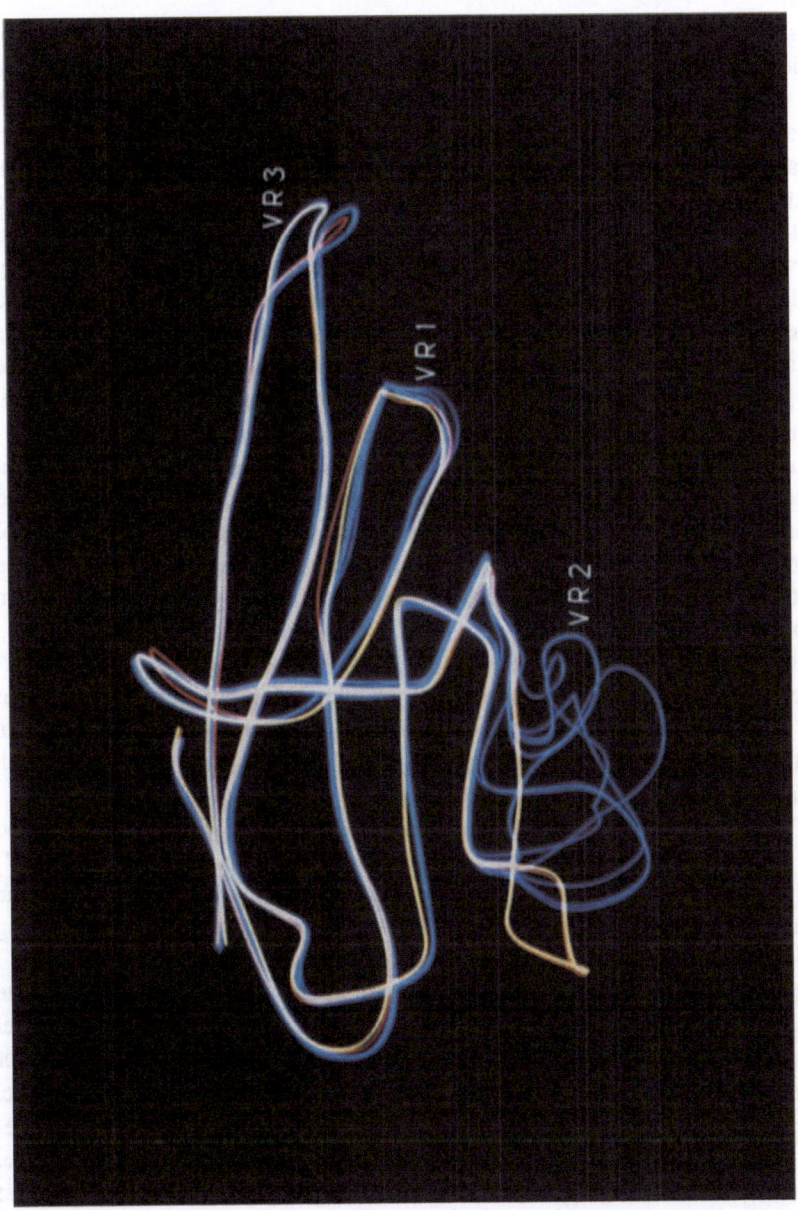

Figure 12. Thin-ribbon overlay of the nine lowest energy refined models of CD4-1: the lowest energy structure is in red, and the next lowest in yellow. Some of the structures are obscured due to their high degree of structural similarity to other structures in the set. This especially true of the red structure which is almost totally obscured by the yellow one in VR2.

intrinsically "smaller" conformational space and are more common, are better suited to this approach.

In contrast, energy-based procedures attempt to scan and evaluate the important regions of the conformational space available to the loop. Typically, conformational searching is carried out by systematic [56,57] or stochastic [58] procedures. Such procedures can generate reasonable structures for loops of up to about nine residues in length [56], and are not limited to loop conformations similar to those already known. Energy-based procedures have difficulty in treating larger loops due to the combinatoric nature of systematic conformational searching: these methods are compute-intensive and do not scale well. To reduce the computational requirements, conformational searching may be divided into two stages. In the first stage, the sidechains of all residues in the loop of interest except glycine and proline are truncated to their C_β carbons, and thus are represented as alanine residues. The conformational space available to these modified loops is then sampled. In the second stage, appropriate sidechains are added, and their conformations are optimized. In contrast to sidechain placement in the core-framework, in loops all sidechains must be treated simultaneously. To insure that appropriate low-energy conformations are obtained robust methods such as simulated annealing can be employed [59,60]. Moreover, since loops occur at the surface of proteins, interaction with the solvent should also be accounted for. Due to the complexity of treating solvent explicitly, several approximate methods have been developed [61,62].

In practice, most methods are neither purely knowledge-based nor purely energy-based. Knowledge-based methods typically rely on energy minimization or molecular dynamics to arrive at a final structure; energy-based methods typically incorporate heuristics into the searching algorithm. A promising approach which combines both methods synergistically has been reported [63].

In our CD4-I studies, knowledge-based approaches were unable to provide good matches for the targeted loops. Instead, an energy-based Monte Carlo approach [64] was used to generate several hundred backbone conformations for each loop in the absence of the other loops. Sidechains for each loop conformation were optimized using simulated annealing including an approximate treatment of solvent [61].

Although considerable effort has been devoted to the development of accurate methods for calculating the structure of loops, in many cases of homology-based model building their structure is less important than that of the remainder of the protein. For example, in the case of enzymes it is the active-site region that is of critical importance. As the target and template enzymes usually belong to the same family, their core-frameworks will most likely be quite similar. In such cases, homology-based model building generally leads to structures of suitable accuracy for many types of molecular modeling studies (see *e.g.* [65]).

3.5. FINAL STRUCTURE REFINEMENT

The loop generation stage provides a range of possible conformations for each loop. If more than one loop must be modeled, combinations of loops must be considered. In refining the structure, the remainder of the protein is allowed to interact with each loop, or set of loops. When extensive sampling such as that described for energy-based procedures has been used, then energy minimization may suffice. In the case of knowledge-based procedures, conformational space may not be adequately represented in the database, and subsequent refinement using molecular dynamics (see *e.g.* [66]) may be required. Often, this refinement will distinguish the lowest energy, and thus most probable, protein structure. For example, final refinement of several combinations of the key three loops in our CD4-I model resulted in a 20 kcal/mol energy difference between the two lowest energy structures. Figure 12 shows a thin ribbon representation comparing the nine lowest energy structures obtained for CD4-I. The red ribbon represents the lowest energy structure and the yellow, the next lowest. Note the similarity of the three loops in the two lowest-energy structures, and the significant divergence in the region of Loop 2 among the higher-energy structures.

3.6. STRUCTURE VALIDATION

Models developed by homology-based model building must then be evaluated for consistency with known properties of proteins and available experimental data. Such measures as solvent accessibility, residue-packing densities, location of hydrophobic or charged/polar residues, electrostatic and solvation free energies have been employed [37,67,68]. In addition, molecular dynamics can be used to explore whether the predicted structure resides in a stable minimum, or whether it resides in a shallow local minimum. Useful experimental data include that obtained from epitope mapping, site-specific mutagenesis studies, and affinity-labelling chemical modification experiments. There does not, however, exist any totally satisfactory approach to structure validation, and each protein must be handled on its own terms. In addition, the process of validation may be part of a larger "iterative cycle" involving prediction, validation, and corrections to the prediction (see Figure 8).

4. Closing Comments

Considerable progress has been made over the last several years in the development of theoretical methods to predict the 3-D structures of proteins. Nevertheless, a reliable and robust *a priori* method is currently unavailable. At present only homology-based methods have produced good structures for a number of proteins (see Table 4 of [69]), although this approach is not without limitations. In this regard, a serious problem is the relatively limited variety of template proteins in important protein families. For example, only one structure of any DNA or RNA polymerase currently exists, namely *E. coli* Pol I [35], and only the atomic coordinates of the α-carbons are available. The need for robust methods for the alignment of template and target protein sequences under conditions of low sequence homology, and the generation of reliable 3-D structures of large loops also remains unmet with a few exceptions. Finally, improved potential energy functions and computationally efficient methods for treating solvent

156

effects are needed. As these needs are met and as the size and diversity of the PDB grows, homology-based modeling of protein structure will be significantly improved in terms of the range of proteins it can treat and in the accuracy with which it can treat them. However, with the growing interest in protein-structure prediction and with the rapidly increasing speed of computers, it is not unreasonable to assume that *a priori* structure predictions may become sufficiently accurate to complement experimental and homology-based model building methods of structure determination before the turn of the century.

References

[1]   Dill, K.A. *Biochemistry* **1990**, *29*, 7133.
[2]   Go, N. *Ann. Rev. Biophys. Bioengineer.* **1983**, *12*, 183.
[3]   Skolnick, J.; Kolinski, A. *Ann. Rev. Phys. Chem.* **1989**, *40*, 207.
[4]   *Prediction of Protein Structure and the Principles of Protein Conformation*, Fasman, G.D., Ed.; Plenum Press: New York, 1989.
[5]   Maggiora, G.M.; Mao, B.; Chou, K.C.; Narasimhan, S.L. in *Methods of Biochemical Analysis, Vol. 35*, Suelter, C., Ed.; John Wiley & Sons: New York, in press.
[6]   Schultz, G.E.; Schirmer, R.H. *Principles of Protein Structure*; Springer-Verlag: New York, 1979.
[7]   Robson, B.; Garnier, J. *Introduction to Proteins and Protein Engineering*; Elsevier: Amsterdam, 1986.
[8]   Gibson, K.; Scheraga, H.A. in *Structure and Expression, Vol. 1: From Proteins to Ribosomes*, Sarma, R.H.; Sarma, M.H., Eds.; Adenine Press: Guilderland, New York, 1988; p. 67.
[9]   Skolnick, J.; Kolinski, A. *J. Mol. Biol.* **1989**, *212*, 787; Sikorski, A.; Skolnick, J. *J. Mol. Biol.* **1990**, *212*, 819; *ibid*, *215*, 183; Skolnick, J.; Kolinski, A. *Science* **1990**, *250*, 1121.
[10]   Wilson, C.; Doniach, S. *Proteins* **1989**, *6*, 193.
[11]   Crippen, G.M.; Snow, M.E. *Biopolymers* **1990**, *29*, 1479.
[12]   Blundell, T.L.; Sibanda, B.L.; Sternberg, M.J.E.; Thornton, J.M. *Nature* **1987**, *326*, 347.
[13]   Greer, J. *Proteins* **1990**, *7*, 317.
[14]   Wagner, C.R.; Benkovic, S.J. *Trends Biotech.* **1990**, *8*, 263.
[15]   Erickson, J.; Neidhart, D.J.; VanDrie, J.; Kempf, D.J.; Wang, X.C.; Norbeck, D.W.; Plattner, J.J.; Rittenhouse, J.W.; Turon, M.; Wideburg, N.; Kohlbrenner, W. E.; Simmer, R.; Helfrich, R.; Paul, D.A.; Knigge, M. *Science* **1990**, *249*, 527.
[16]   Littman, D.R. *Ann. Rev. Immunol.* **1987**, *5*, 561.
[17]   Thornton, J.M.; Gardner, S.P. *Trends Biosci.* **1989**, *14*, 300.
[18]   Ramachandran, G.N.; Ramakrishnan, C.; Sasisekharan, V. *J. Mol. Biol.* **1963**, *7*, 95.

[19]   Levitt, M. *J. Mol. Biol.* **1976**, *104*, 59.

[20]   Chothia, C. *Ann. Rev. Biochem.* **1984**, *53*, 537.

[21]   Alber, T. in *Prediction of Protein Structure and the Principles of Protein Conformation*, Fasman, G.D., Ed., Plenum Press: New York, 1989; chpt. 5.

[22]   Matthews, B.W. *Biochemistry* **1987**, *26*, 6885.

[23]   Chothia, C.; Levitt, M.; Richardson, D. *J. Mol. Biol.* **1981**, *145*, 215; Chothia, C.; Janin, J. *Proc. Nat. Acad. Sci. USA* **1981**, *78*, 4146; Chothia, C.; Janin, J. *Biochemistry* **1982**, *21*, 3955.

[24]   Janin, J.; Chothia, C. *J. Mol. Biol.* **1980**, *143*, 95.

[25]   Cohen, F.E.; Richmond, T.J.; Richards, F.M. *J. Mol. Biol.* **1979**, *132*, 275.

[26]   Cohen, F.E.; Sternberg, M.J.E.; Taylor, W.R. *J. Mol. Biol.* **1981**, *148*, 253; *ibid* **1982**, *156*, 821.

[27]   Chou, K.C.; Nemethy, G.; Scheraga, H.A. *Accts. Chem. Res.* **1990**, *23*, 134.

[28]   Levitt, M.; Chothia, C. *Nature* **1976**, *261*, 552.

[29]   Richardson, J.S. *Adv. Prot. Chem.* **1981**, *34*, 167.

[30]   Leszczynski, J.F.; Rose, G.D. *Science* **1986**, *234*, 849.

[31]   Sutcliffe, M.J.; Haneef, I.; Carney, D.; Blundell, T.L. *Prot. Engineer.* **1987**, *1*, 377.

[32]   Bajaj, M.; Blundell, T.L. *Ann. Rev. Biophys. Bioengineer.* **1984**, *13*, 453.

[33]   Bordo, D.; Argos, P. *J. Mol. Biol.* **1990**, *211*, 975.

[34]   Bowie, J.U.; Reidhaar-Olsen, J.F.; Lim, W.A.; Sauer, R.T. *Science* **1990**, *247*, 1306; Bowie, J.U.; Clarke, N.D.; Pabo, C.O.; Sauer, R.T. *Proteins* **1990**, *7*, 257.

[35]   Ollis, D.L.; Brick, P.; Hamlin, R.; Xuong, N.G.; Steitz, T.A. *Nature* **1985**, *313*, 762.

[36]   Basu, A.; Modak, M.J. *Biochemistry* **1987**, *26*, 1704; Basu, A.; Tirumalai, R.S.; Modak, M.J. *J. Biol. Chem.* **1989**, *264*, 8746.

[37]   Reid, L.S.; Thornton, J.M. *Proteins* **1989**, *5*, 170.

[38]   Collins, J.F.; Coulson, A.F.W. in *Nucleic Acid and Protein Sequence Alignment: A Practical Approach*, Bishop, M.J.; Rawlings, C.J., Eds.; IRL Press: Oxford, 1987; chpt. 13.

[39]   Doolittle, R.F; Feng, D.-F.; Johnson, M.S.; McClure, M.A. *Cold Spring Harbor Symp. Quant. Biol.* **1986**, *51*, 447.

[40]   Manning, M. *J. Pharm. Biomed. Anal.* **1990**, *7*, 1103.

[41]   Sarver, R.W.; Kreuger, W.C. *Anal. Biochem.*, submitted.

[42]   Bussian, B.M.; Sander, C. *Biochemistry* **1989**, *28*, 4271.

[43]   Schulz, G.E. *Ann. Rev. Biophys. Biophys. Chem.* **1988**, *17*, 1.

[44]   Lim, V.I. *J. Mol. Biol.* **1974**, *88*, 873.

[45]   Taylor, W.R. *Prot. Engineer.* **1988**, *2*, 77.

[46]   Furey, W.; Wang, B.C.; Yoo, C.S.; Sax, M. *J. Mol. Biol.* **1983**, *167*, 661.

[47]   Jones, T.A.; Thirup, S. *EMBO J.* **1986**, *5*, 819.

[48]   McGregor, M.J.; Islam, S.A.; Sternberg, M.J.E. *J. Mol. Biol.* **1987**, *198*, 295.

[49]   Summers, N.L.; Carlson, W.D.; Karplus, M. *J. Mol. Biol.* **1987**, *196*, 175.

[50]   Sutcliffe, M.J.; Hayes, F.R.F.; Blundell, T.L. *Prot. Engineer.* **1987**, *1*, 385.

[51]   Ponder, J.W.; Richards, F.M. *J. Mol. Biol.* **1987**, *193*, 775.

158

[52]   Summers, N.L.; Karplus, M. *J. Mol. Biol.* **1989**, *210*, 785.

[53]   Chothia, C.; Lesk, A.M. *J.Mol.Biol.* **1987**, *196*, 901.

[54]   Tramantano, A.; Chothia, C.; Lesk, A.M. *Proteins* **1989**, *6*, 382.

[55]   Blundell, T.; Carney, D.; Gardner, S.; Hayes, F.; Howlin, B.; Hubbard, T.; Overington, J.; Singh, D.A.; Sibanda, B.L.; Sutcliffe, M. *Eur. J. Biochem.*, **1988**, *172*, 513.

[56]   Bruccoleri, R.E.; Karplus, M. *Biopolymers* **1987**, *26*, 1987.

[57]   Moult, J.; James, M.N.G. *Proteins*, **1986**, *1*, 146.

[58]   Shenkin, P.S.; Yarmush, D.L.; Fine, R.M.; Wang, H.; Levinthal, C. *Biopolymers* **1987**, *26*, 2053; Fine, R.M.; Wang, H.; Shenkin, P.S.; Yarmush, D.L.; Levinthal, C. *Proteins* **1986**, *1*, 342.

[59]   Kirkpatrick, S.; Gellatt, Jr., C.D.; Vecchi, M.P. *Science,* **1983**, *220*, 671.

[60]   Nilges, M.; Gronenborn, A.M.; Brunger, A.T.; Clore, G.M. *Prot. Engineer.* **1988**, *2*, 27.

[61]   Hasel, W.; Hendrickson, T.F.; Still, W.C. *Tet. Comp. Meth.* **1988**, *1*, 103.

[62]   Oobatake, M.; Nemethy, G.; Scheraga, H. *Proc. Natl. Acad. Sci. USA* **1987**, *84*, 3086.

[63]   Martin, A.C.R.; Cheetham, J.C.; Rees, A.R. *Proc. Natl. Acad. Sci. USA* **1989**, *86*, 9268.

[64]   Chang, G.; Guida, W.C.; Still, W.C. *J. Am. Chem. Soc.* **1989**, *111*, 4379.

[65]   Weber, I.T.; Miller, M.; Jaskolski, M.; Leis, J.; Skalka, A.M.; Wlodawer, A. *Science* **1989**, *243*, 928; Weber, I.T. *Proteins* **1990**, *7*, 172.

[66]   Bates, P.A.; McGregor, M.J.; Islam, S.A.; Sattenau, Q.; Sternberg, M.J.E. *Prot. Engineer.* **1989**, *3*, 13.

[67]   Lesk, A.M.; Chothia, C.. *Phil. Trans. R. Soc. Ser A* **1986**, *317*, 345.

[68]   Novotny, J.; Rashin, A.A.; Bruccoleri, R.E. *Proteins* **1988**, *4*, 19.

[69]   Barlow, D.J.; Perkins, T.D.J. *Natural Product Reports* **1990**, *7*, 311.

UNDERSTANDING CHEMICAL REACTIVITY THROUGH THE INTERSECTING-STATE MODEL

SEBASTIÃO J. FORMOSINHO
Chemistry Department
University of Coimbra
3049 Coimbra Codex
Portugal

ABSTRACT. A novel theoretical model for the understanding of chemical reactivity is presented. This intersecting-state model allows the estimation of the energy barrier of elementary reactions in terms of reaction energy, force constants and lengths of the reactive bonds, bond order of the transition state, $n^{\ddagger}$, and a parameter λ associated with the energy distribution in the molecular products. For a synchronous reaction $A+B-C \rightarrow A-B+C$ where the resonance effects at the transition state can be neglected $n^{\ddagger}=1/2$; significant resonance effects are present when there are mobile electrons in the reactants and are responsible for higher $n^{\ddagger}$ values. The model has been applied to the study of hydrogen, hydride, proton and methyl transfers, free-energy relationships, sigmatropic shifts and electron transfer reactions and new perspectives for chemical reactivity have been found. It is shown that ISM encompasses several current models of chemical reactivity and constitutes a useful interpretative method for organic reactions.

1. Introduction

For a long time the chemists have tended to believe that the ability of molecules to react is predetermined by their own structural properties, although the encounter of two molecules should add unique features which neither molecule possesses alone. Despite this shortcoming the notion that molecules contain all the information necessary to understand their reactivity has proved extremely useful [1]. For deeper understanding it is necessary to formulate generalizations through conceptual frameworks which will guide our thinking.

The utility of rate constants for understanding chemical reactivity depends largely on interpreting them in terms of reaction energies and molecular structure. Although a full explanation of chemical reactivity is an exercise on the form of potential energy surfaces (PES), such elaborate calculations are often mysterious workings which may fail to unravel the trends of reactivity within families of related molecules. So to understand reactivity one needs to be guided by simple models; simple means mathematically tractable such that the "moving parts" of the model can be open to inspection, and means also clear concepts, full of physical insight [2]. The first level of interpretation for understanding chemical reactivity is intimately connected with the Transition State Theory (TST), which assumes that in a chemical reaction the molecules pass through a configuration of maximum potential energy, called the transition state. Under the assumption that the transition state complexes are in equilibrium with the reactants, TST shows that the rate constants for adiabatic reactions can be expressed by

159

S. J. Formosinho et al. (eds.), Theoretical and Computational Models for Organic Chemistry, 159–205.
© 1991 *Kluwer Academic Publishers.*

$$k = (k_B T / h) c_0^{1-m} \exp(-\Delta G^{\ddagger} / RT) \qquad (1)$$

where $\Delta G^{\ddagger}$ is the free energy of activation, m is the molecularity of the reaction and c_0 is a standard concentration ($c_0 = 1$ mol dm^{-3}); the other symbols have their usual meaning.

The second level of interpretation is not concerned directly with rate constants, but with reaction energy barriers, $\Delta G^{\ddagger}$. For many reactions it has been known that a sufficiently close approximation to the reaction path can be obtained by constructing two intersecting potential energy curves, one for the reactants and another for the products, which associated with a geometrical criterion for the configuration of the activated complexes can provide a fairly good approximation to the energy of the transition state [3]. The purpose of this paper is to show that a theoretical framework can be developed along those lines and in such a way that it can provide an unifying view for chemical reactivity, permitting the identification of several connections among distinct reactions not evident through the current models of chemical reactivity.

2. The Intersecting-State Model

2.1 THE REACTION ENERGY PROFILE

Consider the prototype reaction $A + BC \rightarrow AB + C$ for which the potential energy surface is schematically shown in Figure 1. The course of this reaction corresponds to a movement on the surface as indicated by the minimum energy path shown by the dashed line. Reactants and products are topographically associated with valleys which meet at the saddle point considered here to represent the transition state. An alternative way to find the energy of the transition state is to view the BC and AB motions as independent. This nonadiabatic path is illustrated in Figure 1 by the full straight line: first the BC bond elongates up to the transition state value, $l^{\ddagger}_{BC}$, and then, keeping this length fixed, the distance between the atoms A and B diminishes from infinity to the transition state value, $l^{\ddagger}_{AB}$, virtually through an isoenergetic path. A similar path can be construct from the transition state to the products. Consequently, the overall reaction path for the prototype reaction can be interpreted in terms of independent bond-breaking and bond-forming processes. Thus in the same diagram (Figure 2) one can represent the change in energy for breaking of BC and the change in energy for forming of AB. The true energy reaction path is a continuous one and can be obtained from the two potential energy curves of BC and AB by taking into account the electronic reorganization at the crossing point, i.e., the resonance energy between the initial and final states. For the time being we will neglect the contribution of the resonance energy, ε, at the crossing point of the potential energy curves (the transition state), the so called "diabatic method" of Evans. We will assume only that $\varepsilon > k_B T$, so that the reaction can be considered an adiabatic process.[#]

[#] For a nonadiabatic reaction eq(1) must be multiplied by an electronic transmission factor $\kappa < 1$.

As shown in Figure 2, the horizontal distance between the minima of the potential energy curves, d, represents the sum of the bond extensions to the transition state

$$d = (l_{BC}^{\ddagger} - l_{BC}) + (l_{AB}^{\ddagger} - l_{AB}) \qquad (2)$$

where l_{BC} and l_{AB} are the equilibrium bond lengths of BC and AB and $l^{\ddagger}$ is the bond length at the transition state. The parameter d also represents the distance travelled by the in-flight atom B during the course of the reaction.

The concept of chemical bond order, n, is of great utility to describe quantitatively the course of a chemical reaction [4], because it transforms an infinite path (interatomic distances) to a finite path extending from the reactants, $n=0$, to the products, $n=1$. By

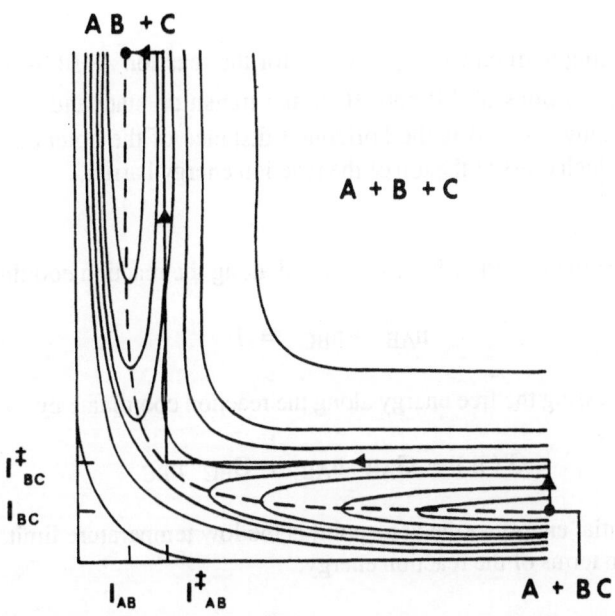

Figure 1. Potential energy surface for the reaction $A + BC \rightarrow AB + C$: ---- minimum energy path; ——— nonadiabatic path equivalent to the intersecting potential energy curve diagram of Figure 2 (reproduced with permission from *Educ. Chem.* **1989**, *26*, 118).

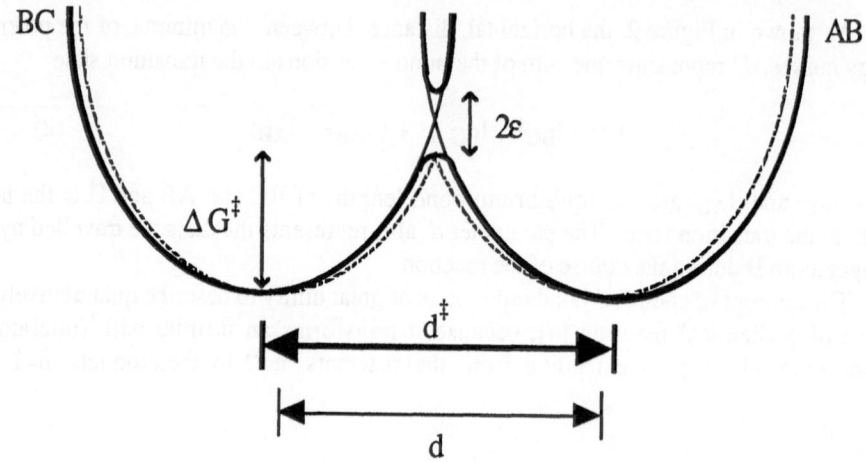

Figure 2. Intersecting harmonic energy curves for the reaction A+BC→AB+C; $d^{\ddagger}$ is the sum of the bond extensions of AB and BC at the transition state and ε is the resonance energy at the crossing point; d is the horizontal distance of the potential energy curves (diabatic curves) which cross at the top of the reaction energy barrier.

assuming that the sum of bond order is conserved along the reaction coordinate, for single bonds,

$$n_{AB} + n_{BC} = 1 \tag{3}$$

Let us start by expressing the free energy along the reaction coordinate by

$$G(n) = G_{AB}\, n_{AB} + G_{BC}\, n_{BC} \tag{4}$$

of which the potential energy at the zero-point is the low temperature limit. This equation can be expressed in terms of the reaction energy,

$$\Delta G = G_{AB} - G_{BC} \tag{5}$$

and

$$G(n) = n\, \Delta G + G_{BC} \tag{6}$$

where

$$n = n_{AB} = 1 - n_{BC} \tag{7}$$

Eq(6) can be simplified by measuring the free energy with respect to the ground state of reactants, $\Delta G(n) = n\, \Delta G$ $(G_{BC}=0)$. This equation represents the "thermodynamic contribution" to the reaction energy barrier. To this one should add an "intrinsic kinetic" contribution which Agmon and Levine [5a] expressed, as in ordinary statistical mechanics, by an additional entropy of mixing, M(n), given by the Shannon function [5b],

$$M(n) = -n_{AB}\, \ln n_{AB} - n_{BC}\, \ln n_{BC} \qquad (8)$$

and, in consequence, one has a generalized free energy function

$$G(n) = n\, \Delta G + \lambda\, M(n) \qquad (9)$$

where λ ($\lambda > 0$) is a parameter with the dimensions of an energy.

One can proceed from such an expression to estimate the energy of the transition states, but this kind of equations are known to have some limitations at very low reaction energies and assume implicitly symmetric potential energy curves for reactants and products [4c]. Nevertheless, they have been found to provide good estimates for the reaction energy profile of many chemical reactions and, consequently, can be a good basis to find a geometrical criterion for the configuration of the activated complexes. This can be open for inspection by comparison with PES calculations for simple elementary reactions.

To find the position of the transition state one has to differentiate $G(n)$ with respect to n and by equating to zero,

$$\partial G(n)/\partial n = \Delta G - \lambda\, \ln\,[\,n/(1-n)\,] = 0 \qquad (10)$$

one can then obtain the bond order at the transition state

$$n(\Delta G)^{\ddagger} = [\,1 + \exp(-\Delta G/\lambda)\,]^{-1} \qquad (11)$$

From eq(7) one gets

$$n(\Delta G)^{\ddagger}_{AB} = [\,1 + \exp(-\Delta G/\lambda)\,]^{-1} \qquad (12a)$$

and

$$n(\Delta G)^{\ddagger}_{BC} = 1 - [\,1 + \exp(-\Delta G/\lambda)\,]^{-1} \qquad (12b)$$

To relate bond order with bond lengths, $\underline{l}$, one can employ the empirical relation proposed by Pauling [6], which has been shown [4b] to be a consequence of assuming the simplest possible description for bond dissociation, i.e., the simplest combination of a repulsive and an attractive function leading to an energy minimum

$$n = \exp\left[-(1-l_0)/a\right] \qquad (13)$$

By substituting eqs(12) into eq(13) one obtains

$$l^{\ddagger}_{AB} - l_{AB} = a \ln\left[1 + \exp(-\Delta G/\lambda)\right] \qquad (14a)$$

$$l^{\ddagger}_{BC} - l_{BC} = -a \ln\left\{1 - \left[1 + \exp(-\Delta G/\lambda)\right]^{-1}\right\} \qquad (14b)$$

Hence, the sum of the bond extensions at the transition state can be obtained by substituting eqs(14) into eq(2),

$$d = a \ln\frac{\left[1 + \exp(-\Delta G/\lambda)\right]}{\left\{1 - \left[1 + \exp(-\Delta G/\lambda)\right]^{-1}\right\}} \qquad (15)$$

In eq(13) a is a constant assumed to be universal, $a = 0.265$ Å. However, this assumption is probably not valid. Agmon [6b] has considered, as a first approximation, a linear dependence of a on l for the generation of reaction coordinates on PES. In fact, if the equilibrium bond length of a molecule B-C or A-B is small its extension to the transition state is expected to be small and if the bond length is large the bond extension is also expected to be large. Thus in eq(15) we will scale a by the sum of the equilibrium bond lengths of AB and BC,

$$d = a' (l_{AB} + l_{BC}) \ln\frac{\left[1 + \exp(-\Delta G/\lambda)\right]}{\left\{1 - \left[1 + \exp(-\Delta G/\lambda)\right]^{-1}\right\}} \qquad (16)$$

where a′ is a dimensionless constant, found to be a′=0.156 [7]. By defining

$$\eta = d / (l_{AB} + l_{BC}) \qquad (17)$$

as the reduced bond extension, then

$$\eta = a' \ln\frac{\left[1 + \exp(-\Delta G/\lambda)\right]}{\left\{1 - \left[1 + \exp(-\Delta G/\lambda)\right]^{-1}\right\}} \qquad (18)$$

Equation (18) is independent of the sign of ΔG (or λ), because $\eta(\Delta G) \equiv \eta(-\Delta G)$; this ensures that microscopic reversibility is obeyed. By expanding eq(18) in a Taylor series, one gets

$$\eta = \eta(0) + (\partial\eta / \partial\Delta G)_0 \, \Delta G + (1/2) \, (\partial^2\eta / \partial^2\Delta G)_0 \, (\Delta G)^2 + ... \quad (19)$$

with

$$\eta(0) = 2 \, a' \, \ln 2 \; ; \quad (\partial\eta / \partial \Delta G)_0 = 0 \; ; \quad (\partial^2\eta / \partial^2\Delta G)_0 = a' / \lambda^2 \; .$$

Therefore, η should show to a good approximation a linear relationship with $(\Delta G)^2$

$$\eta(\Delta G) \approx 2 \, a' \, \ln 2 + (a' / 2) \, (\Delta G / \lambda)^2 \quad\quad\quad (20)$$

2.2. "INCREASED-VALENCE" STRUCTURES FOR TRANSITION STATES

Let us consider now the case where the resonance effect at the transition state is significant. In that respect it is relevant to discuss some aspects of the resonance between Lewis structures. For example, for N_2O none of the Lewis structures used to represent the octet structure of this molecule, including "long-bonds" between pairs of adjacent atoms, such as

$$|N{\equiv}N{-}\underline{\overline{O}}| \leftrightarrow ... \leftrightarrow |\overset{\frown}{N{=}N}{-}\underline{\dot{O}}|$$

can alone account for the similarity of the NN and the NO bond lengths (1.13 and 1.19 Å) to those of triple and double bonds (1.10 and 1.20 Å) respectively. Resonance between the most stable of these structures can account for the observation.

Resonance has an effect on bonding which can be illustrated by referring to valence bond structures ($H^+ H^{\cdot}$) and ($H^{\cdot} H^+$) for one electron H_2^+. The individual structures alone do not have a bonding electron. However, resonance between them generates a 1-electron bond, i.e., $H^+ H^{\cdot} \leftrightarrow H^{\cdot} H^+ \equiv (H \cdot H)^+$. Harcourt [8] has developed a method for writing down all the important Lewis octet structures, together with some indication of the effect on bonding of the resonance between these VB structures, by generating "increased-valence" structures. These structures are particularly relevant for four electrons distributed among atomic orbitals of three atoms and, consequently, they are appropriate to describe transition states, although alternative MO theories can also be used.

To construct a VB structure for a 4-electron 3-centre bonding unit, with a Pauling "3-electron bond" as a component, Harcourt [8] starts by indicating the electron spins (x spin α; o spin β) as shown

$$\overset{\text{o} \quad\; \text{o}}{B \; x \; C} \qquad\qquad \overset{\text{x} \quad\; \text{x}}{B \; o \; C}$$
$$\quad (I) \qquad\qquad\qquad (II)$$

Then an atom A is introduced with one unpaired-electron, whose spin is opposed to that of the electron located in a B-atom atomic orbital, to give

$$\overset{\text{x}}{A} \quad \overset{\text{o}}{B} \overset{\text{o}}{\text{x} C} \qquad \overset{\text{o}}{A} \quad \overset{\text{x}}{B} \overset{\text{x}}{\text{o} C}$$

$$\text{(III)} \qquad\qquad \text{(IV)}$$

If the A- and B-atom atomic orbitals overlap appreciably, we may represent A and B as bonded together to give VB structure (V) in which the electron spins are not specified

$$A - B \cdot \dot{C} \quad \equiv \quad A - B \quad \ddot{C} \quad \leftrightarrow \quad \dot{A} \quad \ddot{B} \quad \dot{C}$$

$$\text{(V)} \qquad\qquad \text{(VI)} \qquad\qquad \text{(VII)}$$

VB structure (V) summarizes resonance between structures (III) and (IV) or is equivalent to invoke resonance between the Lewis structures (VI) and (VII), each with an electron-pair bond and a lone pair of electrons.

These considerations imply that we may generate an "increased-valence" structure (V) from the standard Lewis structure (VI) by delocalizing one electron of the lone-pair of atom C into a vacant bonding BC orbital

$$A - B \overset{\frown}{} \ddot{C} \quad \rightarrow \quad A - B \cdot \dot{C}$$

By doing so, we increase the number of electrons that can participate in the overall bonding.

For N_2O one can apply the same procedure by delocalizing π and π -electrons in the adjacent NO region

$$: N \equiv N \overset{\frown}{\underset{\smile}{-}} \ddot{O} : \quad \rightarrow \quad : N \equiv N \dot{+} \dot{O} :$$

generating an "increased-valence" structure which reveals explicitly the triple-bond character of the NN bond and the double character of the NO bond. Obviously, one could use an older type of increased valence structure $|N \equiv N = \underline{O}|$, but this does not conform with the Lewis rules for first-row atoms, which assumed that only one 2s and three 2p orbitals of these atoms are used for bonding.

2.3. THE RESONANCE ENERGY

As shown in Figure 2, when the resonance energy at the transition state is significant the reaction energy barrier is significantly decreased, although we can always find a parameter d which reproduces the experimental energy barrier. Obviously, such a parameter no longer represents the transition state bond extensions.

As Evans and Warhurst [9] have discussed in detail, strong resonance effects occur when there are mobile electrons which by resonance decrease the energy barrier of the transition state. One way to deal with this problem is to characterize the transition state in terms of a "transition state bond order" $n^{\ddagger}$ which is defined for the thermoneutral situation. For the reaction

$$A + B{\cdot\cdot}C \rightarrow \{A \cdot B \cdot C\}^{\ddagger} \rightarrow A{\cdot\cdot}B + C$$

with conservation of the total bond order (eq(3)), the transition state bond order is $n^{\ddagger} = 1/2$, at the thermoneutral limit, since, at $\Delta G=0$, $n(0)_{AB}=n(0)_{BC}=n^{\ddagger}$. This situation corresponds to the case where one can neglect the effect of the resonance energy.

However, when the atom B possesses mobile electrons a different electronic configuration is possible for the transition state,

$$A + \ddot{B}{\cdot\cdot}C \rightarrow \{A{\cdot\cdot}\ddot{B}{\cdot\cdot}C\}^{\ddagger} \rightarrow A{\cdot\cdot}\ddot{B} + C \qquad n^{\ddagger} = 1$$

which can be characterized by an "increased-valence" structure with a transition state bond order higher than one-half [7,10]. Now we need to relate $n^{\ddagger}$ with the parameter d. This can be carried out if one realizes that a $n^{\ddagger}>1/2$ is equivalent to the "nonconservation of the bond order" at the transition state,

$$n^{\ddagger}_{AB} + n^{\ddagger}_{BC} = m \qquad (21)$$

where m is, in principle, a natural number. For such a generalization it is convenient to divide the bond orders of the fragments AB and BC into subsets i=1 to m, with the partial bond orders $n^{\ddagger}_{AB}(i)$ still obeying eq(3), and

$$n^{\ddagger}_{AB} = \sum_{i=1}^{m} n^{\ddagger}_{AB}(i) \qquad (22)$$

Now, to find the contribution ($d(i)$) of each subset for the overall horizontal displacement of the potential energy curves, d, one has to ensure that for m=1 such a contribution is the sum of the bond extensions given by eq(16). Further, since the sum of the bond extensions tends to infinity when $n \rightarrow 0$, the simplest form that one expects for d is

$$d^{-1} = d(1)^{-1} + d(2)^{-1} + ... + d(m)^{-1} \qquad (23)$$

In fact, by considering all subsets i with zero $n^{\ddagger}(i)$ except the first one, eq(23) leads to a zero contribution for d from all subsets with i>1. By further assuming d(1)=d(2)=...=d(m), then

$$d = d(1) / m \qquad (24)$$

For the mixing entropy one would have a contribution of $\sum \lambda(i) \, M[n(i)]$, where each $M[n(i)]$ obeys eq(8). Then, by adding $\lambda(i)$ as vectors on a multidimensional ortogonal space one has

$$\lambda(m)^2 = \sum \lambda(i)^2 = m \, \lambda(i)^2 \qquad (25)$$

and, consequently, $\lambda(i) = \lambda(1)$ is given by

$$\lambda(1) = (1/m)^{1/2} \, \lambda(m) \qquad (26)$$

By substituting next eqs(24) and (26) into eqs(16)-(18) one gets,

$$\eta_m = \frac{a'}{m} \, \ln \frac{1 + \exp[\sqrt{m} \, \Delta G / \lambda(m)]}{1 - \{1 + \exp[\sqrt{m} \, \Delta G / \lambda(m)]\}^{-1}} \qquad (27)$$

and by expanding in a Taylor series

$$\eta = (2 \, a' \, \ln 2 / m) + (a'/2) \, (\Delta G / \lambda)^2 \qquad (28)$$

where the index m has been omitted in η and λ for simplicity. Since at $\Delta G = 0$ we have $m = 2n^{\ddagger}$, then eq(28) can be rewritten as

$$\eta = \frac{a' \, \ln 2}{n^{\ddagger}} + \frac{a'}{2} \left(\frac{\Delta G}{\lambda} \right)^2 \qquad (29)$$

Equation (29) generalizes eq(20) for situations where the resonance effect is significant and can be characterized by $n^{\ddagger} > 1/2$. However, eq(29) encompasses also the cases where the resonance effect is negligible, with $n^{\ddagger} = 1/2$.

2.4. COMPARISON WITH PES CALCULATIONS

Before pursuing the model further it is useful to compare the values of d estimated through the intersecting-state model (ISM) with the sum of bond extensions at the transition state, $d^{\ddagger}$, given by potential-energy surface calculations, for some elementary reactions. For the reaction $H + H_2 \rightarrow H_2 + H$ ($l_{HH} = 0.7416$ Å), d can be estimated through eq(29) with

$n^{\ddagger}=1/2$, a'=0.156 and ΔG=0. ISM provides a value of d(ISM)=0.320 Å which is ca. 15% lower than the PES calculations, $d^{\ddagger}$(PES)=0.374 Å [11]. The error is much smaller (2%) if instead of considering the length at the minimum of the potential energy curve, one considers the equilibrium bond length at zero-level [7].

Another important reaction to consider is $H_2 + Cl \rightarrow H + HCl$, because it is virtually a thermoneutral reaction and has $n^{\ddagger}$=1 [7]. PES calculations [12a] provide $d^{\ddagger}$ values in the range 0.364 - 0.454 Å which should be compared with the d values estimated through ISM (l_{HH}=0.7416 Å and l_{HCl}=1.2745 Å): d($n^{\ddagger}$=1) = 0.216 Å and d($n^{\ddagger}$=0.5) = 0.432 Å. As expected when the resonance effect is significant, d($n^{\ddagger}$=1) < $d^{\ddagger}$. However, it is gratifying that d($n^{\ddagger}$=0.5) is within the range of the $d^{\ddagger}$ values provided by PES calculations. The agreement is also good (8%) when compared with the Generalized Valence Bond computations of Dunning [12b], which provide good estimates (+0.015 Å) for the equilibrium bond lengths of diatomic molecules such as H_2 and HX; for the reaction under study the sum of the bond extensions is $d^{\ddagger}$=0.40 Å (GVB+1+2).

Although for this reaction the transition state bond order appears to increase due to a strong resonance effect, the "chemical bond order" is conserved along the reaction coordinate, in agreement with the findings of Lendvay [13]. Thus, ISM provides a reasonable estimate of the sum of bond extensions at ΔG=0 through the equation

$$d^{\ddagger}(n^{\ddagger}=0.5) = (2 a' \ln 2) \ (l_{AB} + l_{BC}) \qquad (30)$$

However, for the estimation of the reaction energy barriers one has to take into account the increase in the transition state bond order $n^{\ddagger}>1/2$ due to the resonance effect, and in such cases one might say that $n^{\ddagger}$ is not conserved.

2.5. REACTION DYNAMICS AND THE MIXING ENTROPY PARAMETER

Eq(29) reveals that reactions with $\Delta G \neq 0$ should have higher bond extensions than thermoneutral processes. This view is supported by PES calculations. PES calculations for the elementary reaction $F + H_2 \rightarrow FH + H$ lead to $d^{\ddagger}$ values ranging between 0.426-0.748 Å [12a]. Although these calculations provide too wide a range of values, the ISM estimation for the thermoneutral limit is clearly lower d($n^{\ddagger}$=0.5)=0.358 Å (l_{HF}=0.917 Å); Dunning [12b] estimates $d^{\ddagger}$=0.52 Å.

Let us now examine in more detail the effect of ΔG on η. A strongly exothermic reaction leads, in general, to the formation of vibrationally excited products AB^V. This corresponds to a AB bond extension which is significantly higher than if AB is formed only with translational energy (Figure 3). Therefore λ is smaller for the former situation and higher for the latter. In principle, this argument is also valid for endothermic reactions, because reaction cross sections increase steeply with increasing vibrational energy in the reactant bond [14]. However, one should realize that under current experimental conditions reactants are not prepared in selected vibrational levels, but are populated in a Boltzmann distribution.

Thus λ appears to be related with reaction dynamics. Mass effects and the shape of the inner repulsive wall (sudden or gradual PES) are particularly relevant in this respect [14]. A simple physical rationalization for the mass effect is that the B/C repulsion will force AB to vibrate if it sets in while the A-B bond is extended; but if A-B has reached its minimum bond distance, the B/C repulsion can only give rise to translation. The latter situation occurs if the atom A is light, i.e., if it moves fast [2]. The effectiveness of B/C repulsion in producing vibration in AB depends on the period, t_F, in which the repulsive force is significant. If such a period is long compared to the time of an oscillatory period of AB, t_v, the force will produce a relative motion of B with respect to A, that averages over several AB oscillations, to give no more than a small net change in internal motion (vibration) of AB. In consequence, this simple analysis reveals that a force of significant magnitude that operates over a brief period, $t_F \ll t_v$, will be efficient in translation-to-vibration transfer.

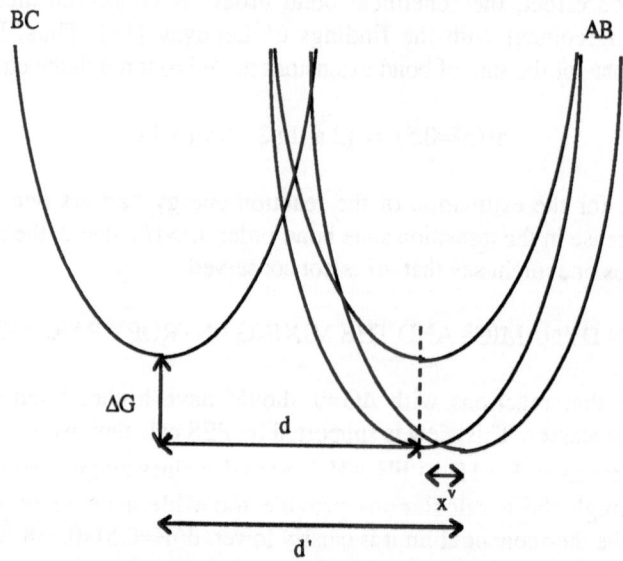

Figure 3. Intersecting energy curve diagram for an exothermic reaction; when AB is formed with translational energy d is independent of ΔG, but if AB is formed with vibrational energy a further bond extension, x^v, occurs and the sum of bond extensions, d', increases.

According to the above considerations and taking into account eq(29), one expects that for exothermic elementary reactions the product vibrational energy (on a reduced scale) should be proportional to ΔG^2. This has been verified experimentally [15] and can be accounted for by a dynamic model of "retreat coordinate" with a gradual repulsive force responsible for separating the products [2]. This view holds essentially for reactions in the vapour phase; in solution vibrational modes of solvent molecules adjacent to the reactive bonds can also accommodate some exothermicity.

2.6. REACTION ENERGY BARRIERS

Within the intersecting-state model, the reaction energy barrier is determined by the shape of the potential energy curves of AB and BC and the geometric criterion for the configuration of the transition state given by eqs(17) and (29).

For Morse oscillators expressed in terms of free energy

$$G_{BC} = D_{BC} \{1 - \exp[-\beta_{BC} (1 - l_{BC})]\}^2 \tag{31a}$$

and

$$G_{AB} = D_{AB} \{1 - \exp[-\beta_{AB} (1 - l_{AB})]\}^2 + \Delta G \tag{31b}$$

where l is the bond length, D_{XY} the dissociation energy of a diatomic molecule XY, and β the corresponding Morse parameter. Thus eqs(31) can be written as

$$1 - l_{BC} = - (1 / \beta_{BC}) \ln [1 - (G_{BC} / D_{BC})^{1/2}] \tag{32a}$$

and

$$1 - l_{AB} = - (1 / \beta_{AB}) \ln \{1 - [(G_{AB} - \Delta G) / D_{AB}]^{1/2}\} \tag{32b}$$

By substituting these eqs in eq(2) one gets

$$d = - (1/\beta_{AB}) \ln\{1 - [(\Delta G^{\ddagger} - \Delta G) / D_{AB}]^{1/2}\} - (1/\beta_{BC}) \ln[1 - (\Delta G^{\ddagger} / D_{BC})^{1/2}] \tag{33}$$

Within the harmonic approximation for the potential energy curves of AB and BC of force constants f_{XY}, one has

$$(1/2) f_{BC} (l^{\ddagger} - l_{BC})^2 = (1/2) f_{AB} [d - (l^{\ddagger} - l_{BC})]^2 + \Delta G \tag{34}$$

and

$$\Delta G^{\ddagger} = (1/2) f_{BC} (l^{\ddagger} - l_{BC})^2 \tag{35}$$

Most chemical reactions are one-bond process, involving the making and the breaking of one bond, or take place in distinct steps, each of one-bond type [16]. Multibond reactions cannot normally be concerted processes, because more energy is needed to break n-bonds as to break one. For example, if in one molecular reagent species the reaction requires the stretching of several chemical bonds of force constants f_i, one can estimate an effective force constant, f, through mechanical coupling by adding vectors which make an angle θ

$$f^2 = \sum (f_i^2 + 2 f_i f_j \cos \theta + f_j^2) \tag{36}$$

172

If the bonds couple in-phase (normal modes) $\theta=0°$ and

$$f = \sum f_i \qquad (37)$$

However, if one is dealing with independent local modes[#] then we take $\theta=90°$ and

$$f = (\sum f_i^2)^{1/2} \qquad (38)$$

It is clear that the local mode behaviour provides a lower energy path than the normal mode one.

Force constants are often estimated from spectroscopic data. Nevertheless, when this is not possible, empirical expressions can be employed; a very convenient one is that of Gordy [17]

$$f_{AB} = \alpha n\,(X_A\,X_B\,/\,l_{AB}^2)^{3/4} + \beta \qquad (39)$$

where α and β are constants, f_{AB} is the force constant of the AB bond, l_{AB} the corresponding length, X_A and X_B the electronegativities of the two atoms and n the bond order of the AB bond.

For a chemical reaction which involves two molecular species either as reactants or products, $AB + CD \rightarrow AC + BD$, the energy barrier is defined for one molecular species. Therefore one can write for the potential energy,

$$2\,E = (1/2)\,f_{AB}\,x_{AB}^2 + (1/2)\,f_{CD}\,x_{CD}^2 \qquad (40)$$

Assuming equal bond stretches, $x_{AB} = x_{CD}$, then the effective force constant is

$$f = (f_{AB} + f_{CD})\,/\,2 \qquad (41)$$

For multibond processes we take the effective bond length as the arithmetical mean of the bond lengths of the corresponding reactive bonds of reactants and products.

3. Comparison of ISM with current models of chemical reactivity

ISM encompasses, as particular cases, several current models of chemical reactivity such as the theory of Marcus [18], the BEBO [4a], Agmon and Levine [5a], Koeppl and Kresge [19] models and two dimensional models such as the ones of Kreevoy and Lee [20] and of Grunwald [21]. It is also in general accord with qualitative electronic theories of chemical

---

[#] The force constant of eq(38) corresponds to an effective bond stretch which is the geometric mean between the stretch for a single bond and the stretch required for all the modes in-phase, at the same potential energy.

reactivity such as the frontier orbital concepts and the orbital-symmetry rules of Woodward-Hoffmann [22], since a forbidden reaction has a zero transition state bond order, $n^{\ddagger}=0$, and an allowed reaction has $n^{\ddagger} \neq 0$.

Pross and Shaik [23] have proposed a qualitative valence-bond (VB) configuration model to describe how reaction energy profiles can be built from VB configurations. Structures for the transition states such as the ones previously presented can be considered within that VB-model. Yates [24] has recently made a comparative study of several intersecting state models, with respect to the problem of photochemical proton transfers, and has concluded that ISM is one of the most general.

3.1. MARCUS THEORY

Let us rewrite eq(34) in terms of the reactant bond extension, x, and the force constants for reactants, f_r, and products, f_p,

$$(1/2) f_r x^2 = (1/2) f_p (d - x)^2 + \Delta G \qquad (42)$$

When $f_r = f_p = f$,

$$x = (d / 2) + (\Delta G / f d) \qquad (43)$$

which, substituted into equation,

$$\Delta G^{\ddagger} = (1/2) f x^2 \qquad (44)$$

leads to

$$\Delta G^{\ddagger} = \Delta G_0^{\ddagger} [1 + (\Delta G / 4 \Delta G_0^{\ddagger})]^2 \qquad (45)$$

This is the equation of Marcus, where $\Delta G_0^{\ddagger}$ is the intrinsic kinetic energy barrier, i.e., the free energy of activation when $\Delta G \rightarrow 0$. This barrier is given by

$$\Delta G_0^{\ddagger} = (1/8) f d^2 \qquad (46)$$

with d independent of ΔG. According to eq(29) and eq(17), this is true when $\lambda >> |\Delta G|$ (in fact $\lambda >> (n^{\ddagger}/2 \ln2)^{1/2} |\Delta G|$). Thus, the theory of Marcus is a particular case of ISM when $f_r = f_p$ and $\lambda >> |\Delta G|$. These are too restrictive conditions for chemical reactions [19,24].

One problem inherent in any Marcus-type equations, based on parabolic potential energy curves, lies in their limiting behaviour at very high endo- or exo-thermicities: for very endothermic reactions the theory leads to $\Delta G^{\ddagger} = \Delta G$ and for very high exothermic processes, which avoid the so called "inverted region", there is a cut-off $\Delta G^{\ddagger} = 0$. Both forms of limiting behaviour have been criticized as being physically unrealistic [24,25]. However, this problem does not arises within ISM: $\Delta G^{\ddagger} \neq 0$ for very exothermic reactions as long as one does not neglect the role of the mixing entropy parameter, λ. The same can be said for the high endothermic limiting behaviour, owing to the square dependence of d on ΔG.

3.2. BEBO MODEL

Within the BEBO model, at the transition state, one can write in terms of free-energy,

$$D_{BC} - D_{BC}\, n_{BC}{}^p = D_{AB} - D_{AB}\, n_{AB}{}^p + \Delta G \tag{47}$$

where the energy coefficient p is given by

$$G(n) = D\, n^p \tag{48}$$

and $\Delta G = D_{BC} - D_{AB}$. With conservation of the total bond order, $n_{AB}=1-n_{BC}=n$, then

$$n^{\ddagger} = [\, 1 + (D_{AB}/D_{BC})^{1/p}\,]^{-1} \tag{49}$$

Since

$$\eta = -a'\,\ln n^{\ddagger} - a'\,\ln(1-n^{\ddagger}) \tag{50}$$

then

$$\eta = a'\ln\frac{[1 + (D_{AB}/D_{BC})^{1/p}]^2}{(D_{AB}/D_{BC})^{1/p}} \tag{51}$$

or

$$\eta = a'\ln\frac{[1 + (1 - \Delta G/D_{BC})^{1/p}]^2}{(1 - \Delta G/D_{BC})^{1/p}} \tag{52}$$

and through expansion in a Taylor series

$$\eta(\Delta G) \approx 2a'\,\ln 2 + a'\,(\Delta G/2\,p\,D_{BC})^2 \tag{53}$$

Thus the BEBO model predicts also a quadratic dependence of η on ΔG, but with a constant intercept, $\eta(0) = 2a'\ln 2$, which corresponds to the ISM case of $n^{\ddagger}=1/2$.

In the same manner the models of Agmon--Levine (eqs(3) to (9)) and the Koeppl--Kresge (eq(34) with a square dependence of d on ΔG) [4c, 5a] are encompassed by ISM, but neglect the role of the resonance effect, i.e. $n^{\ddagger}>1/2$. Nevertheless, we must point out that in the former model λ has a different meaning; λ is equivalent to the intrinsic kinetic energy barrier, $\lambda = \Delta G_0^{\ddagger}/\ln 2$.

3.3. BIDIMENSIONAL REACTION COORDINATE MODELS

Several theoretical models of chemical reactivity take the view that a few elementary reactions are not adequately described by a free energy versus an unidimensional reaction energy profile, and two progress variables are considered: one for the "mean progress" and the other for the "disparity progress" [20,21,26]. This kind of reaction occurs when several processes (bond formation and bond breaking, solvation and desolvation, localization and delocalization of charge) make different progress in the transition state. These two-dimensional models have been questioned, because they bear no resemblance to potential energy surfaces current in molecular dynamics [27]. As we will show such "imbalance reactions" can be described, within ISM, as any other chemical reaction, by the traditional free-energy *versus* one-dimensional reaction energy profile.

Two important modifications to the theory of Marcus, included in ISM, are known to account for some "disparity progress" in chemical reactions: one is the asymmetry of the potential energy curves, $f_r \neq f_p$ [19,24,28] and the other is the linear dependence of d on $(\Delta G)^2$ [19]. Here we will consider another important factor, the siphoning of electronic density at the transition state which makes $n^{\ddagger} > 1/2$.

If one considers the Brønsted coefficient, $\alpha = \partial \Delta G^{\ddagger}/\partial \Delta G$, given by the theory of Marcus (eq(45)), one gets

$$\alpha = (1/2) + (\Delta G / 8\,\Delta G_0^{\ddagger}) \tag{54}$$

Kreevoy and Lee [20] have presented a bidimensional model with

$$\alpha = (1/2) + (\Delta G / 8\,\Delta G_0^{\ddagger}) + \delta/2 \tag{55}$$

where δ is the variable of the disparity progress; the authors admit that δ is the charge of the central atom for the transition state $\{AB^{\delta+}C^{\delta-}\} \longleftrightarrow \{A^{\delta+}B^{\delta-}C\}$. When $\delta = 0$ we have the normal transition state bond order $n^{\ddagger} = 1/2$. Within ISM, for the Marcusian conditions of $f_r = f_p$ and $\lambda \gg |\Delta G|$, from eqs (46) and (54) one can write

$$\alpha = (1/2) + (\Delta G / f\,d^2) \tag{56}$$

or

$$\alpha = (1/2) + [(n^{\ddagger})^2\,\Delta G / (a'\ln 2)^2\,f\,l^2] \tag{57}$$

with $l = l_r + l_p$. When one compares this equation with the one of Kreevoy and Lee ($n^{\ddagger} = 1/2$)

$$\alpha = (1/2) + [(1/2)^2\,\Delta G / (a'\ln 2)^2\,f\,l^2] + \delta/2 \tag{58}$$

one concludes that

$$\delta / 2 = [\Delta G / (a' \ln 2)^2 f l^2] [(n^\ddagger - 1/2)(n^\ddagger + 1/2)] \qquad (59)$$

To account for the disparity, Kreevoy and Lee consider $-1 \le \delta \le 1$. Even within the limiting conditions of the theory of Marcus, the one-dimensional Intersecting-state Model can accommodate the disparity progress of the Kreevoy--Lee model; negative values for the parameter δ correspond to $n^\ddagger < 1/2$, which is possible with some nonadiabatic reactions.

4. Over the Barrier or Through the Barrier?

One of the weaknesses of TST in its usual form is the neglect of quantum mechanical tunnelling; molecules which do not possess enough energy to surmount the reaction energy barrier according to a classical picture, can react to a small extent by 'going' through the barrier.

Villars [29] showed, in the very first paper which discusses the nature of the activation energy of chemical reactions in terms of a bond extension on a potential energy curve, that one is dealing with what has subsequently been called Franck-Condon factors. The same kind of factors has been invoked in the interpretation of nonradiative transitions in large molecules, and subsequently in photochemical reactions [30], in terms of tunnelling of vibrational modes. Thus, a common potential energy barrier can be employed to estimate the rate of a chemical reaction either via thermal activation (eq(1)) or via nuclear tunnelling. For harmonic potential energy curves the rate of tunnelling [30] is given by

$$k_{tun} = \nu \; c_0^{1-m'} \exp[(-2\pi / h)(2\mu \Delta G^\ddagger)^{1/2} \Delta x] \qquad (60)$$

where ν is the frequency of the reactant mode involved in the process, μ the corresponding reduced mass, Δx the width of the energy barrier and the other symbols have their usual meaning. Within the ISM formalism, the barrier width for a tunnelling process from the minimum of the potential energy curve of the reactant in a exothermic reaction is given by

$$\Delta x = d - [2 |\Delta G| / f_p]^{1/2} \qquad (61)$$

In practical terms the empirical value of $n^\ddagger$ can be a useful criterion to assess the occurrence of nuclear tunnelling. For the sake of argument let us consider a family of reactions where the reactants do not possess mobile electrons; one expects a transition state bond order of 1/2. Such a prediction can be confronted with the empirical data, i.e., with the values of d and the value of $n^\ddagger$ which reproduce the experimental $\Delta G^\ddagger$ values. The empirical value of $n^\ddagger$ can be: $n^\ddagger > 0.5$ or $n^\ddagger \approx 0.5$ or $n^\ddagger < 0.5$. When nuclear tunnelling is dominant the corresponding rate constants exceed the rates of thermal activation and, consequently, $n^\ddagger > 0.5$. However, if $n^\ddagger < 0.5$ this suggests that one is dealing with a thermally activated reaction of a nonadiabatic nature or with steric hindrance. Obviously, the present reasoning is applicable to other cases ($n^{\ddagger} \neq 0.5$) where the transition state bond order can be predicted or estimated independently from kinetic data.

5. Hydrogen-Atom Transfer

ISM was initially applied to the study of 25 hydride reactions in the vapour phase [7]. According to eq(33) the value of d was estimated from the experimental activation energy, E_a, the reaction energy, ΔE, and the relevant Morse parameters.[#] No correlation was found between the calculated d values and $(\Delta E)^2$. However, a series of linear correlations were found (Figure 4) when the reduced bond extension, η, was plotted against $(\Delta E)^2$. Families of reactions are found and characterized by a constant $n^{\ddagger}$ and λ; $n^{\ddagger}$ are estimated from the intercepts of the plots in Figure 4

$$n^{\ddagger} \, \eta(0) = 0.108 \tag{62}$$

and λ from the slopes, γ,

$$\lambda = (0.156 / 2\,\gamma)^{1/2} \tag{63}$$

Reactions where the reactants (and products) do not possess nonbonding or antibonding mobile electrons have $n^{\ddagger} = 0.5$ ($\eta(0) = 0.216$). Reactions where the reactants possess pairs of nonbonding electrons, have $n^{\ddagger} = 1$ ($\eta(0) = 0.108$). Examples of such reactions are $X + HA \rightarrow XH + A$, where X represents an halogen atom and A an atom or group of atoms. A transition state bond order of unity can be interpreted in terms of "increased-valence" structures.

It is interesting to point out that such a transition state bond order ($n^{\ddagger} = 1$) can also be rationalized in terms of a molecular orbital energy diagram for a nonlinear molecular species which has features of an electronically excited species. Taking z as the axis of the linear $\{HHX\}^{\ddagger}$ species, it becomes clear that, e.g., a p_x orbital of the atom X which has a nonbonding character, acquires a bonding character in the nonlinear species through interaction with the s-orbital of the terminal of H-atom. So it appears that a nonbonding pair of electrons of X is converted, through a resonance effect, into a bonding pair in the transition state, increasing $n^{\ddagger}$ from 0.5 to 1.

Reactions of $n^{\ddagger} = 1.5$ ($\eta(0) = 0.072$) are found with $H + X_2 \rightarrow HX + X$. In very simplified terms one can view an antibonding pair of electrons of X_2 to be converted into a bonding pair at the transition state, increasing $n^{\ddagger}$ from 0.5 to 1.5. The following "increased-valence" structures of Harcourt can also rationalize the present findings

---

[#] Since such energies were estimated from high temperature data, they correspond entirely to the thermal activation process and have no contribution of quantum mechanical tunnelling.

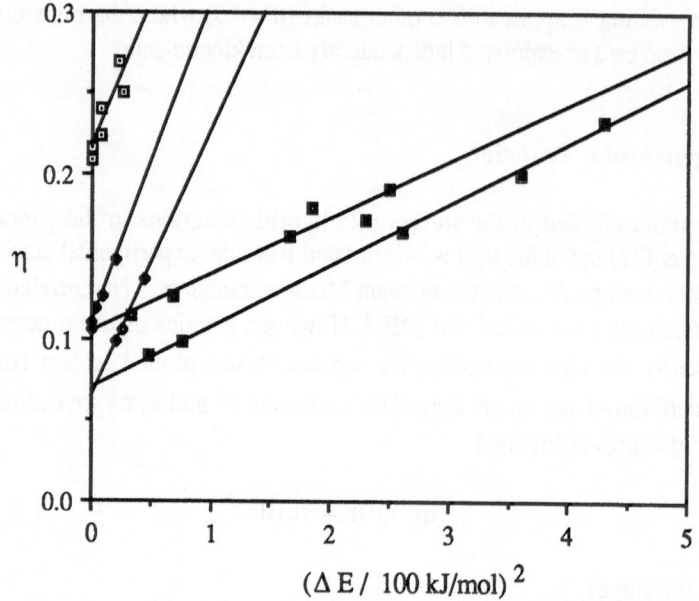

Figure 4. Plot showing the square dependence of η on ΔE for several H-atom transfer reactions; reaction families: i) $H+H_2\rightarrow$; $H+C_nH_{2n+2}\rightarrow$; $CH_3+C_nH_{2n+2}\rightarrow$; ii) $X+CH_nX_{4-n}\rightarrow$ (X=Cl); iii) $X+H_2\rightarrow$; $X+CH_4\rightarrow$; iv) $X+CH_nX_{4-n}\rightarrow$ (X=Br); v) $H+X_2\rightarrow HX+X$, $X+C_nH_{2n+1}X\rightarrow X_2+C_nH_{2n+1}$ (X=I).

$$\ddot{X}\, H\text{-}H \;\rightarrow\; \dot{X} : H\text{-}H \qquad\qquad n^{\ddagger}=1$$

$$X\text{-}X\, H \;\rightarrow\; X\!\div\!X : H \qquad\qquad n^{\ddagger}=1.5$$

To avoid cumbersome diagrams, only the active electrons are illustrated.

There appear to be essentially two types of mixing entropy parameters, corresponding to $\lambda\approx75$ kJ mol^{-1} and $\lambda\approx145$ kJ mol^{-1}. The lowest value is found in many cases where a light atom B in the transition state $\{ABC\}^{\ddagger}$ lies between two dynamically σ*_s

σ*_{p_z}

p_y, p_x   nonbonding

σ*_s

σ*_{p_z}

p_y

p_x

σ_{p_z}

σ_s

σ*_p_z

p_y

p_x

σ_{p_z}

σ_s

$\{H\cdots H\cdots X\}^{\ddagger}$

$\left\{ \begin{array}{c} {}^{\cdots H \cdots} \\ H \qquad X \end{array} \right\}^{\ddagger}$

$n^{\ddagger} = 0.5$ $n^{\ddagger} = 1$

heavy atoms. These families of reactions correspond to the exothermic processes where the products are produced with a high vibrational energy content. For a model of a gradual harmonic force along the retreat coordinate [2] the vibrational energy content, V_{vib}, which can appear in the products is

$$V_{vib} = [m_A / m_B (m_A + m_B)] \ [(\Delta E)^2 / l_{BC}^2 \ \omega^2] \ [1 - \cos(\omega t_c)] \quad (64)$$

where ω is the frequency of the force, t_c the duration of the collision, l_{BC} the bond length of BC close to the transition state value where the repulsive potencial is zero. Eq (64) reveals that V_{vib} is highest when $m_B \ll m_A$. Microscopic reversibility tells us that λ should be virtually the same for the forward and backward reactions; thus the same mass effect is expected for the B and C atoms in the reverse reaction.

6. Hydride transfers

6.1. HYDRIDE TRANSFERS IN SOLUTION

Hydride transfers are a very important category of reactions in organic and biological chemistry, and can be a model for electrophilic reactions, because they can be viewed as transfer of a proton with a pair of electrons between electron deficient sites [31], $A + BH \rightarrow AH^- + B^+$. A large number of theoretical studies have addressed this topic, and the theory of Marcus gives a reasonable representation of the relation between rate and equilibrium constants [32], although several questions remain to be answered, particularly the high intrinsic energy barrier, $\Delta G_0^{\ddagger}$, when compared with proton tranfers.

A convenient set of reactions to be studied within the ISM formalism is between nicotinamide adenine dinucleotide analogues [20] in isopropyl alcohol / water mixtures,

The reactive bonds are C-H in reactant and product, marked in bold. Although the correct values of the force constants and bond lengths of those bonds are not known, one can employ values typical for the CH bond in organic molecules [33]: $f_r = f_p = f_{CH} = 2.9 \times 10^3$ kJ mol^{-1} Å^{-2}, $l_r + l_p = 2 \, l_{CH} = 2.192$ Å. Eq (1) with T= 298 K allows the estimation of $\Delta G^{\ddagger}$ as presented in Table 1; then one can employ eq(35) to estimate the reactant bond extension and from that and the reaction energy values ΔG, eq(34) allows the estimation of d. Finally, η can be calculated through (eq(17)). These relevant data are presented in Table 1 and in Figure 5.

TABLE 1. Bond extensions of hydride transfers of quinolinium ions solution .[a]

R	$\Delta G^{\ddagger}$/kJ mol^{-1}	ΔG/kJ mol^{-1}	d/Å	η
p-CH$_3\phi$CH$_2$	80.9	-6.8	0.4821	0.2199
ϕCH$_2$	80.6	-7.6	0.4825	0.2201
p-FϕCH$_2$	80.5	-8.1	0.4828	0.2202
m-BrϕCH$_2$	80.0	-9.1	0.4828	0.2202
m-FϕCH$_2$	79.6	-9.4	0.4821	0.2199
p-CNϕCH$_2$	79.5	-10.9	0.4838	0.2207
m-CF$_3\phi$CH$_2$	79.5	-11.71	0.4835	0.2206

[a] Ref. 20 ; $f_r = f_p = 2.9 \times 10^3$ kJ mol^{-1} Å^{-2} , $l_r + l_p = 2.192$ Å.

The intercept in Figure 5 leads to $n^{\ddagger} \approx 0.492$ and $\lambda = 86$ kJ mol^{-1}. The transition state bond order is extremely close to 0.5, implying that there is conservation of chemical bond order and resonance effects at the transition state are negligible. The study of other hydride transfer reactions present values of $n^{\ddagger}$ between 0.50 and 0.485, weakly dependent on the ionization energy, I, of the aromatic molecules; $n^{\ddagger}$ decreases slightly with an increase in I [34].

Hydride transfer reactions have a high intrinsic energy barrier (eq(46)), because $n^{\ddagger}$ is reasonably low (≈ 0.5). This becomes obvious when one writes,

$$\Delta G_0^{\ddagger} = (1/8) \ f \ [0.108 \ (l_r + l_p) / n^{\ddagger}]^2 \qquad (65)$$

The reactive center B$^+$ is electron deficient and cannot syphon electronic density into the transition state to increase $n^{\ddagger}$, as found with some H-atom transfers.

Kreevoy and Lee [20] have studied these same reactions within their own theoretical formalism, but have found a "total transition state bond order" which does not agree with the value of unity calculated by ISM. As previously discussed, although the model of Kreevoy and Lee is encompassed by ISM, the meaning of the corresponding molecular parameters is not the same.

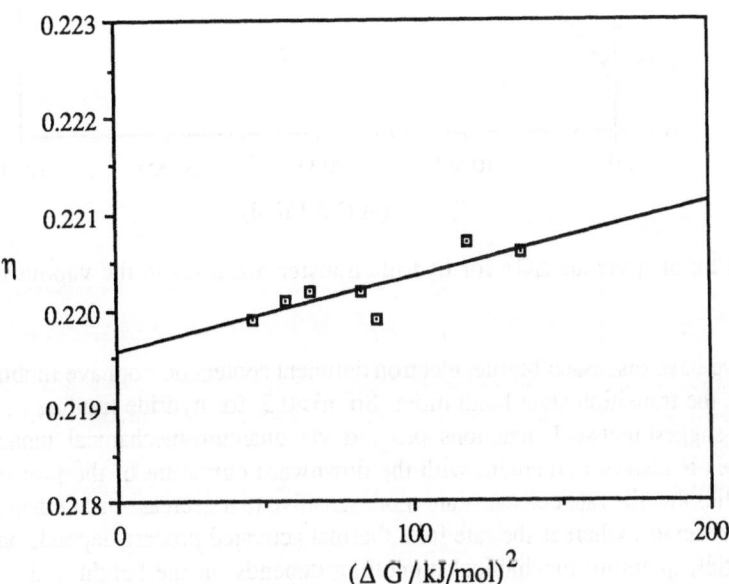

Figure 5. Plot of η versus ΔG^2 for hydride transfers of quinolinium ions in solution (data in Table1).

6.2. HYDRIDE TRANSFERS IN THE VAPOUR PHASE

Meot-Ner (Mautner) and Field [35] have studied several hydride transfers between carbonium ions and hydrocarbons in the vapour phase. The reactions are considerably faster than in solution. The experimental data can be studied under the assumption that one is dealing with a thermal activation process as in liquid solutions. The results taken from ref.34 are presented in Figure 6. The first conclusion that one may draw is that $n^{\ddagger}$ has a considerable high value ($n^{\ddagger}$=0.74), under the present assumptions.

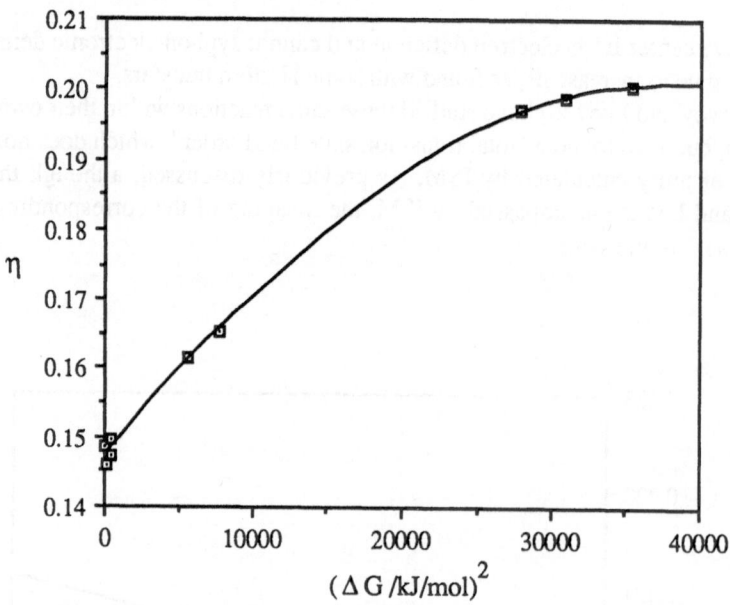

Figure 6. Plot of η versus ΔG^2 for hydride transfer reactions in the vapour phase (refs. 34,35).

As we have discussed before, electron deficient centers do not have mobile electrons to increase the transition state bond order. So $n^{\ddagger}$>0.5 for hydride tranfers in the vapour phase may suggest that such reactions proceed via quantum-mechanical tunnelling. Such a hypothesis is also in agreement with the downward curvature of the plot in Figure 6, which implies that the rate constants are more sensitive to a decrease in ΔG than the thermal activated processes; whereas the rate for a thermal activated process depends on the height of the barrier, quantum-mechanical tunnelling depends on the height and width of the reaction energy barrier. To further explore the tunnelling hypothesis one should estimate the rate for this process (eq(60)) on the ISM energy barrier with the correct value of $n^{\ddagger}$=0.49. Table 2 presents the calculated tunneling rates, with $\nu=10^{13}$ s^{-1} and $\mu=1$ g mol^{-1}, which are in close agreement with experiment. The role of tunnelling in the vapour phase is

TABLE 2 . Quantum mechanical tunnelling rates for hydride transfers in the vapour phase.

ΔH/kJ mol^{-1}	k/mol^{-1} dm^3 s^{-1}	
	exp[a]	tun[b]
0	4×10^8	5×10^8
-80	2×10^{11}	2×10^{11}
-170	8×10^{11}	2×10^{12}

[a] Ref. 35 ; [b] $\nu = 10^{13}$ s^{-1} and $\mu = 1$ g mol^{-1}

in accordance with the suggestions of Kreevoy et al. [20b].

In liquid solutions, the presence of a solvated ion like H_3O^-, drops the tunelling rate by 9 orders of magnitude at $\Delta H=0$ ($k_{tun} = 0.5$ mol^{-1} dm^3 s^{-1}), such that this cannot compete with thermal activation ($k_{therm} = 1.5 \times 10^3$ mol^{-1} dm^3).

7. Proton transfers

7.1. PROTON TRANSFERS IN SOLUTION

When one compares the rate constants of ionization of ketones and sulphones in water which have virtually the same reaction energy (vide Table 3), one concludes that the reactions for sulphones are, intrinsically, two orders of magnitude faster than for ketones. Both kinds of reactions can be studied within the ISM-formalism in the same manner as hydride transfers. The calculated bond extensions, taken from ref.36, are presented in Table 3. The plots of η versus ΔG^2 presented in Figure 7 reveal the existence of two families of reactions, each characterized by n‡ and λ: ketones n‡= 0.56 and λ=300 kJ mol^{-1} and sulphones n‡= 0.70 and λ=127 kJ mol^{-1}. Amines, not presented in Table 3 and in Figure 7, have n‡= 0.85 and λ=87 kJ mol^{-1} [36a].

It has been shown that proton transfer reactions of several kinds of acids can be well interpreted in terms of the ISM. However, in contrast to hydride transfer, proton transfer can occur between electron rich centers and n‡ can be higher than 0.5. In fact n‡ ranges between 0.5 (slow acids) and 1.0 (fast acids) as the following "increased-valence" structures illustrate

$$A\text{-}H \quad B \quad \rightarrow \quad A \cdot H \cdot B \qquad n^{\ddagger} = 0.5$$

$$\ddot{A}\text{-}H \quad B \quad \rightarrow \quad \dot{A}\text{-}H \cdot B \qquad n^{\ddagger} = 0.75$$

$$\ddot{A}\text{-}H \quad B \quad \rightarrow \quad \dot{A}\text{-}H : B \qquad n^{\ddagger} = 1$$

The lowest values are found with ketones and nitroalkanes and the highest values for acids (e.g. HF) where there are reactive centers with mobile electrons and the XH bond has a strong ionic character. Oxygen and nitrogen acids have intermediate $n^{\ddagger}$ values (0.75-0.8).

As far as the mixing entropy parameter, λ, is concerned it shows a decrease with an increase in $n^{\ddagger}$. Owing to the role of the mobile electrons which increase $n^{\ddagger}$, the repulsive wall should have a change which is more sudden when $n^{\ddagger}$ is high than when it is low. For a sudden repulsive barrier, vibrational energy is required to activate the reactive bond of reactants [2,37] and in consequence λ is low. In contrast, a more gradual repulsive wall, expected when $n^{\ddagger}$ is close to 0.5, can be overcome virtually only with translational energy and in consequence λ is high.

TABLE 3. Bond extensions estimated by ISM for the ionization of carbon acids in water[b]

	pK_{HA}[a]	$k/mol^{-1}dm^3s^{-1}$ [a]	$\Delta G^{\ddagger}/kJ\ mol^{-1}$	$\Delta G/kJ\ mol^{-1}$	η
CH_3COCH_3[c]	19.5	8.5×10^{-12}	135.7	110.7	0.202
$CH_2(CO_2Et)_2$	13.3	4.5×10^{-7}	109.0	75.5	0.197
$(CH_3CO)_2CH_2$	8.9	3×10^{-4}	92.9	50.5	0.194
$(PhSO_2)_2CH_2$	11.0	4×10^{-3}	86.6	62.5	0.173
$(EtSO_2)_2CH_2$	12.2	3.1×10^{-4}	92.9	69.3	0.177
$(MeSO_2)_2CHMe$	12.55	2.1×10^{-4}	93.9	71.3	0.177
$(PhSO_2)_2CHMe$	13.76	1.2×10^{-5}	101	78.1	0.181
$(EtSO_2)_2CHMe$	14.4	6.3×10^{-7}	108	81.8	0.189

[a] Data from: Bamford, C.H.; Tipper, C.F.H. *Proton Transfer.Comprehensive Chemical Kinetics;* Elsevier, Amsterdam, 1977, vol.8; pags. 181, 151; [b] T=300 K, $f_r(CH)=2.9 \times 10^3\ kJ\ mol^{-1}\ Å^{-2}$; $f_p(OH)=4.9 \times 10^3\ kJ\ mol^{-1}\ Å^{-2}$; l=2.033 Å; [c] l = 2.056 Å.

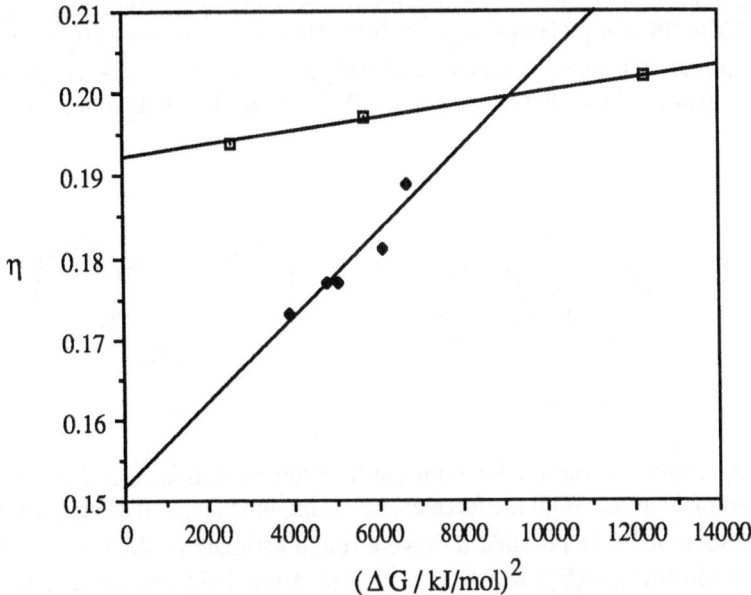

Figure 7. Plot of η versus ΔG² for proton transfers in solutions (data in Table 3).

Hydride transfers have much lower λ, when compared with proton transfer reactions of $n^{\ddagger} \approx 0.5$, probably caused by an even more sudden repulsive barrier. In fact a H⁻-transfer can be seen as a H⁺-transfer followed by the transfer of a pair of electrons.

7.2. PROTON TRANSFER IN EXCITED STATES

Proton transfers in electronically excited states have not been amenable to any reasonable interpretation in terms of the theory of Marcus, in part due to the implicit assumption of the symmetry of the potential energy curves of reactant and product [24,38]. In contrast, ISM provides a simple interpretation of this kind of reactions [39]. The excited-state reactions appear to follow the same basic principles of their ground-state analogues; the transition state bond order does not change appreciably from the ground to the excited state. However, the mixing entropy parameter λ decreases; an enhancement of the dipole moment upon eletronic excitation can increase the suddenness of the repulsive wall of the reaction and decreases λ.

A puzzling problem in proton transfers in excited states is raised by the work of Wan et al. [40]. Proton transfers from the first excited singlet state of aromatic carbon acids such as fluorene (I) are much slower than fluorescence and other deactivation processes, in spite of the tremendous increase in acidity upon electronic excitation: e.g. for fluorene $pK_a(S_0)=30$ to 35 and $pK_a(S_1)=-4$ to -9 [40]. However, suberene (II) in the S_1 state transfers a proton to a water molecule on photolysis in D_2O/CH_3CN, in spite of the fact that its pK_a, estimated by a Forster cycle ($pK_a(S_1)=-7$), is close to that of fluorene.

(I)

(II)

The application of our model to the study of these reactions can shed some light on these intriguing features. With the force constant and bond length data reported in Table 3 and $\lambda\approx250$ kJ mol^{-1} it is possible to have a rough estimate of the range of the proton transfer rates in fluorene(S_1) which, as a typical carbon acid, should have $n^{\ddagger}\approx0.55$; the highest estimated rate is ca. 0.5 mol^{-1} dm^3 s^{-1}, certainly too low to compete with S_1 decay (ca. 10^8-10^9s^{-1}) even in pure water. However, in the proton transfer reaction of suberene is formed a carbanion which is $8\pi(4n)$ system with an antiaromatic character in the ground sate. Consequently, the suberenyl carbanion possesses a pair of electrons not delocalized within the π-system which can increase the bond order at the transition state from a typical $n^{\ddagger}\approx0.5$ to $n^{\ddagger}\approx1$; the relevant "increased-valence" structure is of the kind $\overset{..}{C}\text{-}\overset{..}{C}\text{:}H\cdot B$. With such a value and an even lower λ ($\lambda\approx200$ kJ mol^{-1}) the estimated rate for proton transfer is 0.7×10^8 mol^{-1} dm^3 s^{-1}. For a water concentration of 30 this leads to a rate 2×10^9 s^{-1} which can compete with other S_1 decay processes.

Although fluorene and suberene can both act as carbon acids in S_1 state and have similar pK_as, they are acids of different families; the former is a slow acid and the latter is a fast acid and this can be attributed to the antiaromatic character of suberenyl anion. There are a few other examples of carbon acids which are not slow acids. For example, hydrocyanic acid is virtually a normal acid in the ground state. Our studies [36b] show that the involvement of the $C\equiv N$ bond together with the CH (OH) bond in the reaction coordinate increases $n^{\ddagger}$ and more than compensates (ca. 10^{10} times) the increase in the effective force constant. The "increased-valence" structure $\cdot N\equiv C\text{-}H\cdot B$ leads to $n^{\ddagger}=1.5$; the value estimated from the experimental data is $n^{\ddagger}=1.55$ [36b].

7.3. SOLVENT EFFECTS

Mobile electrons of the reactive centers are able to increase $n^{\ddagger}$ in proton transfer reactions. Since such a pair of electrons also interacts with the solvent molecules, significant solvent effects are expected for these reactions when one changes a solvent of a polar nature (strong interaction with electron pair) by a less polar medium (weaker interaction with electron pair). In consequence, $n^{\ddagger}$ should be lower in polar media and higher in nonpolar media. This is clearly shown in Table 4 where the experimental data of the solvent effects on Brønsted correlations of acid-base catalysis, obtained by Bernasconi and coworkers [41], has been interpreted in terms of ISM [42]. As previously found with other reactions, λ decreases with an increase in $n^{\ddagger}$.

Obviously an increase in solvent polarity also decreases ΔG, so the final outcome of a change in solvent polarity on the chemical reactivity, depends on which factors dominate. This is a problem which will be further explored in the next sections.

TABLE 4. Transition state bond orders, $n^{\ddagger}$, and mixing entropy parameters, λ, for the proton transfer between acetylacetone and RCOO$^-$ in mixtures of dimethylsulphoxide and water.[a]

Me$_2$SO (%)	0	50	90	95
$n^{\ddagger}$	0.598	0.608	0.646	0.667
λ / kJ mol^{-1}	295	223	117	99.4

[a] Ref 42.

7.4. PROTON TRANSFER IN THE VAPOUR PHASE

Proton transfer reactions in the vapour phase have also been studied in terms of ISM [34]. The dominance of the tunnelling process for hydride transfer is also verified with the proton transfers involving O-reactive centers. The empirical transition state bond order ($n^{\ddagger}\approx 1.6$) was found to be higher than the maximum expected value ($n^{\ddagger}\approx 0.8$), and the experimental rate can be well accounted for by the tunnelling of a CH oscillator through an energy barrier estimated by the ISM.

8. Methyl transfers

Methyl transfer reactions, $X^- + CH_3Y \rightarrow XCH_3 + Y^-$ (X and Y are halogen atoms), which are an important class of S_N2 nucleophilic substitutions, have also been studied within the intersecting-state model framework [43]. As will be shown, the overall pattern is similar to that of proton transfers for reactions in solution, but is different for reactions in the vapour phase, because, owing to the masses involved, quantum mechanical tunnelling does not dominate thermal activation.

These reactions have constituted an active field of research, both from the experimental and the theoretical point of view, and although several calculations have been carried out through ab initio methods, to unravel trends of reactivity within a family of related molecules, theoretical models which sacrifice rigour to gain simplicity are often better. In this area the theory of Marcus has been quite popular and Wolfe et al. [44] have even shown that the theory of Marcus calculates energy barriers, from the self-exchange barriers and the reaction energy, in excellent agreement with ab initio calculations. However, as Carrier and Dewar [45] have pointed out, it remains to understand the large variations (ca. 120 kJ mol^{-1}) in $\Delta G_0^{\ddagger}$ and why such reactions should show dramatic solvent effects.

In solution methyl transfers have transition state bond orders which depend on the nature of the reactive centers (nucleophiles) and on the solvent polarity. It was found [46] that the estimated $n^{\ddagger}$ values can be divided into two contributions, $C_n^{\ddagger}$, one of the X and the other of the Y groups,

$$n^{\ddagger} = C_n^{\ddagger} (X) + C_n^{\ddagger} (Y) \tag{66}$$

For example, for the reaction $*FCH_3 + F^- \rightarrow *F^- + CH_3F$, in water, one can estimate a transition state bonder order quite close to 0.5 ($n_w^{\ddagger}(F)=0.54$). The mobile electrons of F interact strongly with the water molecules and are not available to increase $n^{\ddagger}$ at the transition state. A small electronegative group such as F has a higher HOMO which is more available to interact with the H_2O molecules than a large atom such as I, which has a lower HOMO [47]. In fact, the estimated $n^{\ddagger}$ for the reaction $*ICH_3 + I^- \rightarrow *I^- + CH_3I$ in water, is much higher ($n_w^{\ddagger}(I)=0.70$) (vide Table 5).

TABLE 5. Nucleophile transition state bond order contributions, $C_n^{\ddagger}$, for methyl transfer reactions in water and acetone.[a]

$C_n^{\ddagger}$	water	acetone
F	0.27	≈ 0.3
Cl	0.30	0.33
Br	0.32	0.38
I	0.35	0.41

[a] Ref. 46

In a less polar medium the interaction between the nucleophile lone pairs and the solvent molecules decreases and, in consequence, $n^{\ddagger}$ increases. For example, in acetone, for the same reactions $n_{acet}^{\ddagger}(F) \approx 0.6$ and $n_{acet}^{\ddagger}(I)=0.82$. The logical consequence that one may draw from these considerations is that in the vapour phase, owing to absence of solvent molecules, the transition state bond order should be close to 1 and independent of the nature of the nucleophile X. Following the method of Harcourt [8] one can write an "increased-valence" structure which has $n^{\ddagger}=1$, by delocalizing two nonbonding electrons of X (or Y)

$$\ddot{X} \cdot C-\ddot{Y}: \quad \rightarrow \quad :\dot{X}:C-\ddot{Y}:$$

This seems to be confirmed [43], because methyl transfer reactions with nucleophiles such as F, Cl, Br, CH3O, t-BuO, CN, in the vapour phase all fall into a single family with $n^{\ddagger}=0.93$ and $\lambda=580$ kJ mol^{-1}. The transition state bond order is very close to 1, as expected. Nevertheless, Brauman et al. [48] have shown that methyl transfers in the vapour phase proceed via the formation of a weakly stable charge-dipole complex, X$^-$ + CH3Y $\rightleftharpoons$ (XCH3Y)$^-$ $\rightarrow$ XCH3 + Y$^-$. Conservation of the bond order is only valid for a synchronous bond-forming-bond-breaking process. Such a loss of synchronicity due to the formation of a weak complex would lead to a weak decrease in $n^{\ddagger}$, with respect to the maximum value expected ($n^{\ddagger}=1$). Although the formation of a weak intermediate complex leads to a small decrease in $n^{\ddagger}$ that corresponds to a slight increase in $\Delta G^{\ddagger}$, it also leads to a longer period during which the repulsive force is operating [2]. In consequence, very little vibrational energy is released as excess energy in the products and λ is much higher than for H-atom transfer reactions (λ = 145 kJ mol^{-1}). As shown in Table 6, the formation of such intermediate complexes more than compensates the decrease in $n^{\ddagger}$ and is a way to decrease $\Delta G^{\ddagger}$ for methyl transfers which are very exothermic processes in the gas phase (ΔG= -100 to -200 kJ mol^{-1}). In liquid solutions $|\Delta G|$ is quite low and the effect of λ on $\Delta G^{\ddagger}$ is negligible.

TABLE 6. Calculated effects of the parameters $n^{\ddagger}$ and λ on the energy barrier of the elementary process X$^-$ + CH3Y $\rightarrow$ XCH3 + Y$^-$ as a function of the reaction energy.[a]

ΔG/kJ mol^{-1}	λ/kJ mol^{-1}	$n^{\ddagger}$	η	$\Delta G^{\ddagger}$/kJ mol^{-1}
0	145	1	0.108	39
	580	0.93	0.116	45
-50	145	1	0.117	26.9
	580	0.93	0.1167	26
-180	145	1	0.228	104
	580	0.93	0.124	4

[a] f_r = 1.9x10^3 kJ mol^{-1}Å^{-2}, f_p = 3.0x10^3 kJ mol^{-1}Å^{-2}, l = 3.364 Å.

9. Ambident reactions

Whereas methyl transfer reactions proceed faster in nonpolar media, the Menschutkin
(S_N2) reaction $(CH_3CH_2)_3N + CH_3CH_2I \rightarrow (CH_3CH_2)_4N^+ + I^-$ proceeds faster in
polar media. Variations in $n^{\ddagger}$ upon a change on solvent polarity control chemical reactivity
in methyl transfers, whereas for the above mentioned S_N2 displacement the significant
changes of ΔG overcome the effect of $n^{\ddagger}$ and dominate the variations on $\Delta G^{\ddagger}$.

This perspective of $n^{\ddagger}$- and ΔG-controlled reactions constitutes the basis for the
understanding of ambifuncional reagents such as nitrite ion

$$NO_2^- + CH_3I \rightarrow CH_3NO_2 + I^- \qquad n^{\ddagger}\text{- control}$$

$$NO_2^- + BuCl \rightarrow BuONO + Cl^- \qquad \Delta G\text{- control}$$

which has two centers capable of nucleophilic attack, but only one centre of which takes
part in each transition state.

In terms of bond dissociation energies, a gain in polarity of the reactive bonds
during the course of a reaction can lead to a decrease in ΔG [49] and leads also to a
decrease in $n^{\ddagger}$, because the atoms involved are more electronegative; the effects on $\Delta G^{\ddagger}$ are
consequently opposed, but one of them can be dominant. For example, the NO_2^- reaction
with the reactive bonds C-I (reactant) and C-N (products) corresponds to a decrease in
polarity; so it can be more easily controlled by $n^{\ddagger}$. ISM studies on these reactions estimate
$k(CH_3NO_2) / k(CH_3ONO) \approx 30$ [46]. In contrast, the second reaction with C-Cl and C-O
reactive bonds corresponds to an increase in the polarity of the chemical bonds, so it can be
ΔG-controlled; the calculated ratio is $k(BuONO) / k(BuNO_2) \approx 80$. But this is not the whole
story, because other structural factors can also play a relevant role; $n^{\ddagger}$-control is favoured
by a low f_r/f_p ratio and by exothermicity, whereas ΔG-control predominates at higher f_r/f_p
ratios and for more endothermic processes.

10. Pericyclic reactions

We consider now processes characterized by bonding changes taking place through a
concerted reorganization of electron pairs within a closed loop of interacting orbitals, which
have been studied through the ISM [50]. The example to be considered here is the 3,3-
shift in 1,5-hexadiene

To define a curve-crossing model one needs to define an effective force constant and bond
length. The bond order of the cyclic state through which one can admit the reaction to
proceed has $n=7/6=1.17$. From typical bond length data for a C-C and C=C bonds [33]

one can interpolate a value $l_r=l_p=1.46$ Å. By employing the relation of Gordy between force constants, bond orders, bond lengths, and the electronegativity of the bonded atoms (eq(39)), a value of $f=3.3 \times 10^3$ kJ mol^{-1} Å^{-2} is estimated once the equation is calibrated with the data of single and double CC bond stretches.

The bond-breaking-bond-forming process corresponds to $n^{\ddagger}=0.5$. ISM allows the estimation of $\Delta G^{\ddagger}=164$ kJ mol^{-1} which is close to the experimental value, $\Delta G^{\ddagger}=171.4$ kJ mol^{-1} [51]; the empirical $n^{\ddagger}$ value which reproduces the experimental $\Delta G^{\ddagger}$ is extremely close to 0.5 (Table 7) suggesting that this sigmatropic shift is essentially a concerted and synchronous process.

Substituent effects cannot be interpreted in terms of linear free-energy relations (LFER) [27,52]. In principle, one expects that free energy relations (linear or nonlinear) are valid when changes in reactivity are controlled only by changes in ΔG. When ISM is applied to the study of the shift reactions of substituted 1,5-hexadienes [50], it becomes evident that changes in $\Delta G^{\ddagger}$ are not exclusively due to changes in reaction energy (Table 7). Substituents that possess mobile electrons increase $n^{\ddagger}$ and only with methyl derivatives is $n^{\ddagger}=0.5$; we have neglected any effect of λ, because $|\Delta G|$ is quite small.

TABLE 7. Transition state bond orders for 3,3-sigmatropic shifts in 1,5-hexadienes.[a]

Substituents	$\Delta G^{\ddagger}$/kJ mol^{-1}	ΔG/kJ mol^{-1}	η	$n^{\ddagger}$
---	171.4	0	0.221	0.49
3,4-dimethyl	163.0	-18.8	0.221	0.49
2-phenyl	148.4	0	0.205	0.526
2,5-diphenyl	129.6	0	0.192	0.562
3,4-dicyano	133.8	-18.8	0.202	0.535
3,4-diphenyl	129.6	-18.8	0.199	0.544

[a] $f_r = f_p = 3.3 \times 10^3$ kJ mol^{-1} Å^{-2}, $1 = 2.92$ Å; [b] Ref. 51.

11. Free Energy Relationships

Physical organic chemistry has been built around LFER, which is still a matter of active research from experiment to theoretical simulation [53,54]. We ourselves have also addressed these problems under the perspective of the ISM [10,36a,42]. Here we will only deal with the interpretation of the Brønsted relations, the postulate of Hammond and the Reactivity-Selectivity Principle (RSP).

The existance of linear-FER appears to be incompatible with the postulate of Hammond which implies a change on the activated-complex structure by changing the reaction energy. We have shown that virtually linear Brønsted plots (log k / log K or $\Delta G^{\ddagger}$ / ΔG) can be found when the transition state bond order is low (close to 0.5) and $\lambda >> |\Delta G|$ (eq (54) with $\Delta G_0^{\ddagger} > 2\Delta G$) [36a]. This is clearly shown in Figure 8. The plots of $\Delta G^{\ddagger}$ versus ΔG are virtually linear in the range $\Delta G = -50$ to 50 kJ mol^{-1} with $n^{\ddagger}$ = 0.5 and can be linear up to 100 kJ mol^{-1} if there is a mild dependence of $n^{\ddagger}$ on ΔG, as discussed previously. However, when $n^{\ddagger}$ is high (e.g. 0.9) it is clear   from the Figure that the Brønsted plot is strongly nonlinear. The same can also be said when $\lambda \approx |\Delta G|$ [36a].

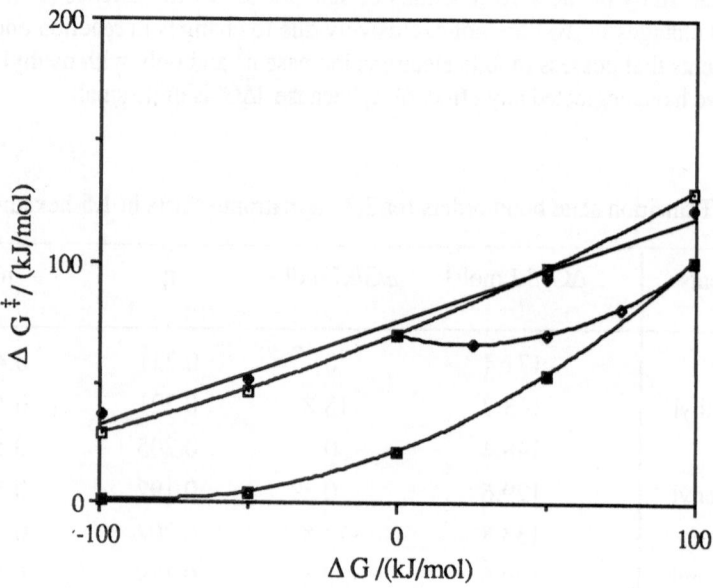

Figure 8. Brønsted plots as a function of the transition state bond order; calculations with the following data: $f_r = f_p = 3 \times 10^3$ kJ mol^{-1} Å^{-2}, l=2 Å, $\lambda >> |\Delta G|$;$\square n^{\ddagger}$=0.5 ; $\blacklozenge$ $n^{\ddagger} = 3 \times 10^{-4}$ ΔG+0.5 (data fit a linear correlation with r=0.98) ; $\lozenge$ $n^{\ddagger} = 3 \times 10^{-3}$ ΔG+0.5 ; $\blacksquare n^{\ddagger}$=0.9.

The present interpretations are in good agreement with the experimental findings of Bernasconi and coworkers [41] on the base catalysis of carbon acid reactions by carboxylic ions. Linear Brønsted plots are observed in water, but become strongly curved in mixtures of Me$_2$SO/water, with the curvature increasing with an increasing on the content of dimethylsulphoxide. As previously discussed, in these apolar media $n^{\ddagger}$ increases and λ decreases (Table 4). ISM shows that the coefficient α can only be identified with the position of the transition state on a reaction coordinate scale of 0 (reactant) and 1 (product), when $f_r=f_p$ and $\lambda>>|\Delta G|$. Otherwise α does not represent the position of the transition state and ISM can even estimate values of $\alpha>1$ and $\alpha<0$ [10,36a].

The postulate of Hammond [55,56] is essentially based on a reasonable intuition: if the reagents are of higher energy than products (exothermic reaction) relatively little change in geometry will be required to reach the transition state, whereas if the reaction is endothermic a higher reorganization in energy is required. As shown in Figure 9 this view is essentially correct if changes in reactivity are controlled by changes in ΔG and $f_r \approx f_p$ and $\lambda>>|\Delta G|$. However, this is no longer true if changes in reactivity are controlled by changes in $\underline{f}$, $\underline{l}$ and/or $n^{\ddagger}$. Even for a thermoneutral reaction the transition state is closer to the products when $f_r<f_p$ and is closer to the reactants when $f_r>f_p$. The same considerations are also applicable to reaction selectivity.

Several chemical reactions verify the RSP, the more reactive species tend to be less selective in their reactivity. Nevertheless, we have shown [10] that this is valid when the reactivity is controlled by changes in ΔG; otherwise exceptions occur and more reactive species can also be more selective than the less reactive ones. Table 8 illustrates the effect of $n^{\ddagger}$ on the selectivity of radicals towards CH bonds. None the less, α is always a good parameter to assess selectivity, for reactions which obey RSP and even for reactions which have a behaviour opposed to such a principle. But literature reports also reaction families where no free-energy relations are observed as, for example, in cation-anion recombinations [57]

$$\log k = \log k_0 + N_+ \qquad (67)$$

where k_0 depends on the identity of the cation and N_+ is a parameter depending on the identity of the nucleophile and the solvent. Such kind of reactions are not anticipated, or even accommodated, by current models of chemical reactivity [57]. The arguments which we have been developing show clearly that eq(67) can be accommodated by ISM when chemical reactivity is dominated by changes in $n^{\ddagger}$ (see also eq(66)). This requires, in general, the maximization of $\partial \Delta G^{\ddagger}/\partial n^{\ddagger}$ and the minimization of $\partial \Delta G^{\ddagger}/\partial \Delta G$. The first condition can be achieved when $f_r \neq f_p$ and the second one requires a fairly low λ and $f_r<f_p$ [58]. Those conditions appear to apply to several cation-anion recombination reactions

$$C_{(solv)}^+ + N_{(solv)}^- \rightarrow C\text{-}N_{(solv)}$$

because the force constant of the C-N bond is significantly higher than those of the solvated species, C^+-solv and N^--solv.

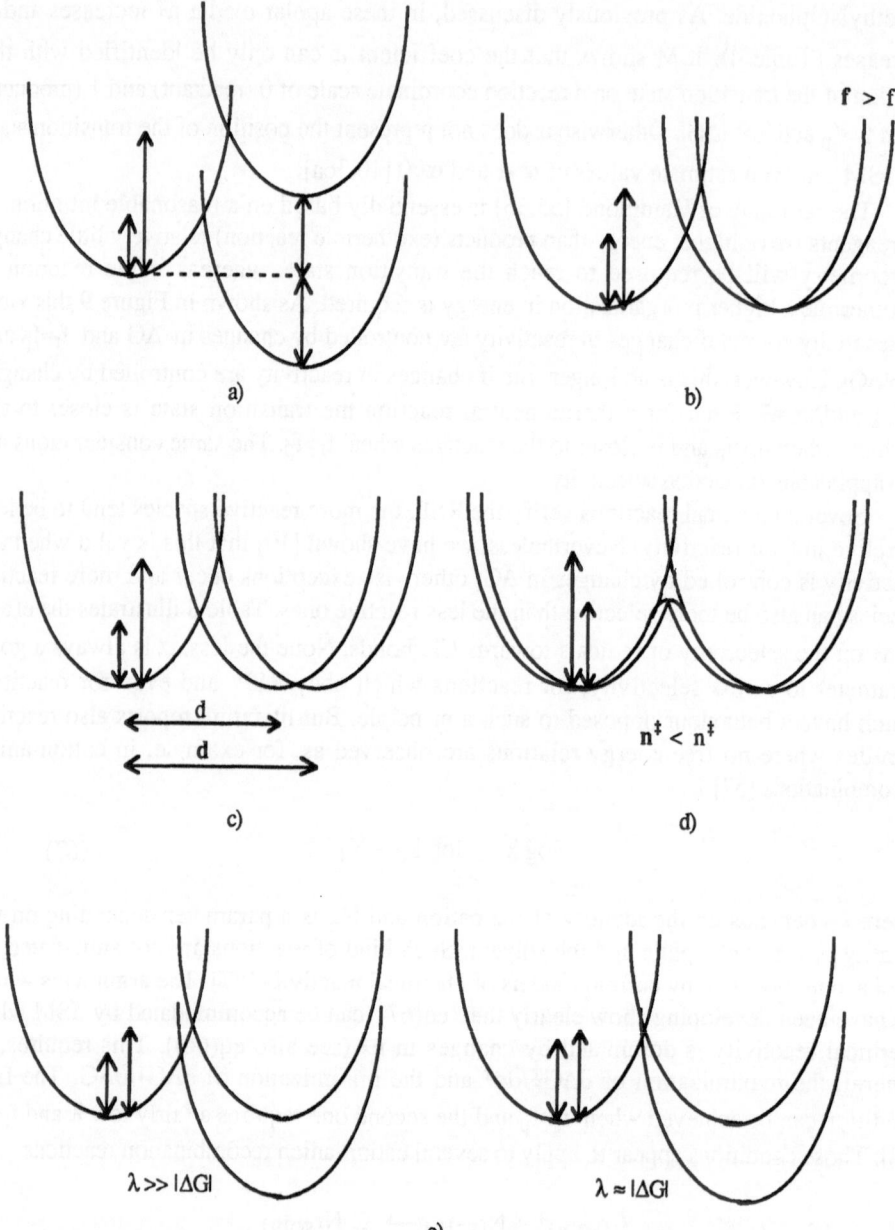

Figure 9. Influence of several structural factors on reaction energy barriers and positions of the transition state on the reaction coordinate: a) reaction energy; b) force constants; c) bond lengths; d) transition state bond order; e) mixing entropy parameter.

Table 8. Selectivity of radicals towards CH bonds; tertiary versus primary hydrogens.

Dominating effect	Radical	ΔE/kJ mol^{-1}	$\Delta E^{\ddagger}$/kJ mol^{-1}	$n^{\ddagger}$	Selectivity
ΔE	F·	-134	5	1	1.4
	Cl·	0	16	1	7.0
$n^{\ddagger}$	Br·	67	75	1	1700
	CH$_3$·	-4	73	0.5	80

12. Electron Transfers

12.1. EXCHANGE REACTIONS

In principle electron transfer is the simplest kind of chemical reaction that one might possibly think of, but in liquid solutions, in spite of extensive experimental and theoretical investigations, it is actually a considerably complex process. Professor Marcus, one of the leading figures on the theories of such reactions, himself recognizes such a complexity: "there are examples "of an old system", $Co^{2+/3+}$ or $Fe^{2+/3+}$ where, despite the current extensive understanding of electron transfers, there remains something to be understood" [59]. In fact, the aquo $Co^{2+/3+}$ and $Fe^{2+/3+}$ pairs have an anomalous behaviour, because the measured self-exchange rates are about 10^7 and 10^3 respectively, faster than those calculated by the theory of Marcus. In contrast, the measured rate for the reaction $Fe(phenanthroline)_3^{2+/3+}$ is ca. 5 orders of magnitude slower than the rate calculated through the same theory [60]. It is this kind of problems that we would like to address now through ISM.

We will start by considering an intramolecular electron transfer reaction in 1,3-dicyanobenzene (DCB) radical anion

which has been studied by e.s.r. and found to be a fast process, $k > 10^{10}$ s^{-1} [61]. Electron transfers are not bond-breaking–bond-forming processes, but a model of intersecting potential energy curves, such as ISM, can be applied to the study of such reactions if the bonds of the oxidized and reduced species, which suffer a significant change in length or force constant, increase in length up to the transition state, i.e., $l^{\ddagger} > l_{ox}$, l_{red} [60,62]. In fact, this requirement appears to be in agreement with potential energy surface (LEPS) concepts [63].

In contrast with this view, in the theory of Marcus [18] one simply minimizes the potential energy of two harmonic oscillators

$$G = (1/2) \; f_{C\equiv N} \; (l^{\ddagger} - l_{C\equiv N})^2 + (1/2) \; f_{C=N} \; (l^{\ddagger} - l_{C=N})^2 \qquad (68)$$

Such a minimization, $\partial G / \partial l^{\ddagger} = 0$, leads to

$$l^{\ddagger} = (f_{C\equiv N} \; l_{C\equiv N} + f_{C=N} \; l_{C=N}) / (f_{C\equiv N} + f_{C=N}) \qquad (69)$$

Equation (69) shows that within the theory of Marcus $l_{C\equiv N} < l^{\ddagger} < l_{C=N}$, and when one substitutes eq(69) into eq(68) obtains an energy barrier

$$\Delta G^{\ddagger} = (1/2) [f_{C\equiv N} \; f_{C=N} / (f_{C\equiv N} + f_{C=N})] \; (l_{C\equiv N} - l_{C=N})^2 \qquad (70)$$

With the relevant force constant and bond length data [33], $f_{C\equiv N} = 1.1 \times 10^4$ and $f_{C=N} = 6.3 \times 10^3$ kJ mol^{-1} Å^{-2}, $l_{C\equiv N} = 1.157$ and $l_{C=N} = 1.34$ Å, eq(70) leads to a barrier $\Delta G^{\ddagger} = 67$ kJ mol^{-1} which leads to a rate constant at room temperature of 20 s^{-1}, ca. 8 orders of magnitude slower than experiment. As Table 9 illustrates, in general, the theory of Marcus overestimates $\Delta G_0^{\ddagger}$ when $l_{red} - l_{ox} \gg 0$ and underestimates it when $l_{red} \approx l_{ox}$.

Table 9. Rates of electron exchanges calculated by the theory of Marcus.[a]

Reactions	$l_{red} - l_{ox}$/Å	k_{Marcus}/k_{exp}
DCB$^{0/-}$	0.183	$\approx 2 \times 10^{-9}$ [b]
AKHD$^{0/+}$	0.146	$4 \times 10^2 - 3 \times 10^4$
Fe(OH$_2$)$_6^{2+/3+}$	0.168	7×10^{-3}
Co(OH$_2$)$_6^{2+/3+}$	0.21	6×10^{-9}
Fe(CN)$_6^{4-/3-}$	0.01	9×10^2
Fe(phen.)$_3^{2+/3+}$	0.0	10^5

[a] Ref. 60; [b] Neglecting the solvent reorganization.

Neglecting any change in the bond lengths of the aromatic ring, ISM leads to the following expression,

$$(1/2) f x^2 = (1/2) f [(0.108 / n^{\ddagger}) l - x]^2 \qquad (71)$$

where $f_r = f_p = (f_{C \equiv N} + f_{C = N})/2$ (eq(41) and $l = l_{C \equiv N} + l_{C = N}$. Because for the meta-derivatives there is not a significant electronic coupling of the CN bonds with the aromatic ring, the transition state bond order is simply taken as the average of the order of the two reactive bonds and $n^{\ddagger} = 2.5$. For these reactions, which are not bond-forming-bond-breaking processes, such $n^{\ddagger}$ value corresponds to the neglect of the resonance effect. This set of equations allows the estimation of the average bond extension, x, of the CN bonds, and the calculated energy barrier (eq(44)) is $\Delta G^{\ddagger} = 12.5$ kJ mol^{-1} which corresponds to a rate constant of 4.5×10^{10} s^{-1} in agreement with the experimental observations. We have neglected any significant contribution, $\Delta G_{out}^{\ddagger}$, of the reorganization of the solvent for the overall energy barrier, by reasons discussed in detail in refs. 60,62 and 63.

Similar calculations [63] were performed for the intramolecular electron exchange of radical anions of benzene-1,3-dicarbaldehyde (3-BDC) in alcohols. The reactive bonds are C-O and C=O and $n^{\ddagger} = 1.5$; the estimated rate constant of 0.9×10^9 s^{-1} is in good agreement with experiment, 0.89-1.5×10^9 s^{-1} [64,65]. The fact that one neglects $\Delta G_{out}^{\ddagger}$, does not mean that the rates of electron transfer are solvent independent. Within ISM significant solvent effects in the energy barrier can be caused by solvents effects on f and/or $n^{\ddagger}$. There is a mild effect (ca. 2-5 times: acetone to methanol) in the rates due to a decrease of force constants with an increase in the polarity of the solvent, but the effects are much more important (up to 10^4 times) [60] due to changes in $n^{\ddagger}$, when there are "mobile electrons" in atoms involved in the reactive bonds. For the latter case the rates increase with an increase in the nonpolar character of the solvent.

A good example to illustrate these features is the intramolecular electron exchange in 1,3-dinitrobenzene (3-DNB) radical anion, which has rates of 1.2-2.2×10^6 s^{-1} in alcohols and in aprotic solvents 2.8-4.2×10^9 s^{-1}[64]. In polar solvents a structure such as

suggests that the bonds relevant for the reaction coordinate are the NO bonds; for the negatively charged group N-O$_2^-$ (l=1.36 Å) and N-O$_2$ (l=1.22Å) [33]. The force constants are not known for such bonds, but can be estimated with the equation of Gordy (eq(39)); this equation once calibrated with the C-O and C=O force constant and bond length data [33] leads to the following data: f=3.6×10^3 kJ mol^{-1} Å^{-2} and f=4.8×10^3 kJ mol^{-1} Å^{-2}, respectively.

There are two NO bonds for each group and they can act as local modes to lead to an effective force constant (eq(38)) of f(NO$_2^-$)=5.1×10^3 kJ mol^{-1} Å^{-2} and f(NO$_2$)=6.8×10^3 kJ

mol^{-1} $Å^{-2}$. So the average force constant for the oxidized and reduced groups is $f=6\times10^3$ kJ mol^{-1} $Å^{-2}$. With these $\underline{f}$ and $\underline{l}$ data and $n^{\ddagger}=1.25$, one can estimate, at room temperature, $\Delta G^{\ddagger}=37.3$ kJ mol^{-1} and $k=2.5\times10^6$ s^{-1} in good agreement with the experimental data in alcohols. This calculation interprets the much lower rates in 3-DNB$^-$ than in 3-BDC$^-$ in terms of an higher force constant and a slightly lower $n^{\ddagger}$.

In nonpolar solvents the lone pairs of the O-atoms interact less strongly with the solvent molecules than in polar media, and an increase in $n^{\ddagger}$ can occur, due to a resonance effect. The maximum value that one can predict through an "increased valence structure" such as

corresponds to $n^{\ddagger}=1.5$; the estimated energy barrier in nonpolar solvents with the same $\underline{f}$ and $\underline{l}$ data is $\Delta G^{\ddagger}=25.9$ kJ mol^{-1} and $k=3\times10^8$ s^{-1}. The estimated rate enhancement is 100 times, ca. an order of magnitude smaller than experiment. For 3-BDC$^-$ the estimated rate of electron exchange in a nonpolar medium, due to an increase in $n^{\ddagger}$ ($n^{\ddagger}=2$) is $k=7\times10^{10}$ s^{-1} too high to be measured by e.p.r. [64].

ISM can also be applied to the calculation of electron exchange reaction rates of coordination compounds and aquo-metal ions [60,62]. Table 10 presents some of the calculated data with the force constants estimated through eqs(38) and (41), with a coordination number of 6. The $n^{\ddagger}$ values are identical to the order of the reactive bonds in reactants and products. In contrast with the theory of Marcus, our model provides calculated rates within an order of magnitude of experiment [60].

TABLE 10. Calculated rate constants for electron exchange reactions by the ISM.[a]

	$l/Å$ [b]	f/kJ $mol^{-1}Å^{-2}$ [c]	$n^{\ddagger}$	k_{ISM}/k_{exp}
$Fe(OH_2)_6^{2+/3+}$	4.08	3.0	1	0.2
$Co(OH_2)_6^{2+/3+}$	4.03	3.0	1	0.9
$Fe(CN)_6^{4-/3-}$	3.83	6.7	1.54	0.3
$Fe(phen.)_3^{2+/3+}$	3.94	2.7	1	0.2
$AKHD^{0/+}$	2.79	3.0[d]	1	0.3-17

[a] Ref. 60, 62;   [b] $l_{ox}+l_{red}$;
[c] $\sqrt{6}$ $(f_{red}+f_{ox})/2\times10^3$;   [d] $(f_{red}+f_{ox})/2\times10^3$.

12.2. REACTION ENERGY EFFECTS

Let us now analyze another kind of electron transfer in organic compounds: reactions between alkylhydrazines (AKHD) and their radical cations [66]. Lone-pair-lone-pair interactions cause these molecules to undergo significant geometric changes upon electron loss. There are small changes in various bond lengths and angles, but the great effect is the N-N change from 1.469Å in the neutral molecule to 1.323Å in the monopositive ion [67]. If one assumes that the various small geometric changes are coupled to this large bond length variation, the reaction coordinate can be considered the N-N bond.

The equation of Gordy calibrated with the C-C and C-N stretching force constants and lengths [33], allows one to estimate for the N-N bond the following data: $f_{ox}= 5.5 \times 10^3$ kJ mol^{-1} Å^{-2} (l_{ox}=1.323Å) and $f_{red}= 3 \times 10^3$ kJ mol^{-1} Å^{-2} (l_{red}=1.469Å). This set of parameters with $n^{\ddagger}$=1 leads to $\Delta G^{\ddagger}$=48 kJ mol^{-1}. This value compares well with the experimental values for these reactions which range between 45-55 kJ mol^{-1} [66]. The increase in $n^{\ddagger}$ due to any resonance effect involving the lone pair of a nitrogen atom is only feasible in the absence of strong steric restrictions [68] which are present in these compounds [67].

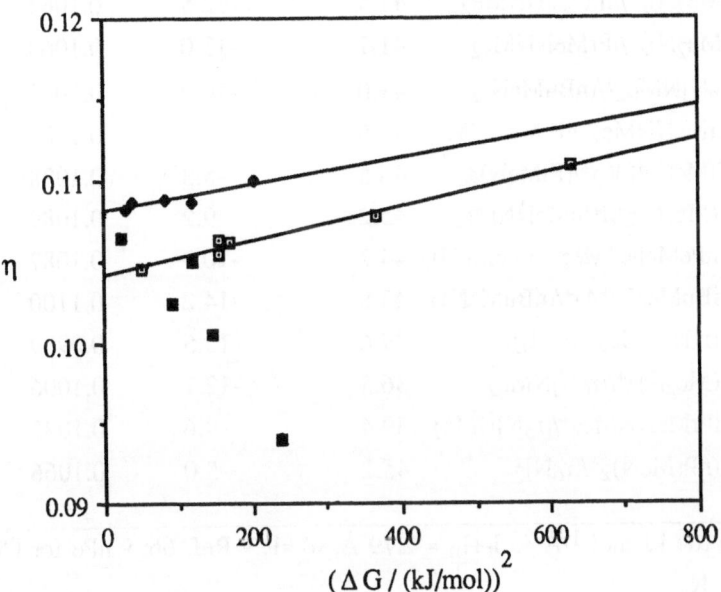

Figure 10. Plot of η versus ΔG^2 for electron transfer reactions between alkylhydrazines and their radical cations (data in Table 10).

Electron transfer reactions between different kinds of molecules have $\Delta G \neq 0$ and ISM predicts that if f, l and $n^{\ddagger}$ are constant, the reduced bond extension should reveal a square dependence on ΔG; η (Table 11) has been calculated with an effective force constant average of the ones presented above, $f=4.3\times10^3$ kJ mol^{-1} Å^{-2} and $l=2.79$Å. Figure 10 illustrates such a dependence for several reactions in the vapour phase [66]. Two reaction series are found virtually with the same λ, $\lambda = 85$ kJ mol^{-1}, and two $n^{\ddagger}$ values both very close to unity, $n^{\ddagger}=1.0$ and $n^{\ddagger}=1.04$. The large change in the N-N length suggests that the products should be formed with a considerable amount of vibrational energy and, in consequence, λ is predicted to have a low value.

TABLE 11. Reduced bond extensions for electron transfer reactions between acyclic hydrazines.[a]

Reaction	$\Delta G^{\ddagger}$/kJ mol^{-1}	ΔG/kJ mol^{-1}	η
1 (Me$_2$N)$_2$$^+$/(nPeMeN)$_2$	36.8	-25.1	0.1109
2 (Me$_2$N)$_2$$^+$/(nPrMeN)$_2$	39.5	-19.2	0.1078
3 (Me$_2$N)$_2$$^+$/(EtMeN)$_2$	40.6	-12.5	0.1055
4 (Me$_2$N)$_2$$^+$/EtMeNNMe$_2$	42.4	-7.1	0.1047
5 (Me$_2$N)$_2$$^+$/nPeMeNNMe$_2$	41.5	-12.5	0.1064
6 (Me$_2$N)$_2$$^+$/iPrMeNNMe$_2$	41.1	-13.0	0.1063
7 [r$_6$N]NMe$_2$$^+$/(nBuMeN)$_2$	41.0	-10.9	0.1051
8 EtMeNNMe$_2$$^+$/iPrMeNNMe$_2$	46.5	-5.4	0.1083
9 EtMeNNMe$_2$$^+$/(EtMeN)$_2$	46.5	-6.3	0.1088
10 (Me$_2$N)$_2$$^+$/iBuMeNNMe$_2$	45.2	-9.2	0.1089
11 nPeMeNNMe$_2$$^+$/(nPeMeN)$_2$	44.2	-10.9	0.1087
12 iBuMeNNM$_2$$^+$/(nBuMeN)$_2$	43.8	-14.2	0.1100
13 [r$_5$N]NMe$_2$$^+$/[r$_5$N]$_2$	29.6	-15.5	0.0939
14 (Me$_2$N)$_2$$^+$/[r$_6$N]NMe$_2$	36.5	-12.1	0.1006
15 iPrMeNNMe$_2$$^+$/[r$_5$N]NMe$_2$	39.4	-9.6	0.1025
16 (nPrMeN)$_2$$^+$/[r$_6$N]$_2$	45.2	-5.0	0.1066

[a] $f = 4.3\times10^3$ kJ mol^{-1} Å^{-2}, $l_r+l_p = 2.79$ Å, $n^{\ddagger}=1$; [b] Ref. 66; [c] nPe for CH$_3$(CH$_2$)$_4$;

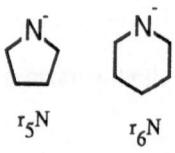

r_5N r_6N

A few reactions of cycloalkylhydrazines do not show such trends, possibly due to variations in $n^{\ddagger}$ and/or λ. For example, for compounds 12 to 16 in Figure 10, η decreases with a decrease in ΔG. As discussed elsewhere [42], this effect can be attributed to a significant change in $n^{\ddagger}$ upon substitution, associated with an increase in the electron affinity of the reactants.

Equation (29) is also verified with several electron transfer reactions between coordinated metal ions [60,69]. The consideration of the role of the mixing entropy parameter can even explain anomalous "cross-reaction" estimates given by the theory of Marcus [60, 70] and shines light on the controversy of the "inverted region" at low ΔG.

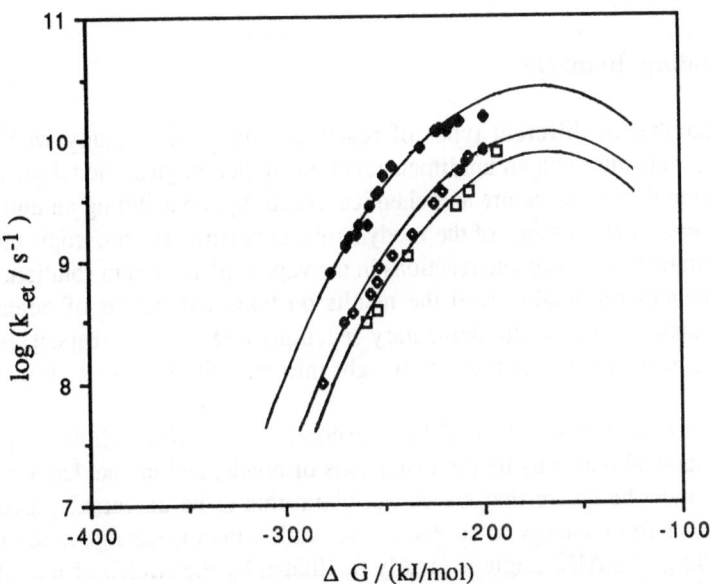

Figure 11. Fitting of the rates of back electron transfer within geminate radical-ion pairs (families of compounds with 1, 2 and 3 aromatic rings) in acetonitrile, as a function of ΔG; —— theoretical curves with the following data: $n^{\ddagger}=1$, $f_r=f_p=155$ J mol^{-1} pm^{-2}, $\lambda=203$ kJ mol^{-1} and l=461 pm 1-ring, 471 pm 2-rings, 475 pm 3-rings (by sequential order from the top) ; experimental points; refs.72 and 73.

The theory of electron transfer reactions is now a highly developed field, encompassing classical, semiclassical and quantum models [71]. Nevertheless, the ISM, a classical model, compares very favourably with other models, namely the quantum mechanical ones. ISM rationalizes a range of free-energy relationships in electron transfer organic reactions, encompassing those depicted by the Marcus inverted region, the Rehm-Weller behaviour, and a "double inverted region" not accounted for by the other models [72]. We have been presenting several examples of reactions where the force constants and bond lengths can be obtained from spectroscopic data. However, there are organic electron transfer reactions which involve stretching and bending vibrations and it is not possible to calculate how these force constants combine to yield the effective force constant for the reaction. In such cases, f_r, f_p, l_r+l_p, and λ have to be treated as adjustable parameters: four is also the number of adjustable parameters in the quantum models. Notwithstanding, studies of families of reactions allow the fitting of such experimental data with parameters entirely consistent with their physical meaning [72]. Data fitting is not completely arbitrary if one selects electron transfer reactions with a reactant in common (only f_r and l are expected to vary) or in different solvents (only λ is expected to vary). This is illustrated in Figure 11 for several return electron transfer reactions within geminate radical ion-pairs of families of compounds with one, two and three aromatic rings [72,73].

13. Concluding Remarks

The large number of different types of reactions amenable to study within the ISM-formalism reveals that such an unidimensional phenomenological model provides a good link between molecular structure and chemical reactivity, constituting an unifying way to rationalize kinetic data in terms of thermodynamic, geometric, spectroscopic and electronic molecular parameters for organic reactions in the vapour phase and in solution. Further, the model reproduces reasonably well the results on bond extensions of potential energy surface calculations of simple elementary reactions and, on that respect, is faithful to quantum chemistry theory, and offers insight into why the so called allowed reactions occur.

The sucess of such simple model it may come as a surprise with the emphasis on the control of chemical reactions by the extensions of bonds, and the neglect of the bending modes. One must be aware that we do not claim this to be universal. Nevertheless, in general, it takes more energy to stretch a bond rather than to change a bond angle and, further, bending of a ABC angle is greatly facilitated by the stretch of the AB (and BC) bonds. On that respect it is relevant to point out, as an example, that the quantum mechanical studies of Mitchell et al. [74] at the 4-31G computational level lead to an average energy of -1 kJ mol^{-1} for the bending and stretch-bend deformation of several nucleophilic methyl transfer reactions; the overall deformation at the transition state is dominated by the stretches of the CX and CY bonds.

ACKNOWLEDGEMENTS

I am grateful to the Instituto Nacional de Investigação Científica for the financial support. This paper has benefit from the comments of Prof. N. Agmon and Dr. B.C. Garrett.

14. References

[1]   Salem, L. *Electrons in Chemical Reactions. First Principles*; John Wiley, New York, 1982;   chap. 6.
[2]   Polanyi, J.C.; Schreiber, J.L. In *Physical Chemistry. An Advanced Treatise;* Jost,W. (Ed.); Academic Press, New York, 1974; chap. 6; Polanyi, J.C. *Disc. Faraday Soc.* **1973**, *55*, 389.
[3]   Evans, M.G.; Polanyi, M. *Trans. Faraday Soc.* **1938**, *34*, 11; Evans, M.G. *ibid.* **1939**, *35*, 824; Marcus, R.A. *Disc. Faraday Soc.* **1960**, *29*, 21; for a recent review on intersecting curve models see ref. 24.
[4]   (a) Johnston, H.S.; Parr, C. *J. Am. Chem. Soc.* **1963**, *85*, 2544; (b) Murdoch, J.R. *J. Mol. Struct. (Theochem.)* **1988**, *163*, 447; (c) Agmon, N. *Int. J. Chem. Kinet.* **1981**, *13*, 333.
[5]   (a) Agmon, N.; Levine, R.D. *Chem Phys. Lett.* **1977**, *52*, 197; (b) Shannon, C.E. *Bell System Technical Journal* **1948**, 27, 379, 623.
[6]   (a) Pauling, L. *J. Am. Chem. Soc.* **1947**, *69*, 542; (b) Agmon, N. *Chem. Phys. Lett.* **1977**, *45*, 343.
[7]   Varandas, A.J.C.; Formosinho, S.J. *J. Chem. Soc. Faraday Trans. 2* **1986**, *82*, 953.
[8]   Harcourt, R.D. *J. Chem. Educ.* **1968**, *45*, 779; *Qualitative Valence-Bond Description of Electron-Rich molecules. Pauling "3-Electron Bonds" and "Increased-Valence" Theory*, Lectures Notes in Chem., Springer-Verlag, vol. 30, 1982, chap. 4, 10-12, 16.
[9]   Evans, M.G.; Warhurst, E. *Trans. Faraday Soc.* **1938**, *34*, 614; Evans, M.G. *ibid.* **1939**, *35*, 824.
[10] Formosinho, S.J. *J. Chem. Soc. Perkin Trans. 2* **1988**, 839.
[11] Varandas, A.J.C. *J. Chem. Phys.* **1979**, *70*, 3786.
[12] (a) Last, I.; Baer, M. *J. Chem. Phys.* **1981**, *75*, 288; Schwenke, D.W.; Tucker, S.C.; Steckler, R.; Brown, F.B.; Lynch, G.C.; Truhlar, D.G. *ibid.* **1989**, *90*, 3110; (b) Dunning Jr., T.H. *J. Phys. Chem.* **1984**, *88*, 2469.
[13] Lendvay, G.; László, B.; Bérces, T.*Chem Phys. Lett.* **1987**, *137*, 175; Lendvay, G. *J. Mol. Struct. (Theochem.)* **1988**, *167*, 331; *J. Phys. Chem.* **1989**, *93*, 4422.
[14] Mok, M.H.; Polanyi, J.C. *J. Chem. Phys.* **1969**, *51*, 145; Polanyi, J.C. *Angew. Chem. Int. Ed. Eng.* **1987**, *26*, 952.
[15] Formosinho, S.J.; Arnaut, L.G. to be published.
[16] Dewar, M.J.S. *J. Am. Chem. Soc.* **1984**, *106*, 209.
[17] Gordy, W. *J. Chem. Phys.* **1946**, *14*, 305.
[18] Marcus, R.A. *J. Phys. Chem.* **1968**, 72, 891.
[19] Koeppl, G.W.; Kresge, A.J. *J. Chem. Soc. Chem. Commun.* **1973**, 371.
[20] (a) Kreevoy, M.M.; Lee, I.H. *J. Am. Chem. Soc.* **1984**, *106*, 2550; (b) Kreevoy, M.M.; Ostovic, D.; Truhlar, D.G.; Garrett, B.C. *J. Phys. Chem.* **1986**, *90*, 3766.
[21] Grunwald, E. *J. Am. Chem. Soc.* **1985**, *107*, 125.

204

[22] Woodward, R.B.; Hoffmann, R. *The Conservation of Orbital Symmetry;* Academic Press, New York, 1970; Fukui, K. In *Modern Organic Chemistry;* Sinanogulu, O. (Ed.); Academic Press, 1965; pag. 49.

[23] Pross, A.; Shaik, S.S. *Acc. Chem. Res.* **1983**, *16*, 363.

[24] Yates, K. *J. Phys. Org. Chem.* **1989**, *2*, 300.

[25] Keeffe, J.R.; Kresge, A.J. In *Investigations of Rates and Mechanisms of Reactions;* Bernasconi, C. (Ed.); John Wiley, New York, 1986; chap. 11.

[26] More O'Ferrall, R.A. *J. Chem. Soc. B* **1970**, 274; Thornton, E.R. *J. Am. Chem. Soc.* **1967**, *89*, 2915; Jencks, W.P. *Chem. Rev.* **1972**,*72*, 705; Bernasconi, C.F. *Acc. Chem. Res.* **1987**, *20*, 301; Albery, W.J.; Kreevoy, M.M. *Adv. Phys. Org. Chem.* **1978**, *16*, 87; Murdoch, J.R. *J. Am. Chem. Soc.* **1983**, *105*, 2159, 2667; Murdoch, J.R.; Magnoli, D.E. *J. Am. Chem. Soc.* **1981**, *103*, 7465, **1982**, *104*, 3792.

[27] Agmon, N. *J. Am. Chem. Soc.* **1984**, *106*, 6960.

[28] Kurz, J.L. *J. Org. Chem.* **1983**, *48*, 5117.

[29] Villars, D.S. *J. Am. Chem. Soc.* **1930**, *52*, 1733.

[30] Formosinho, S.J. *J. Chem. Soc. Faraday Trans. 2* **1974**, *70*, 605; *ibid.* **1976**, 72, 1313; Arnaut, L.G.; Formosinho, S.J.; da Silva, A.M. *J. Photochem.* **1984**, *27*, 185; Arnaut, L.G.; Formosinho, S.J. *ibid.* **1987**, *39*, 13; Formosinho, S.J.; Arnaut, L.G. *Adv. Photochem.* **1991**, *16*, 36.

[31] Watt, C.I.F. *Adv. Phys. Org. Chem.* **1988**, *24*, 57.

[32] See for example, Kreevoy, M.M.; Truhlar, D.G. In *Rates and Mechanisms of Reactions;* Bernasconi, C. (Ed.); John Wiley, New York, 4th ed., 1986; chap. 1.

[33] Gordon, A.J.; Ford, R.A. *The Chemist's Companion;* John Wiley, New York, 1972; pag. 107 and 114.

[34] Formosinho, S.J. *J. Phys. Org. Chem.* **1990**, *3*, 325.

[35] Meot-Ner (Mautner), M.; Field, F.H. *J. Am. Chem. Soc.* **1978**, *100*, 1356.

[36] (a) Formosinho, S.J. *J. Chem. Soc. Perkin Trans. 2* **1987**, 61; (b) Formosinho, S.J.; Gil, V.M.S. *ibid.* **1987**, 1655.

[37] Sathyamurthy, N.; Joseph, T. *J. Chem. Educ.* **1984**, *61*, 968.

[38] Yates, K. *J. Am. Chem. Soc.* **1986**, *108*, 6511.

[39] Arnaut, L.G.; Formosinho, S.J. *J. Phys. Chem.* **1988**, *92*, 685.

[40] McAuley, I.; Krogh, E.; Wan, P. *J. Am. Chem. Soc.* **1988**, *110*, 600; Wan, P.; Krogh, E.; Chack, B. *ibid.* **1988**, *110*, 4073.

[41] Bernasconi, C.F.; Bunnell, R.D. *Isr. J. Chem.* **1985**, *26*, 420; Bernasconi, C.F.; Paschalis, P. *J. Am. Chem. Soc.* **1986**, *108*, 2969.

[42] Arnaut, L.G.; Formosinho, S.J. *J. Phys. Org. Chem.* **1990**, *3*, 95.

[43] Formosinho, S.J. *Tetrahedron.* **1987**, *43*, 1109.

[44] Wolfe, S.; Mitchell, D.J.; Schlegel, H.B. *J. Am. Chem. Soc.* **1981**, *103*, 7692, 7694.

[45] Carrion, F.; Dewar, M.J.S. *J. Am. Chem. Soc.* **1984**, *106*, 3531.

[46] Formosinho, S.J.; Arnaut, L.G. *J. Chem. Soc. Perkin Trans. 2* **1989**, 1947.

[47] Minot, C.; Anh, N.T. *Tetrahedron. Lett.* **1975**, 3905.

[48] Pellerite, M.J.; Brauman, J.I. *J. Am. Chem. Soc.* **1980**, *102*, 5993; **1983**, *105*,

2672; Dodd, J.A.; Brauman, J.I. *ibid.* **1984**, *106*, 5356.

[49] Tykodi, R.J. *J. Chem. Educ.* **1986**, *63*, 107.

[50] Formosinho,S.J. *Tetrahedron.* **1986**, *42*, 4557.

[51] Gajewski, J.J. *J. Am. Chem. Soc.* **1979**, *101*, 4393.

[52] Gajewski, J.J. *Acc. Chem. Res.* **1980**, *13*, 142; Murdoch, J.R. *J. Am. Chem. Soc.* **1983**, *105*, 2660.

[53] See for example Hammond, R.B.; Williams, I.H. *J. Chem. Soc. Perkin Trans. 2* **1989**, 59.

[54] Jencks, W.P. *Chem. Soc. Rev.* **1981**, *10*, 345.

[55] Leffler, J.E. *Science* **1953**, *117*, 340.

[56] Hammond, G.S. *J. Am. Chem. Soc.* **1955**, *77*, 334.

[57] Ritchie, C.D. *Acc. Chem. Res.* **1972**, *5*, 348; *Can J. Chem.* **1986**, *64*, 2239.

[58] Formosinho, S.J., to be published.

[59] Marcus, R.A. In *Understanding Molecular Properties Symp.;* Avery, J. *et al.* (Eds.); Reidel-Dordrecht, Neth., 1986; pag. 229-236.

[60] Formosinho, S.J. *Pure Appl. Chem.* **1989**, *61*, 891.

[61] Rieger, P.H.; Bernal, I.; Reinmuth, W.H.; Fraenkel, G.K. *J. Am. Chem. Soc.* **1963**, *85*, 683.

[62] Formosinho, S.J. *Rev. Port. Quim.* **1986**, *28*, 38.

[63] Formosinho, S.J. *J. Chem. Soc. Perkin Trans. 2* **1988**, 1209.

[64] Grampp, G.; Shohoji, M.C.B.L.; Herold, B.J. *Ber. Bunsenges. Phys. Chem.* **1989**, *93*, 580.

[65] Shohoji, M.C.B.L.; Herold, B.J.; Novais, H.M.; Steenken, S. *J. Chem. Soc. Perkin Trans. 2* **1986**, 1465.

[66] Nelsen, S.F.; Rumack, D.T.; Meot-Ner (Mautner), M. *J. Am. Chem. Soc.* **1987**, *109*, 1373.

[67] Nelsen, S.F.; Cunkle, G.T.; Evans, D.H.; Haller, K.J.; Katfory, M.; Kirste, B.; Kurreck, H.; Clark, T. *J. Am. Chem. Soc.* **1985**, *107*, 3829.

[68] Burrows, H.D.; Formosinho, S.J. *J. Chem. Soc. Faraday Trans. 2* **1986**, *82*, 1563.

[69] Formosinho, S.J. *Rev. Port. Quim.* **1986**, *28*, 48.

[70] Burrows, H.D.; Formosinho, S.J. *Rev. Port. Quim.* **1986**, *28*, 57.

[71] See for example Newton, M.D.; Sutin, N. *Ann. Rev. Phys. Chem.* **1984**, *35*, 437; Marcus, R.A.; Sutin, N. *Biochem. Biophys. Acta* **1985**, *811*, 265.

[72] Arnaut, L.G.; Formosinho, S.J. *J. Mol. Struct. (THEOCHEM)* in press.

[73] Gould,I.R.; Ege, D.; Moser, J.E.; Farid, S. *J. Am. Chem. Soc.* **1990**, *112*, 4290.

[74] Mitchell, D.J.; Schlegel, H.B.; Shaik, S.S.; Wolfe, S. *Can. J. Chem.* **1985**, *63*, 1642.

THE STATES OF AN ELECTRON PAIR AND PHOTOCHEMICAL REACTIVITY

JOSEF MICHL
Center for Structure and Reactivity
Department of Chemistry and Biochemistry
The University of Texas at Austin
Austin, Texas, U.S.A. 78712-1167

ABSTRACT. Photochemical reaction paths tend to pass through biradi-caloid geometries, making a detailed understanding of the electronic states of molecules at such geometries highly desirable. The simplest model that correctly describes their fundamental features, the two-electron two-orbital model, is reformulated in a way that emphasizes the isomorphism between the spaces of the spin functions and the orbital and geminal functions, using angular momentum operators for both. This unifying approach brings to light relations between phenomena that appear quite unrelated otherwise, such as the response of the triplet sublevels to an external magnetic field and the response of the singlet states of a perfect biradical to structural perturbations. It also suggests a specific identification of the three fundamental linearly independent types of perturbations of the states of an electron pair: covalent, polarizing, and magnetizing perturbations.

1. Organic Photochemistry: Why are Electronic States of Biradicaloids of Interest

In this introductory section we argue that an understanding of the nature of the electronic states of an electron pair, and specific-ally, of biradicaloids, is essential for insight into photochemical reactivity.

In solution photochemistry of organic molecules, three kinds of molecular geometries play a special role. They correspond to minima in the lowest excited singlet (S_1) or triplet (T_1) surfaces, or to conical intersections of S_1 with the ground state surface S_0 ("fun-nels"). As electronically excited molecules thermalize their vibra-tional motion after the initial excitation or after a radiationless jump from a higher electronic state to a lower one (internal conver-sion if multiplicity is conserved, intersystem crossing if it is not), they collect at these minima, or pass through the funnels to the ground state.

The three kinds of special S_1 and T_1 potential energy surface regions are (i) "spectroscopic" minima (S), (ii) "excimer" or "exci-plex" minima (E), and (iii) "biradicaloid" minima (B). Their typical role in organic photochemistry can be represented schematically as follows:

S. J. Formosinho et al. (eds.), Theoretical and Computational Models for Organic Chemistry, 207–251.
© 1991 *Kluwer Academic Publishers.*

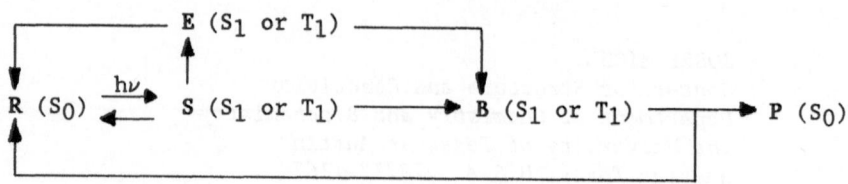

Here, $R(S_0)$ stands for a reactant molecule in the S_0 minimum at its equilibrium ground state geometry, the arrow labeled $h\nu$ indicates vertical excitation by photon absorption or by energy transfer to a spectroscopic minimum S in the S_1 or T_1 state, and the other arrows indicate further fate of this initial vertically excited species. Travel on the S_1 or T_1 surface may take the molecule to an excimer or exciplex minimum E and to a biradicaloid minimum B (an S_0 - S_1 conical intersection is viewed as a limiting case of B). Vertical radiative or non-radiative return from S or E restores R in its ground state (a photophysical outcome). If the return to the S_0 state occurs from B, it is still possible for R to be restored, but there also is some probability of forming a new product P (a photochemical outcome).

While this scheme does not cover all possibilities, it is believed to be representative of a vast majority of organic photochemical processes. For a far more detailed description, see a recent monograph [1].

1.1. Spectroscopic Minima

These minima in S_1 or T_1 are located at geometries close to equilibrium ground state geometries and correspond to the "vertical" or "spectroscopic" excited states. They typically are the first ones in which an excited molecule settles, since the initial excitation normally is vertical. In solution, radiative or radiationless transition to the S_0 surface usually returns the molecules right back to the ground-state reactant and yields no net photochemical change (fluorescence or internal conversion $S_1 \rightarrow S_0$, and phosphorescence or intersystem crossing $T_1 \rightarrow S_0$). In the gas phase the resulting hot ground-state molecule may fragment or isomerize, yielding a net chemical change after all, but in solution thermalization is rapid and such "hot ground-state reactions" seldom occur.

Only in rare instances does an excited molecule reach an S_1 or T_1 spectroscopic minimum of another species by travel on the excited electronic surface, and the return to S_0 then produces the new species as a photoproduct. Such "adiabatic photochemistry" is fairly common in proton transfer reactions but is encountered only infrequently among organic photoreactions that involve more substantial structural reorganization.

A thorough understanding of the spectroscopic minima in S_1 and T_1 is clearly essential for the understanding of the electronic spectra and of the photophysics of the reactant. The existence of these minima is crucial for photochemical processes as well, since they serve as "holding reservoirs", providing the excited molecules with the time needed to escape over small barriers to other minima in S_1 or T_1, located at geometries more strongly differing from that of the reactant. Typically, the latter are "biradicaloid" minima, and return to S_0 is then often followed by the formation of new ground-state species. The travel on the S_1 and T_1 surfaces then represents the chemical transformation proper, i.e., a change in the bonding relations between the individual constituent atoms of a molecule or molecules.

The holding action of the S_1 or T_1 spectroscopic minima is also essential for reaching excimer and exciplex minima and for bimolecular processes in general, since their rates are constrained by the need for diffusion to bring up the requisite ground-state partner. An example of the holding action is a sojourn of a ketone molecule in its long-lived excited triplet state, to which diffusion brings an olefin, with the ultimate formation of an oxetane.

1.2. Excimer and Exciplex Minima

These minima in S_1 and T_1 occur at geometries that correspond to fairly intimate contact between two molecules, each at a geometry close to its ground state equilibrium geometry. The term excimer is used if the two partners are identical (A_2^*) and the term exciplex is used if they are not (AB^*). Most such minima do not have a ground-state counterpart, at least not in solution. This is due to the fact that molecules in their S_0 state are much less "sticky" than electronically excited ones. In the ground state of the molecular pair A_2 or AB, both partners are in their respective S_0 states, and closed shells resist close contact. In the excited state of the pair, one partner is typically in its S_1 or T_1 state, while the other is in its S_0 state (A^*B or AB^*, the excitation may be shared by both to a smaller or larger degree). Alternatively, an electron has been transferred from one partner to the other, so that the excited species corresponds to a contact ion pair $A \cdot \bar{B} \cdot$. In the general case, the locally excited (A^*B, AB^*) and the charge-transfer ($A \cdot \bar{B} \cdot$, $A \cdot \bar{B} \cdot$) wave functions mix, but often, one of the contributions clearly prevails. An open shell – closed shell interaction is frequently attractive, as is the interaction of oppositely charged ions.

In the gas phase, there usually is a very shallow van der Waals minimum at a somewhat larger separation between ground-state partners, but in solution, this is often insufficient for keeping them in close contact, due to their prevalent interaction with the solvent. In some cases, pairwise complexation of ground-state reactant molecules occurs even in solution, as in so-called charge-transfer complexes, and in that case the designation of the excited species as an excimer or an exciplex is not strictly correct: the minimum in S_1 or T_1 then refers to an excited charge-transfer complex.

The minima in this category are usually reached when diffusion brings together one partner in its ground state and the other in its spectroscopic S_1 or T_1 state (other possibilities are less common, e.g., a $T_1 + T_1$ encounter of two molecules in their spectroscopic minima, or ion recombination). Vertical radiative or non-radiative transition to the S_0 surface produces two ground-state partners free to diffuse apart and thus results in no net chemical change. In this sense, excimer and exciplex minima, like spectroscopic minima, are of photophysical rather than photochemical nature, even though a reasonably strong bonding interaction may exist between the partners in the excited state, particularly in S_1.

Nevertheless, the existence of excimer and exciplex minima is crucial for many bimolecular photochemical reactions. They, too, serve as holding reservoirs for travel across barriers along true chemical reaction paths on the excited S_1 or T_1 surfaces. By providing the excited pair of partners A and B with the time needed to overcome reaction barriers in the S_1 or T_1 surface, they permit the system to reach biradicaloid minima at quite different geometries, and after return to S_0, to form new products. Examples of such reaction paths are cycloadditions and, in exciplexes with a high degree of charge transfer, proton transfer processes in which A$\cdot$ acts as a Brønsted base and B$\cdot$ as a Brønsted acid, ultimately yielding a radical pair.

1.3. BIRADICALOID MINIMA

These minima in the S_1 and T_1 states occur at biradicaloid geometries, i.e., those at which the reacting species would have one fewer bond in its ground state than permitted by the rules of valence and in that sense would be a double radical. The "missing" bond can be a localized one, as in the ground state of the linear-chain biradical $\cdot CH_2CH_2CH_2CH_2\cdot$, or a delocalized one, as in the ground state of the cyclic biradical, square cyclobutadiene (C_4H_4). Biradicaloid minima usually correspond to "non-vertical" states in that the corresponding ground state biradicaloid species are hardly ever available for experimentation. These minima are normally reached by travel on the S_1 or T_1 surface starting at spectroscopic minimum, possibly via an excimer or exciplex minimum.

At biradicaloid geometries, the S_0 and T_1 states usually lie fairly close in energy, and either one can lie lower and represent the ground state. This is quite understandable: to reach a biradicaloid geometry from a usual ground-state equilibrium geometry in S_0, the molecule or pair of molecules had to suffer a distortion that destroyed a bond, e.g., by stretching a sigma bond or twisting a pi bond, and the termini of the now absent bond no longer care very much whether they are singlet-coupled or triplet-coupled. Clearly, biradicaloid geometries are energetically quite unfavorable on the S_0 surface, and it is rare for this surface to contain a significant local potential energy minimum at such a geometry. In a simple MO description, a molecule at a biradicaloid geometry contains two approximately non-bonding orbitals occupied by a total of only two electrons in the ground state. It is then also clear why such geometries are geometrically rather favorable on the S_1 and T_1 surfaces and frequently correspond to local minima. As the molecule is distorted towards the more normal geometries the two radical centers begin to develop a stabilizing bonding interaction in the S_0 state. In the simple MO picture, this is represented by the placement of both electrons into the bonding combination offered by the previously approximately non-bonding orbital pair. In the S_1 and T_1 states, the bonding and the antibonding combination are each occupied once, and no energy saving results. Inasmuch as the antibonding orbital is usually destabilized more than the bonding orbital is stabilized, in the S_1 and T_1 state a distortion towards a more normal geometry is actually destabilizing. This simple argument suggests why the S_1 and T_1 states are likely to contain local minima at biradicaloid geometries, while S_0 is not.

A more rigorous definition of a biradicaloid geometry employs the concept of natural orbitals and their occupancies, which are well defined at all levels of quantum mechanical description of molecules in the Born-Oppenheimer approximation: at a biradicaloid geometry, two of the ground state natural orbital occupation numbers are approximately equal to unity (the others, of course, are close to two or close to zero).

Of the three important kinds of minima in S_1 and T_1 surfaces, the biradicaloid minima are clearly the most fundamental for the understanding of light-induced chemical transformations. Return from such a minimum lands the point representing the molecule in a region of the S_0 surface that is rarely accessible by thermal processes in the S_0 state. It is high in energy and permits easy motion towards at least one and possibly several different minima at the more usual geometries at which the missing bond is restored. Thus, at least for some percentage of the initially excited molecules, a chemical transformation to another ground-state species is the likely ultimate outcome. Return through biradicaloid minima is responsible for most organic photochemical reactions, and they could also be called "reactive" minima. Funnels (S_0 - S_1 conical intersections) are expected to occur at biradicaloid geometries as well and can be viewed as extreme cases of biradicaloid minima in S_1 in which the S_0 - S_1 splitting vanishes.

Occupancy of a minimum in S_1 provides an opportunity for a more or less rapid radiationless transition to S_0 in competition with a radiative transition to S_0, a transition to T_1, and adiabation escape over the surrounding barrier to other regions of the S_1 surface (by thermal activation or by tunneling). In contrast, passage through the region of a conical S_0 - S_1 intersection causes an essentially instantaneous nonradiative transition from the S_1 to the S_0 surface, providing little if any opportunity for competing processes. Depending on the overall tilt of the S_1 surface in the region in which the double cone occurs, its S_1 part may but need not correspond to a local minimum in S_1.

1.4. THE NEED FOR THEORY

We conclude that a thorough understanding of the electronic structure of molecules at the three types of geometries at which minima and funnels in the excited state surfaces are most likely to occur is essential for an understanding of the course of organic photochemical reactions. While the spectroscopic minima and the excimer and exciplex minima occur at more or less ordinary molecular geometries, for which qualitative understanding and quantitative methods of computation are widely available and are covered in many elementary and advanced texts, the biradicaloid minima and funnels occur at geometries that are much less familiar from ground-state chemistry and also require a more complicated theoretical treatment, both at the qualitative and the quantitative levels. In the following, we describe a qualitative model suitable for the description of molecular electronic structure at biradicaloid geometries and for the identification of geometries at which S_0 - S_1 conical intersections are likely to occur. For numerical computations, the CAS SCF procedure followed by CI is most appropriate; its use is described elsewhere in this volume.

1.5. THE TWO-ELECTRON TWO-ORBITAL MODEL

In the preceding section we have outlined the motivation for the study of the electronic states of biradicals and biradicaloids. Next, we shall describe a simple standard model that provides a useful approximate description of the low-energy electronic states of these species by focusing on the two least firmly bound electrons. The model considers the configurations built from the two frontier orbitals with their two electrons as the active many-electron space, and treats the rest of the electronic structure as a non-polarizable core. It is often referred to as the 3×3 CI model, or more accurately as the $3 \times 3 + 1 \times 1$ CI model, and this reflects the fact that its singlet configuration space is three-dimensional and its triplet configuration space one-dimensional. Due to its extreme simplicity, this model for the states of an electron pair can be solved algebraically. The results provide considerable insight into the behavior of the electronic wave functions and their energies as a function of molecular structure and outside perturbations. In spite of its simplicity, the model appears to embody most of the essential physics of the problem. It mimics the results of large-scale calculations very successfully and has been used as a unifying concept in a recent treatment of the theoretical aspects of organic photochemistry in a monograph [1].

While the model renders the qualitative nature of the wave functions of the lowest few electronic states very well, it occasionally

inverts the energetic order of nearly degenerate states. Such rever-
sals tend to be important but can be usually anticipated and understood upon more detailed qualitative consideration of the effect of
the electrons attributed to the "fixed core".

The initial systematic introduction of the model into photochemistry is now nearly two decades old [2-5]. More recently, the
model has been elaborated [1,6,7] and its utility for the prediction
and rationalization of geometries at which S_0 - S_1 conical intersections occur has been recognized [1,6-8]. While most of the standard
expositions have concentrated on the orbital and configuration (in
this case, geminal) space as opposed to the spin space part of the
electronic problem, one of the two early reviews treated spin-orbit
coupling as well and formulated useful general rules [2] which have
since found support in *ab initio* calculations [9].

Detailed accounts of the two-electron two-orbital model are
available elsewhere [1-7]. Here, we present a more formalized
pedagogical presentation based on the isomorphism of the spin space
and the geminal space used in the model, in effect extending the
standard "fictitious spin of 1/2" treatment [10] of a two-level one-
particle quantum mechanical system to a "fictitious spins of 0 and 1"
treatment of a four-level two-particle system.

We adopt the usual convention and write basis sets as row
matrices and functions as column matrices, so that the former are
post-multiplied and the latter pre-multiplied by the square matrices
representing operators. Operators are indicated by a caret, vectors
by a bold letter, and tensors by a doubly underlined bold letter.
Cartesian axes are labeled x,y,z, and a general label for any one of
these is u, v or w.

1.6. Organization of the Text

We shall consider first the one-electron aspects of the model.
The spin space and the orbital space are both two-dimensional. We
find an amusing parallel between the three canonical basis sets that
can be used to span the former (adapted to the x, y, and z molecular
axes) and the three that can be used to span the latter (delocalized,
complex, and localized). More general basis sets in either space can
be produced by applying the rotation operator for a particle of spin
1/2. Since we shall eventually deal with the exact solutions both in
the spin space and in the real space (equivalent to full CI), the
choice of the one-electron basis is immaterial. In practice, as we
construct the two-electron wave functions from one-electron wave
functions, we have to choose the latter somehow, and this choice is
frequently dictated by convenience. For instance, the two-dimensional
active space of orbitals may have been defined for us by an open-
shell SCF calculation on the triplet state of a series of related
biradicaloids, which yielded two singly occupied orbitals in each
case, but not necessarily in either the most localized or the most
delocalized form, or even in similar forms for the different mole-

cules. If these orbitals were used as a basis for large-scale singlet CI calculations on these species, and we wish to understand the qualitative features of the results in terms of the simple model, it will clearly be useful to have the ability to transform the one-electron basis set in all of these molecules to some common form, say the most localized orbitals or the most delocalized orbitals.

A general operator in either the spin or the orbital space can be written in terms of the angular momentum operator for a particle of spin 1/2, represented by the Pauli matrices. Casting the Hamiltonian operator in this form provides a natural identification of a perfect biradical as the reference system, and of three linearly independent types of fundamental perturbation: covalent, magnetizing, and polarizing.

We close the one-electron part by considering the spin-orbit coupling operator, which acts simultaneously in the spin and in the orbital space.

Next, we consider the two-electron states and energies. This is essential since the ultimate objective is to find the exact two-electron eigenstates and eigenvalues of the two-electron two-orbital system. This can be done only because the model is so very simple that the solution requires a diagonalization of matrices no larger than 3×3.

The amusing isomorphism of the one-electron spin space and the orbital space is encountered again as we consider two-electron spin wave functions and two-electron orbital (geminal) wave functions. We refer to the two spaces as the spin space and the geminal space. We first construct two-electron basis functions from the one-electron functions. We then use the rotation operators for particles of spin 0 and 1 to describe how the two-electron basis transforms when the one-electron basis is subjected to an arbitrary unitary transformation. Once again, a general operator in either space can be written in terms of identity and the angular momentum operator for a particle of spin 1. Casting the Hamiltonian operator in this form, we again find perfect analogy between the behavior of the spin eigenfunctions in the absence of a magnetic field and the behavior of the geminal eigenfunctions of a perfect biradical. As in the one-electron case, introduction of an outside magnetic field affects the spin part of the total two-electron wave function in exactly the same ways as a perturbation that converts a biradical into a biradicaloid affects its geminal part. We then redefine the perfect biradical, the covalent perturbation, and the polarizing perturbation so as to take proper account of the two-electron part of the Hamiltonian.

Because of the spin-space geminal-space isomorphism found in the simple model, consideration of the behavior of the energies and the wave functions runs along completely parallel tracks in the two cases. This includes the identification of conditions under which the lowest two states are degenerate, of particular interest for photochemistry. Phenomena such as "sudden polarization" and "orbital angular momentum quenching" all fit into a single picture.

Finally, we consider the spin-orbit coupling term in the Hamiltonian, which acts in a space defined as the direct product of the spin and the geminal space. Within the framework of the simple model, we find that in a perfect biradical the triplet state spin-orbit couples only to the highest-energy singlet state S_2.

2. The One-Electron Spin and Orbital Function Spaces

The spin space and the orbital space are isomorphic in that they are each spanned by two orthogonal basis functions.

2.1. Spin Space

The usual choice of the basis functions in the spin space is α and β. These functions are adapted to the z axis in real space in the sense of being the eigenfunctions of the z component of the spin angular momentum operator $\hbar\hat{\sigma}_z/2$, with eigenvalues $+\hbar/2$ and $-\hbar/2$, respectively. The usual choice of spin functions similarly adapted to the orthogonal y and x axes in real space is $(\alpha \pm i\beta)/\sqrt{2}$ and $(\alpha \pm \beta)/\sqrt{2}$, respectively. These are eigenfunctions of the spin angular momentum operators $\hbar\hat{\sigma}_y/2$ and $\hbar\hat{\sigma}_x/2$, respectively.

We shall adopt a slightly different choice of the spin basis functions. In order to guarantee cyclic permutation properties with respect to the indices x, y, and z in the following, we obtain the spin functions adapted to the y axis (α^y, β^y) and those adapted to the x axis (α^x, β^x) from the usual ones, adapted to the z axis $(\alpha^z = \alpha,$ $\beta^z = \beta)$, by a $+2\pi/3$ and a $-2\pi/3$ rotation, respectively, around an axis n with the direction cosines $n_x = n_y = n_z = 1/\sqrt{3}$. Applying the standard rotation operator [10,11] for a particle of spin 1/2,

$$\hat{R}_u^{1/2}(\omega) = \cos(\omega/2) - i\hat{\sigma}\cdot n \, \sin(\omega/2)$$

(1)

$$[\underline{R}_u^{1/2}(\omega)]_{\alpha\beta} = \begin{pmatrix} \cos(\omega/2) - in_z\sin(\omega/2) & -(n_y + in_x)\sin(\omega/2) \\ (n_y - in_x)\sin(\omega/2) & \cos(\omega/2) + in_z\sin(\omega/2) \end{pmatrix}$$

we obtain the three basis sets listed in Table 1, where $\hat{\sigma} = (\hat{\sigma}_x, \hat{\sigma}_y, \hat{\sigma}_z)$. The spin functions α^u and β^u are eigenfunctions of the dimensionless operator $\hat{\sigma}_u$ acting in the spin space, with eigenvalues 1 and -1, respectively ($u = x$, y, or z): $\hat{\sigma}_u\alpha^u = \alpha^u$, $\hat{\sigma}_u\beta^u = -\beta^u$. The representations of the three components $\hat{\sigma}_v$ ($v = x,y,z$) of the operator $\hat{\sigma}$ are the three Pauli matrices. They are shown in Table 2, along with the representation of an operator proportional to the square of the electron spin angular momentum, $\hat{\sigma}^2 = \Sigma_v\hat{\sigma}_v^2$, for all three basis sets given in Table 1 ($v = x$, y, or z). The cyclic permutation properties with respect to x, y, and z are clearly illustrated.

TABLE 1. One-electron basis functions

Basis set:

$a^z = a$ $a^y = [(1 + i)/2](a + ib)$ $a^x = [(1 - i)/2](a + b)$

$b^z = b$ $b^y = [(1 + i)/2](a - ib)$ $b^x = [-(1 + i)/2](a - b)$

Orbital space   *Spin space*

$a = A$, $b = B$ $a = \alpha$, $b = \beta$

TABLE 2. Representations of the operators $\hat{\sigma}_v$ and $\hat{\sigma}^2/3$ in the bases σ^u, β^u or A^u, B^u

	u:	α^u, β^u or A^u, B^u		
v:		x	y	z
$\hat{\sigma}_v$ x		$\begin{pmatrix} 1 & 0 \\ 0 & -1 \end{pmatrix}$	$\begin{pmatrix} 0 & -i \\ i & 0 \end{pmatrix}$	$\begin{pmatrix} 0 & 1 \\ 1 & 0 \end{pmatrix}$
y		$\begin{pmatrix} 0 & 1 \\ 1 & 0 \end{pmatrix}$	$\begin{pmatrix} 1 & 0 \\ 0 & -1 \end{pmatrix}$	$\begin{pmatrix} 0 & -i \\ i & 0 \end{pmatrix}$
z		$\begin{pmatrix} 0 & -i \\ i & 0 \end{pmatrix}$	$\begin{pmatrix} 0 & 1 \\ 1 & 0 \end{pmatrix}$	$\begin{pmatrix} 1 & 0 \\ 0 & -1 \end{pmatrix}$
$\hat{\sigma}^2/3$		$\begin{pmatrix} 1 & 0 \\ 0 & 1 \end{pmatrix}$	$\begin{pmatrix} 1 & 0 \\ 0 & 1 \end{pmatrix}$	$\begin{pmatrix} 1 & 0 \\ 0 & 1 \end{pmatrix}$

2.2. ORBITAL SPACE

Because of the isomorphism between the spin and the orbital space, the orbital functions A^u and B^u can be defined formally as eigenfunctions of the same dimensionless operators $\hat{\sigma}_u$, now acting in the orbital space, $\hat{\sigma}_u A^u = A^u$, $\hat{\sigma}_u B^u = -B^u$. We choose A^z and B^z to represent the most localized possible choice of the two real orbitals, as defined by the usual criterion of minimal interorbital and maximal intraorbital electron-electron repulsion. The reason for this specific choice will become clear later, when we deal with functions in the two-electron (geminal) space. We could have equally well chosen A^z and B^z to represent the most delocalized real orbitals, but any other choice would complicate matters.

As in the spin space, the choice of the localized orbital functions A and B is implied if no superscript is shown.

The formal analogy of the most delocalized "real" orbitals A^x and B^x to the spin functions α^x and β^x, quantized with respect to the x axis, and of the most delocalized complex orbitals A^y and B^y to the spin functions α^y and β^y, quantized along the y axis, is obvious from Table 1 and is emphasized by the notation chosen. The orbitals A^x, B^x have the same complex phase in all space and can therefore be referred to as "real"; we use quotation marks to indicate the distinction. An electron in such an orbital generates no current nor orbital magnetic dipole moment. The orbitals A^y, B^y are essentially complex and carry both a current and an orbital magnetic dipole moment.

In our previous work [1,6] we used the more common basis sets $a = (A-B)/\sqrt{2}$, $b = (A+B)/\sqrt{2}$, $c = (A+iB)/\sqrt{2}$, $c^* = (A-iB)/\sqrt{2}$, but the present usage will bring out more clearly the isomorphism of the spin space and the orbital space.

Note that the three components of the vector operator $\hat{\sigma}$ acting in the orbital space refer to x, y, and z axes located not in ordinary space but rather, in a fictitious three-dimensional space spanned by three possible linearly independent types of one-electron perturbation of a perfect biradical, as defined below. In the two-dimensional one-electron space, a perfect biradical is characterized by a one-electron Hamiltonian operator $\hat{H}_1$ with the properties $<A^u|\hat{H}_1|A^u> = <B^u|\hat{H}_1|B^u> = h_0$ and $<A^u|\hat{H}_1|B^u> = 0$. A more complete definition will be given below when we consider two-electron states.

The equivalence of any two-level problem to the problem of a particle of spin 1/2 in magnetic field is well known [10] and has been used to treat the one-electron part of a two-orbital problem in terms of a fictitious spin of 1/2 in a fictitious magnetic field. The equivalence is obvious from the fact that any 2×2 matrix is fully characterized by its four complex elements and therefore can be written as a linear combination of the unit matrix ($\hat{\sigma}^2/3$) and the three Pauli matrices ($\hat{\sigma}_x$, $\hat{\sigma}_y$, $\hat{\sigma}_z$). The one-electron part $\hat{H}_1$ of a general Hamiltonian operator for the two-orbital model can be expressed as a linear combination of four dimensionless operators associated with $\hat{\sigma}$, and the expansion coefficients h_0, h_x, h_y and h_z are real since $\hat{H}_1$ is Hermitean:

$$\hat{H}_1 = h_0(\hat{\sigma}^2/3) + \mathbf{h} \cdot \hat{\sigma} = h_0\hat{1} + h_x\hat{\sigma}_x + h_y\hat{\sigma}_y + h_z\hat{\sigma}_z \qquad (2)$$

In orbital space, the case $h = 0$ describes a perfect biradical, and in the spin space, it describes the absence of a magnetic field. All three choices of the basis sets α^u, β^u or A^u, B^u, and indeed all other linear combinations of α, β or A, B, then represent eigenfunctions and in that sense are equivalent. Since we can always choose h_0 to represent our energy zero we do not need to be concerned with the $\hat{\sigma}^2$

operator at all. Instead, we can concentrate on the operators $\hat{\sigma}_u$. In orbital space, these represent the various possible types of perturbation of a perfect biradical.

3. One-Electron Part of the Hamiltonian

In the first two subsections, we shall consider only those terms in the one-electron Hamiltonian that act either in the spin space or in the orbital space, but not both. Spin-orbit coupling is considered in the third section.

3.1. SPIN SPACE

We consider the familiar case of spin space first. In this case, the vector h in equation (2) is given by an external magnetic field B, h_0 vanishes, and $\hat{H}_1$ represents the spin Zeeman term in the one-electron Hamiltonian. Using $\hat{\mu}^{(s)}$ for the electron spin magnetic dipole moment operator, we have

$$\hat{H}_1 = -B \cdot \hat{\mu}^{(s)} = (g\beta_e/2)B \cdot \hat{\sigma}^{(s)} \qquad \text{(Zeeman perturbation)} \qquad (3)$$

$$h = (g\beta_e/2)B$$

where we have used a superscript to indicate that the operator $\hat{\sigma}$ operates in the spin space (we shall drop it in the following), and where $\beta_e = |e|\hbar/2mc$ is the Bohr magneton. For a free electron, the dimensionless spectroscopic splitting factor g equals 2.002322. The presence of orbital contributions induced by spin-orbit coupling with the fixed core of the molecule, which is not accounted for explicitly in the two-electron two-orbital model, can be taken into account by treating the g factor as a parameter. In general, it is anisotropic and needs to be treated as a tensor $\mathbf{g}$, so that $\hat{H}_1$ acquires the form $(\beta_e/2)B \cdot \mathbf{g} \cdot \hat{\sigma}$. We shall not introduce this complication explicitly in the following and merely note that the magnetic field vector B should properly be thought of as the vector $B \cdot \mathbf{g}$, and the perturbation vector h in equation (2) as $(\beta_e/2)B \cdot \mathbf{g}$.

The spin eigenfunctions of $\hat{H}_1$ are adapted to the direction of the perturbing field B as the quantization direction. There are three familiar simple cases:

(i) Field directed along z, with $\hat{H}_1$ diagonal in the α^z, β^z representation, and inducing an energy difference $2|h_z| = |g\beta_e B_z|$ between the two spin eigenstates α^z and β^z. The spin state can be said to be magnetized along z.

(ii) Field along y, with $\hat{H}_1$ diagonal in α^y, β^y, and inducing an energy difference $2|h_y| = |g\beta_e B_y|$ between the eigenstates α^y and β^y, with magnetization along y.

(iii) Field along x, with $\hat{H}_1$ diagonal in α^x, β^x, and inducing a Zeeman splitting $2|h_x| = |g\beta_e B_x|$ between the eigenstates α^x and β^x, magnetizing the spin state along x.

The space in which the three components of the spin operator $\hat{\sigma}$ act is just the ordinary Cartesian space in which the magnetic field causing the perturbation is acting.

In a general case, the perturbing vector h is oriented along a direction characterized by a polar angle θ ($0 \leq \theta < \pi$) measured from the z axis and an azimuthal angle ϕ ($0 \leq \phi < 2\pi$) measured from the x axis (counterclockwise when viewed from the $+z$ axis). Then, $\hat{H}_1$ is proportional to $\hat{\sigma}_x \sin\theta\cos\phi + \hat{\sigma}_y \sin\theta\sin\phi + \hat{\sigma}_z\cos\theta$, and its eigenfunctions in the α, β basis set are obtained by rotating the direction h into coincidence with the z axis. They are determined up to arbitrary phase factors, which reflect the manner in which this rotation is accomplished. If the rotation axis is chosen to lie in the xy plane (i.e., the rotation angle is the smallest possible), we have $n_x = -\sin\phi$, $n_y = \cos\phi$, $n_z = 0$, and $\omega = \theta$. Using equation (1), the resulting spin eigenfunctions are

$$(\alpha^h, \beta^h) = (\alpha, \beta) \begin{pmatrix} \cos(\theta/2) & -e^{-i\phi}\sin(\theta/2) \\ e^{i\phi}\sin(\theta/2) & \cos(\theta/2) \end{pmatrix} \tag{4}$$

A more symmetrical set that is often used [10] is $e^{-i\phi/2}\alpha^h$ and $e^{i\phi/2}\beta^h$, but we shall use α^h and β^h. The eigenvalues are $|h| = (g\beta_e/2)|B|$ for α^h and $-|h| = -(g\beta_e/2)|B|$ for β^h. Using $c_v = h_v/|h| = B_v/|B|$ for the direction cosines of h and B in the coordinate axes, we have

$$\alpha^h = [\alpha\sqrt{1 + c_z} + \beta(c_x + ic_y)/\sqrt{1 + c_z}]/\sqrt{2}$$
$$\beta^h = [-\alpha(c_x - ic_y)/\sqrt{1 + c_z} + \beta\sqrt{1 + c_z}]/\sqrt{2} \tag{5}$$

Even in the absence of an external magnetic field, the spin of the electron may be perturbed by nuclear magnets present in the molecule, and by the magnetic field generated by electron motion. Both are of considerable importance in the understanding of biradicals; the latter is described below in the section on spin-orbit coupling.

3.2. ORBITAL SPACE

A consideration of the analogous effects of the operators $\hat{\sigma}_v$ in the orbital space identifies the three linearly independent types of one-electron perturbation of a perfect biradical that we referred to above. The most general type of such perturbation is described by a one-electron Hamiltonian $\hat{H}_1$ of form (2) that now represents the interaction of a fictitious magnetic field B' with a fictitious magnetic moment $-(g'\beta_e/2)\hat{\sigma}$ of a spin 1/2 particle, with h_0 as the energy zero.

$$\hat{H}_1 = h \cdot \hat{\sigma}^{(0)} = (g'\beta_e/2)B' \cdot \hat{\sigma}^{(0)} \tag{6}$$

where the components of h, related to the fictitious magnetic field by $h = (g'\beta_e/2)B'$, are dictated by the matrix elements of the one-electron perturbation of a perfect biradical:

$$h_x = Re\langle A|\hat{H}_1|B\rangle = h_{AB} \qquad \text{("covalent perturbation")}$$

$$h_y = -Im\langle A|\hat{H}_1|B\rangle \qquad \text{(magnetizing perturbation)}$$

$$h_z = (\langle A|\hat{H}_1|A\rangle - \langle B|\hat{H}_1|B\rangle)/2 \qquad \text{("polarizing perturbation")}$$
$$\quad = (h_A - h_B)/2 \tag{7}$$

In the following, we drop the superscript on $\hat{\sigma}$ and do not indicate explicitly the space in which this operator is acting. The fictitious factor g' is merely a computational tool used to make the analogy between the spin and orbital space complete. Its value is arbitrary and cancels out in any final result. Its sign needs to be chosen positive to make the "fictitious spin" analogy complete.

When we abandon the one-electron approximation in the following section and consider the full Hamiltonian, we shall need to modify the definitions of the covalent and the polarizing perturbation. For this reason, the terms are set in quotation marks in expression (7).

The direction of the fictitious magnetic field B' can be described by the direction cosines $c_u = B'_u/|B'| = h_u/|h|$ or the polar angle θ and azimuthal angles given by $\tan\theta = (h_x^2 + h_y^2)^{1/2}/h_z$, $0 \leq \theta < \pi$, and $e^{i\phi} = (h_x + ih_y)/(h_x^2 + h_y^2)^{1/2}$, $0 \leq \phi < 2\pi$.

There again are three simple cases:

(i) An operator $\hat{H}_1 = h_z\hat{\sigma}_z = (g'\beta_e/2)B'_z\hat{\sigma}_z$ is diagonal in A^z, B^z and induces an energy difference $2|h_z| = |g'\beta_e B'_z|$, equal to $|\langle A|\hat{H}_1|A\rangle - \langle B|\hat{H}_1|B\rangle|$, between its eigenfunctions, the localized orbitals A^z and B^z. It could be called the Stark or *polarizing perturbation* operator. When caused by a variation in the chemical structure, it converts a perfect biradical into a *heterosymmetric* [6] *biradicaloid* (for example, going from orthogonally twisted ethylene to orthogonally twisted propene). When caused by an outside electric field, it converts a perfect biradical into a *polarized biradical* (for example, orthogonally twisted ethylene in a field directed along the CC axis).

(ii) An operator $\hat{H}_1 = h_y\hat{\sigma}_y = (g'\beta_e/2)B'_y\hat{\sigma}_y$ is diagonal in A^y, B^y and induces an energy difference $2|h_y| = |g'\beta_e B'_y|$, equal to

$2|Im\langle A|\hat{H}_1|B\rangle|$, between its eigenfunctions, the complex delocalized orbitals A^y and B^y. It could be called the Zeeman or *magnetizing perturbation* operator. It can only be produced by an externally applied magnetic field, but not by a variation in the molecular structure. A structure-related name is therefore not needed, although one might refer to a perfect biradical located in a magnetic field as a *magnetized biradical* (for example, an O_2 molecule in a magnetic field directed along the OO axis). However, we shall see below in the section on spin-orbit coupling that incorporation of internal magnets into the molecule (electron magnetic moments) is described by quite analogous operators proportional to $\hat{\sigma}_y$. The effects of nuclear magnets on orbital magnetic moments are normally too weak to be considered, although the opposite is obviously not true.

(iii) An operator $\hat{H}_1 = h_x\hat{\sigma}_x = (g'\beta_e/2)B'_x\hat{\sigma}_x$ is diagonal in A^x, B^x and introduces an energy difference $2|h_x| = |g'\beta_e B'_x|$, equal to $2|Re\langle A|\hat{H}_1|B\rangle|$, between its eigenfunctions, the "real" delocalized orbitals A^x and B^x. It could be called the bonding-antibonding or *covalent perturbation* operator. It converts a perfect biradical into a *homosymmetric* [6] *biradicaloid* (for example, going from orthogonally twisted ethylene to partially twisted ethylene). This type of perturbation cannot be brought about by the action of an outside field, but only by a variation in the molecular geometry and/or chemical structure.

In a general case, two or all three types of perturbation may be present simultaneously. A biradicaloid characterized by a simultaneous covalent and polar structural perturbation is called *non-symmetric* [6]. The eigenfunctions of the perturbing Hamiltonian $\hat{H}_1$ are obtained from those given in (4) or (5) by the substitution $\alpha \to A$, $\beta \to B$. The eigenvalues are $|h| = (g'\beta_e/2)|B'|$ for A^h and $-|h| = -(g'\beta_e/2)|B'|$ for B^h.

Explicit expressions for the direction cosines c_u in terms of $\hat{H}_1$ are

$$c_x = Re\langle A|\hat{H}_1|B\rangle/[|\langle A|\hat{H}_1|B\rangle|^2 + (\langle B|\hat{H}_1|B\rangle - \langle A|\hat{H}_1|A\rangle)^2/4]^{1/2}$$

$$c_y = -Im\langle A|\hat{H}_1|B\rangle/[|\langle A|\hat{H}_1|B\rangle|^2 + (\langle B|\hat{H}_1|B\rangle - \langle A|\hat{H}_1|A\rangle)^2/4]^{1/2}$$

$$c_z = (\langle A|\hat{H}_1|A\rangle - \langle B|\hat{H}_1|B\rangle)/2[|\langle A|\hat{H}_1|B\rangle|^2 + (\langle B|\hat{H}_1|B\rangle - \langle A|\hat{H}_1|A\rangle)^2/4]^{1/2}$$

(8)

We conclude that in the case of the one-electron Hamiltonian operator acting in the two-orbital space, the fictitious three-dimensional space spanned by the components of $\hat{\sigma}$ has as its dimensions the polarizing perturbation, the magnetizing perturbation, and the covalent perturbation of a perfect biradical. We emphasize, however, that the one-electron treatment described so far merely represents a preliminary step and that definitive definitions of heterosymmetric, homosymmetric and non-symmetric biradicaloids will be given later.

3.3. HÜCKEL THEORY

In the absence of outside fields, there is a simple correlation with quantities familiar from the Hückel theory. The "covalent" term $Re<A|\hat{H}_1|B>$ is identical with the Hückel resonance integral describing the interaction of the localized orbitals A and B, i.e., with the quantity responsible for the existence of covalent bonding in this theory. The "polarizing" term $(<A|\hat{H}_1|A> - <B|\hat{H}_1|B>)/2$ is identical with half the difference of the Coulomb integrals of these two orbitals. The contribution of the resonance integral h_{AB} to the perturbing one-electron Hamiltonian is proportional to $\hat{\sigma}_x$, $\hat{H}_1 = h_{AB}\hat{\sigma}_x$, and that of the Coulomb integral difference is proportional to $\hat{\sigma}_z$, $\hat{H}_1 = [(h_A - h_B)/2]\hat{\sigma}_z$, with $h_0 = (h_A + h_B)/2$ taken to represent the Hückel energy zero.

Calculations on biradicaloids (perturbed biradicals) in which a set of one-electron functions is first obtained by a diagonalization of a suitable one-electron Hamiltonian $\hat{H}_1$ in the absence of a magnetic field therefore normally yield orbitals of energies

$$e_{1,2} = h_0 \pm [h_{AB}^2 + (h_A - h_B)^2/4]^{1/2} \tag{9}$$

and of a form derived by setting $\phi = 0$ and $\theta = \tan^{-1}[2h_{AB}/(h_A-h_B)]$ and replacing α by A and β by B in (4), or by using equation (8) and substituting $-h_{AB}/[h_{AB}^2 + (h_A - h_B)^2/4]^{1/2}$ for c_x, zero for c_y, $(h_A - h_B)/[4h_{AB}^2 + (h_A - h_B)^2]^{1/2}$ for c_z, A for α, and B for β in equation (5). In the general case, these orbitals are distinct from both the most localized set A^z, B^z and the most delocalized set A^x, B^x.

3.4. EXTERNAL FIELDS

The action of outside fields in the orbital space is also described very simply. The contribution to the one-electron Hamiltonian $\hat{H}_1$ that is due to an outside electric field E is $\hat{H}_1 = -E \cdot \hat{m}$, where $\hat{m} = e\hat{r}$ is the electric dipole moment operator. In simple models such as Hückel or PPP, the electron position operator $\hat{r}$ is diagonal in the A, B basis. Using the centroid of electron charge $r_0 = (<A|\hat{r}|A> + <B|\hat{r}|B>)/2$ as the origin of the coordinate system, one then can write

$$\hat{H}_1 = h_z\hat{\sigma}_z = -eE \cdot [(r_A - r_B)/2]\hat{\sigma}_z$$

$$h_z = |e|E \cdot r_{AB}/2 \tag{10}$$

$$r_{AB} = r_A - r_B$$

where $r_A = <A|\hat{r}|A>$ is the charge centroid for orbital A and $r_B = <B|\hat{r}|B>$ is the charge centroid for orbital B. The Stark effect will be the strongest when E is directed along the line joining the two charge centroids and will vanish when E is perpendicular to it. Not surprisingly, then, the effect of an outside electric field directed along $r_A - r_B$ simulates a structural polar effect. In more advanced theories, which avoid the zero-differential-overlap approximation, $\hat{r}$ is not exactly diagonal in the A,B basis, and these results hold only approximately.

The contribution to the one-electron Hamiltonian $\hat{H}_1$ that is due to an outside magnetic field B acting on the electron orbital magnetic moment is $\hat{H}_1 = -B \cdot \hat{\mu}^{(o)}$ where $\hat{\mu}^{(o)} = (e/2mc)\mathbf{l}$ is the electron orbital magnetic moment operator and $\mathbf{l} = -i\hbar(\hat{r} \times \hat{v})$ is the angular momentum operator. Introducing the notation [11] $\hat{v} = \hat{r} \times \hat{\nabla} = (\partial/\partial\xi, \partial/\partial\eta, \partial/\partial\zeta)$, ξ, η, and ζ being the angles of rotation around the x, y, and z axes, respectively, we can write

$$\hat{H}_1 = h_y\hat{\sigma}_y = (\beta_e/2)B \cdot \nabla_{AB}\hat{\sigma}_y \tag{11}$$

$$h_y = (\beta_e/2)B \cdot \nabla_{AB} \tag{12}$$

where the real quantity $\nabla_{AB} = -\nabla_{AB} = 2<A|\hat{v}|B>$ depends on the spatial disposition of the orbitals A and B.

The action of the operator $\hat{v}$ on atomic functions located at the center of the coordinate system is as follows: An s function is annihilated, and for p functions we have $(\hat{v})_z p_x = -p_y$, $(\hat{v})_z p_y = p_x$, $(\hat{v})_z p_z = 0$, etc., by cyclic permutation of indices. The matrix representation of $\hat{v}$ in the basis of the p functions is

$$(\hat{v})_x: \begin{pmatrix} 0 & 0 & 0 \\ 0 & 0 & -1 \\ 0 & 1 & 0 \end{pmatrix} \quad (\hat{v})_y: \begin{pmatrix} 0 & 0 & 1 \\ 0 & 0 & 0 \\ -1 & 0 & 0 \end{pmatrix} \quad (\hat{v})_z: \begin{pmatrix} 0 & -1 & 0 \\ 1 & 0 & 0 \\ 0 & 0 & 0 \end{pmatrix} \tag{13}$$

and this differs only by the factor i from the representation of the dimensionless angular momentum operator J in the x,y,z basis given later (Table 4).

The action of $\hat{v}$ on atomic functions located at other centers in the molecule is more complicated [12]. It needs to be evaluated for the understanding of phenomena such as natural and magnetic circular dichroism and ring currents in cyclic π-electron systems, but is not of immediate interest presently.

3.5. Spin-Orbit Coupling

Unlike the one-electron operators considered so far, which act either in the spin space or in the orbital space but not both, the one-electron spin-orbit coupling operator $\hat{H}_1^{SO}$ acts in both simultaneously [13,14]:

$$
\begin{aligned}
\hat{H}_1^{SO} &= \sum_\kappa (Z_\kappa/|r^\kappa|^3)\; \hat{\mu}^{(o)\kappa}\cdot\hat{\mu}^{(s)} \\
&= (e^2/2m^2c^2)(\hbar^2/4) \sum_\kappa (Z_\kappa/|r^\kappa|^3)\hat{\sigma}^{(s)}\cdot\nabla_{AB}^\kappa\hat{\sigma}_y^{(o)}
\end{aligned}
\tag{14}
$$

where $\hat{\sigma}^{(s)}$ acts in the spin space and $\hat{\sigma}_y^{(o)}$ acts in the orbital space, the sum is over all atoms κ, Z_κ is the atomic number of atom κ, and the superscript κ indicates that nucleus κ is to be taken as the origin in the evaluation. Because of the factor $|r^\kappa|^3$, the effect of atomic orbitals located on atoms other than κ can be neglected in the first approximation, simplifying the evaluation of ∇_{AB}^κ. The operator $\hat{H}_1^{SO}$ acts in a four-dimensional space defined by the direct product of the spin and orbital spaces and spanned by any one of the spinorbital bases $A^u\alpha^v$, $A^u\beta^v$, $B^u\alpha^v$, $B^u\beta^v$.

The effect of the part of the molecule treated as a fixed core in the model, including the inner-shell electrons, can be accounted for by introducing a spin-orbit coupling parameter ζ_κ for each nucleus and writing $\hat{H}_1^{SO}$ in the form

$$
\hat{H}_1^{SO} = \sum_\kappa (\zeta_\kappa/4)\hat{\sigma}^{(s)}\cdot\nabla_{AB}^\kappa\; \hat{\sigma}_y^{(o)}
\tag{15}
$$

Similarity with the operators that we have seen so far becomes particularly obvious by writing this result as

$$
\hat{H}_1^{SO} = \sum_\kappa \hat{H}_1^\kappa = \sum_\kappa \hat{h}_y^{\kappa(s)}\; \hat{\sigma}_y^{(o)}
\tag{16}
$$

$$
\hat{h}_y^{\kappa(s)} = (\zeta_\kappa/4)\hat{\sigma}^{(s)}\cdot\nabla_{AB}^\kappa = h^\kappa\cdot\hat{\sigma}^{(s)}
$$
$$
h^\kappa = (\zeta_\kappa/4)\nabla_{AB}^\kappa
\tag{17}
$$

Comparison of equations (16) and (11) provides a qualitative interpretation. The motion of electrons about each nucleus generates an orbital Zeeman effect in an effective magnetic field, and comparison of equations (17) and (12) shows that this field is due to electron spin: its direction is given by $\hat{\sigma}^{(s)}$ and its strength by the spin-orbit coupling constant ζ_κ [cf. equation (3)]. Contributions from all nuclei need to be added. The spin-space operator $\hat{h}_y^{\kappa(s)}$ is analogous to the electron Zeeman operator of equation (3), for a magnetic field in the direction of ∇_{AB}^κ. It is diagonalized in the same manner as before, yielding the spin eigenfunctions $\alpha^{\nabla\kappa}$ and $\beta^{\nabla\kappa}$ defined by (4) with the choice

$$\tan\theta = [(\nabla_{AB}^\kappa)_x^2 + (\nabla_{AB}^\kappa)_y^2]^{1/2}/(\nabla_{AB}^\kappa)_z, \quad 0 \le \theta < \pi$$

$$e^{i\phi} = [(\nabla_{AB}^\kappa)_x + i(\nabla_{AB}^\kappa)_y]/[(\nabla_{AB}^\kappa)_x^2 + (\nabla_{AB}^\kappa)_y^2]^{1/2}$$

For a single nucleus κ, the eigenvalues are $|h^\kappa|$ for α^∇ and $-|h^\kappa|$ for β^∇. In general, more than one nucleus will have a large ζ_κ value and will have to be considered. Note that the spin eigenfunctions are different for each choice of κ, making a simultaneous diagonalization of the spin function for all values of κ impossible.

The formation of the direct product is particularly simple using the A^y, B^y basis in which the operator $\hat{\sigma}_y^{(o)}$ is diagonal. In the case of a single nucleus κ, the eigenfunctions of $\hat{H}_1^\kappa$ are (i) $A^y\alpha^{\nabla\kappa}$ and $B^y\beta^{\nabla\kappa}$, with the eigenvalue $|h^\kappa|$, and (ii) $A^y\beta^{\nabla\kappa}$ and $B^y\alpha^{\nabla\kappa}$, with the eigenvalue $-|h^\kappa|$.

4. The Two-Electron Spin and Geminal Function Spaces

Our next task is to construct suitable basis sets for the two-electron spin and geminal spaces. Each of the spaces is obtained as a direct product of the one-electron spaces of the first and the second electron, and is spanned by $a(1)a(2)$, $a(1)b(2)$, $b(1)a(2)$, and $b(1)b(2)$. In the case of spin space, $a = \alpha$ and $b = \beta$, in the case of geminal space, $a = A$, $b = B$. The use of permutational symmetry permits a factorization of the four-dimensional space into a direct sum of a three-dimensional subspace whose elements are symmetric with respect to the interchange of electron labels, and of a one-dimensional subspace, whose elements are antisymmetric with respect to such interchange.

The choice of the basis functions in the subspaces has been standardized for the spin space (e.g., [15]) and is shown in Table 3.

The total wave functions in the two-electron two-orbital model are elements in the direct product of the spin space and the geminal space. Since electrons are fermions, the total wave functions need to be antisymmetric with respect to electron label interchange, and are restricted to the six-dimensional space spanned by the basis $T\Theta[x]$, $T\Theta[y]$, $T\Theta[z]$, $S[x]\Sigma$, $S[y]\Sigma$, $S[z]\Sigma$. In molecules composed of light atoms, the first three functions are close in energy. They span the triplet subspace, while the last three span the singlet subspace. The two subspaces interact through the spin-orbit part of the Hamiltonian, as will be discussed in Section 7. For the time being, however, we can restrict our attention to the spin and the geminal spaces separately. Since they are isomorphic, they will be treated jointly much of the time.

TABLE 3. Two-electron basis functions

Basis set:

$$[o] = [o]^z = (1/\sqrt{2})[a(1)b(2) - b(1)a(2)] \qquad \text{antisym.}$$
$$[x] = [x]^z = (-1/\sqrt{2})[a(1)a(2) - b(1)b(2)] \qquad \text{sym.}$$
$$[y] = [y]^z = (i/\sqrt{2})[a(1)a(2) + b(1)b(2)] \qquad \text{sym.}$$
$$[z] = [z]^z = (1/\sqrt{2})[a(1)b(2) + b(1)a(2)] \qquad \text{sym.}$$

Transformation matrix:

$$([o],[x],[y],[z]) = (aa,ab,ba,bb) \begin{pmatrix} 0 & -1/\sqrt{2} & i/\sqrt{2} & 0 \\ 1/\sqrt{2} & 0 & 0 & 1/\sqrt{2} \\ -1/\sqrt{2} & 0 & 0 & 1/\sqrt{2} \\ 0 & 1/\sqrt{2} & i/\sqrt{2} & 0 \end{pmatrix}$$

Adaptation to x and y axes:

$$[o]^x = [o]^y = [o]^z$$
$$[x]^x = [z]^y = [y]^z$$
$$[y]^x = [x]^y = [z]^z$$
$$[z]^x = [y]^y = [x]^z$$

Orbital space: $a = A, b = B$ *1-El. spin space:* $a = \alpha, b = \beta$

Geminal space:
$$[o] = T$$
$$[x] = S[x]$$
$$[y] = S[y]$$
$$[z] = S[z]$$

2-El. spin space:
$$[o] = \Sigma$$
$$[x] = \Theta[x]$$
$$[y] = \Theta[y]$$
$$[z] = \Theta[z]$$

The basis set shown in Table 3 has been constructed from the z-adapted one-electron basis set functions $a^z = a$ and $b^z = b$ (i.e., α,β or A,B). Alternative choices are available, starting with the basis set functions a^y, b^y or a^x, b^z. We shall indicate the choice of the axis u to which the one-electron basis is adapted by a superscript u on the resulting two-electron basis: $[o]^u, [x]^u, [y]^u, [z]^u$. Recalling that the y-axis and x-axis adapted one-electron bases resulted from a rotation of the z-axis adapted basis by $+2\pi/3$ and $-2\pi/3$ around an axis n with direction cosines $n_x = n_y = n_z = 1/\sqrt{3}$, and applying the standard rotation operators [11] for particles of spin 0 and of spin 1,

$$\hat{R}_n^0(\omega) = 1$$

$$\hat{R}_n^1(\omega) = 1 - i n \cdot \hat{J} \sin\omega - (n \cdot \hat{J})^2 (1 - \cos\omega)$$

(18)

$$[\underline{R}_n^1(\omega)]_{[x][y][z]} =$$

$$\begin{pmatrix} 1 - (n_y^2 + n_z^2)(1-\cos\omega) & -n_z\sin\omega + n_x n_y(1-\cos\omega) & n_y\sin\omega + n_z n_x(1-\cos\omega) \\ n_z\sin\omega + n_x n_y(1-\cos\omega) & 1 - (n_z^2 + n_x^2)(1-\cos\omega) & -n_x\sin\omega + n_y n_z(1-\cos\omega) \\ -n_y\sin\omega + n_z n_x(1-\cos\omega) & n_x\sin\omega + n_y n_z(1-\cos\omega) & 1 - (n_x^2 + n_y^2)(1-\cos\omega) \end{pmatrix}$$

we obtain the results listed in Table 3 and note that they satisfy the cyclic permutation symmetry of the indices x, y, and z. The dimensionless operator $\hat{J}(1,2) = [\hat{\sigma}(1) + \hat{\sigma}(2)]/2$ is represented in the $[o],[x],[y],[z]$ basis by the matrices given in Table 4 (note that $\hbar\hat{J}$ is the familiar angular momentum operator). Table 4 also shows that $\hat{J}^2/2$ is equal to the unit operator in the $[x],[y],[z]$ subspace.

TABLE 4. Representations of the operators $\hat{J}_v$, $\hat{J}_v^2$, $\hat{K}_v$, and $\hat{J}^2/2$ in the $[x]^z,[y]^z,[z]^z$ basis[a]

v	$\hat{J}_v$	$\hat{J}_v^2$	$\hat{K}_v$
x	$\begin{pmatrix} 0 & 0 & 0 \\ 0 & 0 & -i \\ 0 & i & 0 \end{pmatrix}$	$\begin{pmatrix} 0 & 0 & 0 \\ 0 & 1 & 0 \\ 0 & 0 & 1 \end{pmatrix}$	$\begin{pmatrix} 0 & 0 & 0 \\ 0 & 0 & 1 \\ 0 & 1 & 0 \end{pmatrix}$
y	$\begin{pmatrix} 0 & 0 & i \\ 0 & 0 & 0 \\ -i & 0 & 0 \end{pmatrix}$	$\begin{pmatrix} 1 & 0 & 0 \\ 0 & 0 & 0 \\ 0 & 0 & 1 \end{pmatrix}$	$\begin{pmatrix} 0 & 0 & 1 \\ 0 & 0 & 0 \\ 1 & 0 & 0 \end{pmatrix}$
z	$\begin{pmatrix} 0 & -i & 0 \\ i & 0 & 0 \\ 0 & 0 & 0 \end{pmatrix}$	$\begin{pmatrix} 1 & 0 & 0 \\ 0 & 1 & 0 \\ 0 & 0 & 0 \end{pmatrix}$	$\begin{pmatrix} 0 & 1 & 0 \\ 1 & 0 & 0 \\ 0 & 0 & 0 \end{pmatrix}$
$\hat{J}^2/2$	$\begin{pmatrix} 1 & 0 & 0 \\ 0 & 1 & 0 \\ 0 & 0 & 1 \end{pmatrix}$	$\hat{K}_x = -(J_y J_z + J_z J_y)$	
		$\hat{K}_y = -(J_z J_x + J_x J_z)$	
Note:	$\hat{J}[o] = 0$	$\hat{K}_z = -(J_x J_y + J_y J_x)$	

[a] Representations in the $[x]^y,[y]^y,[z]^y$ and $[x]^x,[y]^x,[z]^x$ bases are obtained by cyclic permutation of indices.

A general transformation of the two-electron basis set $[x], [y], [z]$ into $[x]^h, [y]^h, [z]^h$ that is induced by the transformation (4) of the one-electron basis set a, b into a^h, b^h follows from the substitution of the rotation parameters $n_x = -\sin\phi$, $n_y = \cos\phi$, $n_z = 0$, and $\omega = \theta$ into expression (18).

$$[o]^h = [o]$$

$$([x]^h, [y]^h, [z]^h) = \tag{19}$$
$$= ([x], [y], [z]) \begin{pmatrix} \cos\theta + 2\sin^2\phi \, \sin^2\frac{\theta}{2} & -\sin2\phi \, \sin^2\frac{\theta}{2} & \cos\phi \, \sin\theta \\ -\sin2\phi \, \sin^2\frac{\theta}{2} & 1 - 2\sin^2\phi \, \sin^2\frac{\theta}{2} & \sin\phi \, \sin\theta \\ -\cos\phi \, \sin\theta & -\sin\phi \, \sin\theta & \cos\theta \end{pmatrix}$$

The transformation matrix in (19) differs from the matrix (A2) given in reference [6] by the factors i and -1 in some of the entries, because the phase factors in the definitions of the two-electron basis functions were chosen differently there and in Table 3. The present choice is more satisfactory and leads to a real transformation matrix.

In terms of the direction cosines c_x, c_y, and c_z of the real or fictitious perturbing field h to which the one-electron basis set has been adapted by the transformation (5), the results for the transformed two-electron basis set are

$$[o]^h = [o]$$
$$[x]^h = [c_z + c_y^2/(1 + c_z)][x] - [c_x c_y/(1 + c_z)][y] - c_x[z]$$
$$[y]^h = -[c_x c_y/(1 + c_z)][x] + [1 - c_y^2/(1 + c_z)][y] - c_y[z] \tag{20}$$
$$[z]^h = c_x[x] + c_y[y] + c_z[z]$$

The transformation properties of the two-electron basis set functions are particularly simple if the field vector h lies in the xz plane (in geminal space, this corresponds to a perturbation of a perfect biradical by structural changes and/or an outside electric field in the absence of magnetic field). Then, $\phi = c_y = 0$, and $[y]$ is invariant:

$$[o]^h = [o]$$
$$[x]^h = [x]\cos\theta - [z]\sin\theta = c_z[x] - c_x[z]$$
$$[y]^h = [y] \tag{21}$$
$$[z]^h = [x]\sin\theta + [z]\cos\theta = c_x[x] + c_z[z]$$

5. The Two-Electron Hamiltonian

We are now ready to derive the form of the full Hamiltonian operator for both electrons. In this section we shall still consider only those terms that act either in the spin space or in the geminal space, but not both. We recall that the use of the model is best justified when the fixed core part of the molecule is taken into account through its action on the properties of the matrix elements of the effective Hamiltonian. This is commonly done for the spin Hamiltonian, but one can adopt the same attitude for the geminal part of the Hamiltonian.

We found earlier that the one-electron part of the Hamiltonian $\hat{H}_1$, acting in the two-dimensional space a,b, can be written as a linear combination of $\hat{\sigma}^2$ and the three components of $\hat{\sigma}$. By a suitable choice of an origin for the one-electron energy scale, we were able to eliminate the need for $\hat{\sigma}^2$ and all structural as well as external-field effects were expressed through $\hat{\sigma}$ alone.

In dealing with the full Hamiltonian $\hat{H}$, acting in the four-dimensional space $[o], [x], [y], [z]$, we encounter an entirely similar situation. $\hat{H}$ has the form

$$\hat{H}(1,2) = \hat{H}_1(1,2) + \hat{H}_2(1,2)$$
$$\hat{H}_1(1,2) = \hat{H}_1(1) + \hat{H}_1(2)$$

(22)

5.1. THE ADDITIVE TERMS

The representations of $\hat{H}_1(1,2)$ in the two-electron space are simply related to those of $\hat{H}_1(1)$ in the one-electron space. Since $\hat{J}(1,2) = [\hat{\sigma}(1) + \hat{\sigma}(2)]/2$, we have

$$\hat{H}_1(1) = h_0 \hat{1}(1) + h \cdot \hat{\sigma}(1)$$
$$\hat{H}_1(1,2) = 2h_0 \hat{1}(1,2) + 2h \cdot \hat{J}(1,2)$$

(23)

Thus, except for a constant term $2h_0 = h_a + h_b$ in the geminal space ($h_0 = 0$ for the spin space), it is clear that the additive "one-electron" part $\hat{H}_1(1,2)$ of the total Hamiltonian operator $\hat{H}$ in either the spin space or the geminal space corresponds to a Zeeman-like interaction of a particle of spin 1 with a field h, either a real magnetic field B, with $2h = g\beta_e B$ or a fictitious field B', with $2h = g'\beta_e B'$.

A general Hermitean operator acting in the antisymmetrized part of the two-electron space would be specified by ten matrix elements, and even after the specification of the state energy zero and the introduction of the operators $\hat{J}_x$, $\hat{J}_y$ and $\hat{J}_z$, six degrees of freedom remain. The addition of the operators $\hat{J}_x^2$, $\hat{J}_y^2$, $\hat{J}_z^2$, and of the operators $\hat{K}_x$, $\hat{K}_y$ and $\hat{K}_z$, defined in Table 4, completes the set. The matrix representations of the operators in the $[x]^Z, [y]^Z, [z]^Z$ basis are given in Table 4; they all are null operators in the $[o]$ space.

5.2. The Inseparable Terms

We next turn our attention to the representations of the inseparable terms $\hat{H}_2(1,2)$. In two-electron spin space, this is the mutual magnetic dipole - magnetic dipole interaction of the two electrons $\hat{H}_d$, and in the geminal space, it is their mutual electric monopole - electric monopole interaction $\hat{H}_e$.

5.2.1 The dipole-dipole interaction

The interaction energy of two magnetic dipoles $\mu^{(s)}(1)$ and $\mu^{(s)}(2)$ is

$$\{[\mu^{(s)}(1)\cdot\mu^{(s)}(2)]r_{12}^2 - 3[r_{12}\cdot\mu^{(s)}(1)][r_{12}\cdot\mu^{(s)}(2)]\}/r_{12}^5$$

where r_{12} is the vector joining their centers and $r_{12} = |r_{12}|$. The magnetic moment due to electron spin is $\mu^{(s)} = -(g\beta_e/2)\hat{\sigma}$ and this leads [14] to an expression for the two-electron "dipole-dipole zero-field splitting Hamiltonian" $\hat{H}_d$, acting in the spin space $\Sigma, \Theta[x], \Theta[y], \Theta[z]$:

$$\hat{H}_d(1,2) = \hat{J}(1,2)\cdot\underline{D}\cdot\hat{J}(1,2) \tag{24}$$

where the symmetric zero-field splitting tensor $\underline{D}$ has the components

$$\underline{D}_{uu} = (g^2\beta_e^2/2)\langle T|(r_{12}^2 - 3u_{12}^2)/r_{12}^5|T\rangle$$
$$\underline{D}_{uv} = \underline{D}_{vu} = (g^2\beta_e^2/2)\langle T|-3u_{12}v_{12}/r_{12}^5|T\rangle, \quad u \neq v \tag{25}$$

Note that the trace of $\underline{D}$ vanishes, and that in the $[o]$ space $\hat{H}_d$ is a null operator. Thus, the spin Hamiltonian leaves the energy of the Σ state identical to the average of the $\Theta[x]$, $\Theta[y]$ and $\Theta[z]$ states.

5.2.2 The monopole - monopole interaction

The point-charge repulsion energy of two electrons is e^2/r_{12}. This gives rise to an equation for the "electron repulsion Hamiltonian" $\hat{H}_e$, acting in the geminal space $T, S[x], S[y], S[z]$. The result agrees with the one given in reference [6], but now is expressed in terms of the operator $\hat{J}$:

$$\hat{H}_e(1,2) = G_0\hat{1}(1,2) + G_1\cdot\hat{J}(1,2) + \hat{J}(1,2)\cdot\underline{G}_2\cdot\hat{J}(1,2) \tag{26}$$

where the scalar G_0, the vector G_1 and the tensor $\underline{G}_2$ jointly characterize the effects of electron repulsion (note that the y components
of G_1 and $\underline{G}_2$ all vanish):

$$G_0 = J_{ab} - K_{ab}$$

$$G_{1x} = [(aa|ab) + (bb|ba)]$$

$$G_{1z} = (J_{aa} - J_{bb})/2$$

$$\underline{G}_{2xx} = 2K_{ab}$$

$$\underline{G}_{2zz} = 2K'_{ab} \qquad\qquad (27)$$

$$\underline{G}_{2xz} = (bb|ba) - (aa|ab)$$

$$\underline{G}_{2zx} = (bb|ab) - (aa|ba)$$

$$G_{1y} = \underline{G}_{2yv} = \underline{G}_{2vy} = 0, \qquad v = x,y,z$$

Here, a and b represent some particular choice of orbitals, such as
A^z, B^z or A^y, B^y or A^x, B^x or a more general one. The Coulomb repulsion
integrals J_{aa}, J_{ab}, and J_{bb}, the exchange repulsion integral K_{ab}, the
hybrid repulsion integrals $(aa|ab)$ and $(bb|ba)$, and the "exchange
repulsion integral" K'_{ab} are defined as follows:

$$J_{ab} = <a(1)b(2)|e^2/r_{12}|a(1)b(2)>$$

$$K_{ab} = <a(1)a(2)|e^2/r_{12}|b(1)b(2)>$$

$$(aa|ab) = <a(1)a(2)|e^2/r_{12}|a(1)b(2)> \qquad (28)$$

$$K'_{ab} = [(J_{aa} + J_{bb})/2 - J_{ab}]/2$$

We now have all the results on hand to write the expressions for
the full two-electron Hamiltonian $\hat{H}$ both in the spin space and in the
geminal space in terms of the operator $\hat{J}$, and to write its representations in the $[o]^z, [x]^z, [y]^z, [z]^z$ basis. Before doing so, however,
we shall simplify the expression for $\hat{H}_2$ by diagonalizing its symmetric part.

5.2.3 Diagonalization of the symmetric part of $\hat{H}_2$

In spin space, this requires a general rotation in the space of the
functions $\Theta[x]$, $\Theta[y]$, and $\Theta[z]$. This can be accomplished by a rotation in the space of one-electron spin functions α^h, β^h, and this is
equivalent to choosing a particular set of molecular axes in real

space. In practice, these axes are frequently obvious from molecular symmetry, but even if they are not, they can always be found easily once $\underline{D}$ is known. Note that $\underline{D}$ can have up to three pairs of equal off-diagonal elements, and these are brought to zero by use of the three degrees of freedom contained in an arbitrary rotation of axes.

We thus do not suffer any loss of generality by assuming that $\underline{D}$ is diagonal and writing $\hat{H}_d$ in the form

$$\hat{H}_d = D_x \hat{J}_x^2 + D_y \hat{J}_y^2 + D_z \hat{J}_z^2 \tag{29}$$

but we have no further freedom in the choice of molecular axes in real space, except for an arbitrary permutation of labels.

It is common to take advantage of the fact that $\sum_u D_u = 0$ and to introduce the parameters D and E,

$$
\begin{aligned}
D &= 3D_z/2 = (3g^2\beta_e^2/4) <T|(r_{12}^2 - 3z_{12}^2)/r_{12}^5|T> \\
E &= (D_x - D_y)/2 = (3g^2\beta_e^2/4) <T|(y_{12}^2 - x_{12}^2)r_{12}^5|T>
\end{aligned}
\tag{30}
$$

in terms of which $\hat{H}_d$ can be written as

$$\hat{H}_d = D(\hat{J}_z^2 - \hat{J}^2/3) + E(J_x^2 - J_y^2) \tag{31}$$

The representation of $\hat{H}_d$ in the $\Sigma, \Theta[x]^z, \Theta[y]^z, \Theta[z]^z$ basis is given in Table 5.

In geminal space, $\underline{G}_2$ already is nearly diagonal, with only one pair of off-diagonal elements, $\underline{G}_{2zx}$ and $\underline{G}_{2xz}$. Full diagonalization therefore requires merely a rotation about the y axis, not in the real space of the molecular framework (this degree of freedom is not available any more), but in the abstract space spanned by the geminal wave functions $S[x]$, $S[y]$, and $S[z]$.

For an arbitrary initial choice of the orbitals a, b, the off-diagonal element $(bb|ba) - (aa|ab)$ may be different from zero. According to (5) and (20), the orthogonal transformation of $S[x]$ and $S[z]$ given in equation (21) that will cause it to go to zero can be brought about by transforming the one-electron functions a and b as in equation (5). As discussed in more detail elsewhere [6], this off-diagonal element vanishes when J_{ab} (and at the same time, K_{ab}) is minimized or maximized. In the former case, $a = A^z$, $b = B^z$, in the latter case, $a = A^x$, $b = B^x$ and we see that the rotation brings the orbital choice into adaptation for either the z or the x axis. In the Appendix section of reference [6], the simple algorithm needed to achieve the diagonalization starting with an arbitrary set of real

orbitals a,b (such as would typically emerge from a numerical computation using standard programs) is described and the invariants of
the transformation are listed.

TABLE 5. Representations of the Hamiltonian in the
$[o]^Z, [x]^Z, [y]^Z, [z]^Z$ basis.

$$\hat{H}_d: \begin{pmatrix} 0 & 0 & 0 & 0 \\ 0 & -D_x & 0 & 0 \\ 0 & 0 & -D_y & 0 \\ 0 & 0 & 0 & -D_z \end{pmatrix}$$

$$\hat{H}_e: \begin{pmatrix} J_{AB}-K_{AB} & 0 & 0 & 0 \\ 0 & J_{AB}+2K'_{AB}-K_{AB} & -i(J_{AA}-J_{BB})/2 & 0 \\ 0 & i(J_{AA}-J_{BB})/2 & J_{AB}+2K'_{AB}+K_{AB} & -i[(AA|AB)+(BB|BA)] \\ 0 & 0 & i[(AA|BA)+(BB|AB)] & J_{AB}+K_{AB} \end{pmatrix}$$

$$\hat{H}^a: \begin{pmatrix} 0 & 0 & 0 & 0 \\ 0 & \underline{V}_{xx} & -iW_z & iW_y \\ 0 & iW_z & \underline{V}_{yy} & -iW_x \\ 0 & -iW_y & iW_x & \underline{V}_{zz} \end{pmatrix}$$

u:

Spin space:	x	y	z
W_u:	$g\beta_e B_x$	$g\beta_e B_y$	$g\beta_e B_z$
$\underline{V}_{uu}$:	$D/3 - E$	$D/3 + E$	$-2D/3$
Geminal space:			
W_u:	γ_{AB}	$\beta_e B \cdot \nabla_{AB}$	$\delta_{AB} + |e|E \cdot r_{AB}$
$\underline{V}_{uu}$:	$2K'_{AB}$	$2(K_{AB} + K'_{AB})$	$2K_{AB}$

[a] Relative to $2h_0 + J_{AB} - K_{AB}$ as the state energy zero.

234

The availability of two fundamentally equivalent choices of orbitals, A^z, B^z and A^x, B^x, also explains our choice of the symbol for the quantity K'_{AB} [6]: if $K_{A^z B^z}$ is the exchange integral between the two z-adapted localized orbitals A^z and B^z, then $K'_{A^z B^z}$, defined by (28), is the exchange integral between the delocalized (x-adapted) orbitals A^x and B^x:

$$K'_{A^z B^z} = K_{A^x B^x}$$
$$K'_{A^x B^x} = K_{A^z B^z}$$

(32)

The analogy between the results for the spin space and the orbital or geminal space is striking. We have already seen that in the one-electron spin space, there is no reason to prefer any one choice of the x,y,z axes in real space in the absence of outside fields or perturbations by the "fixed core" (all α^h, β^h choices are equally good). By the same token, in the one-electron orbital space of a perfect biradical ($h_A = h_B$, $h_{AB} = 0$), there is no reason to prefer any one choice of the x,y,z axes in the fictitious space of covalent (x axis), magnetizing (y axis) and polarizing (z axis) perturbations (all orbital choices A^h, B^h are equally good: localized, partly or fully delocalized, complex).

We see now that in the two-electron spin space, the zero-field-splitting two-electron terms dictate the principal axes x,y,z in the molecular framework. By the same token, in the two-electron geminal space, the two-electron repulsion terms dictate the principal axes in the space of orbital transformations and thus the "principal" orbital choices. The three directions correspond to the most localized orbital choice A^z, B^z, the most delocalized "complex" choice A^y, B^y, and to the most delocalized "real" choice A^x, B^x defined in Table 1. Our insistence on defining the orbitals A^z, B^z as one of the principal choices from the start (the choice of the most localized as opposed to the most delocalized "real" ones was arbitrary), well before electron-repulsion terms were considered, is thus understandable in retrospect.

With the identification of the most localized orbital choice $A = A^z$, $B = B^z$, the two-electron part of the Hamiltonian $\hat{H}_e$ acquires the form

$$\hat{H}_e = (J_{AB} - K_{AB})\hat{1} + [(AA|AB) + (BB|BA)]\hat{J}_x + [(J_{AA} - J_{BB})/2]\hat{J}_z + 2K_{AB}\hat{J}_x^2 + 2K'_{AB}\hat{J}_z^2$$

(33)

The representation of $\hat{H}_e$ in the $T, S[x], S[y], S[z]$ basis is given in Table 5.

At this point, comparison of equations (29) and (33) shows that the similarity of the two-electron spin problem and the two-electron geminal problem is less than perfect. Unlike $\hat{H}_d$, $\hat{H}_e$ is seen to contain off-diagonal terms proportional to J_x and J_z, which couple $S[y]$ with $S[z]$, and $S[x]$ with $S[y]$, respectively.

The analogy is however nearly restored if we redefine the meaning of the requirement that the orbitals of a perfect (unperturbed) biradical be equivalent and non-interacting. We now demand that the structure must not only satisfy (i) $h_A = h_B$ and (ii) $h_{AB} = 0$, but also (iii) $J_{AA} = J_{BB}$ and (iv) $(AA|AB) = (BB|BA)$. Condition (iii) is analogous to (i) and is quite intuitive. Condition (iv) is analogous to (ii) and requires that the repulsion of the overlap density AB with the density AA should equal its repulsion with the density BB. Neither demand is unreasonable if the orbitals A and B are to be truly equivalent. Condition (iv) is perhaps best visualized by recognizing that the condition (ii) is frequently met by A and B belonging to different irreducible representations of a point group of symmetry, and that this same symmetry constraint will make both $(AA|AB)$ and $(BB|BA)$ go to zero individually, since the orbital densities $A(j)A(j)$ and $B(j)B(j)$ will be totally symmetric, like e^2/r_{12}, but the overlap density $A(j)B(j)$ will not $(j = 1,2)$.

With this final definition of a perfect biradical, its two-electron Hamiltonian $\hat{H}_e$ becomes diagonal in the $[o],[x],[y],[z]$ basis, just like the two-electron Hamiltonian $\hat{H}_d$ of an unperturbed two-spin system. The only remaining flaw in the analogy is due to the difference between the electromagnetic interaction of two dipoles (spin space) and two monopoles (geminal space): the former does not make the energy of the $[o]$ state different from the average energy of the $[x]$, $[y]$, and $[z]$ states, whereas the latter does.

Now, finally, we can write in Table 6 our final expressions for those parts of the full Hamiltonian $\hat{H}$ that act separately either in the spin space or in the orbital space, but not in both simultaneously, including both structural and outside field perturbations. The representation of the final form of $\hat{H}$ in the $[o],[x],[y],[z]$ basis is given in Table 5.

It is useful to combine the one-electron and the two-electron terms in W_x and W_z for the geminal space in the absence of outside fields, and we define γ_{AB} and δ_{AB} as follows [6]:

$$\gamma_{AB} = 2h_{AB} + (AA|AB) + (BB|BA) = W_x$$
$$\delta_{AB} = h_A - h_B + (J_A - J_B)/2 \qquad (\text{if } E = 0, \ \delta_{AB} = W_z) \tag{34}$$

The quantity $\gamma_{AB}J_x$ is a generalization of the covalent bonding operator introduced earlier. The value γ_{AB} corresponds to twice the resonance integral between the localized orbitals A and B in semi-empirical theories. The quantity $\delta_{AB}J_z$ is a generalization of the (internal) polarizing operator introduced earlier and is a measure of the electronegativity difference between the localized orbitals A and B. In both cases, electron repulsion effects are now taken into account properly.

TABLE 6. Hamiltonian for the two-electron two-orbital model[a]

$$\hat{H} = \sum_{u=x,y,z} (\underline{W}_u \hat{J}_u + \underline{W}_{uu} \hat{J}_u^2) = \mathbf{W} \cdot \hat{\mathbf{J}} + \underline{W}_{xx} \hat{J}_x^2 + \underline{W}_{yy} \hat{J}_y^2 + \underline{W}_{zz} \hat{J}_z^2$$

	u:			
Spin space:	x	y	z	
$\underline{W}_u$:	$g\beta_e B_x$	$g\beta_e B_y$	$g\beta_e B_z$	(outside magnetic field)
$\underline{W}_{uu}$:	$-D/3 + E$	$-D/3 - E$	$2D/3$	(dipole interaction)
Geminal space:				
$\underline{W}_u$:	$2h_{AB}$		$h_A - h_B$	(internal structure)
	$+ (AA|AB)+(BB|BA)$		$+ (J_A - J_B)/2$	
		$\beta_e \mathbf{B} \cdot \nabla_{AB}$	$+ |e|\mathbf{E} \cdot r_{AB}$	(outside field)
	(covalent)	(magnetizing)	(polarizing)	
$\underline{W}_{uu}$:	$2K_{AB}$	0	$2K'_{AB}$	(monopole interaction)

[a] Relative to $2h_0 + J_{AB} - K_{AB}$ as the state energy zero.

We have noted already that the model Hamiltonian is supposed to contain effective corrections for the presence of the fixed core, so that the interpretation of $\hat{H}_d$ and $\hat{H}_e$ really is not as simple as indicated: $\hat{H}_d$ also contains the effects of spin-orbit coupling with electrons in the "fixed core", while $\hat{H}_e$ also contains the effects of correlation with electrons in the "fixed core". The values determined for the parameters contained in these Hamiltonians by fitting experimental results and assuming that the two-electron two-orbital model is exact therefore do not necessarily agree with values one would obtain from an exact evaluation from the spin and geminal functions of the model using the definitions given here, and an exact fitting of the energies of all six states to the model is indeed in general not possible. In particular the energy of the [y] and [z] states of biradicals tends to be lowered considerably and differentially by correlation with the most loosely held electrons of the fixed core, pushing them below the [o] and [x] states, respectively. The situation has been analyzed in a number of cases. It is fairly well understood [1] and we need not go into detail here.

6. State Energies and Wave Functions for an Electron Pair

Given the Hamiltonian formulated above (Tables 5,6), we are now ready to obtain exact results for state energies and wave functions for the active space of the two-electron two-orbital model in the absence of spin-orbit coupling. It should be noted that in organic molecules spin-orbit coupling within the six states of an electron pair considered in the simple model has a negligible effect on energy differences among the singlet states and is primarily of interest for its effects on the rates of intersystem crossing. It can, however, have a significant effect on energy differences among the triplet levels since they are so much closer to start with.

We shall not elaborate the results for state energies and wave functions in great detail nor provide many examples, since this has been done elsewhere [1,6]. Instead, we shall point out the remarkable similarities between the spin and the geminal spaces, caused by the near identity of the form of their respective Hamiltonians. In order to discuss total energies, we consider the total wave functions, which consist of a product of a spin and a geminal part. At this point, we still neglect spin-orbit coupling.

Within the framework of the simple model, the state energies and wave functions can be found algebraically, since matrices no larger than 3×3 need to be diagonalized. However, the result is sufficiently complicated that it is best to consider explicitly only special cases in which the 3×3 matrix is block-diagonal so that the solutions can be found by inspection and understood in detail. An intuitive feeling for the results obtained in the completely general case can then be developed by using the special cases as starting points.

6.1. PERFECT BIRADICALS IN THE ABSENCE OF OUTSIDE FIELDS

These systems are characterized by diagonal spin and geminal Hamiltonians in the $\Sigma, \Theta[x], \Theta[y], \Theta[z]$ and $T, S[x], S[y], S[z]$ bases, respectively, with the average of the $T\Theta[x]$, $T\Theta[y]$, and $T\Theta[z]$ energies at the origin of our energy scale by definition (equal to $2h_0 + J_{AB} - K_{AB}$). The disposition of the $T\Theta[x]$, $T\Theta[y]$, and $T\Theta[z]$ levels, referred to jointly as the triplet state, bears certain similarities to that of the $S[x]\Sigma$, $S[y]\Sigma$, and $S[z]\Sigma$ levels, which represent the three characteristic singlet states of an electron pair in the two-orbital model. Of course, the difference in scales is huge, on the order of 10^5, since the magnetic dipole interactions between two electrons are much weaker than their electric monopole interactions. Typical differences between the three triplet levels are 10^{-2} - 10^{-1} cm^{-1}, whereas typical differences between the three singlet levels are 10^3 - 10^4 cm^{-1}. Still, the similarity in the mathematical description (Tables 5 and 6) results in very similar behavior, particularly in the way in which the levels respond to a perturbation.

The labels of the three components of the triplet reflect the choice of labels for the principal axes of the D tensor, which is in principle arbitrary. The choice affects the values and signs of the D and E parameters. It has been recommended that they be chosen so that $|D| \geq 3|E|$ and $E < 0$. If D is positive, $T\theta[z]$ is the lowest energy state and is located $-2D/3$ below the energy zero. If $E = 0$, $T\theta[y]$ and $T\theta[x]$ are degenerate and are located at $+D/3$ above zero. This occurs if the x and y molecular axes are equivalent by symmetry. If $E \neq 0$, the degenerate pair is split by $2E$ but its average energy remains at $D/3$. The $T\theta[x]$ and $T\theta[y]$ levels lie higher than the ground state $T\theta[z]$, by $D-E$ and $D+E$, respectively, and the recommended choice of labels places the former above the latter.

There is nothing arbitrary about the label of the singlet state $S[y]\Sigma$, while the labels of the singlets $S[x]\Sigma$ and $S[z]\Sigma$ result from our decision to call the most localized rather than most delocalized "real" orbitals A^z, B^z, so that the latter are A^x and B^x. Then, $S[z]\Sigma$ is the lowest energy singlet and is located $2K_{AB}$ above the energy zero. If $K_{AB} = 0$, the other two states are degenerate and are located $2K'_{AB}$ above zero. If $K_{AB} \neq 0$, the degenerate pair is split by $2K_{AB}$ but its average energy remains at $2K'_{AB} + K_{AB}$. The $S[x]\Sigma$ and $S[y]\Sigma$ levels lie higher than the lowest singlet, $S[z]\Sigma$, by $2(K'_{AB} - K_{AB})$ and by $2K'_{AB}$, respectively.

This perfect analogy between the three levels of the triplet state and the three singlet states is marred only by the fact that the exchange integrals K_{AB} and K'_{AB} cannot be negative so that $S[y]\Sigma$ always lies above $S[x]$, while $T\theta[y]$ and $T\theta[x]$ can lie in either order depending on the choice of labels for the axes. This is only true, however, as long as one adheres to the strict interpretation of the exchange integrals as defined in equation (28), ignoring any possible effects of the fixed core electrons. Correlation with the core electrons actually tends to lower the energies of the $S[y]\Sigma$ and $S[z]\Sigma$ states relative to those of the $S[x]\Sigma$ and the triplet states, and is capable of inverting both the $[y],[x]$ order and the $[z],[o]$ order. If the effect of the fixed core is to be included in the simple model and the parameter K_{AB} is to be determined from experimental data, it may well have a negative value, and this value will depend on whether it is determined from the three singlet energies or from a comparison of the relative energies in the triplet and the singlet manifolds. A similar situation is familiar in the spin problem, where the parameters D and E, as determined from experiment, contain corrections due to interactions with electrons of the fixed core. In certain cases, particularly in transition metal complexes, these "corrections" may actually dominate the dipolar contributions defined by equation (30).

Perfect biradicals in which $K_{AB} = 0$, so that $S[x]\Sigma$ and $S[y]\Sigma$ are degenerate, are called *pair biradicals*, since this situation is reached only if the localized orbitals A and B are quite separate in space, as they would be in a disjoint pair of radicals. Already a moderate distance between two localized radical centers is actually nearly sufficient to reach this limit in practice (e.g., orthogonally twisted ethylene). Pair biradicals are particularly prone to level order inversion by the effect of correlation with electrons of the fixed core, and they tend to have a singlet ground state, with the triplet level slightly above $S[z]\Sigma$. At the same time, in these biradicals $S[y]\Sigma$ tends to dip below $S[x]\Sigma$. Pair biradicals also tend to have very small or vanishing E values, emphasizing the parallel between K_{AB} and E. Of course, as the localized orbitals A and B separate in space, the value of D approaches zero as well, whereas K'_{AB} tends to grow to quite large values.

Perfect biradicals in which $K_{AB} = K'_{AB}$, so that $S[x]\Sigma$ and $S[z]\Sigma$ are degenerate, are called *axial biradicals*, since this situation is frequently reached in the presence of a rotation symmetry axis of order three or more (e.g., the O_2 molecule). This situation implies an equivalence of the localized orbitals A^z, B^z and the delocalized "real" orbitals A^x, B^x with respect to electron repulsion [cf. equation (32)]. It also implies an equivalence of two of the molecular symmetry axes, and thus, $E = 0$.

In axial biradicals, the value of $2K_{AB}$ is at its maximum relative to $2K'_{AB}$, and the singlet-triplet splitting implied by the simple model is so large that the effects of electrons of the fixed core are usually not capable of inverting the order of the lowest singlet and triplet states. Many of the known ground-state triplet biradicals are of this type (e.g., methylnitrene).

6.2. PERTURBED BIRADICALS: BIRADICALOIDS AND A TWO-SPIN SYSTEM IN MAGNETIC FIELD

The analogies between the placement of the three triplet levels and the energies of the three singlets that we have discussed so far may appear a bit far-fetched. The similarity really comes to the fore, however, when we consider the effect of perturbations, since the same drawings can be used to illustrate the behavior of the energies of the three levels in both instances. This is due to the fact that the perturbing Hamiltonian (Table 6) is $\mathbf{W} \cdot \mathbf{\hat{J}}$ in both cases.

Figure 1 shows the familiar response of the three triplet levels, chosen to lie in the order $[z], [x], [y]$, to the perturbation by a magnetic field $\mathbf{W} \cdot \mathbf{\hat{J}} = g\beta_e \mathbf{B} \cdot \mathbf{\hat{J}}$. The three special cases shown correspond to a field directed along the x (left), y (center), or z (right) axis in the molecular framework ($|B_x|$, $|B_y|$, and $|B_z|$,

respectively, are plotted horizontally). When irradiated with micro-
waves of a fixed frequency, say 9.6 GHz, transitions will occur at
different strengths of the magnetic field, giving rise to the familiar triplet EPR pattern [14,15]. It is obvious that for a particular
value of B_y, $T\theta[x]$ and $T\theta[y]$ will be degenerate, and for a particular
value of B_z, $T\theta[x]$ and $T\theta[z]$ will be degenerate, due to the induced
level crossings. For a general orientation of the field B, none of
the levels will be stationary and all crossings will be avoided.

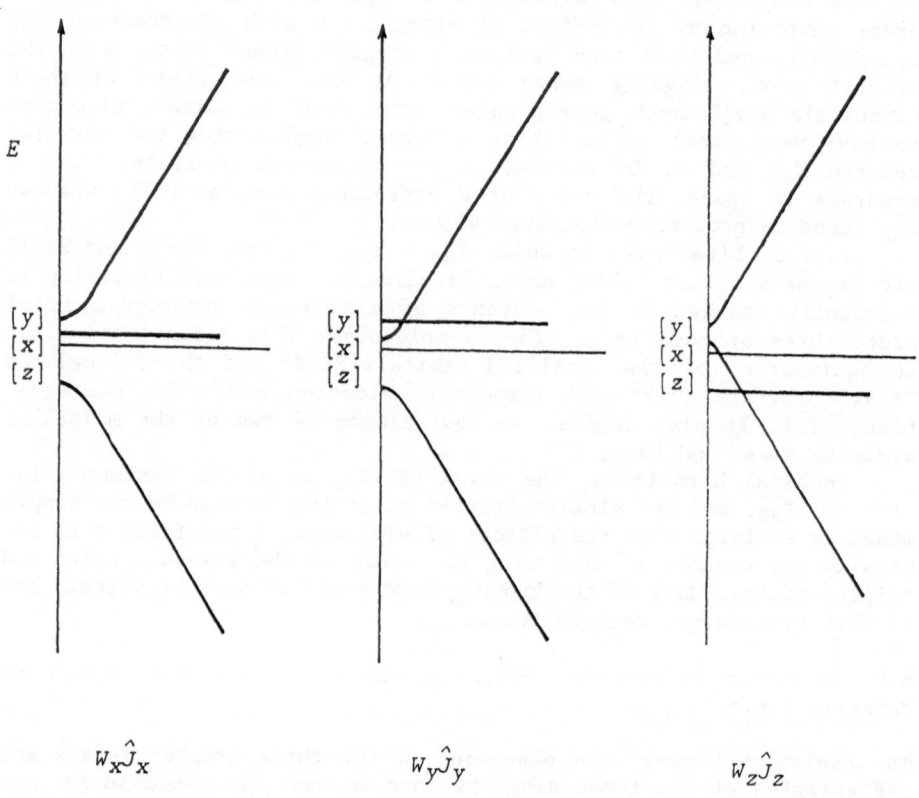

Figure 1. Response of electron pair state energies to the three
linearly independent types of perturbation.

The reasons for this behavior are immediately obvious upon inspection of the Hamiltonian matrix in Table 5. When only one of the components of B, say B_u, is different from zero, the 3×3 matrix factors into 2×2 + 1×1 blocks, with $\underline{V}_{uu}$ remaining in the 1×1 block as an unperturbed eigenvalue and $\Theta[u]$ remaining as an unperturbed eigenfunction. A diagonalization of the 2×2 block yields the familiar hyperbolic curves for the energies and mixes the remaining two basis functions.

For example, for perturbation by a magnetic field along z, the 3×3 block of the Hamiltonian matrix of Table 5 becomes

$$
\begin{array}{l}
[x]: \\
[y]: \\
[z]:
\end{array}
\begin{pmatrix}
\underline{V}_{xx} & -iW_z & 0 \\
iW_z & \underline{V}_{yy} & 0 \\
0 & 0 & \underline{V}_{zz}
\end{pmatrix}
\tag{35}
$$

so that the energies of the perturbed states are

$$
\text{(i)} \quad (\underline{V}_{xx} + \underline{V}_{yy})/2 - (1/2)\sqrt{(\underline{V}_{yy} - \underline{V}_{xx})^2 + 4W_z^2}
$$

$$
\text{(ii)} \quad (\underline{V}_{xx} + \underline{V}_{yy})/2 + (1/2)\sqrt{(\underline{V}_{yy} - \underline{V}_{xx})^2 + 4W_z^2} \tag{36}
$$

$$
\text{(iii)} \quad \underline{V}_{zz}
$$

and the wave functions become

$$
\begin{array}{lll}
\text{(i)} & [x] \longrightarrow & [x]\cos\omega - i[y]\sin\omega \\
\text{(ii)} & [y] \longrightarrow & -[x]\sin\omega - i[y]\cos\omega \\
\text{(iii)} & [z] \longrightarrow & [z]
\end{array}
\tag{37}
$$

$$
\omega = (1/2)\tan^{-1}[2W_z/(V_y - V_x)]
$$

As the strength of the perturbation W_z increases, the energies of the lowest two states approach. They became degenerate at

$$
W_z = \sqrt{(\underline{V}_{xx} - \underline{V}_{zz})(\underline{V}_{yy} - \underline{V}_{zz})} \tag{38}
$$

As the strength of the perturbation increases further, the states exchange their original order, and as it grows beyond all limits, all three wave functions simplify and become the eigenfunctions of $\hat{J}_z$, quantized with respect to the z direction:

(i) $[x] \longrightarrow ([x] - i[y])/\sqrt{2} = b(1)b(2)$

(ii) $[y] \longrightarrow -([x] + i[y]) = a(1)a(2)$ (39)

(iii) $[z] \longrightarrow [z] = [a(1)b(2) + b(2)a(1)]/\sqrt{2}$

In the present case of a spin Hamiltonian, the wave functions in question are $T\Theta[y]$, $T\Theta[x]$, and $T\Theta[z]$. Substitution from Table 5 yields the energies as

(i) $D/3 - \sqrt{E^2 + g^2\beta_e^2 B_z^2}$

(ii) $D/3 + \sqrt{E^2 + g^2\beta_e^2 B_z^2}$ (40)

(iii) $-2D/3$

The functions in the limit of strong perturbation are $\beta\beta$, $\alpha\alpha$, and $(\alpha\beta + \beta\alpha)/\sqrt{2}$, and have z components of the spin angular momentum equal to $-\hbar$, $+\hbar$, and 0, respectively.

The crossing of the lower two levels occurs at magnetic field strength given by

$$B_z = (1/g\beta_e)\sqrt{D^2 - E^2}$$ (41)

The description of the other two cases shown in Figure 1, with magnetic field directed along x and along y, is exactly analogous. We note that for $B_y \neq 0$, the upper two levels cross, and for $B_x \neq 0$, there is no level crossing. It is also obvious from inspection of the Hamiltonian matrix (Table 5) that in the general case, the level crossings will be avoided (i.e., are converted into anticrossings).

Table 5 makes it clear that the same Figure 1 also shows the less familiar response [6] of the three singlet states of a perfect biradical to gradual perturbation by a structural change or by an outside field, again described by the operator $W \cdot \hat{J}$. A structural perturbation converts the biradical to a biradicaloid and eventually, to an ordinary molecule.

For a covalent perturbation, $\gamma_{AB} \neq 0$, $\delta_{AB} = 0$, and the left-hand side of Figure 1 applies. At first, the perturbation produces a *homosymmetric biradicaloid*, and eventually, a covalent bond between A and B. The lowest singlet is stabilized relative to the triplet and becomes degenerate with it when the strength of the perturbation is

$$|\gamma_{AB}| = 2\sqrt{K_{AB}(K_{AB} + K'_{AB})} \tag{42}$$

In the strong perturbation limit, the eigenfunctions become adapted to the x axis, as do the orbitals, which then are the "real" delocalized orbitals A^x and B^x. In this limit, the lowest singlet state is described by a single configuration, with the more stable (bonding) of the two delocalized orbitals doubly occupied. The perturbation increases the gap between the two lowest singlets (the "S_0 - S_1 gap") and induces no singlet level crossings.

In photochemical applications, where minima in the second lowest singlet potential energy surface (S_1) are sought, this is an unproductive type of perturbation of a perfect biradical, and tends to be avoided by photochemical paths. If there is an S_0 - S_1 degeneracy, it occurs at the geometry of the perfect biradical, and only if the biradical is of the axial type.

For a polarizing perturbation, $\gamma_{AB} = 0$, $\delta_{AB} \neq 0$, and the right-hand side of Figure 1 applies. Such a perturbation produces a *heterosymmetric biradicaloid* at first (*weakly* heterosymmetric to the left of the level crossing, *critically* heterosymmetric at the level crossing, and *strongly* heterosymmetric to the right of the level crossing). A very strong perturbation ultimately produces a lone pair - empty orbital ("donor - acceptor") combination. The geminals become adapted to the z axis, as do the orbitals, which become the localized orbitals A^z and B^z. In this limit, the lowest singlet state is again described by a single configuration, with the more stable of the two localized orbitals doubly occupied (lone pair).

Perhaps counterintuitively, if one recalls Hückel theory, this type of perturbation does not initially decrease the singlet-triplet gap ("S_0 - T gap") at all. It does so only after the perturbation reaches the critical strength for the S_0 - S_1 crossing. This occurs when the strength of the perturbation δ_{AB} attains the value δ_0,

$$\delta_0 = 2\sqrt{K'_{AB}(K'_{AB} - K_{AB})} \tag{43}$$

as is readily verified by substitution from Table 5 into the general equation (38).

Perhaps even more counterintuitively relative to the Hückel picture, the polarizing perturbation decreases the S_0 - S_1 gap in a perfect biradical. It therefore is of great interest in the search for minima on the lowest excited singlet surface and for S_0 - S_1 conical intersections. It is likely that many important singlet photochemical processes occur via a return to the lowest singlet state surface at geometries that correspond to a "critically heterosymmetric" biradicaloid [1,6,16] (e.g., photoisomerization of protonated Schiff bases, and certain pericyclic reactions). The donor-acceptor pairs encountered with strongly heterosymmetric biradicaloids also are of interest, particularly in photophysics ("TICT" states [1,6]).

Structural changes capable of introducing a non-vanishing value of δ_{AB} into a biradical are of several types. The most obvious are: (i) An increase of an electronegativity of one of the orbitals (A or B) by substitution or heteroatom replacement, or perhaps by adjustment of a position of an ion or the solvent in vicinity (this is one way of coupling proton motion to excited state relaxation [1,6]). (ii) A change in the electronegativity of one of the orbitals by rehybridization (e.g., pyramidalization at a trigonal carbon atom). (iii) For orbitals delocalized over two or more centers, stabilization or destabilization by introduction of new resonance integrals, say, by transannular interaction (e.g., distorting square cyclobutadiene to a diamond shape). A consideration of these and other structural changes that permit a variation of δ_{AB} is essential for success in the search for critically heterosymmetric molecular geometries and thus for S_0 - S_1 conical intersections.

Simultaneous presence of covalent and polarizing perturbations ($\gamma_{AB} \neq 0$, $\delta_{AB} \neq 0$) leads to *non-symmetric biradicaloids*. Their general features again follow from a consideration of the Hamiltonian matrix in Table 5. For instance, they permit a ready rationalization of the preference of donor-acceptor pair molecules such as NH_2-BH_2 for planarity in the lowest singlet state S_0 and orthogonality in the first excited singlet state S_1 [1,6].

A general wave function of a singlet state S_j of an electron pair is obtained by diagonalization of the matrix of the Hamiltonian $\hat{H}$ given in Table 5:

$$S_j = C_{jx}S[x] + C_{jy}S[y] + C_{jz}S[z] \tag{44}$$

Effects of external fields on the singlet states. As already noted, no structural variation is capable of introducing a perturbation of the $W_y\hat{J}_y$ type, which would correspond to the central part of Figure 1. Such a perturbation is provided by an external magnetic field, ($W_y = \beta_e B \cdot \nabla_{AB}$, Table 5), but the fields available in the laboratory are far too weak for permitting effects of a size comparable with what is easily achievable by a structural manipulation of W_x and W_z.

The effects of easily accessible external electric fields are larger, but still generally very small relative to structural effects. This perturbation is of the type $W_z\hat{J}_z$ and induces the behavior shown on the right-hand side of Figure 1 (e.g., a charge of q = 1.735 |e| located on the CC axis of orthogonally twisted ethylene 1.85 Å from the midpoint of the bond is calculated [1] to bring the lowest two singlet levels S_0 and S_1 to degeneracy).

However, in some cases magnetic and electric fields have large effects: when the molecular structure is such that two of the three singlet levels are nearly or exactly degenerate, even very small perturbations can have dramatic effects on the wave functions.

Two such cases have been recognized for a long time. First, in an axial biradical the $S[x]\Sigma$ and $S[z]\Sigma$ levels are degenerate, permitting even a weak $W_y \hat{J}_y$ perturbation to mix them and to induce an orbital magnetic dipole moment. The magnetic field B required for this needs to have a non-vanishing component along the direction of ∇_{AB}, i.e., along the high-order symmetry axis, in order to make $W_y = \beta_e B \cdot \nabla_{AB}$ different from zero (e.g., in singlet NH, with A and B represented by $2p_N$ orbitals perpendicular to the bond). If the molecule is capable of distortion which removes the exact degeneracy by introducing $\gamma_{AB} \hat{J}_x$ and/or $\delta_{AB} \hat{J}_z$ terms, the magnetizing $W_y \hat{J}_y$ perturbation no longer can mix the $S[x]\Sigma$ and $S[z]\Sigma$ states extensively, and the orbital angular momentum is said to be quenched. If one were to calculate the orbital magnetic dipole moment of the system in an external magnetic field along a path of geometrical distortions that passed through or very close to the point of exact degeneracy (axial biradical), one would obtain essentially no magnetic moment except in the immediate vicinity of the degeneracy point. Such behavior might be called "*sudden magnetization*" and is clearly related to the well known "orbital magnetic moment quenching".

Second, in a pair biradical the $S[x]\Sigma$ and $S[y]\Sigma$ levels are degenerate, permitting even a weak $W_z \hat{J}_z$ perturbation to mix them and to induce an electric dipole moment. The electric field E required for this needs to have a non-vanishing component along the direction of r_{AB}, i.e., along the line joining the charge centroids of the localized orbitals A and B, in order to make $W_z = |e|E \cdot r_{AB}$ different from zero (e.g., in orthogonally twisted ethylene, along the CC line). Of course, a weak perturbation of this type can also be produced by a structural variation (i.e., by a static intramolecular electric field) which causes δ_{AB} to differ from zero (e.g., by pyramidalizing one of the CH_2 groups in the twisted ethylene). If the molecule can be distorted sufficiently in a way that removes the exact degeneracy by introduction of a $\gamma_{AB} \hat{J}_x$ term, the polarizing $W_z \hat{J}_z$ perturbation will no longer be able to mix the $S[x]\Sigma$ and $S[y]\Sigma$ states extensively, and the electric dipole moment will be largely quenched. The electric dipole moment induced in the system by an external (E) or internal (δ_{AB}) electric field, calculated along a path of geometrical distortions that passes through or very close to the point of exact degeneracy (pair biradical), is generally small, except in the immediate vicinity of the degeneracy point. Such behavior is well known as "*sudden polarization*" [1,6,17].

Since the $S[y]$ and $S[z]$ states are not degenerate for any choice of K_{AB} and K'_{AB} realizable in practice, we do not have a third analogous phenomenon, "sudden bond formation". There also is no external field that would act like a $W_x \hat{J}_x$ perturbation and mix the $S[y]$ and $S[z]$ states. This type of covalent perturbation is available from structure variation (γ_{AB}), but the conversion of a strong covalent bond to a dissociated one is always gradual. This circumstance is at the heart of the difficulty in reaching an agreement on a quantitative definition of a biradicaloid: how strong does a bonding interaction between two radical centers need to be before it is no longer useful to talk about a biradical or biradicaloid but instead, about a system with a weak (or strong) covalent bond?

Inspection of Figure 1 suggests one additional possibility for a "sudden" behavior. In any perfect biradical other than an axial one, which has a degenerate lowest singlet already, it is possible to induce S_0 - S_1 degeneracy by introduction of a polarizing perturbation δ_{AB} of just the right size (δ_0) to reach a critically heterosymmetric biradicaloid. At this point, a state with a wave function given by $S[z]$ is degenerate with one whose wave function contains roughly equal parts of $S[x]$ and $S[y]$ and is therefore highly polarized. Even if γ_{AB} does not vanish exactly, so that the crossing is avoided somewhat, one can expect a very abrupt change in the dipole moments of the lowest two singlet states as structurally determined δ_{AB} or an external electric field E are varied in a way that takes the system through the critical point. Highly non-linear polarizability, bistability, and other interesting phenomena can be expected when such nearly critically heterosymmetric biradicaloids are examined in the laboratory.

7. Spin-Orbit Coupling in a Two-Electron System

We are finally ready to examine the action of the spin-orbit coupling operator $\hat{H}^{SO}(1,2)$ in the spin-geminal product space spanned by $T\theta[x]$, $T\theta[y]$, $T\theta[z]$, $S[x]\Sigma$, $S[y]\Sigma$, and $S[z]\Sigma$. This operator consists of an additive part $\hat{H}_1^{SO}(1,2)$ due to the interaction of electron spin with the magnetic field generated by its own orbital motion, and of an inseparable part $\hat{H}_2^{SO}(1,2)$ due to the interaction of electron spin with the magnetic field generated by the motion of another electron:

$$\hat{H}^{SO}(1,2) = \hat{H}_1^{SO}(1,2) + \hat{H}_2^{SO}(1,2)$$
$$\hat{H}_1^{SO}(1,2) = \hat{H}_1^{SO}(1) + \hat{H}_1^{SO}(2) \tag{45}$$

where $\hat{H}_1^{SO}$ is the operator defined in equations (15) - (17). It is customary to neglect the effects of $\hat{H}_2^{SO}(1,2)$ in semiquantitative treatments although they need to be kept in rigorous calculations, and we shall presently neglect $\hat{H}_2^{SO}(1,2)$ as well.

Working out the matrix elements of $\hat{H}_1^{SO}(1,2)$ and using the definition of the vector h^κ in equation (17), we obtain

$$<T\theta[u]|\hat{H}_1^{SO}(1,2)|S[y]\Sigma> = -<S[y]\Sigma|\hat{H}_1^{SO}(1,2)|T\theta[u]>$$

$$= \sum_{\kappa}(\zeta\kappa/2)(\nabla^{\kappa}_{AB})_u = \sum_{\kappa}h^{\kappa}_u = h^{SO}_u \qquad (46)$$

while all other matrix elements vanish. The representation of $\hat{H}_1^{SO}(1,2)$ in the above basis then is

$$
\begin{array}{l}
T\theta[x]: \\
T\theta[x]: \\
T\theta[x]: \\
S[x]\Sigma: \\
S[y]\Sigma: \\
S[z]\Sigma:
\end{array}
\left(
\begin{array}{cccccc}
0 & 0 & 0 & 0 & h_x^{SO} & 0 \\
0 & 0 & 0 & 0 & h_y^{SO} & 0 \\
0 & 0 & 0 & 0 & h_z^{SO} & 0 \\
0 & 0 & 0 & 0 & 0 & 0 \\
h_x^{SO} & h_y^{SO} & h_z^{SO} & 0 & 0 & 0 \\
0 & 0 & 0 & 0 & 0 & 0
\end{array}
\right) \qquad (47)
$$

where $h^{SO}_u = \sum_{\kappa}h^{\kappa}_u$, and h^{SO} can be referred to as the spin-orbit coupling vector.

In a perfect biradical, the lowest two singlet states, S_0 and S_1, thus do not spin-orbit couple to the triplet in the present approximation. The highest singlet $S[y]\Sigma$ will couple to all three components of the triplet if none of the components of the vector $\sum_{\kappa}h^{\kappa}$ vanish. Symmetry may demand otherwise. For instance, in an axial biradical, in which $S[x]$ and $S[z]$ are degenerate, the vector ∇^{κ}_{AB} lies in the high-order axis of rotational symmetry and only the corresponding component of the triplet state will interact with $S[y]\Sigma$.

In perturbed biradicals, the $S[y]\Sigma$ character is shared among two or all three singlet states as shown in equation (44) and the spin-orbit coupling matrix elements for each of these will then be simply proportional to the coefficient of the $S[y]\Sigma$ part of the singlet wave function. For a singlet wave function S_j defined in equation (44), the matrix elements of spin-orbit coupling are

$$<T\theta[u]|\hat{H}_1^{SO}|S_j> = C_{jy}h_u^{SO} \qquad (48)$$

The strength of the coupling thus depends on three factors: the weight of the $S[y]$ contribution to the singlet state wave function, the spatial disposition of the orbitals A and B, and the atomic number of the atoms on which the orbitals reside, which dictates the size of ζ_K. An actual rate of intersystem crossing will depend on other factors as well, in particular, the density of states and Franck-Condon factors.

Inspection of Table 5 and Figure 1 reveals that in a homosymmetric biradicaloid or a molecule in which the electron pair forms an ordinary covalent bond, the $S[y]$ character is shared between the lowest and the highest of the three singlet states, S_0 and S_2. Spin-orbit coupling between the ground state S_0 and the triplet state should thus be favorable. In contrast, the coupling between S_1 and the triplet should not. It has indeed been recognized for a long time [2] that in orthogonally twisted ethylene, spin-orbit coupling between the triplet and the lowest singlet S_0 is very small and that intersystem crossing is likely to occur at geometries that are only partially twisted (homosymmetric biradicaloids), even though they are less favorable energetically in the triplet state. Also, the $S_1 \rightarrow T$ intersystem crossing in simple olefins is notoriously slow.

By the same token, ring closure in triplet chain biradicals, such as 1,4 or 1,5 biradicals, should preferentially occur at geometries at which a covalent perturbation γ_{AB} is present, i.e., where the atomic orbitals of the two radical centers overlap at least to some degree. Strong overlap would be very unfavorable energetically in the triplet state, and even relatively weak overlap may cause the appearance of an activation energy to the process - in order to reach the geometries at which intersystem crossing is really fast, a little extra energy is needed. Once intersystem crossing occurs at such a geometry, the singlet molecule is looking into an 80 kcal/mol abyss and will almost inevitably close the ring, without having any time for bond rotations or other activities. Thus, ring closure from the triplet state may appear to be concerted in spite of its spin-forbidden nature.

In a weakly heterosymmetric biradicaloid, the triplet state does not spin-orbit couple to the lowest singlet state S_0 and couples to the two excited singlets S_1 and S_2 instead. In a strongly heterosymmetric biradicaloid, i.e., a donor-acceptor pair (such as a twisted aminoborane NH_2-BH_2), the triplet should spin-orbit couple to the lowest singlet S_0 and not the next higher one, S_1. In such molecules, partial untwisting will again be needed to achieve efficient spin-orbit coupling, and in that sense they should be like olefins.

Under conditions of rapid relaxation among the three triplet levels, the singlet-triplet coupling is frequently characterized by the "total spin-orbit coupling strength" for the S_j state, defined by

$$SOC_j = (\sum_u |<T\theta[u]|\hat{H}^{SO}|S_j>|^2)^{1/2} \qquad (49)$$

In our approximation, this is equal to the product of the absolute magnitude of C_{jy} and the length of the spin-orbit coupling vector h^{SO}:

$$SOC_j = |C_{jy}| \; |h^{SO}| = |C_{jy}| \; |\sum_{\kappa} h^{\kappa}| \qquad (50)$$

The vector nature of the addition of the contributions from the individual nuclei κ in the construction of the total spin-orbit coupling vector h^{SO} is important. It is often assumed that introduction of a heavy atom into a molecule cannot but increase the spin-orbit coupling strength (the "heavy-atom effect"). This is clearly wrong. If the additional contribution is a vector that is opposed to the resultant of those already present, the value of SOC_j may actually decrease upon the introduction of a heavy atom. We believe to have identified an experimental case of such behavior recently [18].

The spin-orbit coupling matrix elements are so small in ordinary organic molecules that they have no significant effect on energy differences between the singlet levels. However, since the three triplet levels are so closely spaced together, any differential effects of their coupling to the singlet state are likely to affect the size of the D and E parameters. The spin-orbit contribution to these zero-field splitting parameters appears to be small in molecules such as aromatic hydrocarbons, but it is not obvious that it is also negligible in other light-atom molecules, and in biradicals in particular. Indeed, in the O_2 molecule, where the $S[y]\Sigma$ state lies only 1.64 eV above the T states, spin-orbit coupling provides the dominant contribution [15b].

The change in the energy, $E[u]$, of the $T\Theta[u]$ sublevel of the triplet state can be estimated from second-order perturbation theory. Using $E(T)$ for the average triplet energy, we obtain

$$\Delta E[u] = \sum_j c_{jy}^2 (h_u^{SO})^2 / [E(T) - E(S_j)] \qquad (51)$$

For instance, in the usual case of $D > 0$, $E < 0$ (level order $\Theta[z]$, $\Theta[y]$, $\Theta[x]$ in the order of increasing energy), spin-orbit coupling will change the values of the D and E parameters as follows:

$$D' = D + \sum_j [E(T) - E(S_j)]^{-1} c_{jy}^2 \{(h_z^{SO})^2 - [(h_y^{SO})^2 + (h_x^{SO})^2]\} \qquad (52)$$

$$E' = E + \sum_j [E(T) - E(S_j)]^{-1} c_{jy}^2 [(h_y^{SO})^2 - (h_x^{SO})^2]$$

8. Summary: Electronic States of an Electron Pair

In our quest for an understanding of the electronic states of biradicals and biradicaloids we have achieved a unification of quite a few apparently unrelated aspects of chemistry, such as the adaptation of orbitals to perturbations, geminal transformations induced by orbital transformations, zero-field splitting in triplets, S_0 - S_1 conical intersections, sudden polarization, orbital angular momentum quenching, intersystem crossing as a function of structure, twisting in TICT states, bistability, non-linear polarizability, and others. Most of the results are not new, but we believe that the organization based on angular momentum operators is, and it is hoped that the unification will eventually lead to new insights.

Perhaps the most significant aspect of the simple model used is the generality of its treatment of the states of an electron pair, arguably one of the most important building blocks of chemical structures, and the powerful concepts of covalent, polarizing, and magnetizing perturbations, and of critical biradicaloids. We believe that the resulting guidance will continue to be useful in the search for minima in the lowest excited singlet surface S_1, and for its conical intersections with the ground state singlet surface S_0.

Acknowledgement

I am grateful to the National Science Foundation for support (CHE 9000292). I thank the faculty of the Chemistry Department at the Technion in Haifa, Israel, for their warm hospitality during the initial stages of my work on this project (Manson Visiting Professorship). It is a pleasure to express my gratitude to Prof. Vlasta Bonačić-Koutecký of the Free University, Berlin, Germany, for years of collaboration on the theory of biradicals and biradicaloids, and to acknowledge a useful discussion with Prof. Joe Paldus of the University of Waterloo, Ontario, Canada.

References

[1]   Michl, J.; Bonačič-Koutecký, V. *Electronic Aspects of Organic Photochemistry*, John Wiley and Sons, Inc.: New York, 1990.

[2]   Salem, L.; Rowland, C. *Angew. Chem., Int. Ed. Engl.* **1972**, *11*, 92.

[3]   Michl, J. *Mol. Photochem.* **1972**, 4, 257.

[4]   *Diradicals*; Borden, W. R., Ed.; Wiley: New York, 1982.

[5]   Salem, L. *Electrons in Chemical Reactions*; Wiley: New York, 1982.

[6]   Bonačič-Koutecký, V.; Koutecký, J.; Michl, J. *Angew. Chem., Int. Ed. Engl.* **1987**, *26*, 170.

[7]   Michl, J.; Bonačič-Koutecký, V. *Tetrahedron* **1988**, 44, 7559.

[8]   Bonačič-Koutecký V.; Michl, J.; Köhler, J. Chem. Phys. Lett. **1984**, *104*, 440. Bonačič-Koutecký, V.; Michl, J. *J. Am. Chem. Soc.* **1985**, *107*, 1765.

[9]   Caldwell, R. A.; Carlacci, L.; Doubleday, C. E., Jr.; Furlani, T. R.; King, H. F.; McIver, J. W., Jr.; *J. Am. Chem. Soc.* **1988**, *110*, 6901.

[10] Cohen-Tannoudji, C.; Diu, B.; Laloë, F. *Quantum Mechanics*; Wiley: New York, 1977.

[11] Altmann, S. L. *Rotations, Quaternions, and Double Groups*; Clarendon Press: Oxford, 1986.

[12] Warnick, S. M.; Michl, J. *J. Am. Chem. Soc.* **1974**, *96*, 6280.

[13] Richards, W. G.; Trivedi, H. P.; Cooper, D. L. *Spin-Orbit Coupling in Molecules*; Clarendon Press: Oxford, 1981.

[14] McGlynn, S. P.; Vanquickenborne, L. G.; Kinoshita, M.; Carroll, D. G. *Introduction to Applied Quantum Chemistry*; Holt, Rinehart and Winston: New York, 1972.

[15] (a) Carrington, A.; McLachlan, A. D. *Introduction to Magnetic Resonance*; Harper and Row: New York, 1967. (b) Weltner, W., Jr. *Magnetic Atoms and Molecules*; Van Nostrand Reinhold: New York, 1983.

[16] Bernardi, F.; De, S.; Olivucci, M.; Robb, M. A. *J. Am. Chem. Soc.* **1990**, *112*, 1737.

[17] Bonačič-Koutecký, V.; Bruckmann, P.; Hiberty, P.; Koutecký, J.; Leforestier, C.; Salem, L. *Angew. Chem., Int. Ed. Engl.* **1975**, *14*, 575.

[18] Fisher, J. J.; Michl, J., unpublished results.

AB-INITIO MODELLING OF CHEMICAL REACTIVITY USING MC-SCF AND VB METHODS

Michael A. Robb
Department of Chemistry
King's College London
Strand
London WC2R 2LS

F. Bernardi
Dipartimento Chimico 'G. Ciamician'
Universita di Bologna
Bologna

ABSTRACT

Our first objective is to illustrate how *thinking in VB language* solves many of the objectivity problems in the practical application of MC-SCF theory to problems in chemical reactivity. Thus the essential elements of MC-SCF theory will be reviewed with the objective of conveying sufficient information that the practical and conceptual aspects of MC-SCF theory can be understood. We shall also show that there exists a completely rigorous transformation to Valence Bond Space. Thus we will illustrate that one has the option of *thinking* in VB space (a natural one for the chemist) but doing all the *numerical* work in MC-SCF space.

Our second second objective to illustrate how MC-SCF data on chemical reactivity can be visualized and conceptualized using this VB model. A Heisenberg VB hamiltonian is obtained from an MC-SCF computation in terms of coulomb and exchange parameters that have the same interpretation as in the HL theory of the H_2 molecule. This data can then be fitted and visualized using contour diagrams.

I Introduction

We believe that theoretical investigations of chemical reactivity ought to have 2 facets i) a numerical investigation of structure and energetic problems and ii) Modelling and visualization of the numerical results. Thus one must be concerned with the numerical computation of the equilibrium structures, transition structures, reaction paths and barriers that are associated with various competing mechanisms. Here the theoretical method chosen must be capable of representing the competitive reaction paths with a balanced accuracy. Often SCF (including UHF methods) cannot achieve this balance where homolytic bond cleavage takes place or when bi-radical centres are created. In this case computations at the MC-SCF level are needed. After the numerical investigation is finished, one must attempt to rationalize the results for a spectrum of

253

S. J. Formosinho et al. (eds.), Theoretical and Computational Models for Organic Chemistry, 253–288.
© 1991 *Kluwer Academic Publishers.*

related reactions with a model that gives us some "understanding".

Over the last few years we have been involved in a collaborative project that has involved the study of chemical reactivity using MC-SCF theory coupled with the use of a VB model for interpretive purposes. Our experience (see ref. 1-8 for example), indicates that MC-SCF theory will evolve to become a routine technique for the study of reactivity. However, the MC-SCF method is often criticized for its possible lack of objectivity and the fact that one requires some considerable experience and insight before MC-SCF codes can be used in a routine way. In these notes we will show that the VB model [9-11] we have been using to give us *understanding* provides just the required insight necessary to use the MC-SCF method in an effective manner. Thus simple VB models have proved to be very important not only because they provide a basis for "understanding" chemical reactivity but also, because they provide the basis for the types of decisions (selection of the active orbitals or the choice of starting geometry in a transition structure computation) that one must make in performing the MC-SCF computations themselves.

In these notes we have 3 objectives. Firstly we shall review the essential elements of MC-SCF theory. Our objective here is to convey only sufficient information that some of the practical and conceptual difficulties with MC-SCF theory can be understood. Secondly, we wish to show that there exists a completely rigorous transformation to Valence Bond Space. Thus we have the option of *thinking* in VB space (a natural one for the chemist) but doing all the *numerical* work in VB space. Thirdly, we shall illustrate how Heisenberg VB hamiltonian can be parametrized from MC-SCF wavefunctions. This technique enables us to "model" reactivity and visualize the results.

2 MC-SCF Theory for Pedestrians

The essential elements of MC-SCF theory will be now be reviewed with the objective of conveying sufficient information that the practical and conceptual aspects of the theory can be understood. There are now many good reviews on the subject and ref. 12 and 13 are quite helpful. Our treatment is based upon ref. 13-16.

At the most elementary level MC-SCF theory is very simply related to ordinary SCF theory. In SCF theory, one has only two types of orbitals occupied and unoccupied. In MC-SCF theory there are 3 types illustrated in Figure 1.The core orbitals are always doubly occupied and are similar to the closed shell doubly occupied orbitals in SCF theory. It is the Valence or Active orbitals that play the crucial role in MC-SCF theory. The reference CI expansion contains configurations with (in general) all possible occupancies of the valence orbitals (complete active space SCF - CASSCF). Thus in SCF one has a single configuration of the electrons, while in MC-SCF the orbitals are optimized for an linear combination of electronic configurations. One must choose the partition between active/inactive and active/virtual at the outset of the computation. It is this choice that leads to the possibility of a lack of objectivity in MC-SCF computations. From the outset, it must be recognized that in MC-SCF theory, one is not attempting to compute the true dynamic electron correlation energy; rather, one is seeking a wavefunction that will yield a balanced representation of the potential energy surface in the region of products, reactants and transition state. One needs to be able to demonstrate that one has achieved this and one needs to be able to choose the active orbital space so that this objective is in fact achieved.

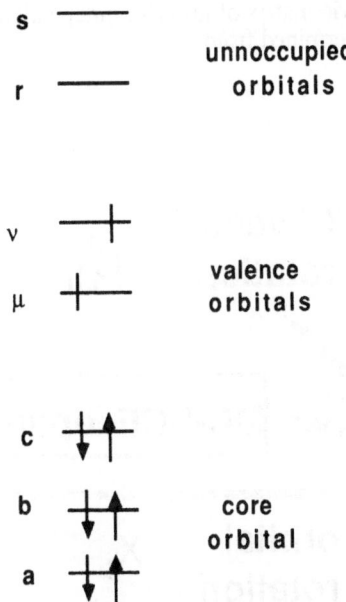

Figure. 1 Three types of orbitals in MC-SCF theory.

Let us now briefly summarize the essentials of MC-SCF theory. It is helpful to take an entirely different approach to the usual formulation of closed shell SCF theory. In fact, it proves to be useful to think of the MC-SCF process in a similar way to geometry optimization. Thus we shall view the orbital and CI coefficient variables that occur in MC-SCF in the same way as the internal geometrical variables in a geometry optimization. Technically in order to do this we must assume that we have an orthogonal set of starting orbitals Φ_0 and an orthogonal set of CI vectors $|K\rangle$. The MC-SCF CI expansion (for state κ) is written as

$$\Psi^\kappa = |0\rangle + \Sigma_K C^\kappa{}_{0K} |K\rangle \qquad [1]$$

where $|0\rangle$ is the eigenvector of interest obtained by solving the CI problem using the starting set of orthogonal orbitals Φ_0 and the $|K\rangle$ are the remaining eigenvectors of the CI problem. The final MC-SCF wavefunction can be seen to correspond to a *step* in the space of the *variables* X_{ij} (for orbital rotation among the Φ_0) and $C^\kappa{}_{0K}$ (for CI vector rotation between $|0\rangle$ and $|K\rangle$) as shown in Fig. 2. The final orbitals Φ are obtained from the starting orthogonal Φ_0 set by unitary transformation as

$$\Phi = U \Phi_0 \qquad [2]$$

where the unitary transformation U is written in terms of its parameters X_{ij} as

$$U = \exp X \qquad [3]$$

where **X** is a skew-symmetric matrix of n(n-1)/2 independent parameters. Similarly the final CI wavefunction is determined from

$$A = \exp C \qquad\qquad [4]$$

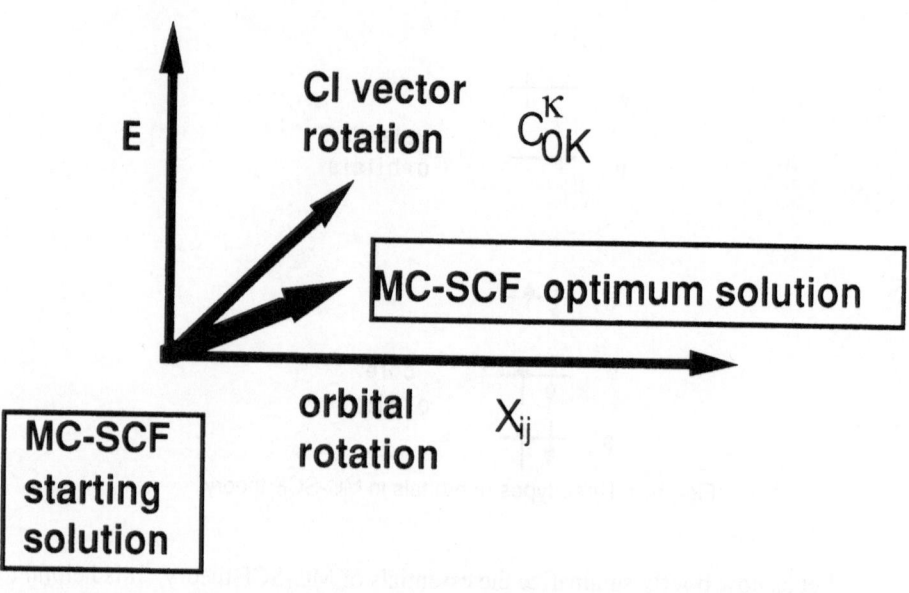

Figure 2 Variables in the MC-SCF process

Thus the variables in **U** and **A** play the same role that the Cartesian co-ordinates do in a geometry optimization while the X_{ij} and $C\,^\kappa_{0K}$ are like bond angles or bond lengths. Consequently, the MC-SCF energy to second order can now be written in terms of the parameters X_{ij} and $C\,^\kappa_{0K}$ as

$$E_{\text{MC-SCF}} = E_0 + E^X\,X + X\,{}^t E^{XX}\,X + C^t\,E^{cc}\,C + C^t\,E^{cx}X + X\,{}^t\,E^{xc}\,C$$

$$[5]$$

where the arrays **X** and **C** have been written in a linear form in eq. 5. For example

$$X = \begin{bmatrix} X_{21} \\ X_{31} \\ X_{32} \end{bmatrix}$$

$$[6]$$

The quantities E^X etc. denote the first and second derivatives of the energy with respect to orbital rotations. Thus for example,

$$E^C = \partial E / \partial C_{oK} \qquad [7]$$

$$E^X = \partial E / \partial X_{ij} \qquad [8]$$

$$E^{XX} = \partial^2 E / \partial X_{ij} \partial X_{kl} \qquad [9]$$

Minimization of the energy gives the usual Newton-Raphson equations

$$\begin{bmatrix} E^{XX} & E^{XC} \\ E^{CX} & E^{CC} \end{bmatrix} \begin{bmatrix} X \\ C \end{bmatrix} = - \begin{bmatrix} E^X \\ E^C \end{bmatrix} \qquad [10]$$

In fact, as we have formulated the problem E^{CC} has the simple form

$$\{E^{CC}\}_{KL} = \delta_{KL}(E_0 - E_K) \qquad [11]$$

so that the parameters C can be eliminated. However we leave the equations on this form to illustrate the structure of the theory. Clearly we have the minimum energy when

$$\begin{bmatrix} E^X \\ E^C \end{bmatrix} = 0 \qquad [12]$$

Thus the energy optimization process can be thought of in much the same way as geometry optimization.

The reader may be surprised that we have apparently have no Fock matrix type eigenvalue problem and no concept of orbital energies as in ordinary SCF theory. In fact the E^X in eq. 5 do depend on two Fock type operators

$$F^1 = h^C + \sum_{\nu\mu} (J_{\nu\mu} - 1/2\, K_{\nu\mu}) \gamma_{\nu\mu} \qquad [13]$$

and

$$F^2 = \sum_{i\chi} |i\rangle \left\{ \sum_{\nu\mu\eta} (\langle i| h^C |\eta\rangle \gamma_{\eta\chi} + [i\nu|\mu\eta] \Gamma_{\chi\nu\mu\eta}) \right\} \langle \chi | \qquad [14]$$

where h^C is the usual closed shell inactive Fock operator, $J_{\nu\mu}$ and $K_{\nu\mu}$ coulomb and exchange operators, and $\gamma_{\nu\mu} / \Gamma_{\chi\nu\mu\eta}$ the one and two electron density matrices. The $\gamma_{\nu\mu}$ and $\Gamma_{\chi\nu\mu\eta}$ depend upon CI expansion coefficients. The summations over $\chi\nu\mu\eta$ include the active orbitals and over i for either inactive or virtual orbitals. Thus the Fock type operators for MC-SCF theory look similar to those of closed shell SCF theory except that the $J_{\nu\mu}$ and $K_{\nu\mu}$ are weighted by the CI expansion coefficients via the $\gamma_{\nu\mu}$

/ $\Gamma_{\chi\nu\mu\eta}$. The specific formula for E^X depends upon the type of orbital rotation and is defined by the Fock type matrix given below

	core	valence	virtual
core	Undef.	$F^1 - F^2$	F^1
valence	$F^1 - F^2$	$F^2_{\mu\nu} - F^2_{\nu\mu}$ Undef. for CAS	F^2
virtual	F^1	F^2	Undef.

$$F =$$

[15]

Thus

$$E^X_{valence/virtual} = \frac{\partial E}{\partial X_{valence/virtual}} = F^2_{valence/virtual}$$

[16]

so that if the MC-SCF wavefunction is optimized with respect to valence/virtual rotations the corresponding part of the Fock matrix is diagonal. But note that for example

$$\frac{\partial E}{\partial X_{core/core}} = 0$$

[17]

since the energy of an MC-SCF wavefunction is invariant to core/core rotations. As a consequence the Fock operator $F_{core/core'}$ is not defined. Thus we have the freedom to choose *any* rotation of the core orbitals that we find useful. Thus we could choose a set of orbitals so that

$$F^1_{core/core'} = \varepsilon_{core}\delta_{core/core'} \tag{18}$$

where ε_{core} is an *orbital energy*. In a similar fashion the MC-SCF energy is invariant to virtual-virtual rotations. Finally, notice that

$$E^X_{valence/valence'} = \frac{\partial E}{\partial X_{valence/valence'}} = F^2_{valence/valence'} - F^2_{valence'/valence} \tag{19}$$

In the case of a CAS MC-SCF then the energy is invariant to valence-valence rotations ie

$$E^X_{valence/valence'} = 0 \tag{20}$$

It is this feature that enables us to transform to VB space and we return to this important point in the next subsection.

Now let us briefly digress to extract conventional closed shell SCF theory from this formalism so that the ideas we have developed are clearer. For the closed-shell RHF case we have only a single reference so the blocks E^{CC}, E^{CX}, etc are omitted from the energy expression (eq.5) and eq. 10. Thus the equation system reduces to

$$E^X + E^{XX} X = 0 \tag{21}$$

which has the formal solution

$$X = -(E^{XX})^{-1} E^X \tag{22}$$

If we take E^{XX} to be diagonal and approximate it as

$$(E^{XX})_{ij,ij} = \varepsilon_i - \varepsilon_j \tag{23}$$

then the X_{ij} are found as

$$X_{ij} = -(E^X)_{ij}/(\varepsilon_i - \varepsilon_j) \tag{24}$$

However, provided X_{ij} is small enough so that $\tan 2 X_{ij} \approx X_{ij}$, then we can recognize eq. 24 as the formula for a 2X2 Jacobi plane rotation. Then we can determine U rather than X by solution (diagonalization) of

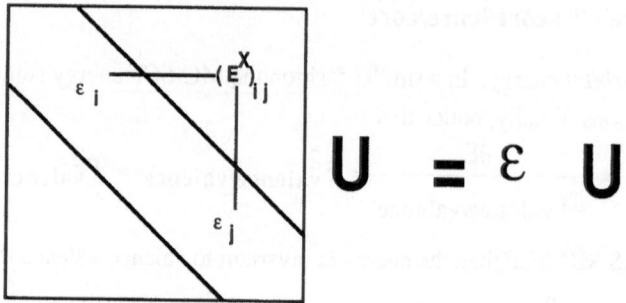

$$U = \varepsilon\ U$$

[25]

When one evaluates the expressions for the $(E^X)_{ij}$ one simply obtains the usual Hartree-Fock operator matrix elements since F^1 and F^2 become identical to the usual SCF Fock operators when the valence orbitals become doubly occupied (ie $\gamma_{\nu\mu}=2.0\delta_{\nu\mu}$, $\Gamma_{\chi\nu\mu\eta}=2.0\delta_{\chi\nu}\,\delta_{\mu\eta}-\delta_{\chi\mu}\,\delta_{\nu\eta}$). The ε_i are arbitrary but are usually taken to be the diagonal elements of the Hartree-Fock operator. In fact eq. 21 is just the second-order approximation and the eq. 24/25 are a first-order approximation to the conventional SCF equations. The only difference from the conventional textbook formulation at this stage is that we have formulated the problem in an orthogonal basis rather than the usual AO basis.

In fact we have used a pseudo 1st order approach to the solution of the MC-SCF equations with some success. One uses a <u>diagonal</u> approximation to E^{XX} and solves

$$E^{XX}\ X = -\ E^X$$

[26]

the C in eq. 10 having been obtained by solving the CI eigenvalue problem directly via the augmented eigenvalue problem

$$\begin{bmatrix} E_0 & E^C \\ E^C & E^{CC} \end{bmatrix} \begin{bmatrix} 1 \\ C \end{bmatrix} = E_{MC\text{-}SCF} \begin{bmatrix} 1 \\ C \end{bmatrix}$$

[27]

This assumes that the orbital rotation /CI vector coupling E^{CX} is negligible which is often the case.

What are the practical difficulties with using MC-SCF? Essentially they arise either from the strong coupling CI vector with orbital rotations or with the definition of the core/valence/virtual partition. In eq. 11 when $(E_0 - E_K)$ is small (such as when excited state and ground state become close together) the CI vector and orbital rotation become pathologically coupled but these situations are quite rare. Most difficulties are associated with the correct choice of core/valence/virtual partition. We shall point out some of the mathematical/technical problems here and return to the chemical/conceptual ones later. Since the orbital rotations are given as the solutions of eq. 22 then $E^{XX}=\partial^2 E/\partial X_{ij}\,\partial X_{kl}$ must be non-singular. For a CAS MC-SCF this will be true until $\gamma_{\nu\nu}$ approaches 0 or 2. For example, if a valence orbital becomes doubly occupied then the energy is invariant to rotations between the *core* orbitals and this *particular valence* *orbital*. Thus $\partial^2 E/\partial X_{ij}\partial X_{ij}=0$ for i=*core* j=*particular valence orbital*. Thus one of the central difficulties with MC-SCF is avoiding this problem with a sensible choice of valence space. Now let us consider on other problem - the MC-SCF solution obtained with moderate sized active spaces is routinely a local rather than a global minimum! At first sight this may seem rather surprising but a simple example clarifies the problem. Suppose we are trying to describe the dissociation of $H_2C=O$ into H_2 and CO with 4 σ / σ^* type active orbitals (two a_1 and two b_2) with the two π / π^* type b_1 orbitals inactive/virtual and 4 electrons. One has no difficulty finding a local minimum in this case. However, because of the large dynamic $(\pi)^2$ / $(\pi^*)^2$ correlation of the CO type b_1 orbital, the replacement of a pair of a_1 or b_2 orbitals in the active space with a pair of CO π type b_1 orbitals leads to an undesired solution of lower energy. Of course the dilemma disappears upon enlargement of the reference space but this makes the computation more expensive.

Attempts have been made to make the choice of core/valence/virtual orbital partition into a black box. In particular the unrestricted natural orbitals can provide a good starting point for MC-SCF as as suggested by Pulay[17]. We now explore this point in some detail since it also gives us some additional insight into MC-SCF theory. As discussed by Pulay[17] in his paper one can understand the problem with a two electron example. In UHF wavefunction for a two electron localized bond, one seeks the optimum energy of a wavefunction of the form

$$\psi^{UHF}= |\Phi_a\Phi_b|$$

[28]

At the equilibrium geometry Φ_a and Φ_b are the same and one has a closed shell SCF wavefunction. For some stretched geometry the orbitals Φ_a and Φ_b will be different and in general become localized onto atomic sites. However, ψ^{UHF} is not a pure spin state and is in fact a *component* of an MC-SCF wavefunction as we will now demonstrate. Firstly, we can project onto a pure singlet spin state

$$\psi^{PUHF}= |\Phi_a\Phi_b| - |\Phi_a\Phi_b|$$

[29]

Let us define the delocalized and othonormalized orbitals $\chi_{a/b}$

$$\chi_{a/b} =\frac{1}{\sqrt{2\pm2\lambda}} (\Phi_a\pm\Phi_b)$$

[30]

with λ corresponding to the overlap

$$\lambda = \langle\Phi_a|\Phi_b\rangle$$

[31]

The correct spin projected wavefunction ψ^{PUHF} is thus transformed into an MC-SCF wavefunction

$$\psi^{MC-SCF}= C_1 |\chi_a\overline{\chi_a}| - C_2 |\chi_b\overline{\chi_b}|$$

[32]

Where C_1 and C_2 are given as

$$C_{1/2} = \frac{1 \pm \lambda}{\sqrt{2 \pm 2\lambda^2}}$$

[33]

The $\chi_{a/b}$ are natural orbitals and have an occupancy of

$$\sigma = 1 \pm \lambda$$

[34]

As the bond dissociates homolyticaly λ goes to zero and the two occupancies approach 1. As the overlap gets large, λ goes to 1 and one of the orbitals becomes doubly occupied and we go smoothly to the SCF solution. In a true MC-SCF computation $C_{1/2}$ are variables. In the UHF they are determined completely by λ. Generalizing these ideas, the proposal of Pulay is

      1)The UHF natural orbitals are good starting orbitals for MC-SCF

      2)The MC-SCF active space should contain the fractionally occupied UHF natural orbitals (with occupation numbers between 0.02 and 1.98)

A simple example is appropriate. In Fig 3 we show the transition state geometry for the transition structure between Benzene and Dewar Benzene. The active space corresponds to the 6 π orbitals of Benzene. In Table 1a we show the MC-SCF convergence starting from the UHF natural orbitals. In 3 iterations one has convergence to 10^{-4}. The natural orbital occupation numbers of the UHF and MC-SCF wavefunctions are given in Table 1b. Notice the *pairing* of the UHF natural orbitals implied by eq. 34.

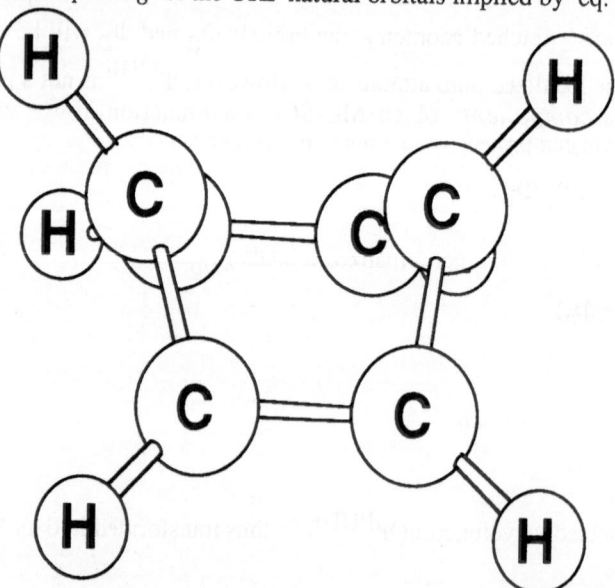

Fig 3. Benzene -Dewar Benzene 6 orbital CAS Transition State Geometry

TABLE 1a UHF/MCSCF Energy   Benzene -Dewar Benzene 6 orbital CAS

Transition State Geometry

UHF energy	MC-SCF energy
-227.7425	-227.8337 (0)
	-227.8369 (3)
	-227.8369 (8)

TABLE 1b UHF/MCSCF Natural Orbital occupation Numbers Benzene -Dewar

Benzene 6 orbital CAS Transition State Geometry

UHF-NO	MC-SCF-NO
1)1.8365	1.8818
2)1.7378	1.8521
3)1.5075	1.8236
4)0.4924	0.1794
5)0.2622	0.1362
6)0.1635	0.1268

Let us conclude this section with a very brief discussion of MC-SCF analytical derivatives. We shall follow the prescription of Almlof and Taylor [13] which illustrates the power of working in the space of orbital rotations.   The most difficult problem is the identification of the variables.If we imagine that we have the MC-SCF solution at geometry q_i and we require the evaluation of the MC-SCF at at $q_i + \Delta q_i$ then we have the steps illustrated in Box 1.

It then becomes obvious how to write the expression for the gradient

$$dE/q_i = [\partial E/\partial q_i + (\partial E/\partial T)(\partial T/\partial q_i)] + (\partial E/\partial A)(\partial A/\partial q_i)$$

[36]

We can now begin to appreciate why the MC-SCF has been formulated using the X and A variables. Since one is using a CI optimized wavefunction $\partial E/\partial A = 0$ and the second term in eq 36 vanishes. The term $(\partial E/\partial T)(\partial T/\partial q_i)$ is the response of molecular orbitals to nuclear distortion. It has two parts one that arises from orbital reoptimization $X \partial E/\partial X$ and one that arises from orbital re-orthogonalization $Y \partial E/\partial Y$ in Box 1

$$(\partial E/\partial T)(\partial T/\partial q_i) = (\partial E/\partial X)(dX/dq_i) + (\partial E/\partial Y)(dY/dP)(dP/dq_i)$$

[37]

where we use the vector P to denote the parameters of the atomic orbital basis (co-ordinates of the Gaussian centres and the exponents). As discussed by Almlof and Taylor [13], in the case where the basis function are fixed to the nuclei, the derivatives of the basis functions, $P_i = dP/dq_i$ is unity for all functions attached to the nucleus i and zero otherwise.

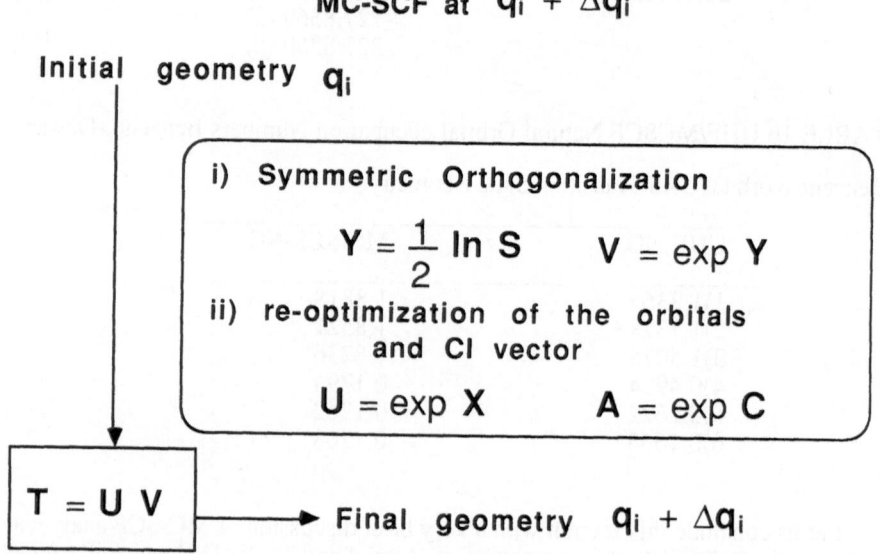

MC-SCF at $q_i + \Delta q_i$

Initial geometry q_i

i) **Symmetric Orthogonalization**

$$Y = \frac{1}{2} \ln S \qquad V = \exp Y$$

ii) **re-optimization of the orbitals and CI vector**

$$U = \exp X \qquad A = \exp C$$

T = U V → **Final geometry $q_i + \Delta q_i$**

Box 1 Steps involved in performing the MC-SCF at one geometry using the orbitals of a previous geometry

The expression for the gradient for q_a takes on the deceptively simple form

$$E_a = E^a + E^C C_a + E^X X_a + E^Y Y_p P_a \qquad [38]$$

where E^a is just the MC-SCF energy expression evaluated with derivative integrals $(\partial E/\partial q)$, $E^C = \partial E/\partial C_{oK} = 0$ since the CI vector optimized, and $E^X = \partial E/\partial X = 0$ since the orbitals are optimized. The

$$X_a = \partial X/\partial q_a \qquad [39]$$

are the response of the molecular orbitals to the nuclear distortion (which would require solution of the coupled perturbed MC-SCF equations: CP-MCSCF) and are not required. Only the Lagrangian

$$E^Y = \partial E/\partial Y \qquad [40]$$

and derivatives of the overlap integrals

$$Y_p = -1/2 \; \partial S / \partial P \qquad\qquad [41]$$

are needed, so that the gradient has the simple form

$$E_a = E^a + E^Y Y_p P_a \qquad\qquad [42]$$

It can be seen that the gradient expression for MC-SCF is no more computationally difficult than SCF theory. Thus we now have established the *numerical* part of our apparatus for the study of reactivity problems.

3 Analysis of MC-SCF wavefunctions in Valence Bond (VB) Space

3.1 INTRODUCTORY REMARKS

It should be clear from the preceding discussion, that the computation of MC-SCF wavefunctions and analytic energy derivatives is now well developed and computations can be carried out routinely. An MC-SCF program [15,16] is now included in GAUSSIAN 90 [18] However, it is clear that the method will never be a "black box" like SCF theory. One must have a "model" for the choice of the active orbital space It turns out that VB theory provides just such a model. The VB method gives us a clear objective basis for choosing the active space via Evans' diabatic surface model. Clearly, in order to *Think VB* one must show how the MC-SCF wavefunction can be transformed to VB space and examine the MC-SCF results in VB space. In fact, as we shall discuss, the transformation to VB space gives us a numerical representation of the so called Heisenberg hamiltonian. This Hamiltonian is a function of the simple coulomb and exchange integrals that are familiar from the Heitler-London VB treatment of H_2.

An overview of the theoretical basis for the relationship between MC-SCF and VB theory for 2 electron systems is given in Box 3 and 4. We shall briefly discuss this before moving to a more general exposition. In MC-SCF, if a complete active space is used for the CI expansion (CASSCF), then the energy is invariant to linear transformations of the active orbitals among themselves. Thus (eq. 19/20 and Box 2), the specific choice of the parameters $X_{valence/valence}$' is arbitrary. A number of possible choices for the valence orbitals can be made (Box 2) and these include natural orbitals, localized MO [19] and non-orthogonal orbitals [20]. Such transformations clearly change the CI expansion coefficients and, of course, the interpretation of the CI configurations themselves and we summarize the situation in Box 3. For a 2 electron bond such as that of H_2 χ_a and χ_b will just be the bonding σ_g and anti-bonding σ_u orbitals and the wavefunction corresponds to the $(\sigma_g)^2$ and $(\sigma_u)^2$ configurations (top of Box 3). When the orbitals are localized to form η_a and η_b (Box 3 middle) the

MC-SCF⇔Valence Bond

(a) Valence Orbital Transformations

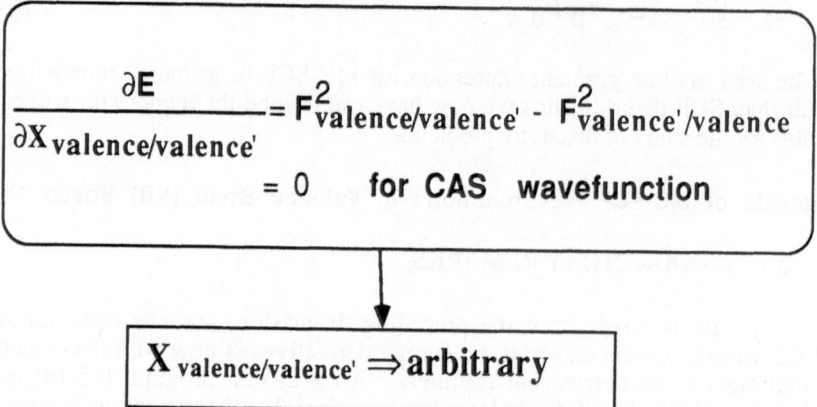

$$\frac{\partial E}{\partial X_{valence/valence'}} = F^2_{valence/valence'} - F^2_{valence'/valence}$$

$$= 0 \quad \textbf{for CAS wavefunction}$$

$$X_{valence/valence'} \Rightarrow \textbf{arbitrary}$$

$$X_{valence/valence'} = -X_{valence'/valence} \Rightarrow \textbf{Orthogonal Orbitals}$$

a) Natural Orbitals
b) Localized Orbitals (Orthogonal VB)

$$X_{valence/valence'} \neq -X_{valence'/valence}$$
$$\Rightarrow \textbf{non-orthogonal Orbitals}$$
$$\Rightarrow \textbf{non-orthoganal VB}$$

Box 2 MC-SCF VB transformation of Valence Orbitals

MC-SCF⇔Valence Bond
b) Wavefunction

---

a) Natural orbitals

$$\psi^{MC\text{-}SCF} = C_1 \, |\chi_a \overline{\chi_a}| - C_2 \, |\chi_b \overline{\chi_b}|$$

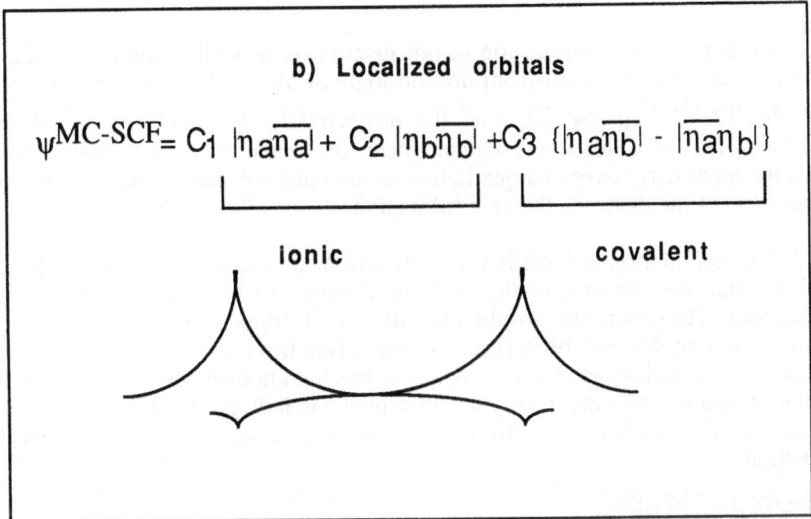

b) Localized orbitals

$$\psi^{MC\text{-}SCF} = C_1 \, |\eta_a \overline{\eta_a}| + C_2 \, |\eta_b \overline{\eta_b}| + C_3 \, \{|\eta_a \overline{\eta_b}| - |\overline{\eta_a} \eta_b|\}$$

ionic   covalent

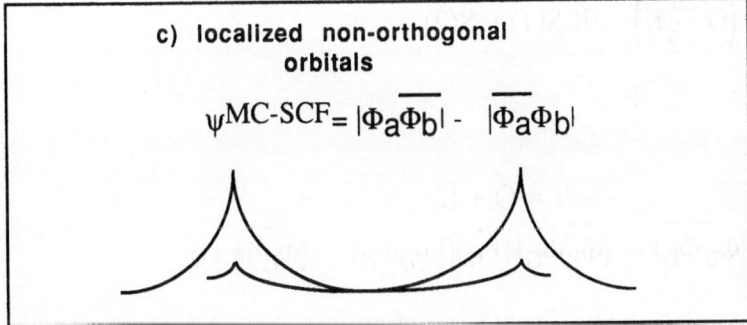

c) localized non-orthogonal orbitals

$$\psi^{MC\text{-}SCF} = |\Phi_a \overline{\Phi_b}| - |\overline{\Phi_a} \Phi_b|$$

Box 3 Two electron wavefunctions as function of various valence orbital transformations.

wavefunction becomes an orthogonal VB wavefunction with ionic or charge transfer configurations and covalent contributions. Note the *negative* tails on the orbitals that are required to keep the orbitals orthogonal. The tails on the orbitals and the charge transfer configurations are intimately related. When we allow the orbitals to be non-orthogonal (Box 3 bottom) we can find orbitals where the charge transfer contribution is zero. Now the *distorted atomic orbitals* have positive tails. These positive tails are just a manifestation of charge transfer. The important message is

For as CAS MC-SCF where the number of orbitals is equal to the number of electrons, it is always possible to transform to a VB wavefunction with covalent terms only. The corresponding valence orbitals will be non-orthogonal (and in general like distorted AO). The role of orbitals distortion/non-orthogonality and charge-transfer is interchangeable.

Finally, let us relate this discussion to our discussion of UHF natural orbitals. Of course our wavefunction built from non-orthogonal MC-SCF orbitals has the same form as the Ψ^{PUHF} in eq. 29. And the projected UHF wavefunctions and VB wavefunctions for a two electron system are identical. When we move to more than two electrons the ideas carry over in a qualitative fashion and we shall summarize the most important parts of the theory in the next subsection.

A VB formulation of CASSCF wavefunctions is attractive because as chemists we believe that we can attach physical significance to the various VB covalent configurations. However, we would also like to introduce an other aspect. VB wavefunctions can always be written as eigenfunctions of what is known as a Heisenberg spin hamiltonian. The parameters of this hamiltonian have a simple physical interpretation and this provides a very useful way of Modelling MC-SCF results. Let us introduce the idea with a 2 electron 2 orbital system again. Using a covalent wavefunction

$$\psi = |\Phi_a \overline{\Phi_b}| - |\overline{\Phi_a} \Phi_b|$$

[43]

we can write the energy as the expectation value of a *spin* hamiltonian H_s

$$\widehat{H_S} = \left(Q - \frac{1}{2}K\right) - 2K\,\widehat{S(1)} \cdot \widehat{S(2)}$$

[44]

to give

$$E = Q + K$$
$$= \langle \frac{1}{\sqrt{2}}\{|\Phi_a\overline{\Phi_b}| - |\overline{\Phi_a}\Phi_b|\}\widehat{H_S}\{|\Phi_a\overline{\Phi_b}| - |\overline{\Phi_a}\Phi_b|\}\rangle$$

[45]

The parameters Q and K are the usual coulomb and exchange integrals

$$Q_{ab} = \langle \Phi_a \,\widehat{h}|\Phi_a\rangle + \langle \Phi_b \,\widehat{h}|\Phi_b\rangle + [\Phi_a\Phi_a \frac{1}{r_{12}}|\Phi_b\Phi_b]$$

[46]

$$K_{ab} = 2 \langle \Phi_a | \Phi_b \rangle \langle \Phi_a | \hat{h} | \Phi_b \rangle + [\Phi_a \Phi_b \frac{1}{r_{12}} | \Phi_a \Phi_b]$$

[47]

For our purposes, it is simpler to work in matrix notation. Taking as basis states

$$|\overline{\Phi_a \Phi_b}|$$

and

$$|\overline{\Phi_b \Phi_a}|$$

the matrix representation of the Heisenberg hamiltonian is just

$$\hat{H}_s = \begin{bmatrix} Q & K \\ K & Q \end{bmatrix}$$

[48]

Why should we be interested in such abstractions? An example shown in Box 4 for 4 orbitals and 4 electrons clarifies the main ideas. For a 4 orbital exchange reaction we change the bonding from I to II. The total energy in the VB scheme is just a combination of the K_{ij} between the various sites. Clearly if we can obtain these exchange integrals from an MC-SCF computation we can explain the origin of energetic effects in terms on these simple parameters. Further, if these K_{ij} can be fitted to simple functions of distance we have the possibility of an analytic representation so that we can visualize the energy and its components. Referring to the second part of Box 4, we also have the possibility of diabatization of the energy into contributions from reactant-like or product-like configurations. The energy of the diabat corresponding to the reactant is given as α_R and for the product α_P while β is the resonance energy. The energy is obtained from the two level secular equation. Since the explicit energy expressions for the various components can be written in terms of the K_{ij} we can have an analytic representation of these components for vizualization as we will describe later.

The generalization of these ideas to more than 2 electrons is summarized in Box 5. We shall now demonstrate that when localized MC-SCF orbitals are used we can partition the CI hamiltonian. The projection onto the space of covalent configurations yields an effective hamiltonian that has the property that it reproduces a subset of the MC-SCF eigenvalues explicitly. The final step involves expressing this effective hamiltonian in terms of the Heitler-London Q and K integrals via a Heisenberg hamiltonian

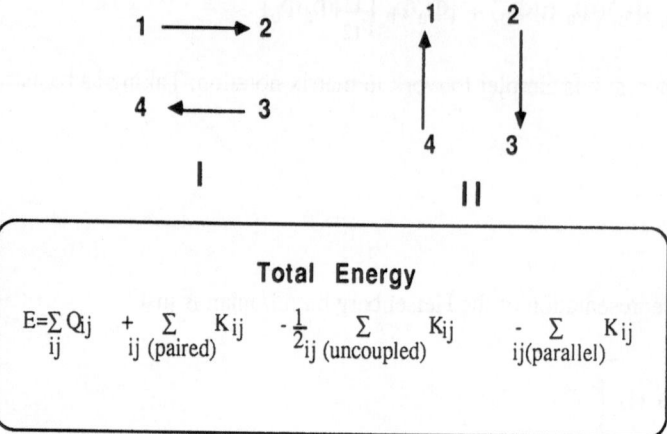

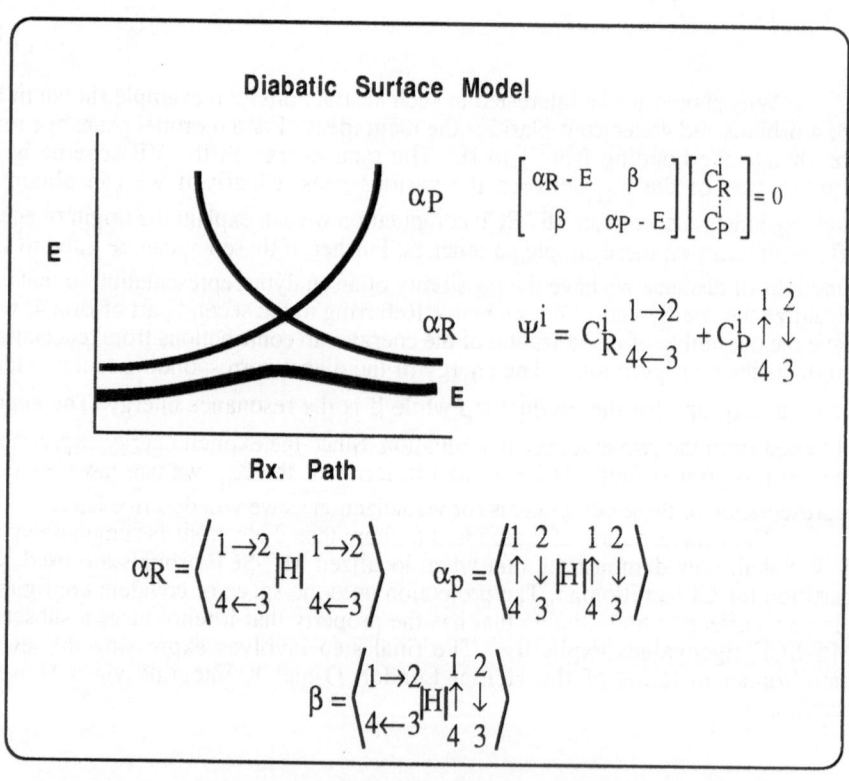

Box 4. 4 orbitals 4 electrons: Total Energy (Coulomb +Exchange)
and Diabatic Surface Model

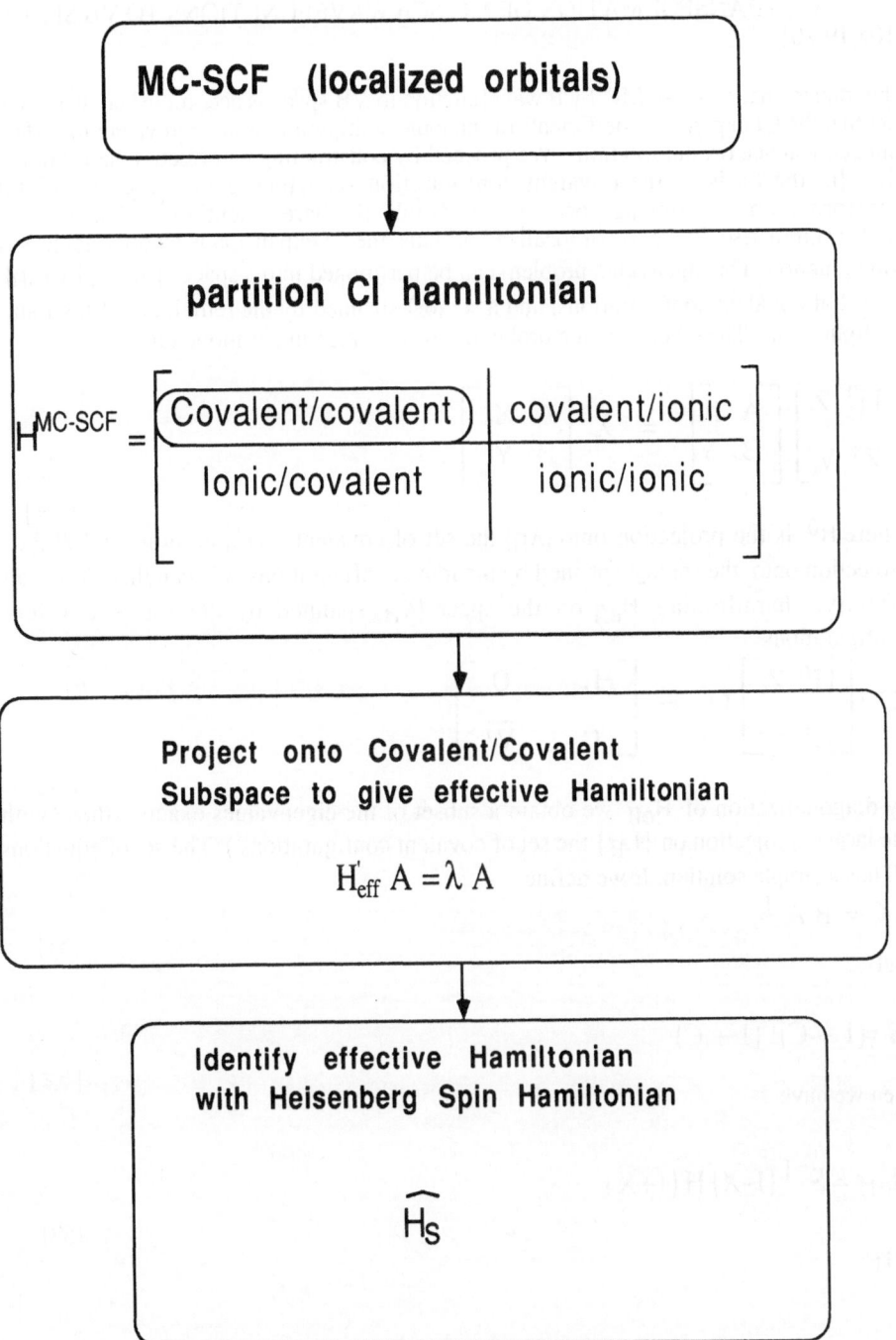

Box 5. Overview: Transformation of MC-SCF wavefunctions to VB Space

3.2) TRANSFORMATION OF MC-SCF WAVEFUNCTIONS TO VB SPACE [10a,19,20]

The transformation of an MC-SCF wavefunction to VB space is accomplished using the fact that the CI expansion coefficients of the ionic configurations are equivalent to certain valence valence orbital rotations. We proceed by constructing an effective hamiltonian [10a]in the basis of the covalent configurations and then generating the orbital transformation. For our purposes we need only the bare essentials which we now briefly summarise. In a basis of localized orbitals, the CI expansion is an orthogonal VB wavefunction. The eigenvalue problem can be partitioned into a space spanned by {Λ_R}, the set of covalent configurations, and a set {ϕ_S} spanned by the remainder of the ionic configurations. The CI eigenvalue problem can be written in partitioned form as

$$\begin{bmatrix} H^0 & Z \\ Z^\dagger & W \end{bmatrix} \begin{bmatrix} A & X \\ B & Y \end{bmatrix} = \lambda \begin{bmatrix} A & X \\ B & Y \end{bmatrix}$$

[49]

where H^0 is the projection onto {Λ_R} the set of covalent configurations and W the projection onto the set {ϕ_S} spanned by the ionic configurations. We can then define an effective hamiltonian H_{eff} on the space {Λ_R} spanned by the set of covalent configurations

$$U^{-1} \begin{bmatrix} H^0 & Z \\ Z^\dagger & W \end{bmatrix} U = \begin{bmatrix} H_{eff} & 0 \\ 0 & \overline{W} \end{bmatrix}$$

[50]

by diagonalization of H_{eff} we obtain a subset of the eigenvalues exactly (those with the largest projection on {L_R} the set of covalent configurations). The set of equations 50 has a simple solution. If we define

$$C = B A^{-1}$$

[51]

and

$$S = (1 + C)^\dagger (1 + C)$$

[52]

then we have

$$H_{eff} = S^{-1} (1-X) H (1+X)$$

[53]

with

$$X = \begin{bmatrix} 0 & -C^\dagger \\ C & 0 \end{bmatrix}$$

[54]

We immediately recognize the role of the metric S in equation 53 (bi-orthogonality) and thus we can interpret

$$H_{eff}^{NO} = (1-X) H (1+X)$$

[55]

as the effective hamiltonian in a non-orthogonal basis. This can be symmetrically orthogonalized to give

$$H_{eff} = S^{-1/2}(1-X) H (1+X) S^{-1/2}$$

[56]

From the definition of X in eq. 54, we see that H^{NO}_{eff} is just the hamiltonian in a non-orthogonal basis defined as

$$\widetilde{\Lambda_R} = \Lambda_R + \sum_S C_{SR} \, \Phi_S$$

[57]

Thus the solutions of

$$H_{eff}' A = \lambda A$$

[58]

namely

$$\psi_\lambda = \sum_R A_R \widetilde{\Lambda_R}$$

[59]

are a non-orthogonal VB expansion based upon the non-orthogonal (viz. eq. 55]) covalent configurations Λ_R'.

It is consistent to regard the Λ_R' as if they had been constructed from non-orthogonal localized orbitals. This follows from the fact that we could write eq. 57 as

$$\widetilde{\Lambda_R} = \exp\left(\sum_{ij} m_{ij}\widehat{E_{ij}}\right) \Lambda_R$$

[60]

The m_{ij} are the parameters of a linear transformation that can be determined from the C_{SR} in eq. 57 . The operators E_{ij} are defined so that orbital i is replaced by orbital j

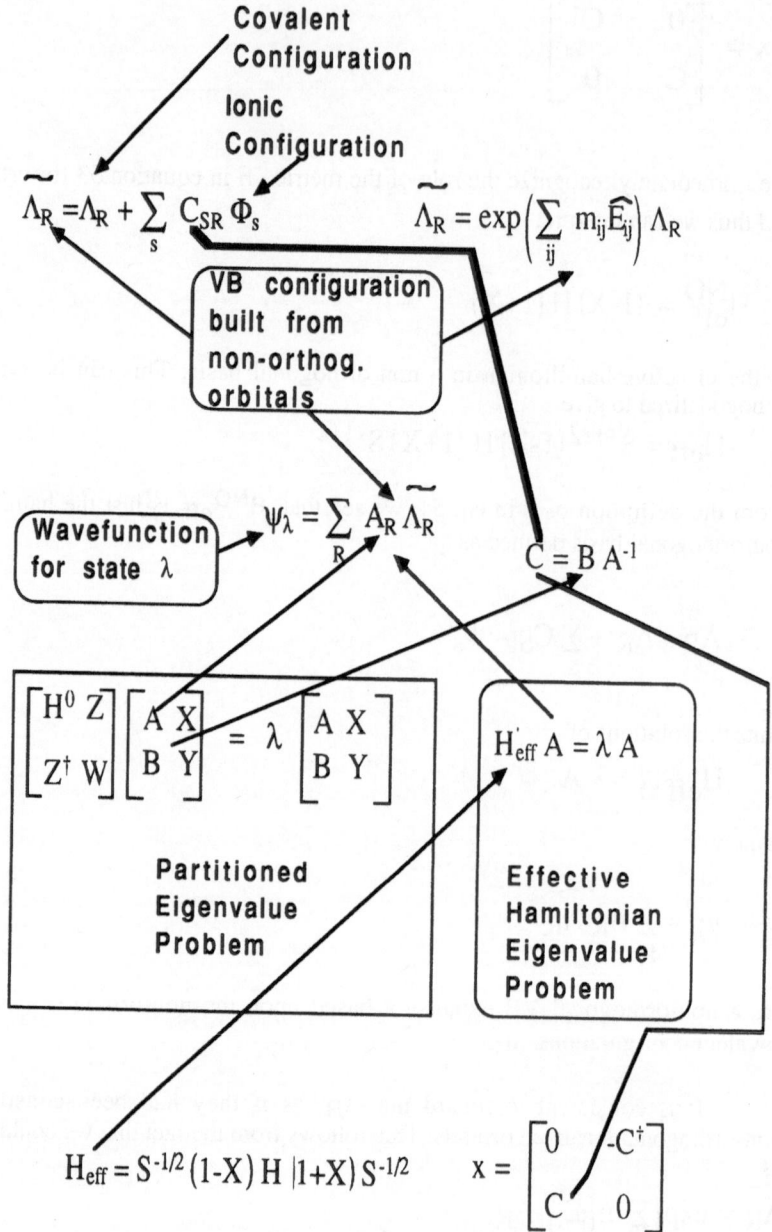

Covalent Configuration

Ionic Configuration

$$\tilde{\Lambda}_R = \Lambda_R + \sum_s C_{SR} \Phi_s \qquad \tilde{\Lambda}_R = \exp\left(\sum_{ij} m_{ij} \hat{E}_{ij}\right) \Lambda_R$$

VB configuration built from non-orthog. orbitals

Wavefunction for state λ

$$\Psi_\lambda = \sum_R A_R \tilde{\Lambda}_R \qquad C = B A^{-1}$$

$$\begin{bmatrix} H^0 & Z \\ Z^\dagger & W \end{bmatrix} \begin{bmatrix} A & X \\ B & Y \end{bmatrix} = \lambda \begin{bmatrix} A & X \\ B & Y \end{bmatrix} \qquad H'_{eff} A = \lambda A$$

Partitioned Eigenvalue Problem

Effective Hamiltonian Eigenvalue Problem

$$H_{eff} = S^{-1/2} (1-X) H \, |1+X) \, S^{-1/2} \qquad x = \begin{bmatrix} 0 & -C^\dagger \\ C & 0 \end{bmatrix}$$

Box 6 Overview of transformation of MC-SCF wavefunction to VB Space everywhere it occurs in Λ_R. Now from eq. 60 and 57 we have

$$\sum_S B_S \Phi_S = \sum_R A_R \, [\exp(M) - 1] \lambda_R$$

[61]

where

$$M = \left(\sum_{ij} m_{ij} \widehat{E_{ij}} \right)$$

[62]

If one pre-multiplies eq. 55 by

$$\sum_R \widehat{E_{ij}} A_R \, \langle \Lambda_R |$$

we have

$$\sum_{RS} A_R \, \langle \Lambda_R \widehat{|E_{ji}|} \Phi_S \rangle \, B_S = \sum_{RT} A_R \langle \Lambda_R \widehat{|E_{ji}|} (\exp(m)-1) | \Lambda_T \rangle \, A_T$$

[63]

When the expansion of the exponential is truncated one has a set of linear equations that can be solved for the parameters (m_{ij}) of the transformation to non-orthogonal orbitals.

The interconnection between the basic mathematical relation is summarized in Box 6 and we now attempt to point out the significance of the various relationships. Equations 51 and 54 are both equivalent definitions of $\Lambda\tilde{\,}_R$, a VB configuration built from non-orthogonal orbitals. In eq. 57 (lhs of Box 5) the non-orthogonality is represented in terms of ionic configurations ϕ_S in eq. 60 the orbitals distort via exp $(\Sigma_{ij} m_{ij} E_{ij})$. The wavefunction Ψ_λ for state λ (eq 59) can be obtained from either the partitioned eigenvalue problem eq 49 (as a combination of covalent + ionic configurations) or from the effective hamiltonian problem eq 58 (as a combination of configurations built from non-orthogonal orbitals (viz. eq. 61). *Note that the effective hamiltonian eq. 58* **exactly** *reproduces a subset of the eigenvalues of the full partitioned eigenvalue problem of eq. 49.*

3.3 COMPUTATION OF H_s FROM H_{eff}

It remains now to show how the Heisenberg Hamiltonian can be constructed in general given the effective Hamiltonian. The Heisenberg hamiltonian is defined in the following way;

1)The matrix elements of H_s are evaluated *symbolically* as if the orbitals were *orthogonal.*

2) The coulomb and exchange terms that occur are replaced by the Q_{ab} or K_{ab} of equations 46/47

eg.

$$[\Phi_a \Phi_b \, |\frac{1}{r_{12}}| \Phi_a \Phi_b] \Rightarrow K_{ab} = 2 \, \langle \Phi_a | \Phi_b \rangle \langle \Phi_a \, \widehat{|h|} \, \Phi_b \rangle + [\Phi_a \Phi_b \, | \frac{1}{r_{12}} \, | \Phi_a \Phi_b]$$

[64]

For 4 electrons and 4 orbitals, there are 6 covalent determinants as shown below.

1) $|3\ 4\ \overline{1}\ \overline{2}\ |$ 2) $|2\ \overline{4}\ \overline{3}\ 1\ |$ 3) $|2\ \overline{3}\ \overline{4}\ 1\ |$

4) $|1\ 4\overline{2}\ \overline{3}\ |$ 5) $|1\ \overline{3}\ \overline{2}\ 4\ |$ 6) $|1\ \overline{2}\ \overline{3}\ 4\ |$

TABLE 2 Symbolic (non-redundant §) form of Heisenberg hamiltonian H_S for 4 orbitals and 4 electrons.

	1	2	3	4	5	6
1	$Q_1-K_{12}-K_{34}$		K_{23}	$-K_{24}$	$-K_{13}$	$-K_{14}$
2		$Q_2-K_{13}-K_{24}$	K_{34}	K_{12}		
3			$Q_3-K_{14}-K_{23}$			

§ The remaining matrix elements are equal to those given. Matrix elements between pairs of determinants that differ by two spin interchanges (eg. 1,6 or 2,5 or 3,4) are equal. Thus $H_{12} = H_{56}$, $H_{15} = H_{63}$ etc

Thus the general form of H_S can be written as

$$\langle K|\widehat{H_S}|L\rangle = \delta_{KL}a^{KL}\,Q + \sum_{ij} b_{ij}^{KL}\,ij$$

[65]

where K and L index the configurations and ij runs over the orbitals. The a^{KL} and b_{ij}^{KL} are simply numerical coefficients that can be computed from the usual rules for the evaluation of matrix elements over determinants built from orthogonal orbitals. Now we simply make the identification

$$(H_{eff})_{KL} = \langle K|\widehat{H_s}|L\rangle$$

[66]

The Q and K_{ij} thus become the *unknowns* in the equations which can be solved by least squares fitting. This is necessary because in general the number of $Q + K_{ij}$ is greater than the number of independent KL. In the case of 2,3 and 4 particles the rank of the coefficient matrix is one so no approximation is implied. Because of non-orthogonality, $(H_{eff})_{KL}$ will have matrix elements for KL where b_{ij}^{KL} is zero.

4 Applications

4.1 SOME REMARKS ON THE CHOICE OF THE ACTIVE ORBITAL SPACE IN MC-SCF COMPUTATIONS

We have just discussed the transformation of MC-SCF wavefunctions to VB space. We believe that if one *Thinks VB* then one now has a model that can be used in order to select the active space for a given transition structure reactivity problem. Thus with some experience in the interpretation of MC-SCF results in VB language much of the objectivity problems of MC-SCF go away. Here we shall summarize the main criteria involved in the choice of the active space.

1) if practical, the number of active orbitals should be equal to the number of active electrons. However, if doubly occupied orbitals are only involved in "dative" bond making/breaking then this rule can be relaxed (however this obviously prejudices the result)

2) MC-SCF is not in general a method for the dynamic electron correlation. Thus doubly occupied orbitals which have a large contribution to the dynamic correlation should remain inactive and the dynamic electron correlation computed in some other way. This may have the effect that ones MC-SCF solution is a local rather than a global minimum .

3) the active apace for a reactivity problem should be chosen so that it can represent product and reactant diabatic surfaces with the same level of confidence.

For us rule 3 is the most important principle. A simple example serves to illustrate. One of the first examples we have studied [1] was the 1,3 supra sigmatropic shift.

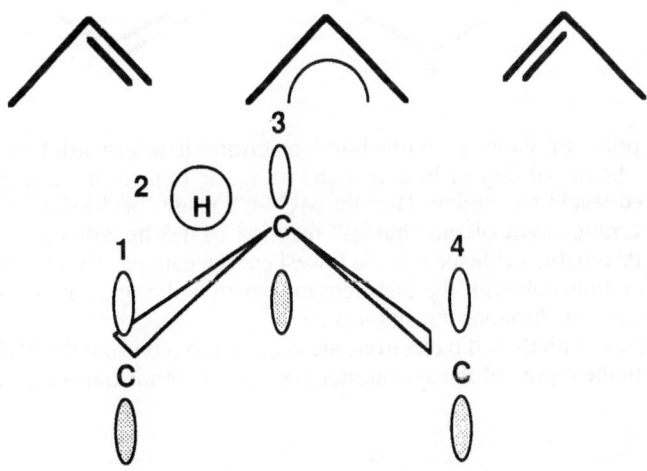

For this example the reactants correspond to 1-2 and 3-4 singlet spin coupling and the product (trimethylene diradical) to 2-3 spin coupling with 1 and 4 a loosely coupled singlet diradical. While SCF theory cannot describe the 2-3 spin coupling with 1 and 4 a loosely coupled singlet diradical, the MC-SCF correctly predicts a diradical trimethylene minimum and a 1,2 transition structure. Many other examples can be found elsewhere.

4.2 DIABATIC REPRESENTATION OF THE POTENTIAL ENERGY SURFACE FOR THE COPE REARRANGEMENT

There has been a controversy concerning the mechanism of the chair Cope rearrangement for many years (see ref. 7 and papers cited therein) Conflicting experimental and theoretical studies provide evidence to support both a synchronous mechanism with an "aromatic" transition state

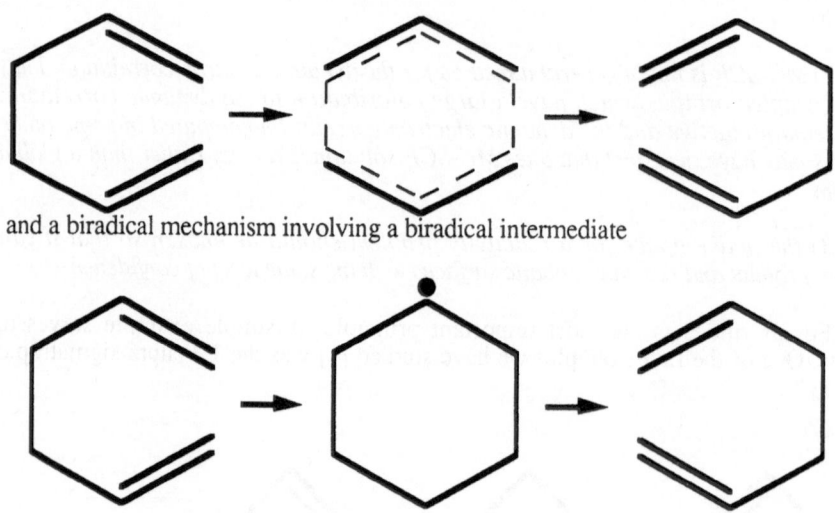

and a biradical mechanism involving a biradical intermediate

From a theoretical point of view, in multi-bond reactions it is essential to use a wavefunction where the possibility of biradical and aromatic transition states can be treated with a balanced level of accuracy. Thus the MC-SCF results of Morokuma et al [21] on the Cope rearrangement of the "model" reaction of 1,5 hexadiene are very convincing and provide reliable evidence that the lowest energy pathway for the "model" reaction is the synchronous one with the biradical intermediate lying 22 Kcal mole^{-1} higher in energy than the synchronous transition state.

Clearly 6 valence orbitals and 6 electrons are required to represent the MC-SCF or VB wavefunction in the region of the synchronous or asynchronous transition state.

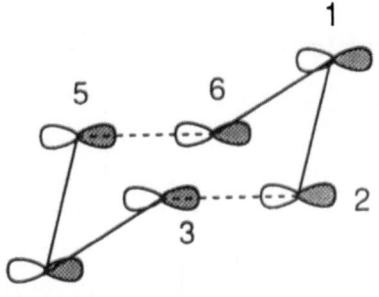

It turns out that the K_{ij} from the MC-SCF computation can be well represented [7] by functions of the form

$$K_{ij} = a \exp (br_{ij})$$

|67|

and the coulomb interaction between non-bonded centres by

$$Q_{ij} = \frac{n\varepsilon}{n-m} \left\{ \frac{m}{n} \left[\frac{r_0}{r_{ij}} \right]^n - \left[\frac{r_0}{r_{ij}} \right]^m \right\}$$

|68|

and the σ component of a double bond by

$$Q_{ij} = D_e [\, 1 - \exp(-a \{r_{ij} - r_0\})\,]^2$$

|69|

In Figure 4 below we show the potential energy surface obtained from the Heisenberg hamiltonian. It reproduces the main features of the ab-initio surface: the synchronous transition state, the two asynchronous transition states and the bi-radical minimum. The main VB components of the MC-SCF wavefunction will be the two Kekule structures and the Dewar Structure

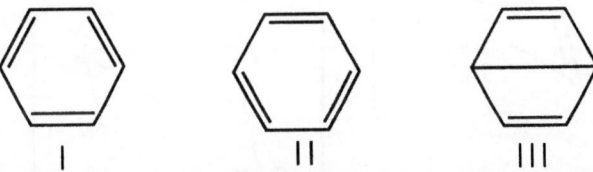

Since we have an analytical representation of the Heisenberg hamiltonian we can plot the energies of each of these components of the energy. The result is shown in Figure 5. Remarkably, each of the transition states sits on a ridge between two diabatics and each of the minima lies on a single diabatic.

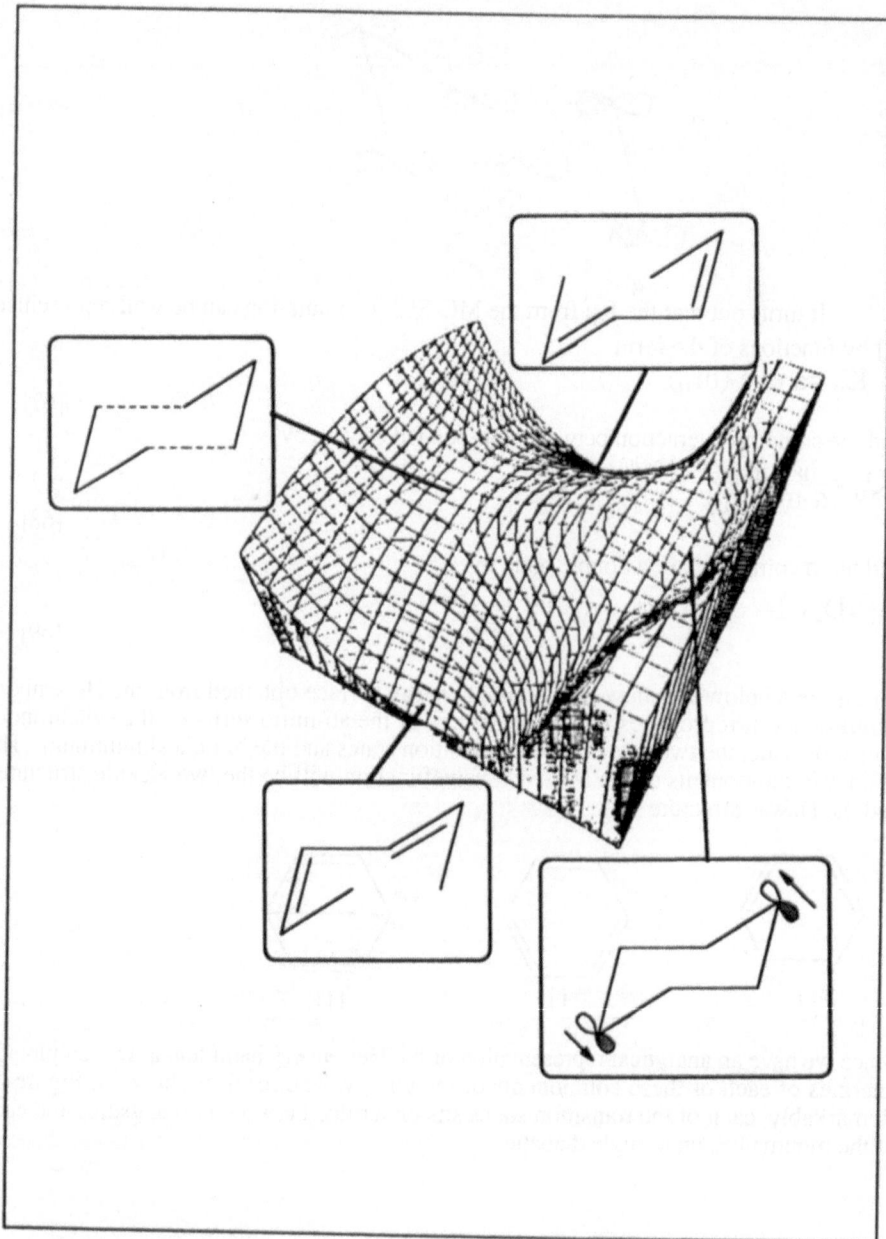

Figure 4 Potential Energy Surface of Cope Rearrangement computed from Heisenberg hamiltonian.

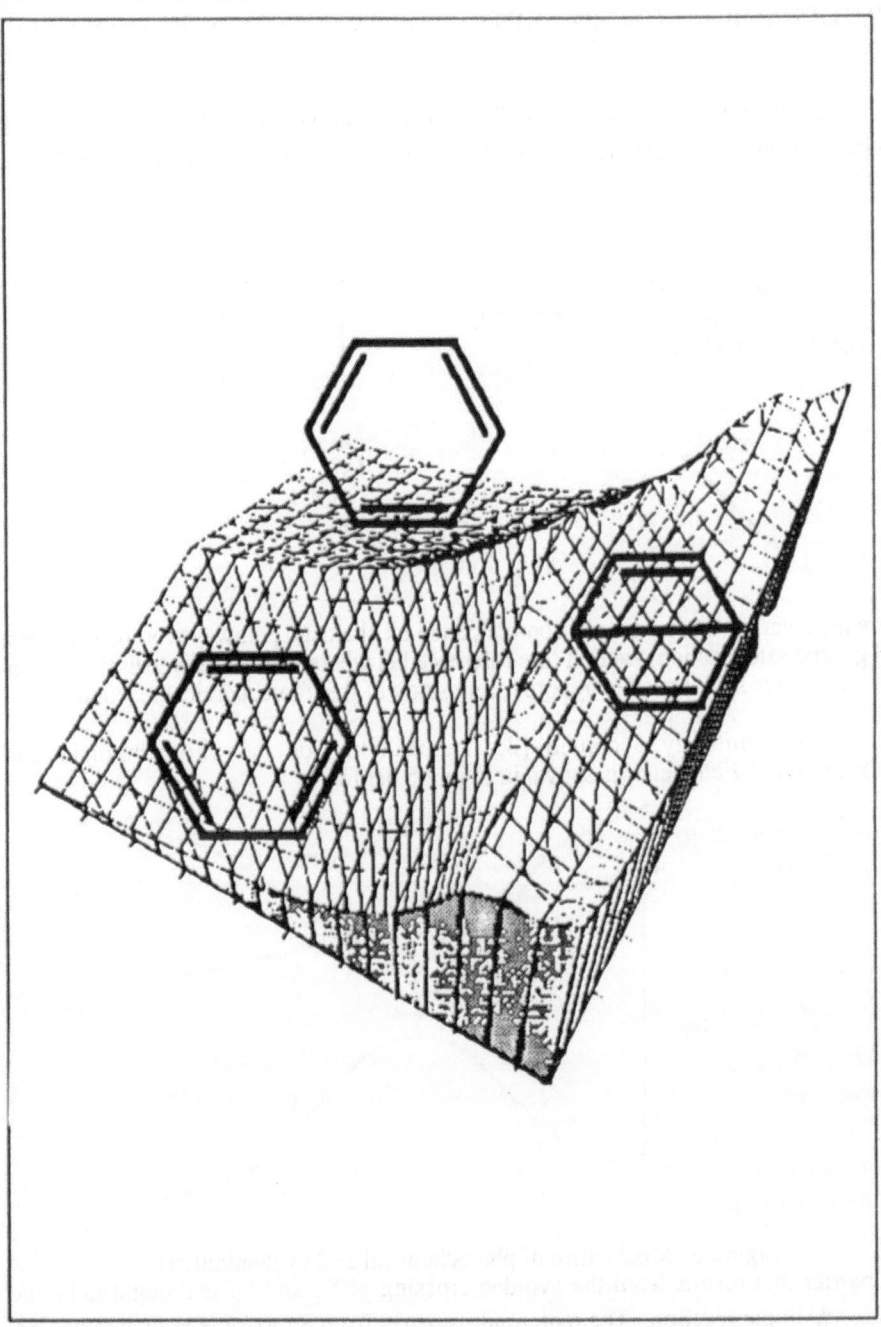

Figure 5. Diabatic surfaces for the Cope Rearrangement

4.3 ANALYTIC REPRESENTATION OF A CONICAL INTERSECTION IN THE PHOTOCHEMICAL CYCLOADDITION OF TWO ETHYLENE MOLECULES

The choice of reference space for the 2+2 cycloaddition of two ethylene molecules is obvious. There are 4 ethylenic π orbitals and 4 π electrons involved as illustrated below

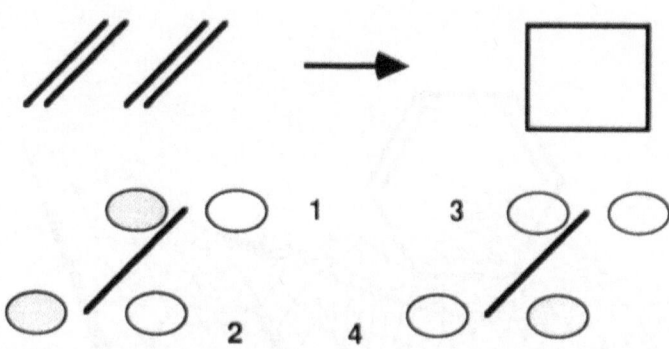

There are a variety of concerted / non-concerted mechanistic pathways for the forbidden [22] ground sate reaction reaction (see reference [1] where an exhaustive study has been done). Here we shall consider a more recent study of the photochemical reaction.

The commonly accepted [23] mechanism for the photochemical 2+2 cycloaddition of 2 ethylenes is shown in Figure 6 below

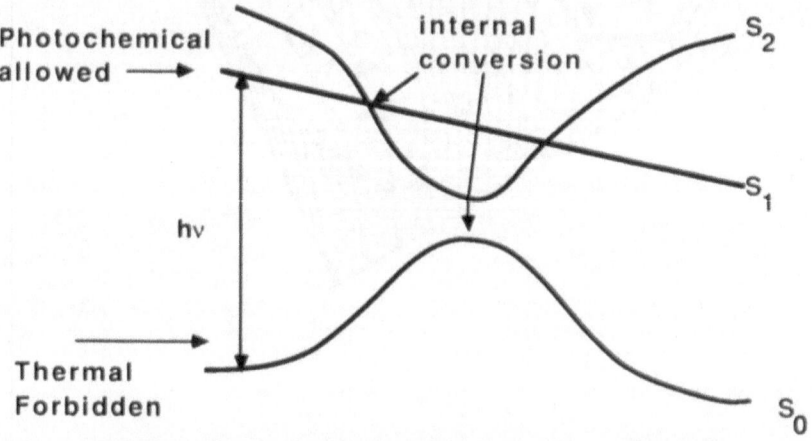

Figure 6. Mechanism of photochemical 2+2 cycloadditions.

The barrier that results from the avoided crossing of S_1 and S_2 is thought to be the *bottleneck* in the reaction. The radiationless decay from S_2 to S_0 is known to be very fast. This result is curious since the rate of decay from S_2 to S_0 will be very slow if the *gap*

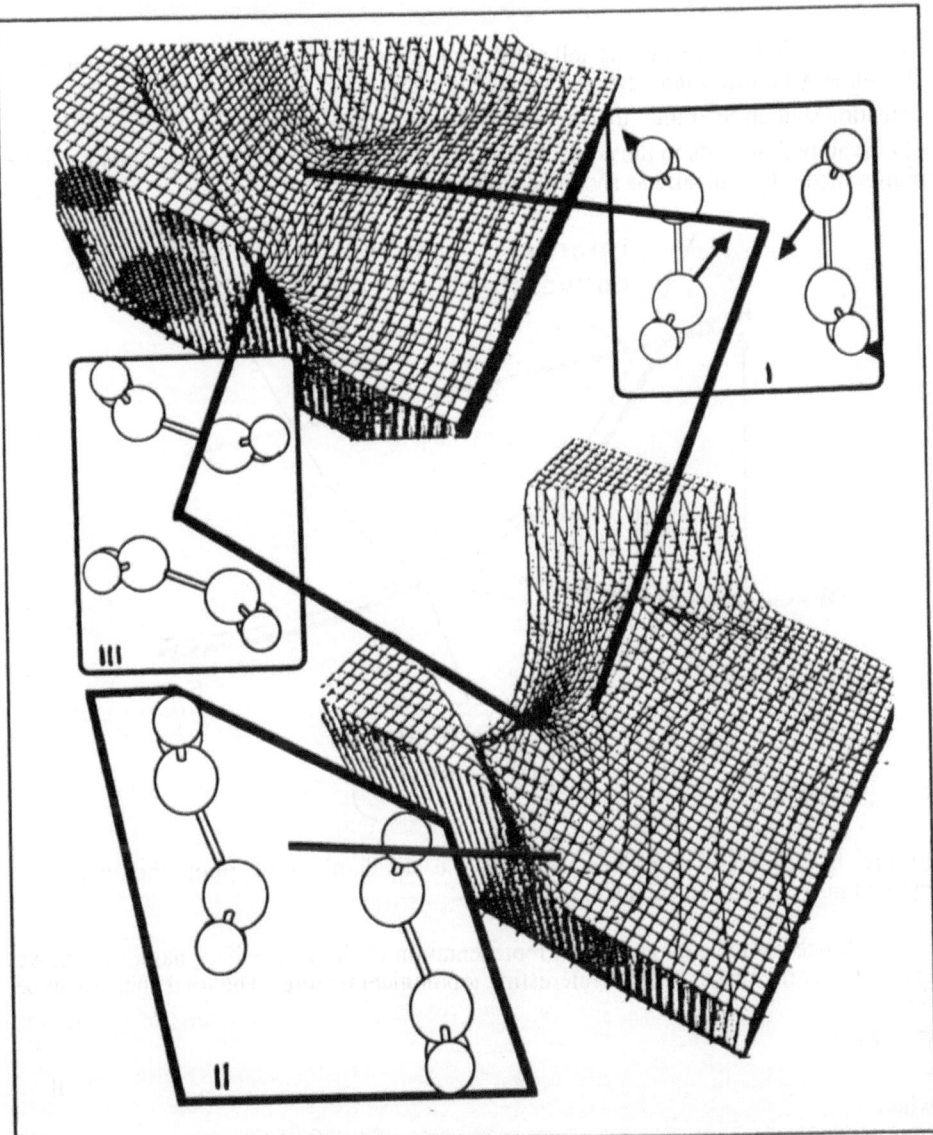

Figure 7. Conical intersection (III) on the 2+2 potential energy surface. The upper surface is the excited state S_2 and the lower surface is the ground state S_0. The left-right coordinate is the distance between the two ethylenic fragments. The bottom-top coordinate is the coordinate that connects the trans bi-radical (II) with the 2_s+2_s critical point (I). The 2_s+2_s critical point (I) is a transition state in S_2. The arrows indicate the negative direction of curvature that leads to the conical intersection.

is more than a few kcal mol^{-1}

MC-SCF computations tell a different story. The surfaces computed with fitted Heisenberg hamiltonians are shown in Figure 7. The 2_s+2_s critical point (I) is a transition state in S_2 rather than a minimum. The arrows indicate the negative direction of curvature that leads to the conical intersection[24] (III). Thus the mechanism can be represented schematically as shown in Figure 8 below.

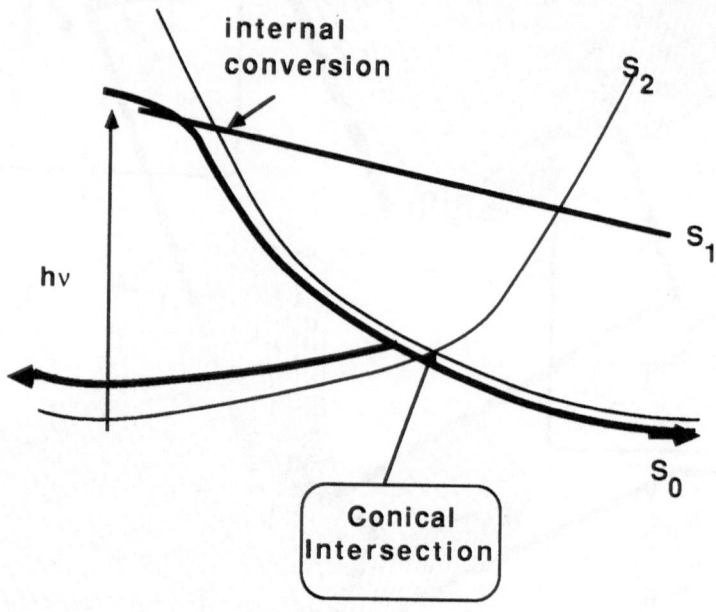

Figure 8 Schematic representation of the mechanism of photochemical 2+2 cycloadditions via a conical intersection

Since we have an analytical representation of the Heisenberg hamiltonian, we can discover the reason for this interesting topological feature. The total energy can be written as

$$E = Q \pm T$$

[70]

where

$$T = \{(K_P - K_X)(K_R - K_X) + (K_P - K_R)^2\}^{1/2}$$

[71]

and

$$K_R = K_{12} + K_{34} \quad K_X = K_{14} + K_{23} \quad K_P = K_{13} + K_{24}$$

[72]

Thus when

$$K_P = K_X, \quad K_R = K_X, K_P = K_R$$

[73]

the energy of ground and excited states are equal and we have a conical intersection. The plot of T is shown below in Figure 9.

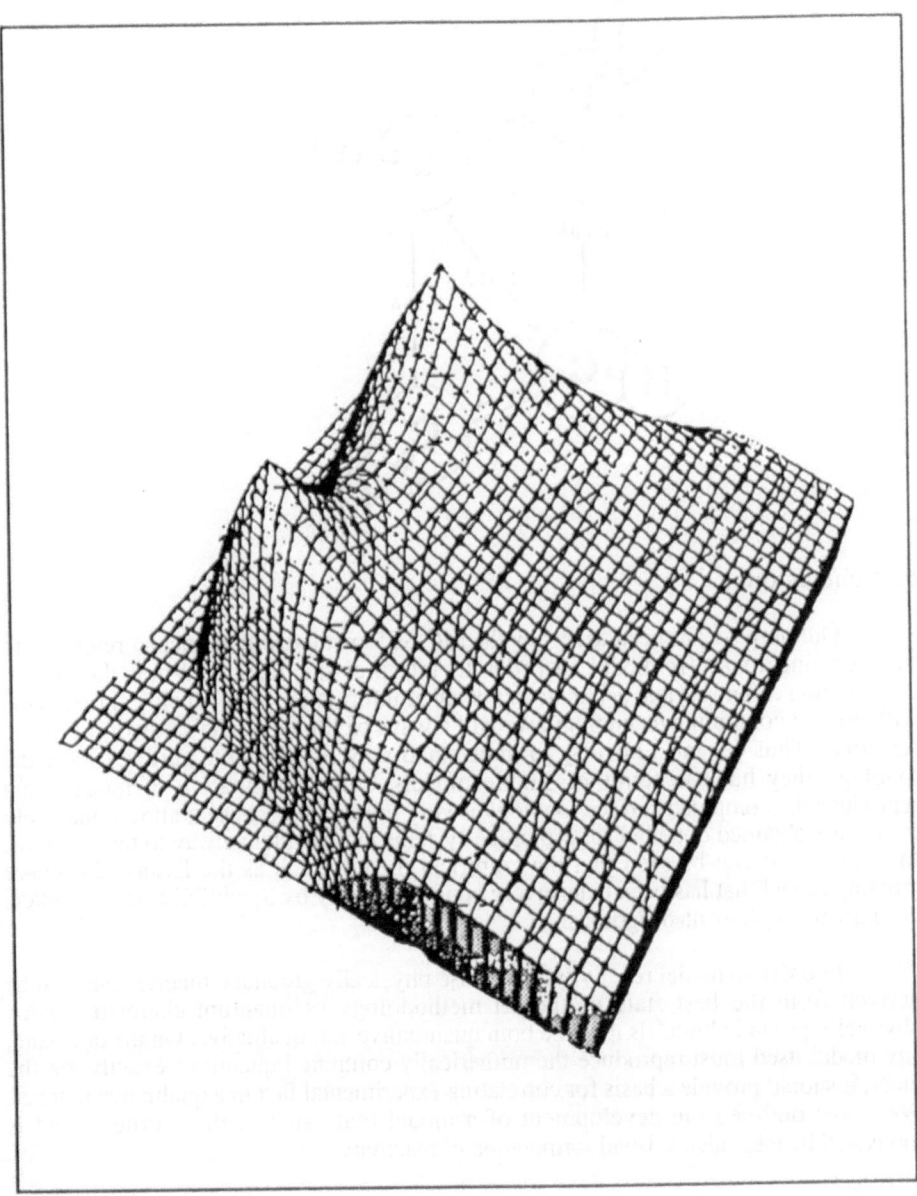

Figure 9. The total exchange energy of 2+2 cycloaddition reaction on the same grid as Figure 7.

The singularity occurs at the geometry

5. Conclusions

Our purpose has been to review the MC-SCF method with particular reference to transforming to a VB model in order to interpret the results. While theoretical computation can now yield accurate and detailed information about chemical reactivity, with present computing technology, we will always be limited to the study of prototype reactions. Thus we must also try to understand why the potential surfaces have the topology they have in terms of simple models. The examination of models that reproduce the computed surface topology is very important because it allows one apply the results obtained obtained for prototype systems in a qualitative way to real systems. In particular it can be seen that the empirical models such as the Evans [25] curve crossing model that has been extensively applied recently by Shaik[26] and co-workers can now be implemented rigorously.

In order to model reactivity, must use physically grounded theories that can be derived from the best state of the art methodology of quantum chemistry. Such physically grounded models must be both quantitative and qualitative. On the one hand, any model used must reproduce the numerically computed quantities exactly, on the other, it should provide a basis for correlating experimental fact in a qualitative manner. We have outlined the development of a model that satisfies these criteria and is motivated by the Valence Bond formulation of reactivity .

Acknowledgements
All of the development of MC-SCF derivatives was performed in collaboration with H.B. Schlegel (Detroit). These codes have been interfaced with the GAUSSIAN system of programs. The applications work has been carried out in collaboration with

A. Bottoni, M. Bearpark, S. De M. Olivucci J. Mc Douall and G. Tonachini. The development of transformation methods from MC-SCF to VB space was carried out with J. McDouall. The work on conical intersections was carried out by M. Olivucci.This work has received the generous support of NATO, the EEC and the SERC (UK). Some of the molecular structures in the manuscript have been drawn using the MODEL program written by H.B. Schlegel.

REFERENCES

[1] Bernardi, F. ; Bottoni, A. ; Robb, M.A. ; Schlegel, H. B. ; Tonachini, G. *J.Am.Chem.Soc.* **1985**, *107*, 2260-2264.

[2] Bernardi, F.; De, S.; Olivucci,M.; Robb,M.A. *J.Am.Chem.Soc.* **1990**, *112*, 1737-1743.

[3] Bernardi, F. Bottoni, A..; Field, M.J. ; Guest, M.F. ; Hillier, I.H. Robb, M.A.; Venturini, A. *J.Am.Chem.Soc.* **1988**, *110*, 3050-3055.

[4] McDouall, J.H.W. ; Robb, M.A. Niazi, V. Bernardi, F. Schlegel, H.B. *J. Am.Chem.Soc.* **1987**, *109*, 4642.

[5] Tonachini, G.; Schlegel,H.B.; Bernardi,F. Robb,M.A *J. Am.Chem.Soc.* **1990**, *112*, 483-490.

[6] Bernardi, F.; Bottoni, A.; Robb,M.A Venturini, A *J. Am.Chem.Soc.* **1990**, *112*, 2106-2114.

[7] Bearpark,M. ; Bernardi,F.; Olivucci,M. ;Robb,M.A. *J. Am.Chem.Soc.* **1990**, *112*, 1732-1736.

[8] Bernardi, F. ; Bottoni, A. ; Olivucci, M.; Robb,M. A.; Schlegel , H.B. ;Tonachini G. *J. Amer. Chem.Soc.* **1988** *110*, 5993-5995

[9] Bernardi, F; Olivucci ;M. Robb, M.A. *Research on Chemical Intermediates*, **1989**, 217-249

[10] (a.)Bernardi, F. ; Olivucci, M. ; McDouall, JJW. ; Robb,M.A. J.Chem. Phys. 16. **1988**, *89* ,6365 (b.) Robb, M.A. ;Bernardi,F. in J. Bertran and I.G.Csizmadia (ed.) *New Theoretical concepts for Understanding Organic reactions* (Kluwer Academic Publishers) **1989**, 101-146

[11]Bernardi, F. ;Robb, M.A. *Adv. Chem. Phys.***1987**, *67* 155-248

[12]J. Olsen,J.;Yeager, D.L; Jorgensen, P. *Adv. Chem. Phys.***1983** *54*

[13]Taylor P. Almlof, J. *Intern. J. Quantum Chem.* **1985**,*27* 743

[14] Levy B. *Intern. J. Quantum Chem* **1970** *4* 297

[15]Eade, R.H.A. ;Robb, M. A. *Chem. Phys. Lett.***1981** *83* 362

[16]Schlegel H.B.; Robb, M. A. *Chem. Phys. Lett.***1982** *93* 43

[17] Pulay,P.;Hamilton,T.P. *J.Chem. Phys.* **1988** *88* 4926

[18]**Gaussian 90**, Frisch, M. J. ; Head-Gordon, M. ; Trucks, G. W. ; Foresman, J. B. ; Schlegel, H. B.; Raghavachari, K.; Robb, M.; Binkley, J. S. ; Gonzalez, C. ; Defrees, D. J. ; Fox, D. J. ; Whiteside, R. A. ; Seeger, R. A. ; Melius, C. F. ; Baker, J. ; Martin, R. L. ; Kahn, L. R. ; Stewart, J. J. P. ; Topiol, S. ; and Pople, J. A. ; Gaussian, Inc., Pittsburgh PA.

[19] McDouall J. J. W. ; Robb M. A., *Chem. Phys. Lett.* **1987** *132* 319

[20] McDouall J. J. W. ; Robb M. A., *Chem. Phys. Lett.* **1987** *142* 131

[21]Morokuma, K.;Borden, W.T.;Hrovat D.A.; *J. Amer. Chem. Soc.* **1988** *110* 4474

[22] Woodward, R.B.; Hoffmann, R. *Angew. Chem.Int.Ed.Engl.* **1969**, *8*, 781-853.

[23](a).Turro, N.J, *Modern Molecular Photochemistry*, Benjamin Publishing, USA, **1978** (b) Michl,J. ; Bonacic-Koutecky, V. *Electronic Aspects of Organic Photochemistry* , Wiley, New York, **1990**

[24] (a)Salem, L., *Electrons in Chemical Reactions: First Principles* , Wiley, New York, **1982** (b)Tully, J.C., ;Preston, R.K. *J. Amer. Chem. Soc.* **1971**, *55*, 562 (c)Von Neumann,J. ; Wigner,E.,*Physik. Z.* **1929**,*30*, 467 (d)Teller, E, *J. Phys. Chem.* **1937**,*41*, 109 (e)Herzberg,G. ; Longuet-Higgins,H.C., *Trans. Faraday Soc.* **1963**,*35*, 77 (f)Herzberg,G.*The Electronic Spectra of Polyatomic Molecules* Van Nostrand, Princeton **1966** pp442 (g)Mead,C.A. ; Truhlar,D.G.,*J.Chem.Phys.* **1979**, *70*, 2284 (h) Mead,C.A., *Chem.Phys.* **1980**, *49*, 23 (i)Keating,S.P, ; Mead,C.A., *J.Chem.Phys.***1985**, *82*, 5102 (j) Keating,S.P, ; Mead,C.A. *J.Chem.Phys.***1987**, *86*, 2152

[25](a) Evans, G.; Polanyi, M. *Trans.Far.Soc.* **1938**, *34*, 11.   (b) Evans, G.; Warhurst, E. *Trans Far.Soc.***1938**, *34*, 614. (c)

[26](a) Pross, A.; Shaik, S,S *Acc. Chem. Res.* **1983** *16*, 363 (b) Shaik, S,S, *Prog. Phys. Org. Chem.* **1985**, *15*, 198-337

THE SUPRA-SUPRA MECHANISM OF FORBIDDEN AND ALLOWED CYCLOADDITION REACTIONS: AN ANALYSIS USING A VB MODEL

Fernando Bernardi
Dipartimento di Chimica "G.Ciamician"
dell'Universita di Bologna,
Via Selmi 2, 40126 Bologna, Italy and the

Massimo Olivucci and Michael A.Robb
Department of Chemistry; King's College,
London, Strand,
London WC2R 2LS, U.K.

ABSTRACT:

The MC-SCF potential energy surfaces for the 2+2 cycloaddition of two ethylenes, the 4+2 cycloaddition of butadiene and ethylene and the 1,3 dipolar cycloaddition of fulminic acid(HCNO) and acetylene are analyzed in terms of the coulomb and exchange energy of Heitler-London VB theory as described in the companion paper [1]. We demonstrate that the electronic origin of the reaction barrier is the same for the 3 reactions. However, while the global shape of the exchange energy (net bonding effects) is broadly similar, the very different mechanisms (different surface topology) arises from the behaviour of the quasi-classical coulomb energy. The behaviour of the coulomb contribution is easily rationalized using simple qualitative concepts.

1 Introduction

The study of organic reactivity has undergone a complete revolution in the past decade. The availability of standard quantum chemistry codes with analytic gradients and frequency capabilities has enabled the investigation of molecular structures and transition states for real chemical reactivity problems. (For a survey of recent developments the reader is referred to two recent volumes of Adv. in Chem. Phys. ref [2,3]). The more recent availability of methods such as MC-SCF has extended these possibilities to bi-radicaloid systems and excited states.

However equally important is the availability of physically grounded models that can provide *understanding* of chemical reactivity. In the past, theories of reactivity have been based on empirical structure-reactivity relationships (viz. linear free energy relationships) or qualitative theoretical concepts (viz. Woodward-Hoffmann approach or the frontier orbital method)[4-11]. However, there is a different approach, which is potentially more fruitful. In this approach one uses physically grounded models that can be obtained from the best state of the art methodology of quantum chemistry. These physically grounded models must be both quantitative and qualitative. On the one hand, any model used should reproduce the numerically computed quantities exactly, on the

289

S. J. Formosinho et al. (eds.), Theoretical and Computational Models for Organic Chemistry, 289–313.

other, it should provide a basis for correlating experimental facts in a qualitative manner. In the preceding chapter in this volume we have discussed a model that satisfies these criteria [12-16] and is motivated by the Valence Bond formulation of reactivity first used by Evans [17-18]. The model has been implemented [16] in the context of effective hamiltonian theory [19] to reproduce the results of MC-SCF, CI and perturbation theories exactly. However, the most difficult concept required for its qualitative application is the Valence Bond treatment of H2.

The objective of this article is to illustrate, in some detail,the applicability of the model described in the previous chapter [1] to a problem of chemical interest; namely the concept of forbidden and allowed cycloadditions. We shall consider 3 prototype cycloaddition reactions a) the forbidden cycloaddition reaction of ethylene+ethylene, (b) the allowed cycloaddition (Diels Alder) reaction of butadiene+ethylene and (c) the allowed 1,3 dipolar cycloaddition of fulminic acid+acetylene..

We begin by briefly reviewing the mechanistic information obtained from the ab-initio MC-SCF computation of the potential energy surface and then illustrate how these results can be rationalized using a VB model.

2. Review of mechanistic results obtained at the MC-SCF level

Let us briefly review the mechanistic information that is available from our previous MC-SCF computations (cycloaddition of two ethylenes [20] ,as an example of a forbidden reaction and the cycloaddition of butadiene and ethylene [21] and that of fulminic acid plus acetylene [22] as examples of allowed reactions).These results were obtained with ab-initio MC-SCF techniques using minimal STO-3G [23] and extended 4-31G [24] basis sets.

Our MC-SCF results show that ,while in the allowed cycloadditions both the concerted and two-step paths exist,with the concerted mechanism energetically favoured, in the forbidden reaction only two-step paths exist and the critical points associated with the concerted paths lie at an higher energy than those associated with the two-step paths.Consequently to understand the difference between forbidden and allowed cycloadditions it is important to understand the differences associated with the supra-supra approaches in the two cases.

We shall illustrate our discussion using 3d potential energy diagrams that have been designed to reproduce the ab-initio MC-SCF data. We shall use E and two geometrical variables: a distance between the two reacting fragments r and an angle α) illustrated in fig.1

2.1) CYCLOADdITION OF TWO ETHYLENES

In fig .2 we show the potential energy diagram in the r/α space illustrated in fig.1.In this subspace the synchronous supra-supra reaction path passes over a local maximum and there is a synchronous channel involving a coplanar syn diradicaloid structure. When rotation about the CC bond is considered, the syn diradicaloid structure turns out to be a local maximum. True transition states exist only for two reaction paths associated with

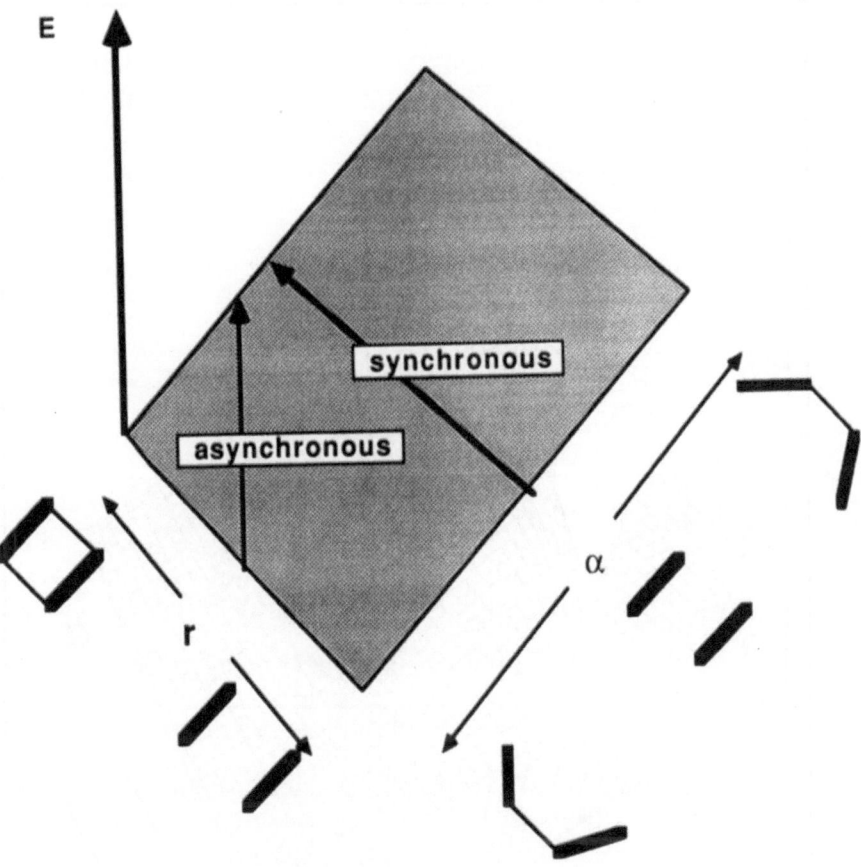

Fig.1 Schematic representation of 3^d energy diagrams which will be used to discuss synchronous vs asynchronous bond formation in cycloaddition reactions.

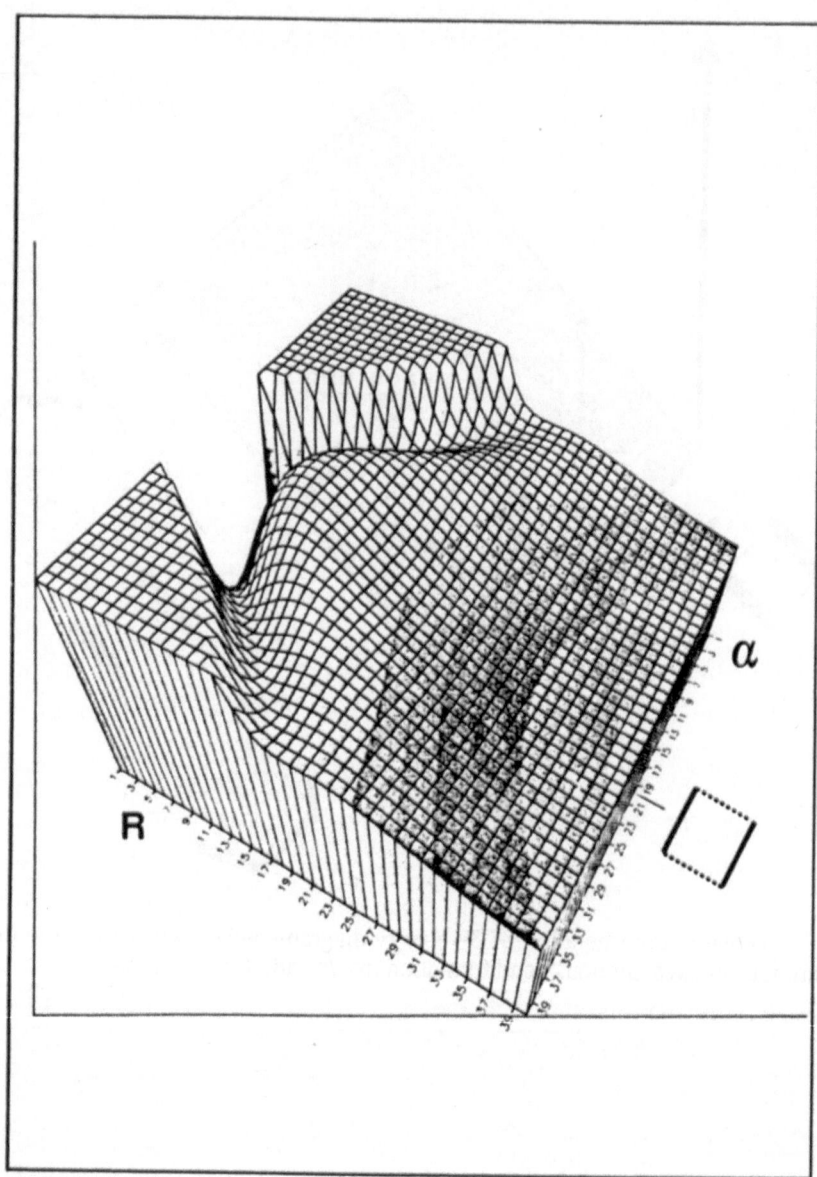

Fig.2 Potential energy surface for the 2s+2s cycloaddition of two ethylene molecules. One axis (diagonal top left to bottom right) is the interfragment distance r and the other axis (diagonal top right to bottom left) the angle α defined in Fig. 1. Each division on the r axis corresponds to an increment of 0.1a_0 (.053 Å) and the first division corresponds to r=3.5a_0 (1.85Å). Each division on the a axis corresponds to an increment of 3.5 degrees and the first division corresponds to α =20 degrees.

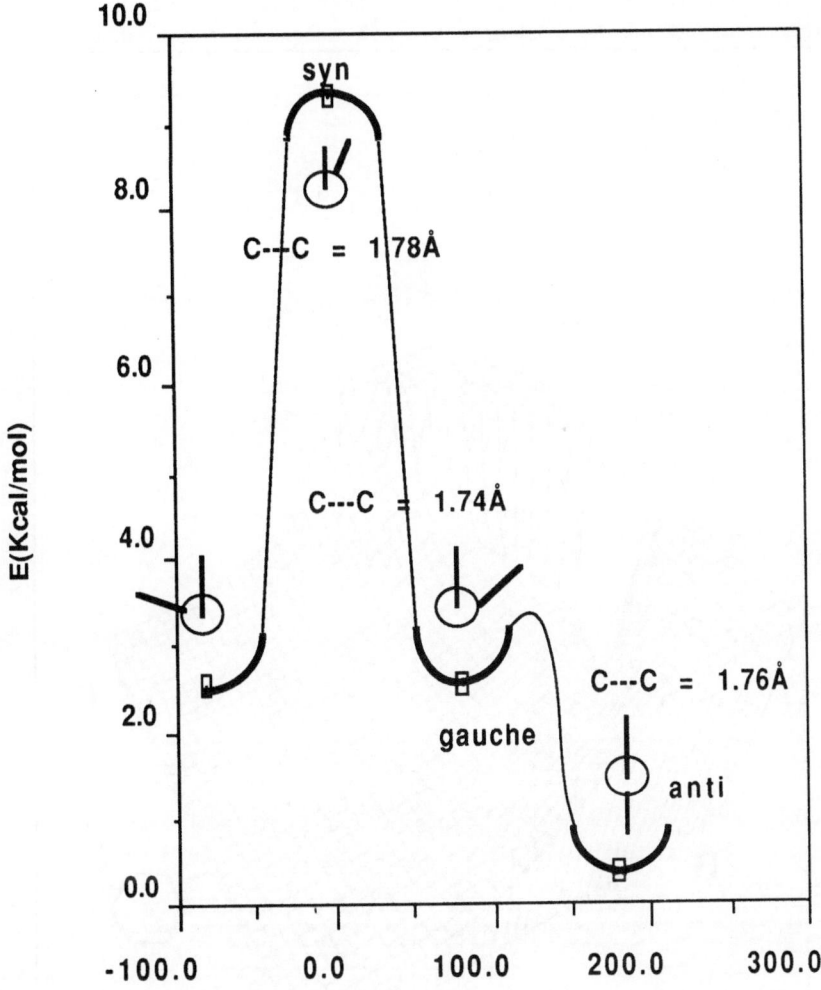

Fig. 3 Energy profile for the syn diradicaloid structure (C-C) rotational angle

294

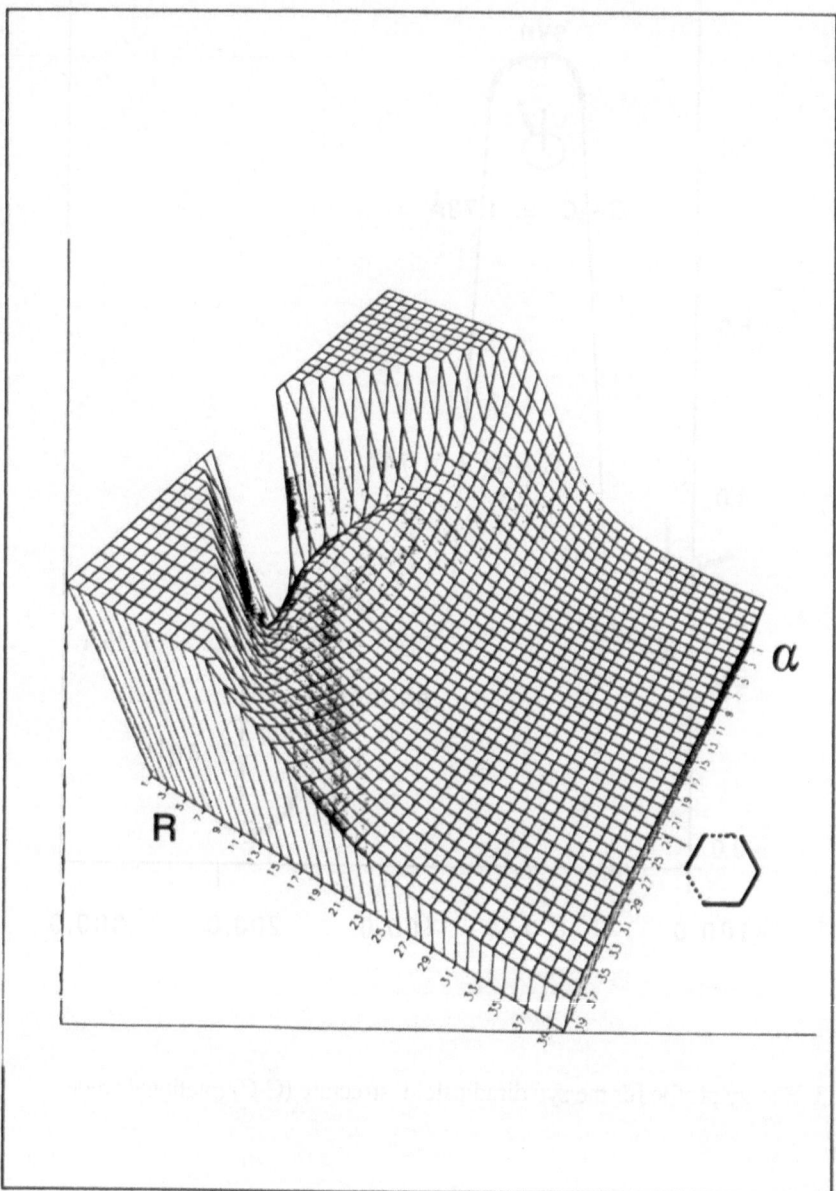

Fig.4 Potential energy surface for the Diels Alder reaction of cis butadiene and ethylene
molecules. The two axes are defined as in Fig. 1 Each division on the r axis corresponds
to an increment of .2a_0 (.105Å) and the first division corresponds to r=3.0a_0 (1.58 Å).

Each division on the α axis corresponds to an increment of 3.5 degrees and the first
division corresponds to α = 20 degrees.

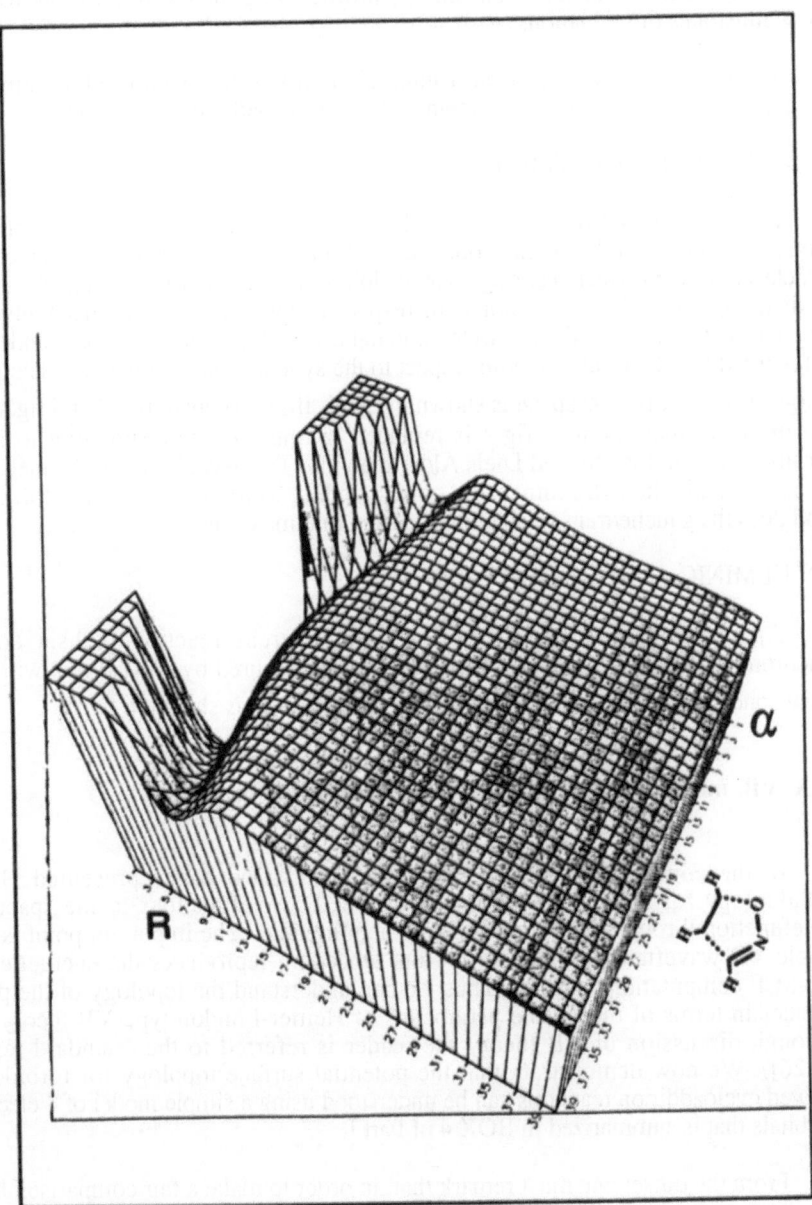

Fig.5 Potential energy surface for the 1,3 dipolar cycloaddition of Fulminic acid (HCNO) and acetylene molecules. The two axes are defined as in Fig. 1 Each division on the r axis corresponds to an increment of .2a_0 (.105Å) and the first division corresponds to r=3.0a_0 (1.58 Å). Each division on the α axis corresponds to an increment of 3.5 degrees and the first division corresponds to α = 20 degrees.

gauche and trans approaches . The energy profile along the CC rotation shown in fig 3 below illustrates this situation.

A critical point has been found also along the supra-antara approach : characterization of this point has shown that it is a second order saddle point.

2.2 BUTADIENE + ETHYLENE

For this allowed reaction, at the 4-31G level ,there are three different reaction paths,i.e. a concerted synchronous path which leads to cyclohexene,a concerted asynchronous path which leads again to cyclohexene via a syn-gauche attack and which is disfavoured by 2.2 Kcal/mol with respect to the synchronous mechanism and a two-step path which leads to vinylcyclobutane via an anti-gauche attack and which is disfavoured by 11 Kcal/mol with respect to the synchronous mechanism. The potential energy surface in the r/α space is shown in fig. 4. If we compare fig. 2 and fig.4 we see that the local maximum in fig.2 is replaced by the *col* corresponding to a proper transition state in the allowed Diels Alder reaction. The asynchronous channels remain. When one adds the extra dimension corresponding to rotation about the forming C-C bond does the gauche/trans transition states and minima occur.

2.3 FULMINIC ACID + ACETYLENE

Again for this allowed reaction two different reaction paths,a concerted synchronous path and a two-step path which is disfavoured by 7 Kcal/mol with respect to the concerted mechanism. The surface in the r/α space is shown in fig. 5

3. A VB model for Supra-Supra cycloaddition reactions

In the companion article in this volume [1] we have presented [16] the methodology for the transformation of MC-SCF wavefunctions to the space of VB wavefunction through the use of effective hamiltonians. The important point is that the simple VB wavefunction obtained in this procedure reproduces the energetics of the MC-SCF computations exactly. Thus we can understand the topology of the potential surfaces in terms of the simple parameters of Heitler-London type VB theory (For a thorough discussion of VB theory the reader is referred to the standard textbooks [25,26]). We now demonstrate that the potential surface topology for forbidden and allowed cycloaddition reactions can be understood using a simple model of 4 electrons in 4 orbitals that is summarized in BOX 4 of Part I.

From the outset one must remark that, in order to make a fair comparison between synchronous (*aromatic*) pathways and possible asynchronous pathways, it is essential to be able to treat bi-radicaloid and and quasi-closed shell structures in a balanced way. In practise this implies that MC-SCF and multi-reference CI is necessary. With SCF , one will always obtain the result that the synchronous pathway is preferred! We shall use as benchmark the cycloaddition of two ethylene molecules where the use 4 active orbitals and 4 active electrons is obviously correct. It is not immediately obvious that this is possible for the other two reactions. For example, the Diels Alder reaction of butadiene and Ethylene involves 5 resonance structures built from 6 active orbitals. However he transition structure region is dominated by the two resonance structures

In other words, the central double bond in the butadiene fragment is *passive* in the region of the transition state. Similarly, in the cycloaddition of fulminic acid (HCNO) to acetylene, the N lone pair behaves Similarly. Thus the use of a 4 orbital 4 electron model is justified.

Heitler-London type VB theory uses only "perfect-pairing" type "structures. (The effects of ionic structures are included via the delocalization of the atomic orbitals). Thus for a model two-bond exchange reaction there are only two structures that need to be considered.

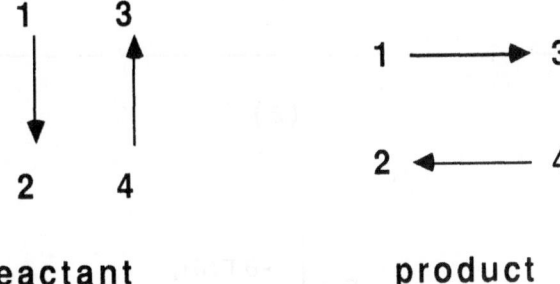

reactant   **product**

We shall use the two schemes outlined in Part I; namely a diabatic surface decomposition (Part 1 Box 4) and a decomposition into coulomb Q and exchange energies T. In the diabatic surface model the energy is decomposed into α_R is the "diabatic" energy of the reactant configuration , α_P the "diabatic" energy of the product configuration and the resonance β. The repulsive bond-breaking diabatic curve α_R coincides with the reactant adiabatic energies at the reactant geometries and the attractive bond-making diabatic curve α_P coincides with the product energies at the product geometries. The eigenvalues of the two level problem are

$$E = \frac{1}{2}[\alpha_R + \alpha_P] \pm \sqrt{(\alpha_R - \alpha_P)^2 + 4\beta^2}$$

[1]

Alternatively, we can decompose the energy into Coulomb and Exchange contributions. It is convenient to write the exchange energy in terms of product K_P ,reactant K_R and nonbonded K_X contributions defined as

$$K_R = K_{12} + K_{34}$$

[2a]

$$K_P = K_{13} + K_{24}$$

[2b]

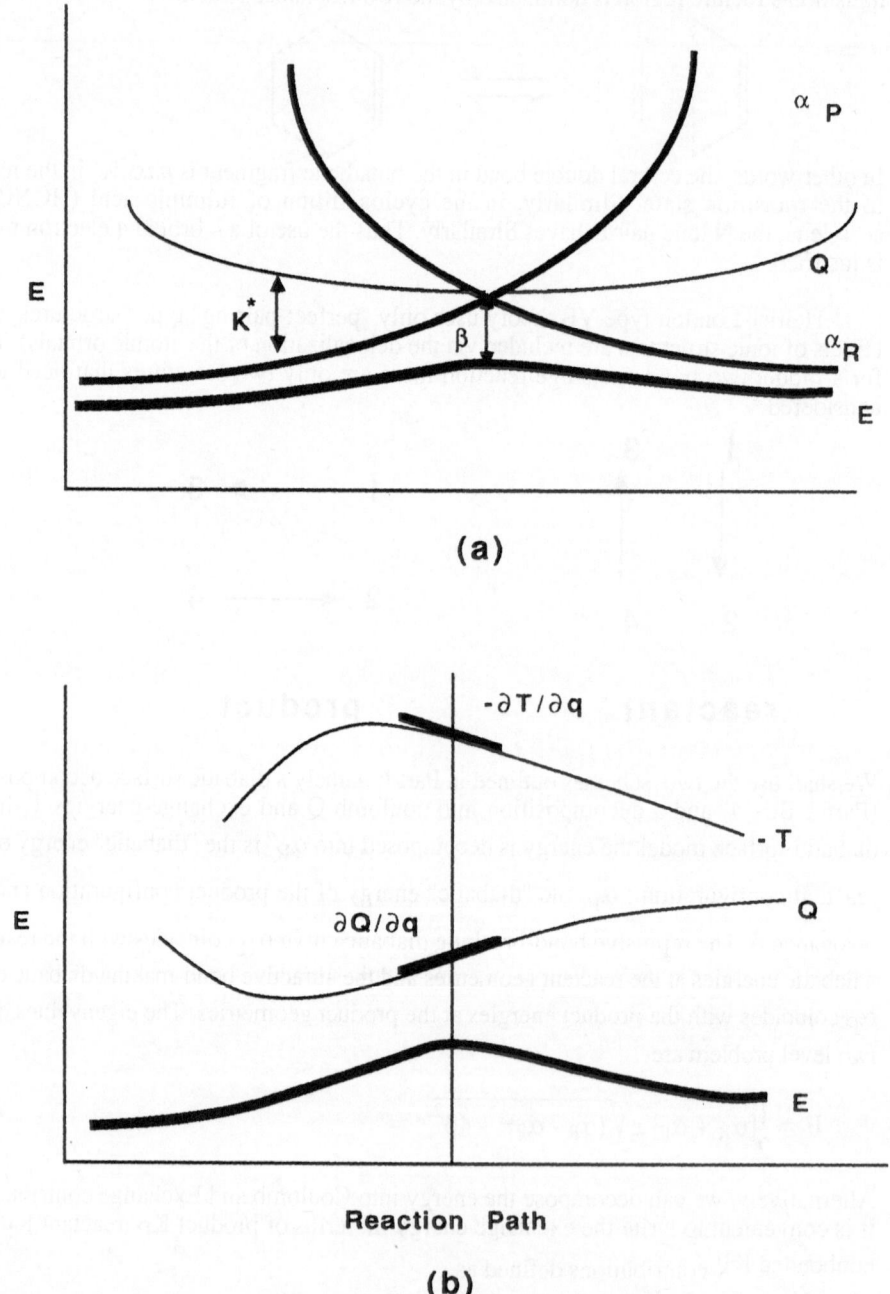

(a)

Reaction Path

(b)

Fig.6 Schematic Behaviour of Diabatic surfaces (a) and Q/T (b)

$$K_X = K_{14} + K_{23}$$

[2c]

The total energy can now be written as
$$E = Q \pm T$$

[3]

where T is the total exchange energy

$$T = \sqrt{(K_P - K_X)(K_R - K_X) + (K_P - K_X)^2}$$

[4]

We can combine these two approaches by writing (from eq 1.)
$$T = \sqrt{((K^*)^2 + \beta^2)}$$

[5]

In eq. 5 K^* is the *bond exchange* . It is a combination of the product K_P ,reactant K_R and nonbonded K_X contributions that depends upon the nature of the spin coupling. It is given approximately by

$$K^* \approx K_R - K_P$$

[6]

The relationship of these quantities along a typical reaction co-ordinate is shown schematically in Fig. 6.
The quantity Q is very flat until the interfragment distance is small (unless highly polar reactants are involved). The bond exchange K decreases (as the exchange energy or bonding of the reactants decreases and the exchange energy of the product configuration increases) until the region of the transition state where it becomes zero. The precise form of K^* and β depends upon the details of the choice of spin functions |16|; however, in the region of transition state K_R and K_P have similar magnitudes and thus K^* is approximately zero so that T and β are approximately equal. Thus T or β is approximately equal to $K_R - K_X$ (or $K_P - K_X$) in the region of the transition state.

Any critical point is associated with the obvious condition that $\partial E/\partial q_i = 0$ for all geometric variables. Thus if we view the potential surfaces using the quantities Q and -T then we have two possible conditions for a critical point on the total energy surface

i) $\partial Q/\partial q_i = -\partial T/\partial q_i$ (ie. Q and -T must have exactly opposite slope as illustrated in fig. 6 above)
ii) $\partial Q/\partial q_i = -\partial T/\partial q_i = 0$ (ie both must have zero slope)

Further, the nature of the critical point (ie. the signs of the $\partial 2E/\partial q_i^2$) depends upon the curvature of the Q and -T surfaces at the transition structure. Thus the "nature" of a critical point is dependent on the sign of $\partial 2Q/\partial q_i^2 - \partial 2T/\partial q_i^2$ (These simple conditions are to be contrasted with the diabatic decomposition where the position of the critical point and its nature depends on the behaviour of ridge of intersection of two diabatics and the resonance energy). For the problems under consideration, along the concerted path,

$\partial Q/\partial\alpha = -\partial T/\partial\alpha = 0$ by symmetry (rigorously for the 2s+2s and Diels Alder reactions and approximately for the 1,3 dipolar cycloaddition), and the transition structure will occur along r when the slopes of Q and -T with respect to r are opposite or zero.

The quantities Q and T have simple physical interpretations. The quantity Q is essentially classical in origin and will be dominated by steric effects (bond stretching and angle strain), electrostatic effects (dipole-dipole interactions) and non-bonded repulsions. The term Q can be thought of as the energy of the system if all the bonding interactions between the reactive sites are "switched off". Thus, for example the Q contribution arising from an ethylenic fragment behaves as if there were no p bond and so has a minimum at around the CC single bond length. In contrast, the contribution to Q arising from the interaction of two methylene groups on different ethylene groups will be repulsive due to the non-bonded repulsions. The quantity -T, on the other hand, is purely quantum mechanical in origin and represents the net bonding energy. The behaviour of the exchange integrals K_{ij} that occur in T can be rationalized using the Heitler-London expression for 2 electrons

$$K_{ij} = [ij|ij] + 2\ s_{ij} <i|h|j> \hspace{3cm} [7]$$

where $[ij|ij]$ is the usual two electron exchange repulsion integral, $<i|h|j>$ is the one electron exchange integral and s_{ij} is the overlap integral between orbitals i and j. K_{ij} will be dominated by the attraction term and thus is negative. These exchange integrals are simple functions of the distance between the reactive sites i and j by virtue of s_{ij}.

The numerical values of Q and the individual exchange integrals that occur in T, K and β are obtained by transformation of an MC-SCF computation to VB space (Part 1 and ref. [16]). To obtain a global view of the surface we have fitted the individual coulomb and exchange integrals obtained from the MC-SCF effective hamiltonian computations to simple functions of the distances between the reactive atomic sites . We emphasize that we are interested in the topology of the surfaces only. Barrier heights etc. are very sensitive to the basis sets used and the inclusion of dynamic electron correlation effects. The surface topology (number and nature of critical points) for these 3 reactions is the same for STO-3G and 4-31G basis so we have computed the effective Hamiltonians at the STO-3G level. We have repeated a few of the computations at the 4-31G level. The numerical values of the exchange integrals are almost identical (see ref. 16) and the only effect is to make Q slightly more repulsive at short range which has the effect of making the diradicaloid regions of the surface flatter [20,21].

4. Analysis of potential energy surfaces for Supra-Supra cycloaddition in terms of i) Diabatic surfaces and ii)Q and T

Our objective is to illustrate how the surface topology seen in Fig. 1,4 and 5 can be understood in terms of the VB model discussed in the previous section.

Before discussing the the potential energy surfaces we must make some comments on the r/α choice of variables. The reaction coordinate corresponds approximately to the variable r. At the transition state there is a negative direction of curvature with respect to this variable.For a true transition state the potential surface is positively curved with respect to the remaining variables. However certain plausible reaction paths may infact pass over local maxima where the potential energy surface is negatively curved in one of the directions that is orthogonal to the reaction coordinate. Indeed this is exactly what

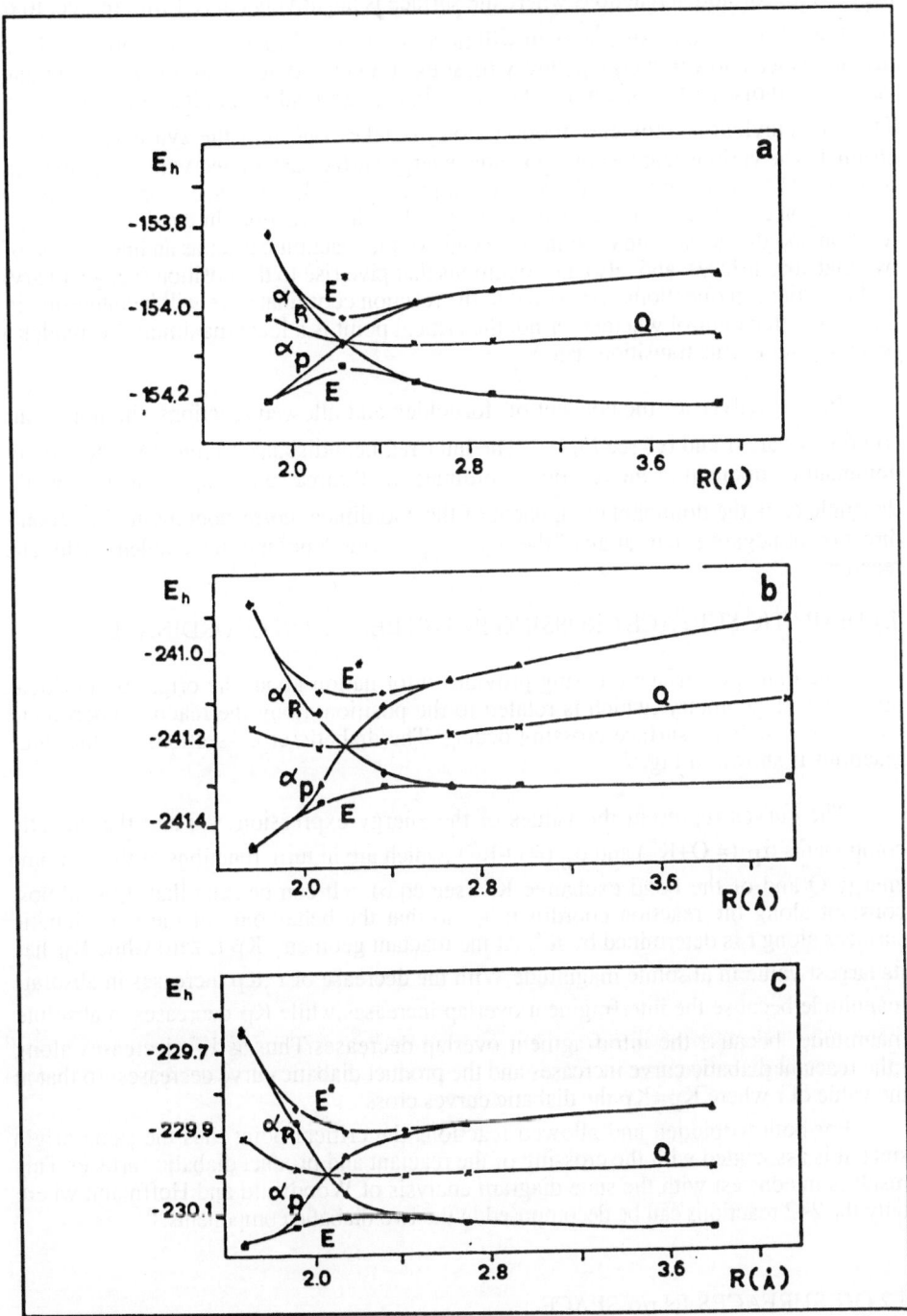

Fig. 7 Diabatic Surface Cross-sections along r for ethylene + ethylene (a), Fulminic Acid + acetylene (b) and Butadiene + ethylene (c)

happens for forbidden reactions where the surface is negatively curved with respect to α (see Fig. 2). Of course, such a path will never correspond to a real path followed by a reaction since a lower energy pathway must exist. On the other hand ,motion along one variable orthogonal to a given reaction coordinate can lead to an alternative pathway.

Thus for ethylene + ethylene, motion along α takes one into the syn asynchronous channel. The various features of a potential energy surface associated with a reaction can be understood by examining the surface in many possible subspaces corresponding to the reaction coordinate and variables orthogonal to this direction. In the direction of the reaction coordinate we can examine the origin of the reaction barrier as an intersection of two diabatic surfaces and also the conditions that give rise to the position (i.e. geometry) of the barrier . In directions orthogonal to the reaction coordinate ,we will be interested in the factors that control whether or not the critical point is a local maximum (forbidden) or a minimum (true transition state).

In the analysis of the concept of forbidden and allowed reactions ,the important coordinates are r and α (see Fig.1) .The interfragment distance r represents the largely dominant component of the reaction coordinate in all three supra-supra reactions,while the angle α is the dominant component of the coordinate corresponding to the second direction of negative curvature of the supra-supra critical point of the ethylene+ethylene reaction.

4.1 DIABATIC SURFACE CROSSING IN THE REACTION COORDINATE

The diabatic surface crossing provides information about the origin of a critical point and its geometry ,which is related to the position along the reaction coordinate r,where the diabatic surface crossing occurs. The diabatic cross-sections for the three reactions is shown in Fig. 7 .

The curves represent the values of the energy expression [1] and the diabatic components α_R (= Q+K*) and α_p (= Q-K*) ,which are in turn functions of the coulomb energy Q and of the bond exchange K* (see eq.6). It can be seen that Q is almost constant along the reaction coordinate r , so that the behaviour of the two diabatic surfaces along r is determined by K*. At the reactant geometry Kp is zero while K$_R$ has its largest value in absolute magnitude.With the decrease of r ,Kp increases in absolute magnitude because the interfragment overlap increases,while K$_R$ decreases in absolute magnitude because the intrafragment overlap decreases.Thus,as K* decreases along r,the reactant diabatic curve increases and the product diabatic curve decreases ,so that at the value of r where K$_R$=Kp the diabatic curves cross.

For both forbidden and allowed reactions the critical point has the same origin since it is associated with the crossing of the reactant and product diabatic surfaces This result is in contrast with the state diagram analysis of Woodward and Hoffmann where only the 2+2 reactions can be decomposed in the two diabatic components.

4.2 Q/T SURFACES IN r/α SPACE

The principal objective of the diabatization of the potential energy surface just discussed is to demonstrate the common origin of the reaction barrier in the 3 reactions. However a more detailed rationalization of the potential surface topology in terms of the diabatic surface model is complicated by the fact that a)the transition structure may not correspond exactly to the diabatic surface intersection and b)the exchange energy enters any qualitative argument in both K^* and the resonance energy β. Thus in order to discuss the surface topology in a complementary way that can be more easily generalized in a qualitative fashion the parameters Q and T can be used. In this way the quasi-classical coulomb and quantum mechanical exchange effects can be cleanly separated.

The surfaces for Q and T for the 3 reactions are presented in Fig. 8 through 13. The numerical value gradient and curvature of these surfaces, in the variable R and α, at the concerted transition structure are given in table 1.

Table 1. Gradient and Curvature§ of the potential energy surfaces at the "concerted transition structure" for 2_s+2_s cycloaddition reaction of two ethylene molecules, the Diels Alder reaction of Butadiene and ethylene and the 1,3 Dipolar cycloaddition of Fulminic acid and Acetylene.

	2_s+2_s	Diels Alder	1,3 Dipolar
$\partial Q/\partial r$	0.008	0.049	0.044
$-\partial T/\partial r$	-0.008	-0.049	-0.044
$\partial Q^2/\partial r^2$	0.212	0.060	0.073
$-\partial T^2/\partial r^2$	-0.884	-0.272	-0.139
$\partial E^2/\partial r^2$	-0.671	-0.211	-0.066
$\partial Q^2/\partial \alpha^2$	0.025	-0.011	0.047
$-\partial T^2/\partial \alpha^2$	-0.101	0.012	-0.030
$\partial E^2/\partial \alpha^2$	-0.076	0.008	0.017

$\S$ The curvature $\partial E^2/\partial q_i^2$ was evaluated analytically, while the $\partial Q^2/\partial^2 q_i$, $\partial T^2/\partial q_i^2$ have been approximated by finite difference.

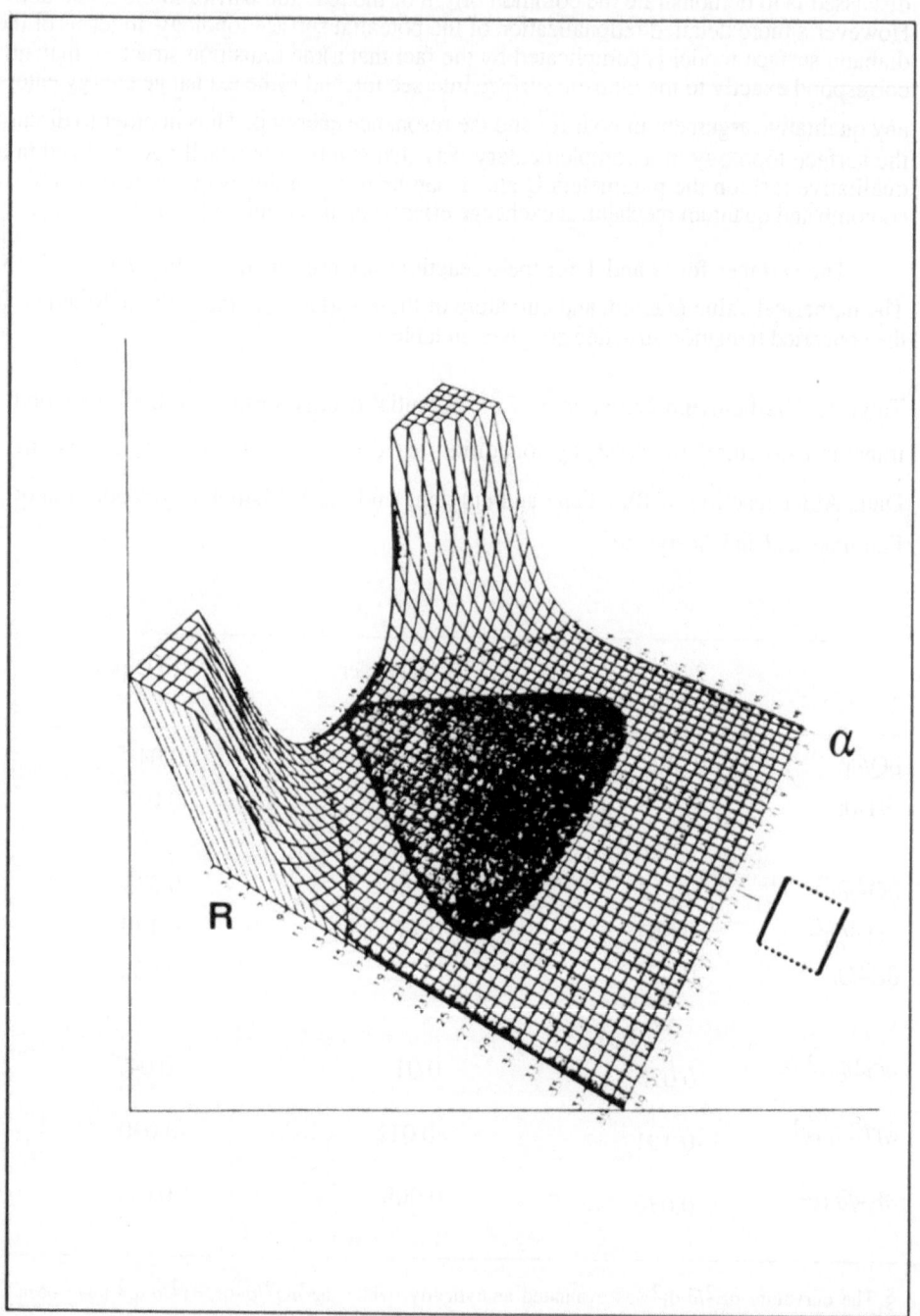

Fig. 8. Coulomb potential energy surfaces (Q) for the 2s+2s cycloaddition of two ethylene molecules. The two axes are defined as in Fig. 2.

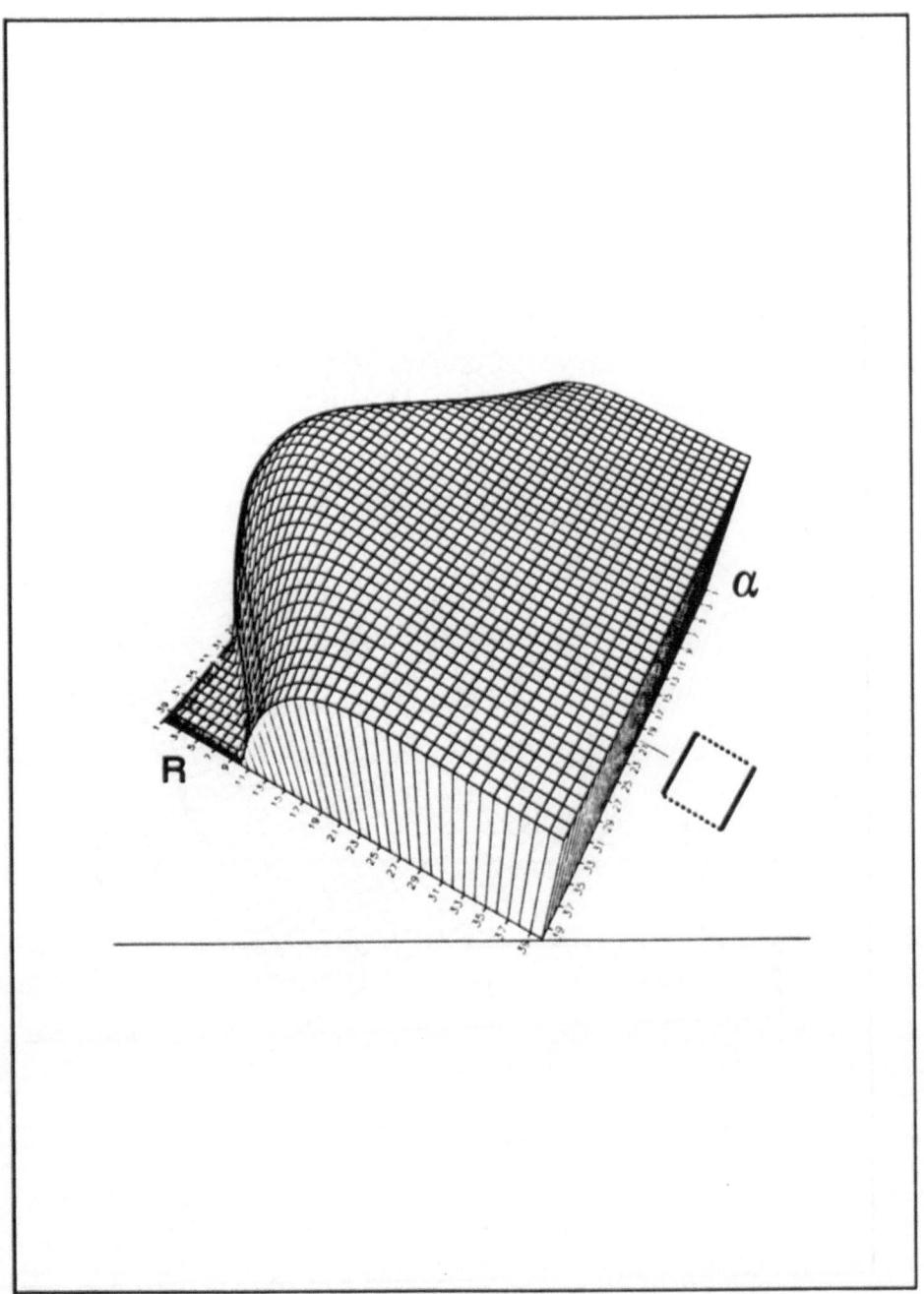

Fig. 9. Exchange energy surfaces (-T as defined in eq.4) for the 2s+2s cycloaddition of two ethylene molecules. The two axes are defined as in Fig. 2.

306

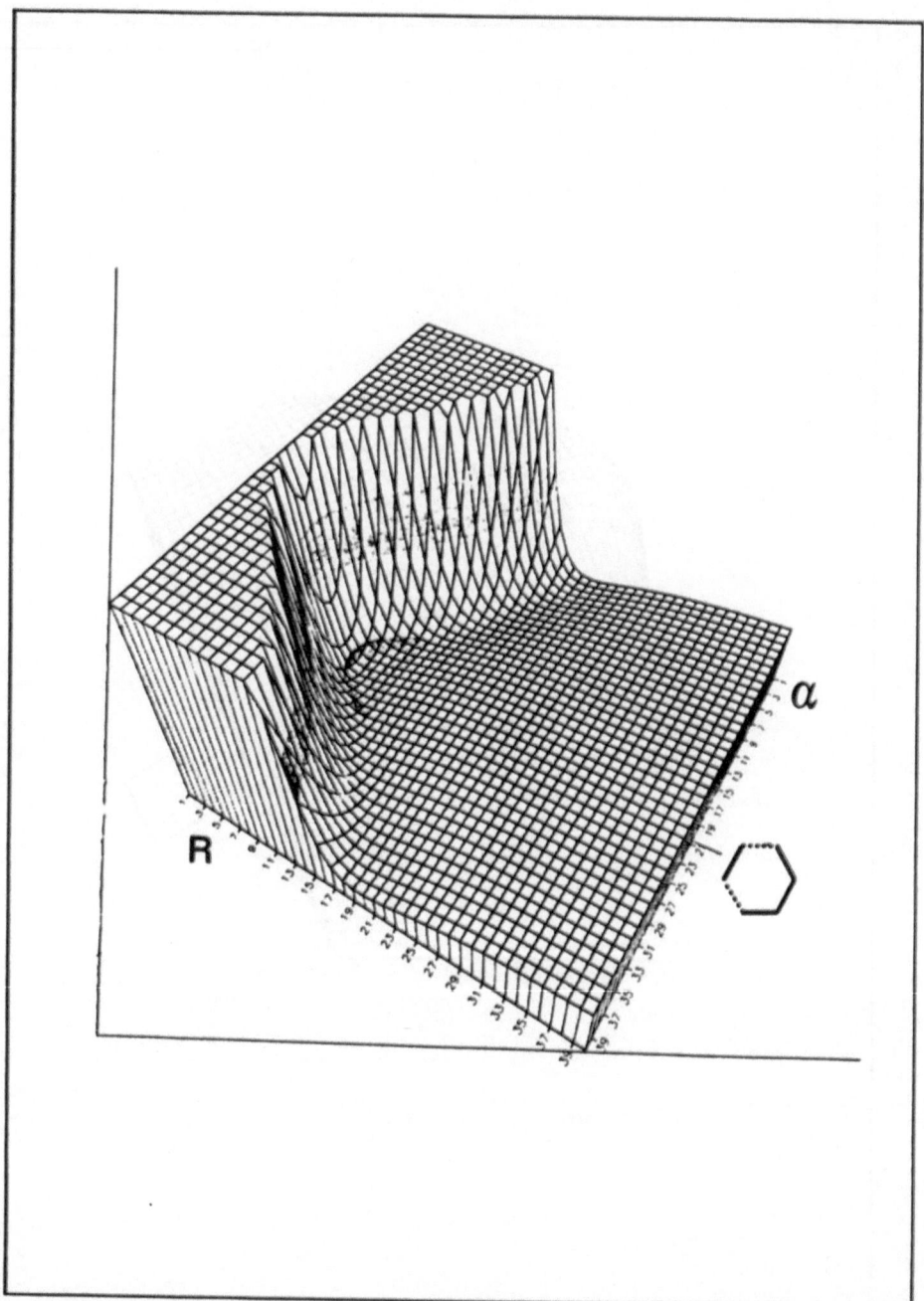

Fig. 10. Coulomb potential energy surfaces (Q) for the Diels Alder reaction of cis butadiene and ethylene molecules. The two axes are defined as in Fig. 4.

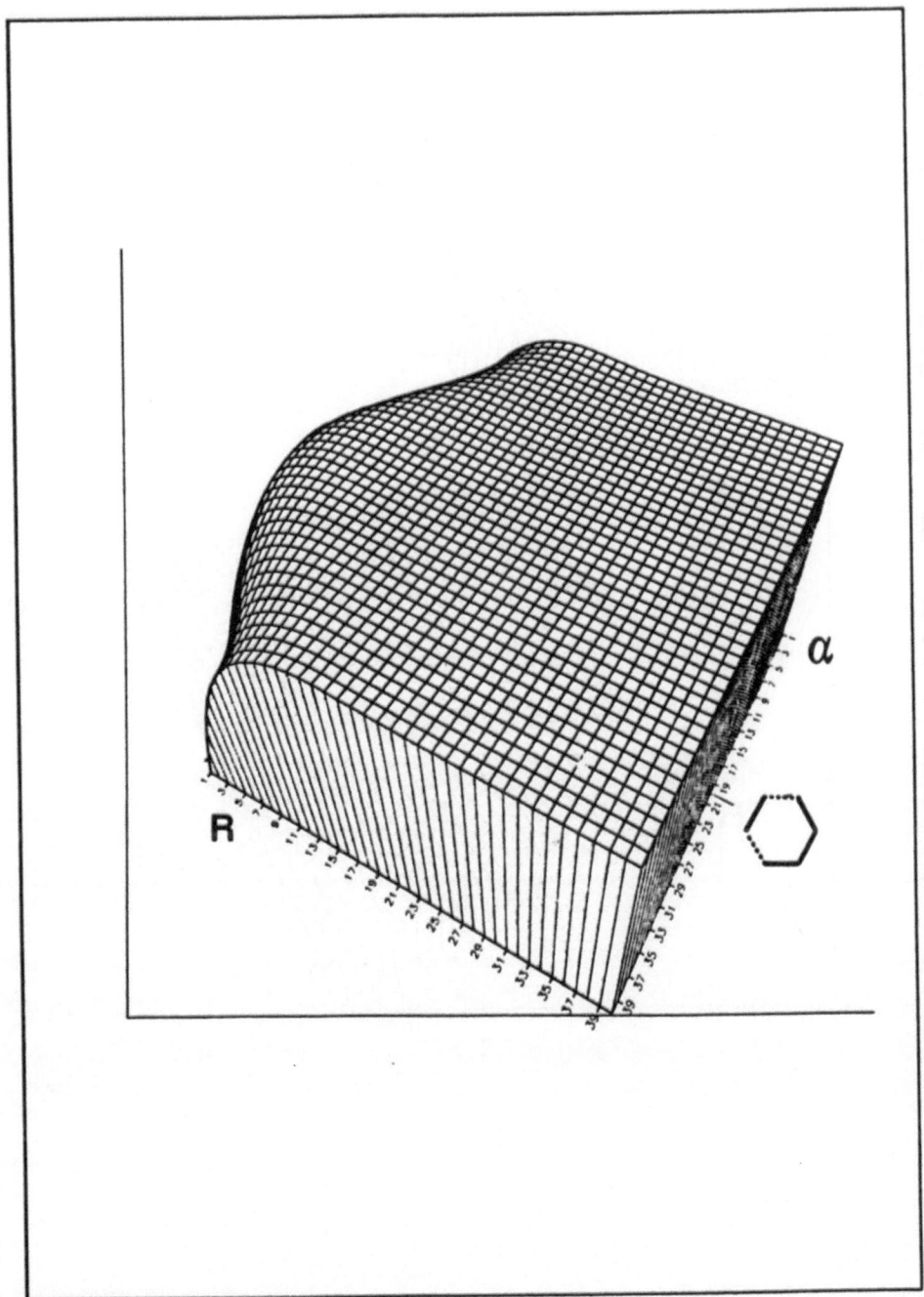

Fig. 11. Exchange energy surfaces (-T as defined in eq.4) for the Diels Alder reaction of cis butadiene and ethylene molecules. The two axes are defined as in Fig. 4.

308

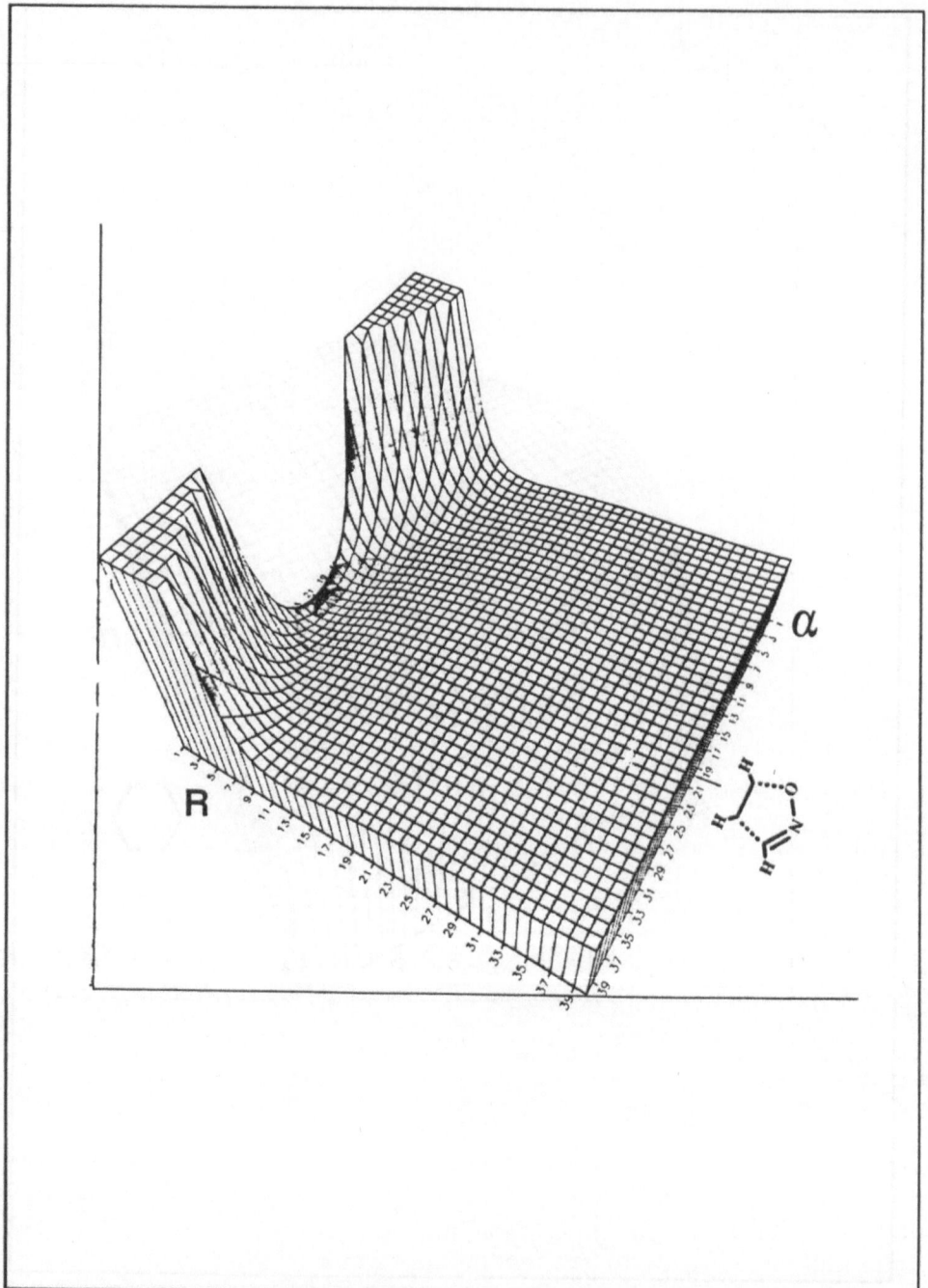

Fig. 12. Coulomb potential energy surfaces (Q) for the 1,3 dipolar cycloaddition of
Fulminic acid (HCNO) and acetylene molecules. The two axes are defined as in Fig. 5.

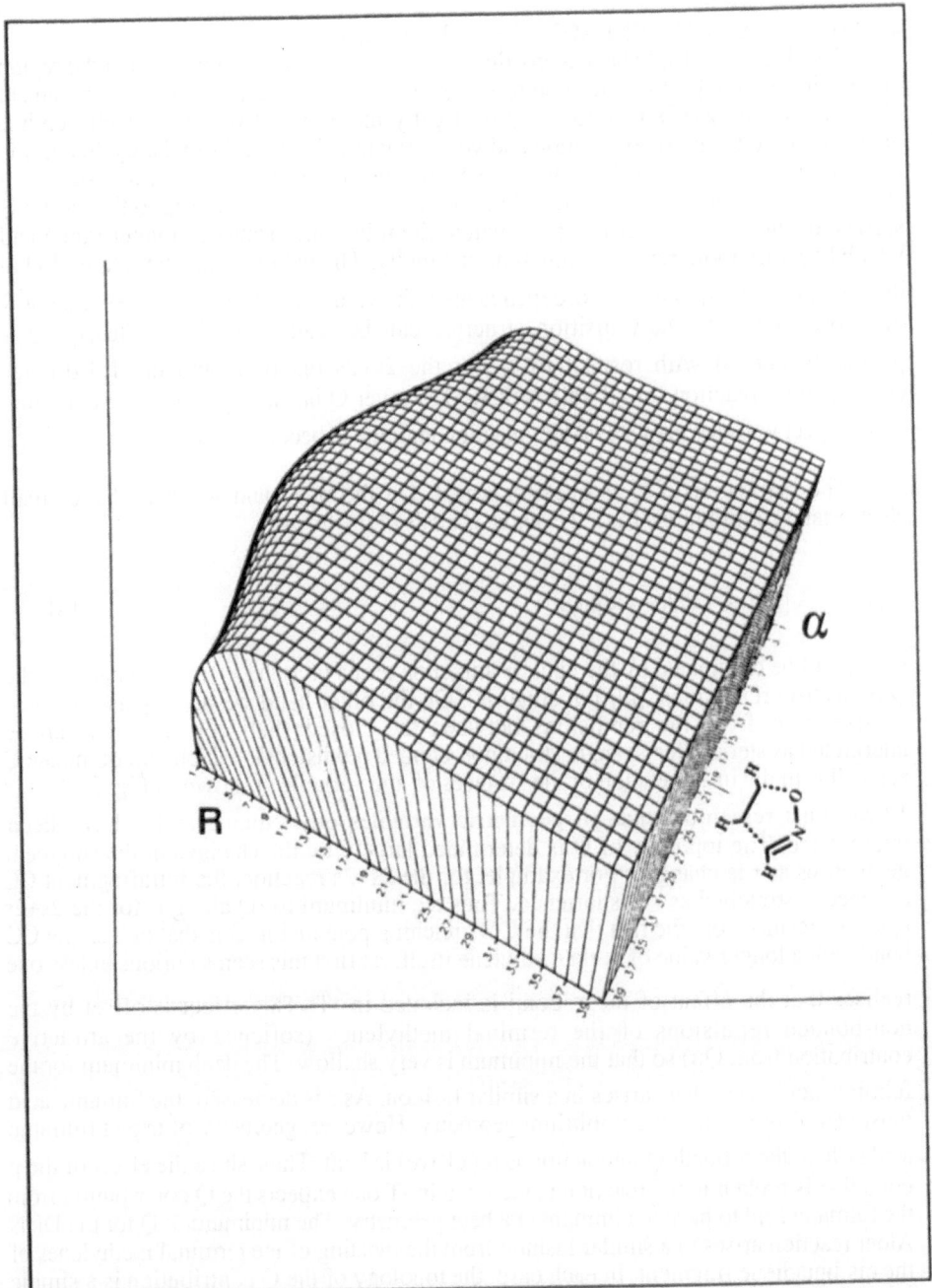

Fig. 13. Exchange energy surfaces (-T as defined in eq.4) for the 1,3 dipolar cycloaddition of Fulminic acid (HCNO) and acetylene molecules. The two axes are defined as in Fig. 5.

The essential features of the surfaces can be summarized as follows.

4.2.1)THE COULOMB ENERGY (Fig. 8,10 and 12)

The 3 surfaces for Q have a broadly similar topology with a minimum in the region of the critical point for the concerted reaction. Q is very flat except at small interfragment distances for the 2s+2s reaction.   Q is slightly more attractive (ie the well depth is greater) for the Diels Alder reaction and very attractive for the 1,3 dipolar cycloaddition reaction. However, the surface of Q is soft (in the sense that the minimum occurs at short range) for the 2s+2s and the 1,3 dipolar cycloaddition as opposed to the harder Q surface of the Diels Alder reaction where the minimum occurs at longer range and which becomes more repulsive much more rapidly. The nature of the curvature of Q in the direction corresponding to α depends upon the value of r. The data at the value of R that corresponds to the transition structure can be read in Table 1. Clearly, Q is positively curved with respect to α for the 2s+2s reaction and the 1,3 dipolar cycloaddition reaction for all values of R. However Q has a slight negative curvature with respect to α for the Diels Alder reaction until the R becomes quite small.

The behaviour of Q is quasi-classical. On the one hand we must have small electrostatic contributions from the active orbitals of the form

$$Q_{ij} = [ii|jj] + <i|h|i> + <j|h|j> \qquad [8]$$

which will be dominated by the nuclear attraction terms in $<i|h|i>$ at large r and by the two-electron repulsions, $[ii|jj]$ at small R . On the other hand we have quasi-classical contributions from the repulsions between the closed shell cores,(which can be interpreted as steric effects, angle strain,non-bonded repulsions) which will dominate Q at small r. In the interpretation of the surfaces of the variable Q shown in Fig. 8 , 10 and 12,we must remember that all geometric variables other than r and α have been interpolated. The topology of Q is determined largely by the changes in the fragment geometries as r is changed. For example, for the 2s+2s reaction, the intrafragment CC distance is stretched as r is shortened. Thus the minimum for Q along r, for the 2s+2s reaction, results from the fact that the CC stretching potential in Q is that of a single CC bond with a longer value of r_e than ethylene itself. At first this seems curious unless one realises that the effect of the π bond is included in -T. This effect is offset by the non-bonded repulsions of the terminal methylenes (softened by the attractive contribution from Q_{ij}) so that the minimum is very shallow. The deep minimum for the fulminic acid+ acetylene arises in a similar fashion. As r is decreased, the fulminic acid must bend from its linear equilibrium geometry. However, geometry of triplet fulminic acid (where the π bonding interaction is repulsive) is bent. Thus, since the effect of the p bond that is broken in the reaction is included in -T one expects the Q contribution from the fulminic acid to have a minimum at a bent geometry. The minimum in Q for the Diels Alder reaction arises in a similar fashion from the twisting of the terminal methylenes of the cis-butadiene fragment. In each case the topology of the Q contribution is a simple manifestation of the skeletal deformations of a molecule where the bonding interactions have been "switched off".

4.2.2)THE TOTAL EXCHANGE ENERGY (Fig. 9,11 and 13)

The surfaces for -T also have a broadly similar shape for the 3 reactions. For each reaction, -T has a critical point on the concerted path (a local maximum) and becomes attractive more rapidly along an asynchronous co-ordinate. In particular, -T has a pronounced maximum in the region of the critical point for the 2s+2s reaction. Note that the curvature of the surface evaluated at the concerted reaction path with respect to the direction α depends critically upon the value of R. At large R, the curvature of -T is positive respect to the direction α but as one approaches the critical point in -T it becomes negatively curved.

The behaviour of -T is representative of the net bond breaking/making energy. It is controlled simply by the distances between the reactive sites (ie s_{ij} in eq. 7). On the reactant side of the surface we are breaking π bonds so that the surface is weakly repulsive and on the product side we are forming new σ bonds so that the surface is very strongly attractive. For a fixed value of r if we change the value of α from 90 degrees, then we increase the magnitude of Kp and K_X (since the related s_{ij} vary exponentially) while the value of K_R remains approximately fixed. We can distinguish the following situations:

i)Firstly at large r the terms K_X and Kp will be small so that -T is approximately -$[K_R$_$Kp]$. and will have a small positive curvature along α .

ii)For values of R in the transition state region and before the crossing -T is approximately -$[K_R$-$K_X]$ and along α its curvature can be either slightly negative or with a small minimum,depending on the eigenvector dominated by α.

iii)For the other values of r (i.e. after the transition state)-T is approximated by -$[Kp$-$K_X]$ and its curvature along a will be negative .

4.2.3)THE NATURE OF THE CRITICAL POINTS (TS or LOCAL MAXIMUM) IN TERMS OF $\partial E^2/\partial R^2$, $\partial Q^2/\partial R^2$,$\partial T^2/\partial R^2$

The position of the critical point on the concerted synchronous pathway for the 2s+2s reaction occurs at a position which is also a critical point in -T and in Q (the values of $\partial Q/\partial R$ and -$\partial T/\partial R$ in Table 1 are both almost zero). Since Q is very flat this is the only possibility. Thus the maximum in -T coincides approximately with the local maximum on the energy surface shown in Fig. 2. The position of the critical point with respect to r for the 1,3 dipolar cycloaddition reaction occurs just before the minimum in Q (ie on the attractive part of Q) while -T is still decreasing in magnitude. Thus for the 1,3 dipolar cycloaddition reaction the transition state occurs only just before the critical point in -T where the curvature in -T is negative with respect to the direction α (see also Table 1). In contrast, the geometry of the critical point for the Diels Alder reaction is considerably in advance of the critical point in -T where the curvature of -T in the direction α is positive. The channel on the asynchronous diradical pathway for the 2s+2s reaction results from the coincidence of the steeply repulsive Q and strongly attractive -T components of the energy. In the Diels Alder reaction the Q becomes steeply repulsive much earlier than -T becomes attractive along the asynchronous reaction path and the condition for a critical point ($\partial Q/\partial q_i = -\partial T/\partial q_i$) cannot be met. The situation is similar for the 1,3 dipolar reaction.

From the preceding discussion we see that, while the general shapes of the Q and T surfaces for the 3 reactions are similar, the condition for a critical point (slopes of Q and -T opposite or equal to zero) and the nature of the critical point (transition state or local maximum) are delicately intertwined and yield a very different topology on the surface that corresponds to the total energy. The critical points on the concerted synchronous pathway for the 2s+2s reaction and the 1,3 dipolar cycloaddition reaction both occur where the curvature of -T with respect to the direction α is negative and the curvature of Q is positive. In the 2s+2s critical point, the curvature of -T is dominant and this point is a local maximum with respect to the directions α and r. However, the dominant positive curvature of Q in the direction α turns the critical point for the 1,3 dipolar cycloaddition reaction into a true transition state. If the curvature of Q was smaller with respect to the direction α, the curvature would be dominated by -T and the critical point for the 1,3 dipolar cycloaddition reaction could become a local maximum. In contrast, the transition structure for the Diels Alder reaction occurs where the curvature of -T in the direction α is positive and the curvature of Q in the direction α is slightly negative and the positive curvature of -T dominates. Thus from a global point of view -T favours the asynchronous pathway in all cases and it is the behaviour of Q that determines the geometry where the transition state occurs and its nature. If Q is very "soft" then we have the local maximum with two asynchronous "channels" (2s+2s reaction) dominated by the topology of -T. If Q is "hard" so that the transition state occurs well before the maximum in -T where the curvature of -T with respect to α is still positive then the concerted mechanism is favoured (Diels Alder). The 1,3 dipolar reaction falls into the delicate region in between where Q is quite soft and -T is negatively curved with respect to α . Here, the dominant positive curvature of Q with respect to α causes the preference for the concerted mechanism.

5 Conclusions

For 3 different cycloaddition reactions we have been able to show that while the electronic origin of the ridge that separates reactants and products is the same, the different surface topology is controlled by the behaviour of the quasi-classical coulomb energy of simple Heitler-London VB theory. The net bonding energy is described by the total exchange energy -T. The surface of -T is broadly similar for the 3 reactions and tends to favour the asynchronous pathway. The similarity of -T is not too surprising since it is controlled mainly by the distances between the atomic sites involved in bond breaking/making. However, the asynchronous pathway can only exist if Q is sufficiently soft so that the sharply attractive regions of -T become accessible.

Q is quite soft for the 2s+2s reaction because it is dominated by the single bond CC stretching potential of an ethylene molecule where the p bond is absent. This potential has its minimum at a bond length that is obviously longer that that of a CC double bond. Thus the repulsive non-bonded interactions of the terminal methylenes are offset by the stretching of the CC bond length in the ethylenic fragment which is attractive in the Q component of the energy. Q is harder in the other two reactions since the corresponding attractive components in Q arise from angle deformation and torsion.

References

1. *Ab-initio Modeling of Chemical Reactivity using MC-SCF and VB Methods*, M.A. Robb and F. Bernardi, (in L.G. Arnaut, S. Formosinho and I.G.Csizmadia ed. Theoretical and Computational Models for Organic Chemistry, Kluwer Academic Publishers, this voume)
2. Ab initio methods in Quantum Chemistry I, Ed. H.P. Lawley *Adv. Chem.Phys.* **1987**, 67
3. Ab-initio methods in Quantum Chemistry Il, Ed. K.P. Lawley *Adv.Chem.Phys.* **1987**, 69.
4. Woodward, R.B.; Hoffmann, R. *Angew. Chem.Int.Ed.Engl.* **1969,** 8, 781-853.
5. Epiotis, N.D. *Theory of organic reactions; Springer Verlag:* Heidelberg, **1978**.
6. Epiotis, N.D. *Lect. Notes Chem.* **1982**, 29.
7. Epiotis, N.D. *Lect. Notes Chem.* **1983**, 34.
8. Shaik, S.S., *J.Am.Chem.Soc.* **1981**, 103, 3692-3701.
9. Pross, A.; Shaik, S.S. *Acc.Chem.Res.* **1983**, 16, 363.
10. Shaik, S.S. Prog.*Phys.Org.Chem.***1985**, 15, 197.
11. Pross, A. *Adv.Phys.Org.Chem.* **1985**. 21, 99.
12. Bernardi, F.; Robb, M.A. *Mol.Phys.* **1983**, 48, 1345-1355
13. Bernardi, F.; Robb, M.A. *J.Am.Chem.Soc.* **198**4, 106, 54-58
14.Bernardi, F.; Olivuccci M.; Robb, M.A.; Tonachini, G. J. *Am.Chem.Soc.* **1986**, 108, 1408-1415.
15. Bernardi, F.; Robb, M.A. *Adv.Chem.Phys.* **1987,** 67, 155-248.
16.Bernardi, F. ; Olivucci, M. ; McDouall, JJW. ; Robb,M.A. *J.Chem. Phys.* **1988**, 89 ,6365
17. Evans, G.; Polany, M. *Trans.Far.Soc.* **1938**, 34, 11.
18. Evans, G.; Warhurst, E. *Trans Far.Soc.* **1938**, 34, 614.
19. Durand, P.; Malrieu, JP. *Adv.Chem.Phys.* **1987,** 67, 321-412.
20.Bernardi, F. ; Bottoni, A. Robb, M.A. Schlegel, H. B. ; Tonachini, G. *J.Am.Chem.Soc.* **1985**, 107, 2260-2264.
21.Bernardi, F. Bottoni, A..; Field, M.J. ; Guest, M.F. ; Hillier, I.H. Robb, M.A.; Venturini, A. *J.Am.Chem.Soc.* **1988**, 110, 3050-3055.
22.McDouall, J.H.W. ; Robb, M.A. Niazi, V. Bernardi, F. Schlegel, H.B. *J. Am.Chem.Soc.* **1987**, 109, 4642.
23.Hehre, W.J. ; Stewart, R. F. ; Pople, JA.*J. Chem.Phys.* **1969**, 51, 2657-2664.25.
24.Ditchfield, R.; Hehre, W.J.; Pople, JA.*J. Chem.Phys.* **1971**, 54, 724-728.
25 McWeeny, R.; Sutcliffe, B. *Methods of quantum mechanics*; Academic: New York, **1969**.
26.Eyring, H.; Walter, J.; Kimball, G. *Quantum Chemistry*; Wiley: New York, **1944**.

EXCITED STATE PROTON TRANSFER REACTIONS

Noam AGMON

*Department of Physical Chemistry
and The Fritz Haber Research Center
The Hebrew University
Jerusalem 91904, Israel*

*"Let us descend to the garden
Things seen from there one does not see from here"*
An Israeli folk song.

ABSTRACT. Proton dissociation in solution is traditionally described via a two step kinetic mechanism of complex formation and proton escape. This mechanism leads to an ultimate exponential decay of the acid concentration. In contrast, we find that the fluorescence signal from an excited, negatively charged hydroxypyrene derivative in solution has a non-exponential tail, attributed to reversible geminate recombination in the excited state. The theory of reversible diffusion influenced reactions predicts a $t^{-3/2}$ asymptotic decay, in quantitative agreement with experiment.

The addition of inert salts screens the Coulombic attraction thus diminishing geminate recombination. As a result, the non-exponential tail and the steady-state quantum yield decrease with increasing ionic strength. This contrasts with the classical theories of Brönsted and Bjerrum which predict zero salt effect on the dissociation direction. We model the screening effect by a "Naive Approximation", assuming zero screening beyond the separation of the nearest, oppositely charged ion. For interpreting pH effect on the proton transfer reaction, we approximate the many-body problem by introducing a bimolecular boundary condition in the diffusion equation. This is compared with one dimensional stochastic simulations.

S. J. Formosinho et al. (eds.), Theoretical and Computational Models for Organic Chemistry, 315–334.
© 1991 *Kluwer Academic Publishers.*

1. Introduction

Proton transfer is a fundamental process in both chemistry [1-3] and biology [4]. In particular, proton dissociation namely, proton transfer to solvent, from aromatic dye molecules in their excited electronic state [5] can be easily studied by virtue of their strong fluorescence signal [6]. The older fluorescence measurements did not possess time resolution: It was only possible to obtain steady-state quantum yields under conditions of constant illumination [6]. The conventional interpretation of the experimental data assumed a chemical kinetic scheme, such as [3]

$$R^*OH \; \rightleftarrows \; R^*O^- \ldots H^+ \; \rightarrow \; R^*O^- + H^+ \qquad (1)$$

In this scheme the proton first dissociates to form a geminate ionic complex and then separates from the parent anion. In the time domain, the rate-equations for this two-step mechanism leads to bi-exponential decay of the excited acid (R^*OH) population.

At low pH values, when additional protons are present, the separation step becomes reversible and one observes homogeneous proton recombination. The reaction under these conditions undergoes a transition from unimolecular (correlated pairs) to a bimolecular (or pseudo-unimolecular) reaction. The rate of this recombination reaction is expected to diminish with increasing concentration of inert salt, which screens the Coulombic attraction between the proton and the anion. In fact, the classical Brönsted-Bjerrum theory of salt effects puts all of the effect in the recombination reaction while predicting zero salt effect on the dissociation direction [7].

Time resolved measurements of ultrafast proton dissociation in the excited state are challenging these classical views [8-13]. Our picosecond fluorescence experiments have shown that the kinetic scheme, eq 1, does not accurately describe the kinetic data. Firstly, the R^*OH fluorescence decays asymptotically as a power law, rather than exponentially [10]. The area under the non-exponential tail contributes to the observed quantum yield. In addition, there is a marked salt effect on the elementary dissociation process:

The relative R^*OH/R^*O^- quantum yield decreases with increasing ionic strength [11].

In order to account for the above behavior we have replaced the conventional rate equations by a Smoluchowski equation, describing diffusion in the potential of mutual interaction. In this picture one introduces explicitly the R^*O^-/H^+ distance distribution, thus obtaining a partial (instead of an ordinary) differential equation for the time evolution of proton dissociation. The fact that the observed reaction is *reversible* in the excited-state, has promoted the development of the theory of reversible diffusion influenced reactions [14], as an (almost obvious) extension of the theory of (irreversible) diffusion influenced reactions [15, 16].

The aim of this lecture is to provide a qualitative description of reversible proton transfer reactions in the excited-state, using the extended theory of diffusion influenced reactions. The complete equations and numerical procedures may be found in the literature [10-14]. Major results include (i) the asymptotic power-law decay and the evidence for diffusive kinetics [10]; (ii) The salt effect [11] and the "Naive Approximation" for the screening function [17, 11] and (iii) an extension [18] of the theory for approximating the effect of competing geminate and homogeneous proton recombination expected at low pH values.

2. The Experimental System

Our work [9-13] has centered on the dye molecule shown in Figure 1. This molecule is officially called 8-hydroxypyrene 1,3,6-trisulfonate and colloquially known as pyranine or HPTS. We also denote it by ROH. Dissolved as its trisodium salt, it is already triply charged in its ground state, where the pK^0 value of the OH group is around 8. In the first excited singlet state the hydroxy group becomes highly acidic [5], with a pK^* value of 1.4.

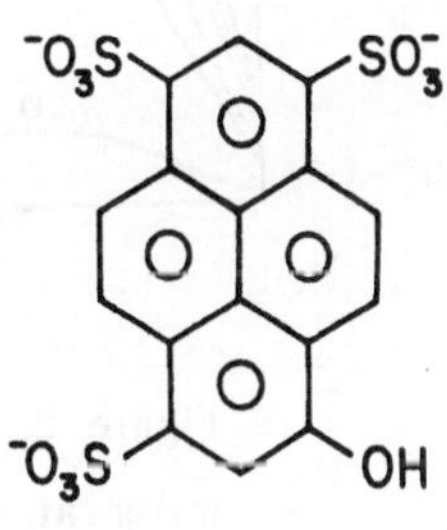

Figure 1. Pyranine.

HPTS shows a strong fluorescence signal, with an absolute quantum yield close to unity (hence it is a good laser dye). In water at pH=6, R*OH fluorescence peaks around 445 nm and is about 1/20 of the R*O⁻ fluorescence signal, which peaks at 510 nm. A steady-state (constant illumination) fluorescence spectrum [13] is shown in Figure 2. The small R*OH/R*O⁻ quantum yield, 1/23, is due to the fast proton dissociation. As demonstrated by Weller [5], the fluorescence spectrum can be titrated by the addition of strong acids. This effect, shown in Figure 2, indicates that the reverse reaction of proton recombination in the excited state cannot be negligible [5].

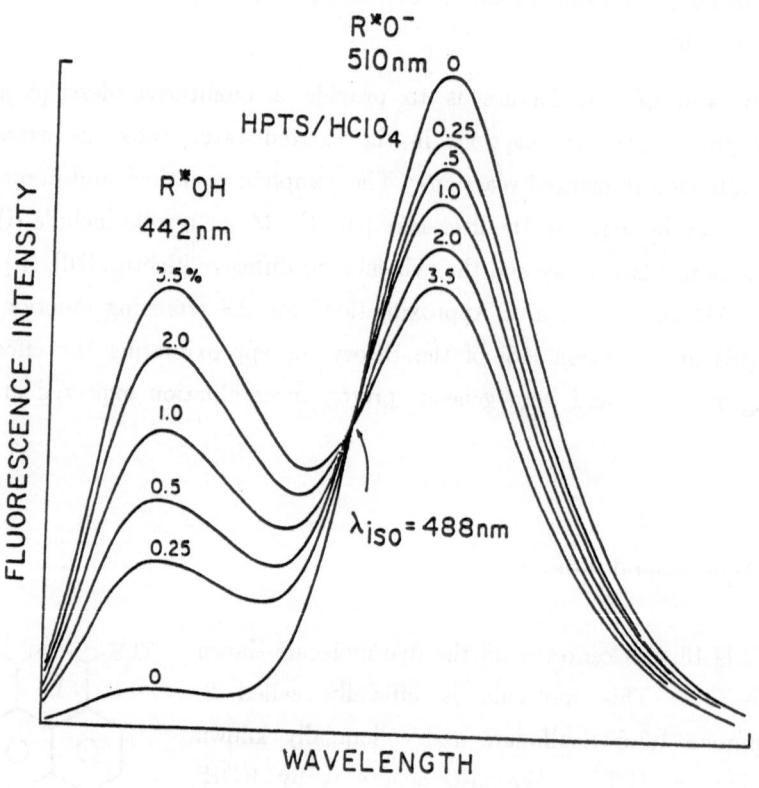

Figure 2.   Steady-state fluorescence spectra of HPTS in water at room temperature [13].   Curve labelled 0 is at neutral pH. Other spectra are at low pH and are labelled by the volume fraction of 70% perchloric acid added.   Note the isoemissive point at 488 nm.

Early time resolved measurements could not follow the excited-state kinetics within the short HPTS fluorescence lifetime (τ_f=5-6 ns). Laser pulses of about 20 ns have been used in conjunction with absorption spectroscopy (HPTS absorption peaks at 403 nm and its anion at 454 nm) to determine the ground state recombination rate coefficient [19] and the effect of fluorescence quenching on the ground-state recombination yields [20]. As expected, the exothermic ground-state recombination reaction ($pK^0 \approx 8$) occurs at a nearly diffusion-controlled rate, $1.9 \cdot 10^{11} M^{-1} s^{-1}$ [19]. The dependence of this recombination rate on ionic strength is shown in Figure 3: Increased screening of the proton-anion Coulombic attraction decreases the recombination rate. The effect is milder than predicted by the Debye-Hückel theory [7, 21], which could fit experiment only by invoking the empirical Davies correction [19].

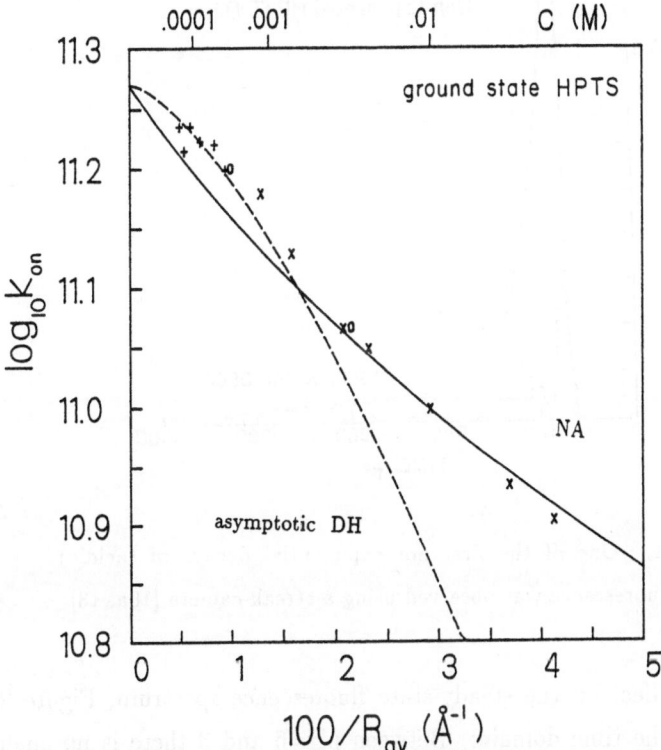

Figure 3. Steady-state protonation rate coefficients for ground-state HPTS [19], as a function of $HClO_4$ (+'s) and $KClO_4$ (X's) concentration, c (or the average radius of a sphere containing one salt ion, R_{av}), compared with the asymptotic Debye-Hückel expression and the Naive Approximation [11c], see below.

It was not until the development of picosecond fluorescence spectroscopy that the fate of the excited acid could be followed in the time domain. From the nearly exponential decay profile it was verified [9] that in the excited-state the hydroxylic proton is ejected at an ultrafast rate (100 ps). It has occurred to the researchers [9] that, since the fluorescence spectrum can be titrated (Figure 2), there should be an observable recombination process also for the geminate, correlated pair. Indeed, when they looked closely at the decay of the R*OH fluorescence, shown in Figure 4, they discovered a long time, non-exponential, tail which they have attributed to reversible rebinding of the geminate proton, occurring (almost) without quenching of the excited-state [9].

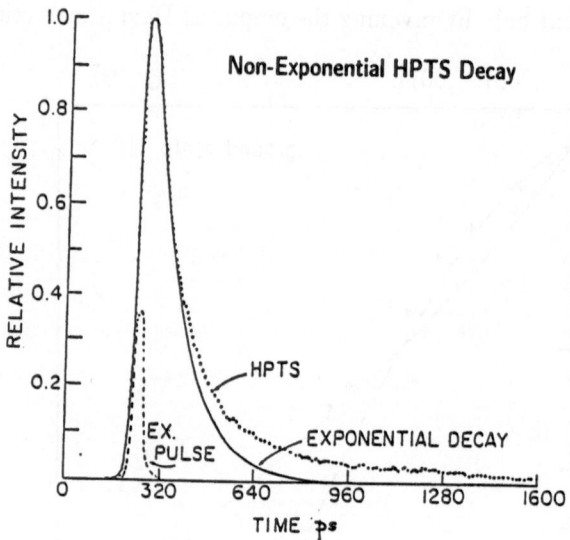

Figure 4. One of the first non-exponential decays of (acidic) HPTS fluorescence was observed using a streak-camera [10a, 13].

The pH effect on the steady-state fluorescence spectrum, Figure 2, can also be seen in the time domain: Between pH=6 and 3 there is no change in the time resolved data, but at lower pH values the long time tail increases (Figure 5). This corresponds to the pH range for which the R*OH peak in Figure 2 increases in amplitude. The increase in the tail in strongly acidic solvents shows that it is due to recombination with protons, while its pH independence at neutral pH values shows that the tail under these conditions is not due to homogeneous protons. This provides an additional support to the interpretation of the geminate tail as arising from geminate recombination.

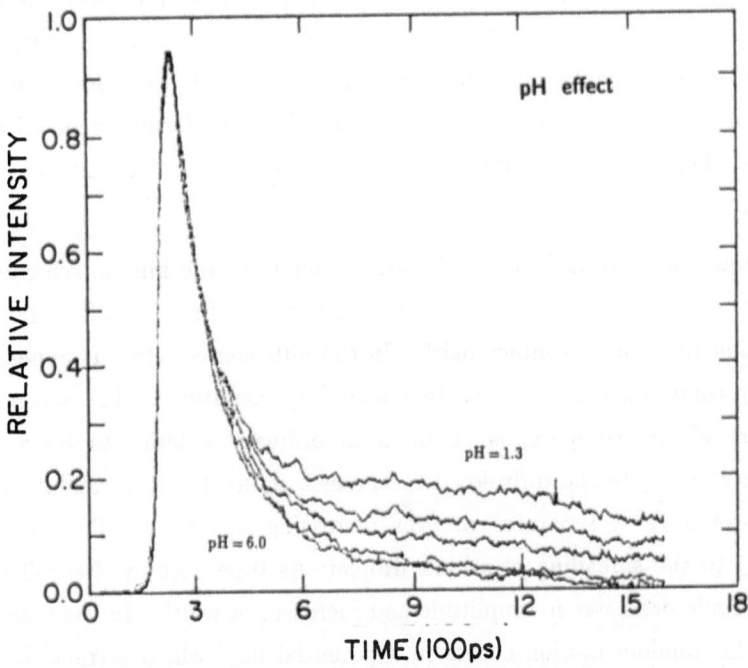

Figure 5. pH dependence of HPTS (acidic form) transient fluorescence signal at 430nm [10a]. pH values (top to bottom) are: 1.3, 1.7, 2.0, 3.6 and 6.0.

3. The Long Time Tail and Proton Diffusion

The traditional analysis of proton transfer reactions [3, 5-7] is by a kinetic scheme e.g., eq 1, and the ensuing chemical rate equations describing the time dependence of bulk concentrations. This is insufficient for explaining the long time tail. The minimal level of complexity has to involve the spatial diffusion of the proton in the electrostatic field of the anion, as depicted by the time-dependent Smoluchowski equation [15], but with a boundary condition (the "back-reaction" boundary condition) which describes reversible reactions [10].

Additional complications to the kinetics besides proton diffusion are possible. These could include vibrational relaxation of the excited pyrene derivative, dielectric relaxation of the solvent (water) and heat transfer. These processes in water occur on timescales which are typically shorter than 10 ps, hence they are over by the time the proton is ejected (100 ps). Rotational diffusion of the HPTS anion (hence of the rebinding site) could also play a role. In water at room temperature it takes some 150 ps. Hence before the proton has had a chance to recombine, the rebinding site is uniformly smeared around a sphere. One is left with the relative proton-anion translational diffusion to consider.

How does a diffusional mechanism differ from the kinetic scheme in eq 1? Both are two-step mechanisms with an identical first step: The reversible dissociation to form a "contact pair". In the diffusion equation it appears as a boundary condition (the back-reaction boundary condition). The second step, separation of the complex, is assumed in ordinary kinetics to be a single, elementary step. In the diffusional mechanism it involves the random motion of the proton as it separates to ever increasing distances. This motion is analogous to the spreading of an ink droplet: As time goes by, its bell shaped density profile decreases in amplitude and increases in width. In the case of the proton, this random motion occurs in a potential field which attracts it to the quadruply charged HPTS anion, resulting in enhanced geminate recombination, detectable through the non-exponential tail in the fluorescence decay. The diffusion equation is therefore generalized to a spherically symmetric Smoluchowski equation. Its numerical solution, convoluted with an instrument response function of about 100 ps, is compared in Figure 6 [10c] with single photon counting data points [multiplied by $\exp(t/\tau_f)$].

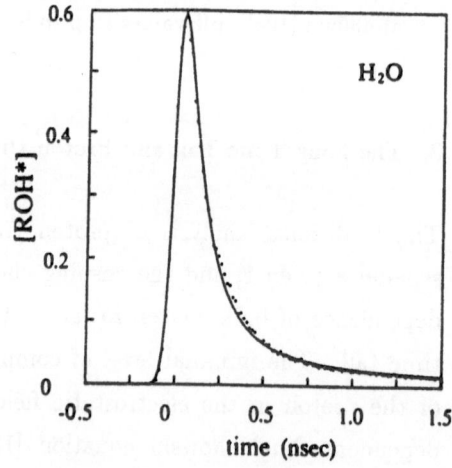

Figure 6. HPTS fluorescence decay.

The solution of the Smoluchowski equation for a Coulomb potential, $-R_D/r$, depends on the "Debye radius", $R_D \equiv |z_1 z_2| / (\epsilon k_B T)$, whose magnitude (nearly $30\,\text{Å}$ for HPTS in water) is determined by the charges, z_1 and z_2, the temperature, T, and the dielectric constant ϵ of the solvent. The relative diffusion coefficient, D, approximately equals the proton diffusion coefficient, which is nearly an order of magnitude faster than any other ion. This is due to the Grotthuss mechanism of proton diffusion by a chain of proton transfers between neighboring water molecules. All the above quantities are experimentally determinable. The only adjustable parameters are the two rate constants for the first kinetic step: The dissociation rate parameter, κ_d, and the recombination rate parameter, κ_r. The effect of some of the above parameters on the probability for observing a bound pair is shown in Figure 7.

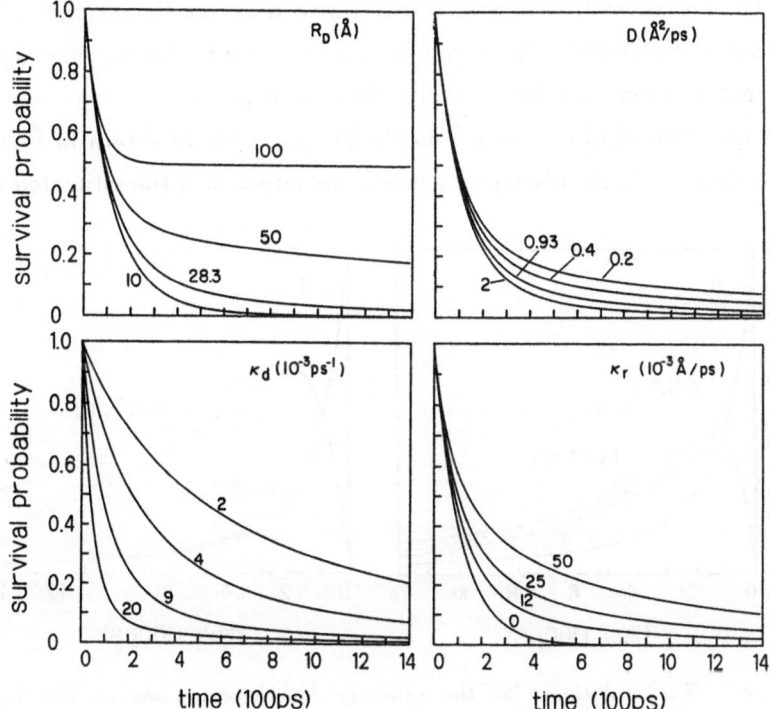

Figure 7. Effect of various parameters on the solution of the spherically symmetric (3d), time-dependent Smoluchowski equation with back reaction [10a]. In each panel one parameter is varied while the others are held constant. In all panels, the third curve from the top is identical and corresponds to the observed HPTS signal, multiplied by $\exp(t/\tau_f)$.

324

The kinetic scheme, eq 1, can be solved analytically. This solution involves one additional rate parameter for the separation of the ion pair. Using the steady-state solution of the Smoluchowski equation, one may express this rate coefficient in terms of R_D and D. As seen in Figure 8a, the resulting solution (dashed curve) does not fit the transient data (full curve). These two curves possess however the same area, thus predicting the same quantum yield [10b]. If the third rate coefficient is treated as an adjustable parameter we obtain the bi-exponential fit shown in Figure 8b [10b]. It agrees with the transient data up to about 1ns but, because it decays exponentially rather than as a power law (see below), it underestimates the area under the curve by about 30%. The relative R^*OH/R^*O^- quantum yield that we obtain from the ratio of the two peaks in the steady state fluorescence spectrum (Figure 2) is 0.043. From the area under the transient decay (divided by the fluorescence lifetime, $\tau_f = 5.2$ ns) we find 0.037. This small discrepancy may be due to some degree of unaccountable quenching, but is still within the experimental error bars. The relative quantum yield obtained from the biexponential fit shown in Figure 8b drops to about 0.03, the discrepancy now being larger than the estimated error.

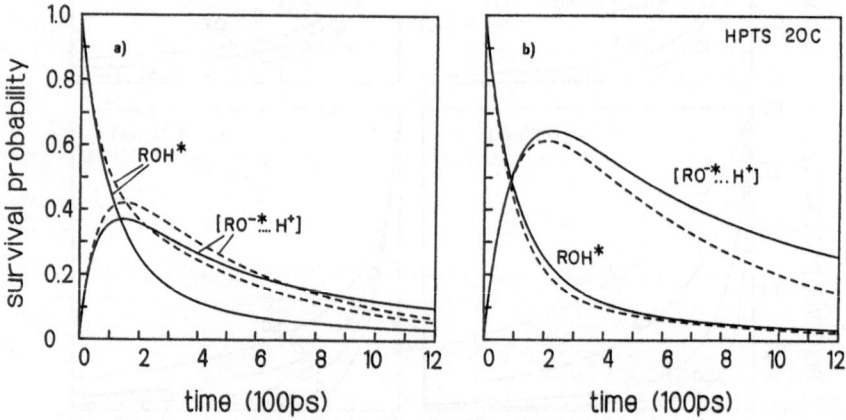

Figure 8. Two solutions for the ordinary kinetic equations of the Eigen mechanism [3], eq 1 (dashed curves), as compared with the exact numerical solution for the Smoluchowski equation (full curves). Both models have the same κ_d and κ_r, but different values for the complex separation rate constant in the kinetic scheme were employed: (a) giving the same area or (b) the same initial transient behavior as compared with the exact solution for the R^*OH decay [10b].

The smallness of the quantum yield predicted from the bi-exponential decay is one indication that the long-time behavior is not exponential. In the diffusional scheme one can show [14] that the asymptotic decay of the bound state should be a power law. In a three dimensional space it is expected to behave as $t^{-3/2}$, with a prefactor which depends on R_D, D, κ_d and κ_r. As Figure 7 demonstrates, this factor is enhanced by increasing R_D and κ_r and diminishes with increasing D and κ_d. To test the asymptotic behavior, the fluoresence signal has been followed [10c] over a large time range using the time correlated single photon counting technique. The transient decay is multiplied by $\exp(t/\tau_f)$, to correct for the finite fluorescence lifetime, τ_f. The data points, shown on a log-log scale in Figure 9, covers nearly three orders of magnitude in intensity. The full curve is the exact numerical solution to the Smoluchowski equation, while the dashed line is the asymptotic $t^{-3/2}$ behavior (a power law becomes a straight line in a log-log scale).

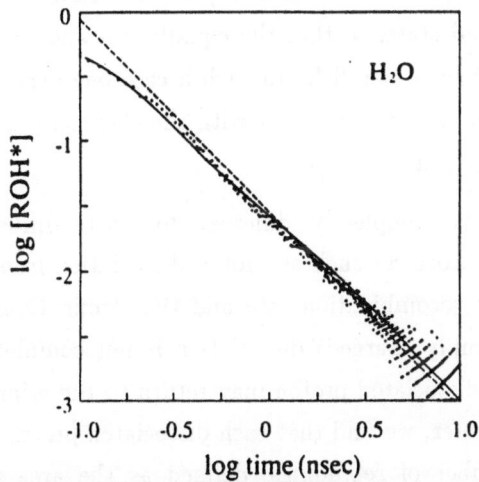

Figure 9. The asymptotic behavior of HPTS fluorescence decay (for an infinite fluorescence lifetime): Same data as in Figure 6, only on a log-log (base 10) plot [10c].

Figure 9 is a direct experimental proof for the inadequacy of the rate equation, as opposed to the diffusional approach. A bi-exponential decay, even when it involves one additional adjustable parameter, fits only up to about 1ns (half of the time scale shown). Approximately at this time the transient data approaches the asymptotic behavior which it subsequently follows for more than an order of magnitude in intensity. We expect the range of this power-law behavior to be limited only by homogeneous recombination with the very small concentration of protons available at the neutral pH values of the experiment.

4. Salt Effects on Steady State and Transient Kinetics

For dissociation reactions in which the reverse, recombination step is slow, dissociation is virtually complete once the pair has separated to the contact distance. Under these conditions the magnitude of the interaction in the dissociated pair, in particular the screening of this interaction by ions, would not affect the dissociation rate. In terms of the Hammond postulate [22] and its extensions [23] an exothermic dissociation process would have its transition state close to the bound state, so that the equilibrium and recombination rate coefficients would change in parallel. In such a case one expects no salt effect on the dissociation reaction, in agreement with the classical picture of Brönsted and Bjerrum for kinetic salt effects [7].

The situation is completely different for fast, diffusion influenced, reversible dissociation processes such as proton dissociation from excited HPTS [11]. Due to the fast recombination rate and the strong Coulomb attraction (HPTS anion is quadruply charged) dissociation is not complete once the OH bond is cleaved: The dissociated proton may return to the origin of its random walk and rebind. In water, we find that each dissociated proton rebinds at least once [10a]. The number of rebindings, defined as the area under the non-exponential tail divided by the area under the initial exponential decay, is determined by the magnitude of the proton/anion attraction. Screening of this interaction by inert salts is expected to decrease geminate recombination and tilt the balance towards the dissociated state. A particularly large effect is expected in low dielectric solvents, where the Coulomb attraction is especially large. As Figure 10 shows, there is a dramatic effect of added salt ($NaNO_3$) on the steady state fluorescence spectrum of HPTS in a 50/50%(v) mixture of methanol/water: With increasing salt concentration the acidic peak (at 432nm in this solvent) decreases while the basic peak (508nm) increases in amplitude. Note also the nice isoemmisive point obtained (482nm), implying that only two states (acidic and basic) are involved and that the salt has practically no quenching effect on the excited HPTS.

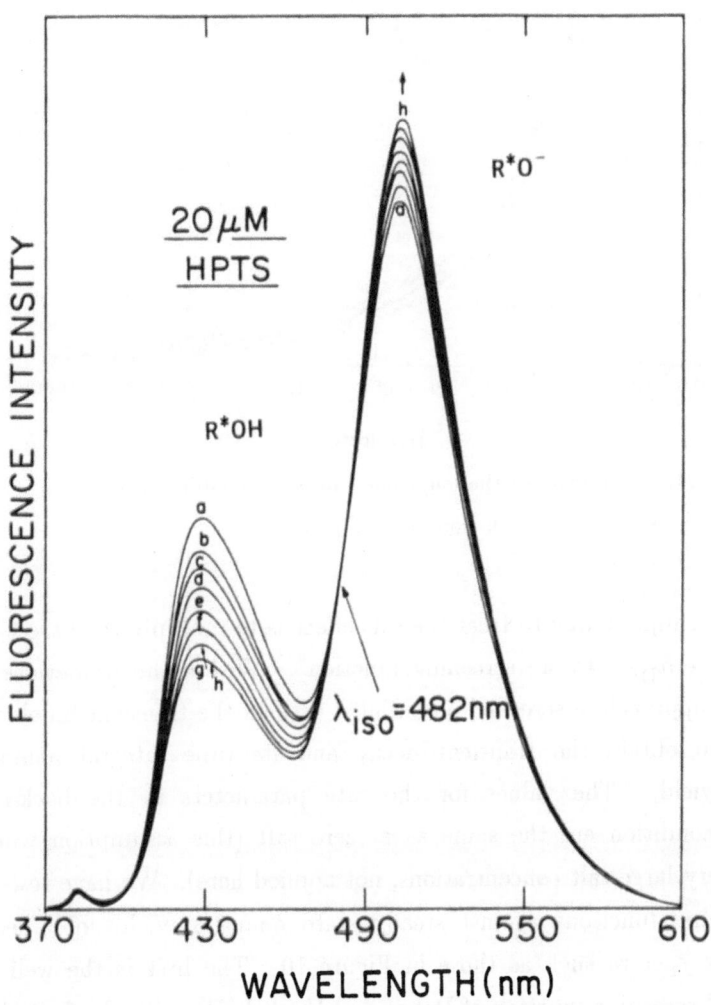

Figure 10. Steady state fluorescence spectra of HPTS in 50/50% (by volume) methanol/water as a function of inert salt ($NaNO_3$) concentration [11c]. Curves a-h correspond to $0, 5, 14, 27, 50, 83, 130$ and $222\,mM$ $NaNO_3$, respectively.

The decrease in the R*OH quantum yield at this salt concentration range is entirely due to the suppression of the non-exponential tail by the screening effect which diminishes the geminate recombination process. This is confirmed by direct time resolved measurements [11b] shown in Figure 11.

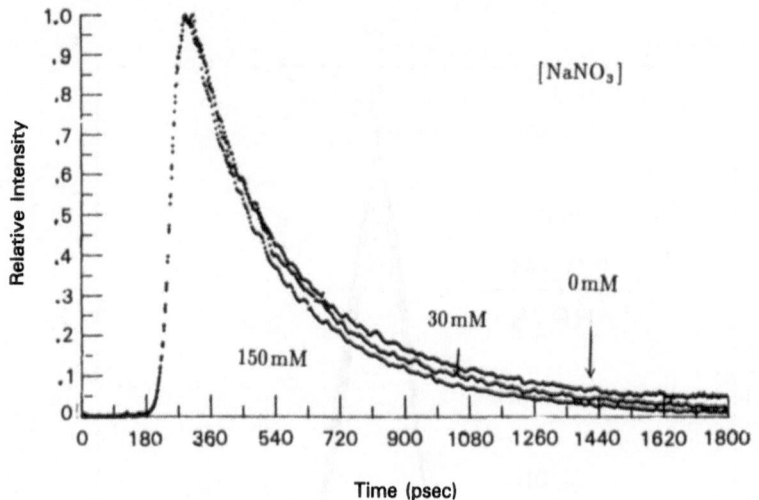

Figure 11. Supression of the long-time tail, by the addition to water of an inert salt (NaNO$_3$). Streak-camera data [11b].

The simplest way to treat the salt effect is to multiply the bare Coulomb attraction, $-R_D/r$, by a "screening function". Within the framework of our diffusional approach, a screened potential is used in the transient Smoluchowski equation to obtain the transient decay and its time integral namely, the quantum yield. The values for the rate parameters in the back-reaction boundary condition are the same as at zero salt (this assumption will break down at very large salt concentrations, not applied here). We have tested [11c] two screening functions against steady state quantum yields obtained from fluorescence spectra such as those in Figure 10. The first is the well known exponential screening function of Debye and Hückel [21], with the finite ion size correction. It is shown as the dashed curves in Figure 12. As concluded earlier from the fit to the ground-state recombination rate coefficient (Figure 3), the Debye-Hückel potential overestimates the screening effect.

The full curves in Figure 12 represent a "Naive Approximation" (NA) [17], which assumes no screening up to the average distance of the nearest cation, R_{av}, and full screening (zero potential) at larger distances. Hence

$$V_{NA}(r) = -(R_D/r - R_D/R_{av}), \qquad r \leq R_{av} \qquad (2)$$

and zero otherwise. Taking $4\pi c R_{av}^3/3 = 1$, where c is a uniform salt concentration, we conclude that the quantum yield in the NA is proportional to $c^{-1/3}$, rather than exhibiting the $c^{-1/2}$ dependence predicted by the Debye-Hückel theory [7, 21]. This agrees with experiment above about 20 mM salt.

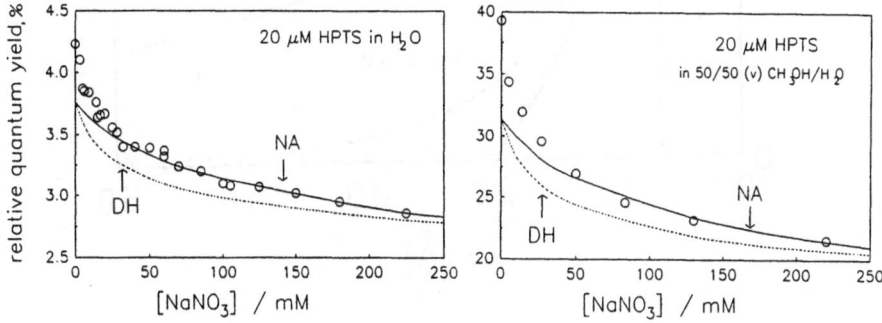

Figure 12. Salt effect on the relative R^*OH/R^*O^- quantum yield in two different solvents: Water (left panel) and a 50/50% (by volume) mixture of methanol/water [11c]. Circles are experimental data obtained from the relative height ratio of the two peaks in the steady-state fluorescence spectrum e.g., Figure 10. Dashed and full curves correspond to the Debye-Hückel expression (with finite ion-size correction) [21] and the Naive Approximation [17, 11c], respectively. Both models employ the zero-salt kinetic parameters.

The difference between the two screened potentials is illustrated by Figure 13. Both predict a remarkably similar value at the contact distance, a. The main difference between the potentials is in their width rather than their depth. The narrower Naive Approximation potential restricts the proton to be closer to the parent anion thus leading to more geminate recombination and larger probabilities for the bound (acidic) state.

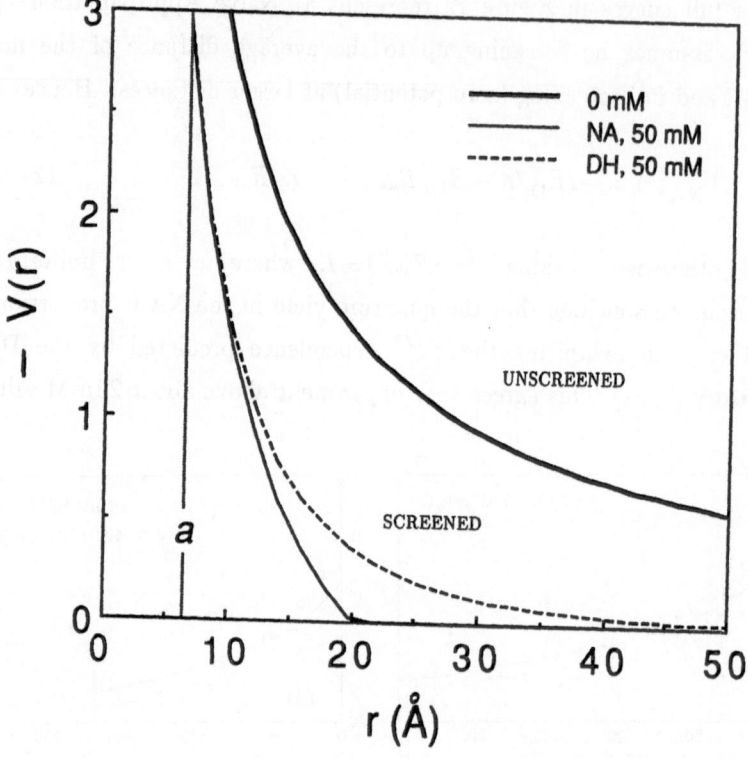

Figure 13. The screened Coulomb potentials [11c]: The Naïve Approximation (full curve) is compared with the Debye-Hückel potential (dashed curve), both at $c=50\,$mM. The unscreened potential (wide full curve) is also shown for comparison.

5. The pH Effect as a Many-Body Problem

As seen in Figure 5 above, the addition of homogeneous protons results in enhanced proton/anion recombination and therefore an enhanced long time tail. Once the geminate proton dissociates, it competes with the homogeneous protons over the binding site. This many body effect may be treated within the framework of reversible diffusion influenced reactions.

The complete analysis involves the solution of a multidimensional diffusion (Smoluchowski) equation in the coordinates of all of the available protons. At high proton concentrations this becomes impractical and one searches for simple but useful approximations. One approximation, using convolution relations derived for isolated pairs, is given in [14]. An even simpler idea is to extend the back-reaction (reversible) boundary condition into a "bimolecular boundary condition", by multiplying the rate coefficient for rebinding, κ_r, by the probability that the binding site is unoccupied by any of the other protons [18]. A comparison of this approximation with one dimensional stochastic simulations [18a] for free diffusion and an initially unbound site is shown in Figure 14. It is seen that the approximation is excellent when dissociation is fast, but deteriorates in the limit of irreversible recombination.

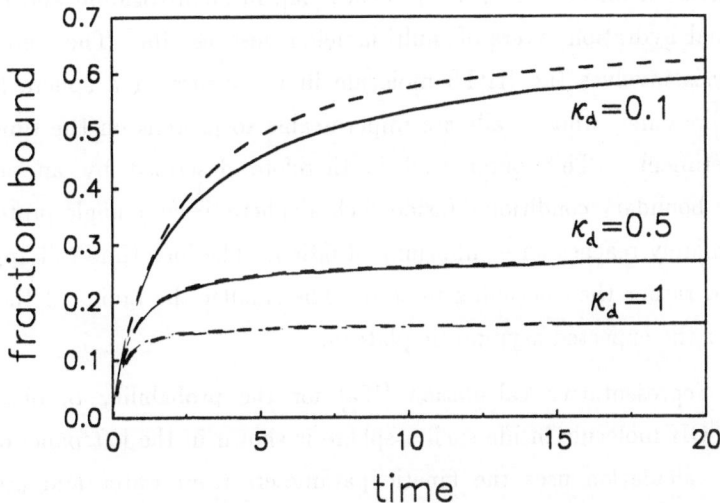

Figure 14. The bimolecular boundary condition for the Smoluchowski equation (dashed curves) compared with an exact simulation (full curves) [18a]. The simulation involves 100,000 realizations of unbiased (zero potential), one dimensional random walks involving 20 walkers on 100 lattice sites. The walkers are initially distributed randomly over all sites except the binding site. At later times one walker (at most) can occupy the binding site. The diffusion rate is equal to the rate of entering the site (κ_r), and both are set to unity. The rate for leaving the site (κ_d) varies.

6. Diffusion in Restricted Geometries

The discussion so far centered on proton diffusion in an infinite space. Hence, a spherically symmetric diffusion (Smoluchowski) equation in three dimensional space has been employed in the data analysis. An inner boundary condition (at the contact distance) has been imposed to describe reaction, but no outer boundary condition. Almost all of the interesting biological applications [4] involve proton diffusion in cavities and restricted geometries. These may include the inner volume of an organelle, the water layers between membranes or pores within a membrane.

HPTS fluorescence has been employed in the study of proton diffusion within inverted micelles [24], in liposomes [4a], in apomyoglobin and the inter-membranal hydration layers of multi-lamellar vesicles [4b]. The simplest case for analysis involves the HPTS molecule in the center of a sphere (inverted micelle, liposome) whose walls are impermeable to protons on the timescale of the experiment. This outer wall is therefore described by an additional reflective boundary condition. Inside such a sphere, even a single proton/anion pair ultimately reaches an equilibrium situation: The long-time tail approaches a plateau, rather than decaying to zero. The smaller the radius of the sphere, the higher the expected asymptotic plateau.

A representative calculation [12a] for the probability of observing a bound R^*OH molecule inside such a sphere is shown in the left panel of Figure 15. the calculation uses the kinetic parameters from water and assumes a sphere radius of $250\overset{0}{A}$. The right panel shows a preliminary experiment [4a] monitoring HPTS inside liposomes (curve labelled B) as compared with its fluorescence trace in pure water (curve A). It is seen that the long-time tail is indeed enhanced when the fluorophor is located inside a liposome. A more quantitative determination of liposome size distribution will allow a quantitative comparison between experiment and theory.

It is important to find good model systems to test the applicability of the theory in restricted geometries. Once the method is calibrated against the radius (or volume) of the cavity, the HPTS fluorescence decay trace could be used as a "microscopic ruler" to determine cavity size [12a].

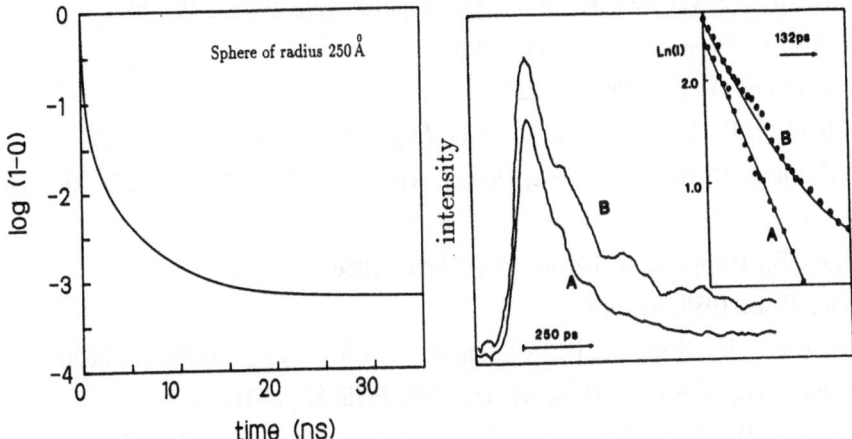

Figure 15. Reversiblle proton dissociation inside a finite cavity. Left panel [12a] is a representative calculation for the probability of observing a bound pair (i.e., assuming an infinite radiative lifetime, τ_f). Right panel [4a] compares transient HPTS fluorescence in water (trace A) with its signal when located inside a liposome (trace B). The insert shows the same data on a semi-logarithmic scale.

7. Acknowledgements

I am indebted to my colleagues and collaborators: A. Blumen, D. Huppert, E. Pines and A. Szabo, whose talents made this exposition possible. This work is supported in part by grant number 86-00197 from the US-Israel Binational Science Foundation (BSF), Jerusalem, Israel. The Fritz Haber Research Center is supported by the Minerva Gesellschaft für die Forschung, München, BRD.

8. References

[1] Bell, R. P. *The Proton in Chemistry*, 2nd ed.; Chapman and Hall: London, 1973.

[2] Caldin, E. F.; Gold, V. *Proton Transfer Reactions*; Chapman and Hall: London, 1975.

[3] Eigen, M. *Angew. Chem. Int. Ed.* 1964, 3, 1.

[4] (a) Gutman, M. *Meth. Biochem. Anal.* 1984, 13, 1.

334

(b) Gutman, M.; Nachliel, E. *Biochim. Biophys. Acta* 1990, 1015, 391.

[5] Weller, A. *Prog. React. Kinet.* 1961, 1, 189.

Z. Phys. Chem. N. F. 1958, 17, 224.

[6] Ireland, J. F.; Wyatt, P. A. H. *Adv. Phys. Org. Chem.* 1976, 12, 139.

[7] Perlmutter-Hayman, B. *Prog. React. Kinet.* 1972, 6, 239 and refs. therein.

[8] Kosower, E. M.; Huppert, D. *Ann. Rev. Phys. Chem.* 1986, 37, 127.

[9] Pines, E.; Huppert, D. *Chem. Phys. Lett.* 1986, 126, 88;

J. Chem. Phys. 1986, 84, 3576.

[10] (a) Pines, E.; Huppert, D.; Agmon, N. *J. Chem. Phys.* 1988, 88, 5620.

(b) Agmon, N.; Pines, E.; Huppert, D. *ibid.* 1988, 88, 5631.

(c) Huppert, D.; Pines, E.; Agmon, N. *J. Opt. Soc. Amer. B* 1990, 7, 1545.

[11] (a) Pines, E.; Huppert, D. *J. Am. Chem. Soc.* 1989, 111, 4096.

(b) Pines, E.; Huppert, D.; Agmon, N. In *Ultrafast Phenomena VI, Springer Verlag Series Chem. Phys.* 1988, 48, 517. *(c) J. Phys. Chem.* 1990, 95, xxxx.

[12] (a) Agmon, N. *J. Chem. Phys.* 1988, 88, 5639. (b) *ibid.* 1988, 89, 1524.

[13] Pines, E. *Ph. D. Thesis;* Tel-Aviv University: Tel-Aviv, 1989.

[14] Agmon, N.; Szabo, A. *J. Chem. Phys.* 1990, 92, 5270.

[15] Rice, S. A. *Diffusion Limited Reactions,* In *Comprehensive Chemical Kinetics;* Bamford, C. H.; Tipper, C. F. H.; Compton, R. G., Eds.; Elsevier: Amsterdam, 1985; vol. 25.

[16] Gösele, U. M. *Prog. React. Kinet.* 1984, 13, 63.

[17] Agmon, N. *Chem. Phys. Lett.* 1987, 141, 122.

[18] (a) Schnörer, H.; Blumen, A.; Agmon, N., in preparation.

(b) Szabo, A.; Zwanzig, R., unpublished.

[19] Förster, Th.; Völker, S. *Chem. Phys. Lett.* 1975, 34, 1.

[20] Hauser, M.; Haar, H.-P.; Klein, U. K. A. *Ber. Bunsenges. Phys. Chem.* 1977, 81, 27.

Haar, H.-P.; Klein, U. K. A.; Hauser, M. *Chem. Phys. Lett.* 1978, 58, 525.

[21] Debye, P.; Hückel, E. *Physik. Z.* 1923, 24, 185.

[22] Hammond, G. S. *J. Am. Chem. Soc.* 1955, 77, 334.

[23] Agmon, N. *J. Chem. Soc. Faraday II* 1978, 74, 388.

Int. J. Chem. Kinet. 1981, 13, 333.

[24] Bardez, E.; Goguillon, B.-T.; Keh, E.; Valeur, B. *J. Phys. Chem.* 1984, 88, 1909.

AN EXPLORATORY STUDY TO CORRELATE EXPERIMENTAL AND THEORETICAL ACIDITIES OF ORGANIC MOLECULES

C. ÖGRETIR and N. KANISKAN
Chemistry Department
Faculty of Arts and Sciences
Anadolu University
Eskisehir, Turkey

ABSTRACT. After a brief survey of the fundamental concept, a brief account of acidity and basicity and the application of Molecular Orbital Calculations to heteroaromatic compounds will be reviewed. Subsequently, correlations between experimentally obtained acidity constants (i.e. pK_a values) and *ab initio* relative energy values will be examined. Some representative molecules will be used to demonstrate the "intramolecular" and "intermolecular" correlations.

1. Introduction

The ionization constants of acids (K_a) and bases (K_b) are a measure of relative tendencies to lose or gain protons in solutions. Hence, they enable us to arrange a series of compounds in order of ascending acidic strength.

In practice one prefers to use the pK_a value instead of ionization constant, K_a (1a). In 1923 Brönsted extended the use of acidic ionization constants to bases defining K_b as well as pK_b (1b). Consequently, we may infact be able to use pK_a values to measure basicities of various nitrogen atoms in complex heterocyclic molecules.

$$K_a = \frac{[H^+][A^-]}{[HA]} \qquad pK_a = -\log_{10} K_a \qquad (1a)$$

$$K_b = \frac{[H^+][B]}{[HB^+]} \qquad pK_b = -\log_{10} K_b \qquad (1b)$$

$$pK_a + pK_b = K_w = 14 \qquad (1c)$$

In heterocyclic chemistry, this kind of information has many useful applications. We may outline some examples as follows:

(i) In order to ensure that one has either the conjugate acid or the conjugate base in a desired concentration range one must do the necessary calculations using the pK_a value of the compound. For example take the quinoline system with a pK_a value of 4.94.

S. J. Formosinho et al. (eds.), Theoretical and Computational Models for Organic Chemistry, 335–353.
© 1991 *Kluwer Academic Publishers*.

$$K_a = 1.15 \times 10^{-5}$$

conjugate acid + H$^+$

conjugate base

The following equation allows one to calculate the concentration of the conjugate base at varying pH values.

$$\% \text{ conjugate base} = \frac{100}{1 + \text{antilog}(pK_a - pH)}$$

The following Table indicates what these percentage values are at 1 or 2 pH units up and down of the pK_a value of quinoline.

	pH=pK$_a$-2	pH=pK$_a$-1	pH=pK$_a$	pH=pK$_a$+1	pH=pK$_a$+2
% conjugate base	1.00	9.10	50.00	90.90	99.00
% conjugate acid	99.00	90.90	50.00	9.09	0.99

Consequently, with suitably chosen buffers one can guaranty a high percentage of one ionic form or the other.

(ii) The determination of UV spectra is one of the most important component of the determination of ionization constants (K_a). Different ionic species such as A and AH$^+$ can have different spectra. Therefore, a spectroscopic study must be carried out before ionization constant can be determined (Figure 1).

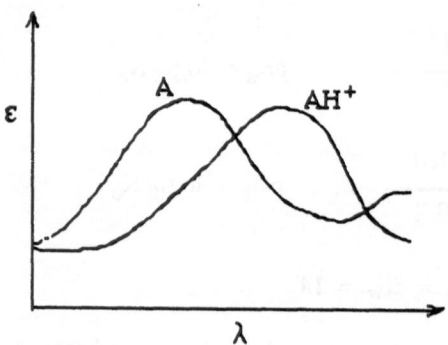

Figure 1. A schematic illustration of different ionic species with different electronic spectra.

Thereafter, buffers that are suitable for isolating each ionic species in spectroscopically pure form can be chosen.

(iii) Another important application involves the determination of the position of tautomeric equilibria.

Basicity measurements have been extensively used to determine the predominant structure of potentially tautomeric compounds. If pK_a and pK_a' are the values for the individual forms, the equilibrium constants ($K_{a\ taut}$) for the tautomerization can be obtained from Eqs. 2a and 2b.

$$pK_{a\ taut} = pK_a - pK_a' \tag{2a}$$

$$K_{a\ taut} = 10^{-pK_{a\ taut}} \tag{2b}$$

Since the alkylation does not greatly alter the pK_a or pK_a' values, the approximate values for tautomerization constants can be determined through partial methylation (Scheme 1).

SCHEME 1

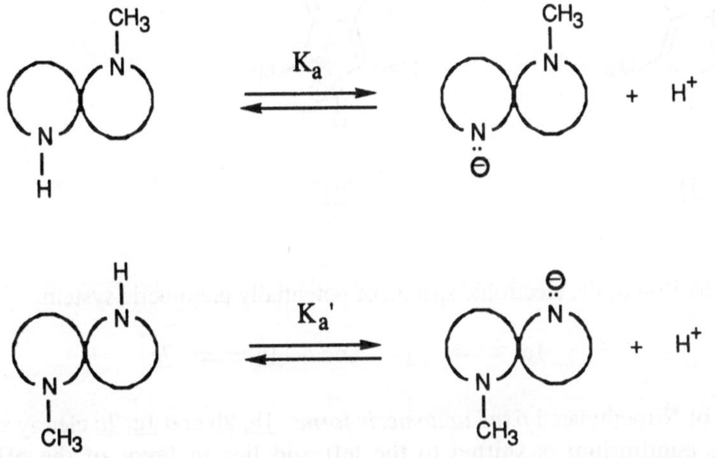

Let us consider the following system:

(a)	R = R′ = H .
(b)	R = H, R′ = Me
(c)	R = Me, R′ = H
(d)	R = R′ = Me

Comparison of the electronic spectra of potentially tautomeric systems

1a ⇌ 2a   and   1c ⇌ 2c

with those of N-methylated *fixed tautomeric forms*, 1b, 2b and 1c, 2c clearly show that the tautomeric equilibrium is shifted to the left and lies in favor of the 6H structures. Measurement of the pK_a values for the monoprotonated derivatives of the *fixed tautomeric form* model compounds 1b, 2b as well as 1d, 2d (Table 1) provided a quantitative measure for the equilibrium and established that for both 5,7-dimethyl (a) and the 1,4,5,7-tetramethylpyrrolo[3,4-d]pyridazines (c) the 6H tautomeric form is favored to the extend of *ca* 540:1, compared with 2H structure which indicated that the 6H structure is more stable than the 2H structure [1].

TABLE 1. pK$_a$ Values for Various Protonated
Pyrrole[3,4-d]pyridazines.

compound	formation of monocation
1a $\rightleftharpoons$ 2a	7.18 ± 0.03*
1b	7.06 ± 0.03
2b	9.79 ± 0.10
1c $\rightleftharpoons$ 2c	7.75 ± 0.02*
1d	7.02 ± 0.07
2d	10.60 ± 0.09

* "average" value measured on the tautomeric mixture.

2. MO Theory and Heterocyclic Systems

There are several simple introductions to the molecular orbital treatment of aromatic molecules [2], but a brief account is included here, mainly to outline the special problems associated with heteroaromatic molecules.

In the simpler types of molecular orbital approximation, the π-electrons are assumed to move independently in molecular orbitals that can be represented as linear combinations of the atomic π-orbitals. When these molecular orbitals are used as the basis for an approximate calculation, the energy of molecular orbitals and the distribution of the π-electron in each depend on the values of certain integrals. These integrals are of two types, termed Coulomb and Resonance integrals: both have the dimensions of energy and their signs are negative indicating stabilization. The Coulomb integrals are characteristic of an atomic π-orbital in a given molecular environment and are a measure of the effective electronegativity of that atom towards π-electrons. The resonance integrals are characteristic of a π-bond between two adjacent atoms and are a measure of the stability that a localized π-bond would have if formed between them. Neither the Coulomb integrals nor the Resonance integrals are usually evaluated by integration; the final calculations are based on certain simplifying assumptions concerning the relative values of the different Coulomb integrals and the relative values of different Resonance integrals.

The unsubstituted aromatic hydrocarbons offer the simplest subjects for such calculations since all the carbon atoms are usually assigned the same Coulomb integral (α) and all C-C bonds are usually assigned the same Resonance integral (ß). The π-electron energy is then a function of α and ß; for benzene, for example, this energy is $6\alpha + 8ß$.

In extending this treatment to heteroaromatic molecules, the modification concerns appropriate Coulomb integral for the heteroatom. It is convenient to express this Coulomb integral (α_x) in terms of α and ß, the standard integrals associated with the aromatic hydrocarbons. Most of the calculations have involved nitrogen compounds and there has been considerable variation in the Coulomb integral used for the neutral nitrogen atom such as in pyridine; the values used have ranged from $\alpha + 2ß$ to $\alpha + 0.2ß$, the lower values being the more recent [3].

Another problem concerns the Coulomb integral (α_c) of the carbon atom adjacent to the heteroatom. These carbon atoms have been considered to be more negative than the carbon atoms in benzene because of the inductive effect of the heteroatom as realized through the σ-bonds. This has been accommodated in the calculation by making the appropriate Coulomb integrals more negative than α. It is also convenient to express the value of α_c in terms α and ß and so the Coulomb integrals used for the heteroatom and for the adjacent carbon atoms may be defined by two parameters h and h´ according to Eqs. 3 and 4:

Heteroatom	$\alpha_x = \alpha + hß$	(3)
Adjacent carbon atoms	$\alpha_c = \alpha + h´ß$	(4)

The Resonance integral of the π-bond between the heteroatom and carbon is another possible parameter in the treatment of heteroaromatic molecules. The carbon-heteroatom Resonance integral $ß_{C-X}$ can also be expressed in terms of ß of the C-C bond (Eq. 5):

$$ß_{C-X} = kß \qquad\qquad (5)$$

For nitrogen compounds the uncertainty in choosing k is probably unimportant, partly because more detailed calculations have suggested that this Resonance integral is very similar to that for a C-C bond [4] (i.e. k is not very different from unity) and partly because the relative values for the reactivity indicies at different positions are not very sensitive to change in this parameter.

The semi-empirical quantum-mechanical treatment of the structure and reactivity of homo- and heteroatomic systems in the last 30 years or so has, by making appropriate assumptions and simplifications, been giving moderate to good correlations with measured experimental parameters data such as

1. Bond length
2. Dipole moment
3. UV absorption

and using computed

4. *ab initio* data

Unfortunately, the most widely quoted parameters, the calculated π-electron densities, have only very recently become related to experimentally measured quantities. Of course, values obtained by calculation have naturally varied according to the details of the theory used and the assumptions made. Although values consistent with chemical reactivity have been obtained in most cases, this has not been so in several other cases. For a time, π-electron density was believed to be linked with reactivity to electrophilic and nucleophilic reagents, but as it became clear that this is not necessarily so, a search for a better analysis of reactivity was made which led to calculations of polarizability, frontier electron density of *atom localization* energy.

Indeed, in the discussions of reactivity, qualitative estimates of the relative stabilities of reaction intermediates can be employed as models for the relative stabilities of transition states leading to them.

Calculated values of π-electron density obtained by the method of atom-localization energy, are still subject to some error and changes in accord with the chosen method, as it can be seen from Table 2.

TABLE 2. Calculated π-Electron Densities for Pyridine.

N	2	3	4	method	year
1.14	0.98	0.98	0.95	HMO	1958
1.20	0.92	1.00	0.95	HMO	1959
1.01	1.005	1.002	0.975	*ab initio*	1967
1.14	0.98	0.98	0.95	*ab initio*	1968

It is probably fair to say that MO treatment of systems with heteroatoms is still somewhat crude and beset with difficulties. It is very important to make clear that which method is being used in a calculation and what assumptions are being associated with the chosen method.

3. Correlation Attempts

In principle we may compare the relative basicity of two nitrogen atoms located at two sites of the same molecule (Intramolecular Correlation) as well as the basicity of two nitrogen atoms located in two different molecules (Intermolecular Correlation). These correlations can be achieved both at the *semi-empirical* and at the *ab initio* level of theory.

3.1. SEMI-EMPIRICAL APPROACH

The early work of Conant and Wheland [5] provided an order of relative acidities for a number of hydrocarbons and Wheland [6] showed that this order was consistent with HMO theory. Wheland made an assumption that σ-bond energy changes were constant for a series of compounds and he determined the acidity differences by the π-energy changes, $\Delta E_\pi = E_\pi(ArCH_2^-) - E_\pi(Ar-H)$ for the proton donation process, $ArCH_3 \longrightarrow ArCH_2^-$.

Shorly after Mc Ewen [7] derived a list of approximate pK_a values for a series of hydrocarbons, relative orders of acidity of other hydrocarbons have been derived from metalation reactions by Morton [8] and Gilman [9]. Shatenshtein [10] used the rate of base-catalysed proton abstraction as a measure of acidity and found a reasonable correlation between the Mc Ewen's pK_a values and logk for the exchange the acidic hydrogens in liquid deuteroammonia. With such a correlation it has been established that kinetic acidity is in good agreement with thermodynamic acidity. Streitwieser [11] demonstrated that a plot of the estimated pK_a values against the ΔE_π values calculated from HMO theory shows a good linear correlations (Figure 2) for the carbanions that are expected to be coplanar.

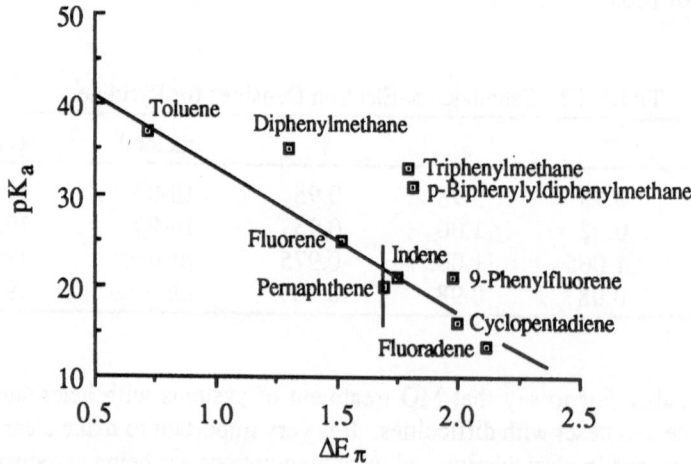

Figure 2. Correlation of hydrocarbon acidity with computed ΔE_π values.

One of the very first attempts to apply MO theory to heterocyclic compounds was done by Longuet-Higgins [12]. He has shown how simple approximation techniques give a satisfactory accounts of observing acidities and basicities. What he has done actually is that taking into account the π-energy difference between the neutral and the protonated molecule such as quinoline (3) and quinolinium (4) which are both derived from naphtalene.

As a first approximation if the only change is that of Coulomb integral of one carbon, r, then it may be thought of that the π-energy change for the equilibrium is going to be as

$$\Delta E_\pi \cong q_r (h_{N^+} - h_{N^.}) \beta \qquad (6)$$

The data in Table 3 show that simple heterocyclic compounds differ in basicity only by 0.66 pK_a units.

TABLE 3. Basicity of Some Nitrogen Containing Bases [11].

compound	pKa (aqueous)
pyridine	5.23
quinoline	4.94
acridine	5.60
isoquinoline	5.14

Consequently it has been suggested that if Eq. 6 is a reasonable approximation then a plot of pK_a versus electron density, q_r, in which r is the ring carbon replaced by nitrogen should give a linear correlation [11]. That point was indeed confirmed in our own studies and a nice correlation between pK_a values of some substituted imidazo[4,5-f]quinolines (5a) and the electron densities calculated by HMO Programme (Table 4 and Figure 3) was obtained. The only point which scatters considerably from the linearity in the case of deprotonation is that 2-pyridyl and that is probably due to the possibility of intramolecular hydrogen bonding.

TABLE 4. π-Electron Densities (q_N) and Acidity Constants (pK_a) of 2-Substituted 3H-Imidazo[4,5-f]quinoline (5a) for the First Protonation and Deprotonation.

5a

molecule	R	q_{N6}	pK_{a_1} (prot. at N6)
I	hydrogen	0.9484	6.83
II	phenyl	1.0649	4.97
III	4-methylphenyl	0.9297	5.24
IV	4-methoxyphenyl	0.9282	6.05
V	2-pyridyl	1.2118	1.88
VI	3-pyridyl	1.0668	3.28
VII	4-pyridyl	1.0388	4.50

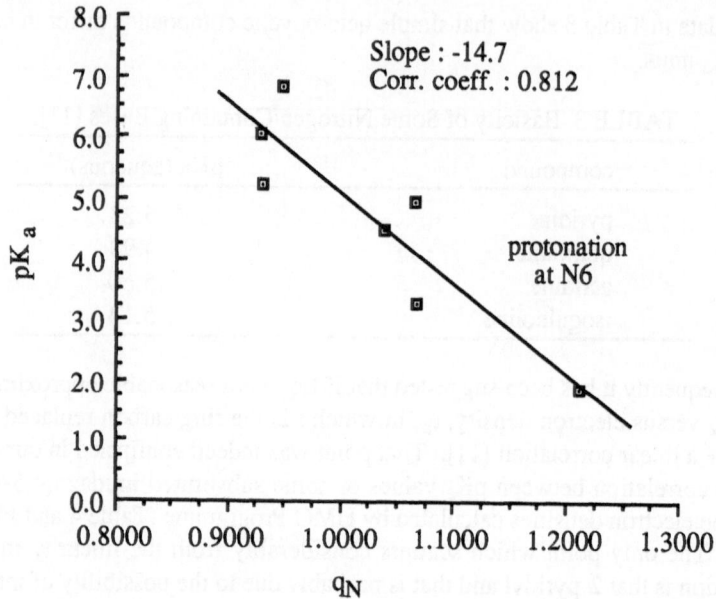

Figure 3. The graph of pK_a versus to q_N for 2-substituted 3H-imidazo[4,5-f]quinolines.

3.2. AB INITIO APPROACH

We are now in a position to extend these studies to several representative heteroaromatic compounds involving the more vigorous *ab initio* MO theory.

(i) Intramolecular correlations can be important in those molecules which contains competative heteroatoms for protonation. Imidazoquinolines (5) are good examples for this group. The availibility of two *aza* nitrogens in those molecules brings the question of which ring is the first for protonation.

5b

Among the other possible forms we have concentrated our work primarily on imidazo[4,5-f]quinolines starting from the synthesis of 5a. The acidity constants of those studied molecules were determined using spectrophotometric technique [13].

By using *ab initio* programmes it is possible to calculate the favorable form in terms of energies and electron densities. In our work using Monstergauss programme at STO-3G level without geometry optimization we demonstrated that the first protonation of imidazo[4,5-f]quinolines (5b) takes place at N6 which is located in pyridine ring. That conclusion came out from the calculated energies. As it can be seen from the Figure 4 the energy difference between the two alternative protonation is found to be 8.75 kcal/mol. The point that the first protonation takes place in pyridine ring has been also confirmed by UV data of our recent work and in a previous work [14]. The similarity of pK_a values of molecule 6 to 7 and molecule 8 to 9 suggest that the first protonation indeed takes place at N6 which is located in pyridine ring.

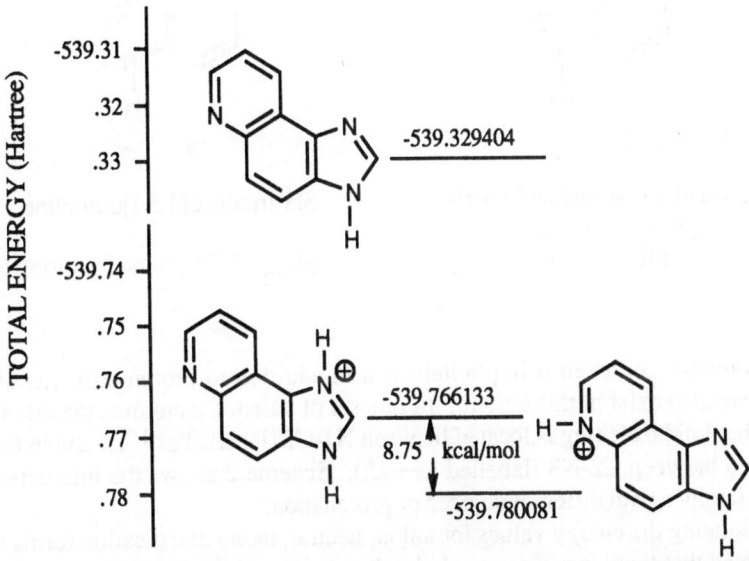

Figure 4. Energy values for neutral and two alternative protonation forms of imidazo[4,5-f]quinoline.

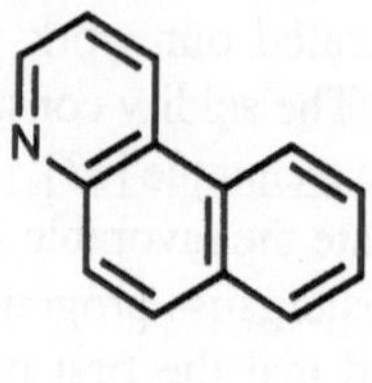

6

[5,6]Benzoquinoline

$pK_a = 5.5$

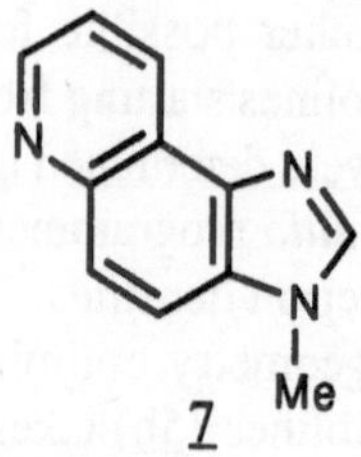

7

3-Methylimidazo[4,5-f]quinoline

$pK_a = 4.55$

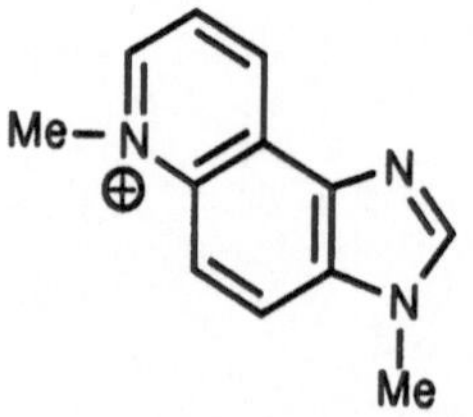

8

3,6-Dimethylimidazo[4,5-f]-
quinolinium cation
$pK_a = 1.86$

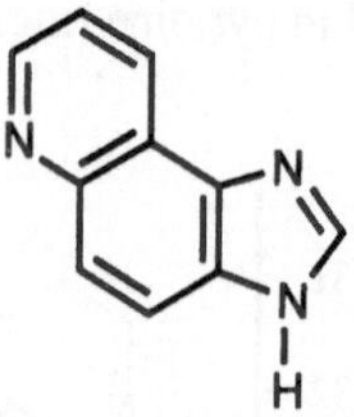

9

3H-Imidazo[4,5-f]quinoline

$pK_{a_2} = 2.83$ (second protonation)

It is recognized even if implicitly that in addition to proton tautomerism valence tautomerism also exist in this system. In one set of valence tautomers the double bond in five member imidazole ring is located between N1-C2 (labelled as $^{1,2}\Delta$) and in the other set it is located between C2-N3 (labelled as $^{2,3}\Delta$). Scheme 2 shows the interdependence of these two families with different degrees of protonation.

Calculating the energy values for anion, neutral, mono and dication forms it was also demonstrated that there is a nice correlation between computed gas phase proton affinities and pK_a values determined in aqueous solution for imidazo[4,5-f]quinolines (Table 5 and Figure 5).

SCHEME 2

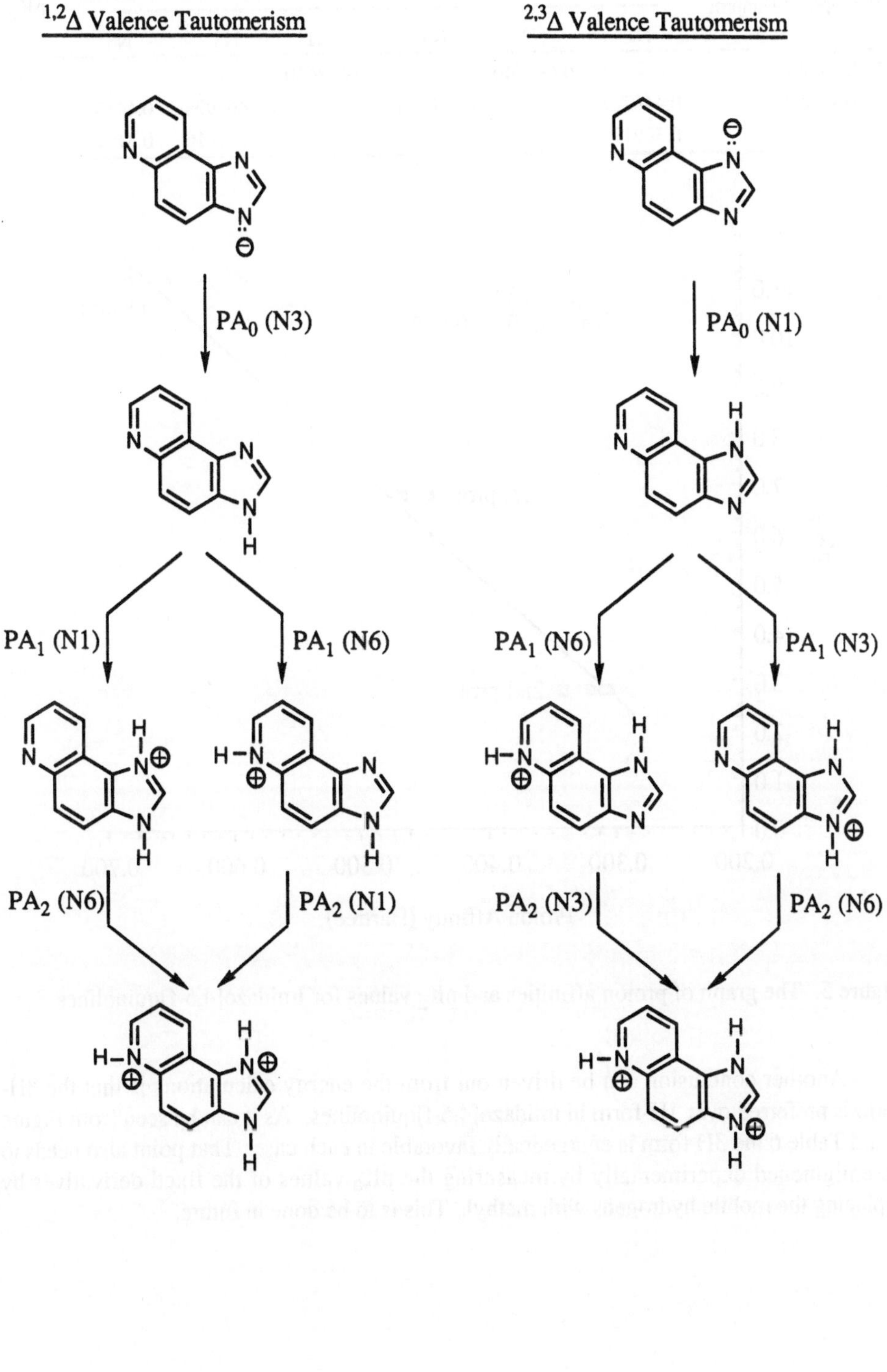

$^{1,2}\Delta$ Valence Tautomerism $^{2,3}\Delta$ Valence Tautomerism

348

TABLE 5. Calculated Proton Affinities and pK_a Values for Imidazo[4,5-f]quinolines.

$-\Delta E = E_{prot} - E_{unprot}$	$1,2\Delta$ or 3H			$2,3\Delta$ or 1H			pK_a
	N1	N3	N6	N1	N3	N6	
PA_0 (deprot.)	-	0.683740	-	0.653630	-	-	11.17
PA_1 (first prot.)	0.446729	-	0.450677	-	0.467998	0.445379	6.83
PA_2 (second prot.)	0.309190	-	0.305240	-	0.341810	0.319190	2.83

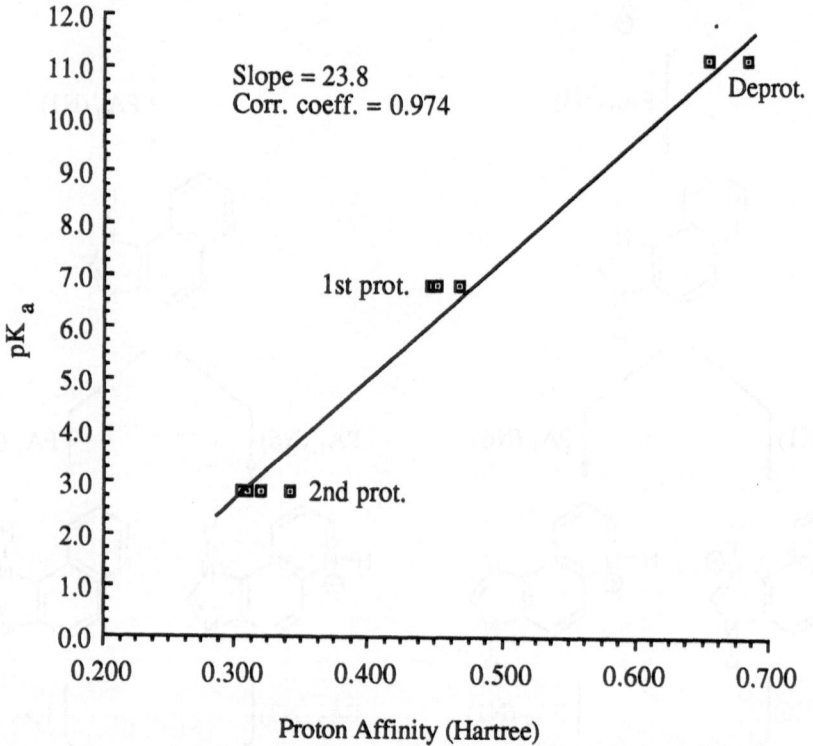

Figure 5. The graph of proton affinities and pK_a values for Imidazo[4,5-f]quinolines.

Another conclusion can be driven out from the energy calculations is that the 3H-form is preferred over 1H-form in imidazo[4,5-f]quinolines. As it can be seen from Figure 6 and Table 6 the 3H form is energetically favorable in each case. That point also needs to be enlightened experimentally by measuring the pK_a values of the fixed derivatives by replacing the mobile hydrogens with methyl. This is to be done in future.

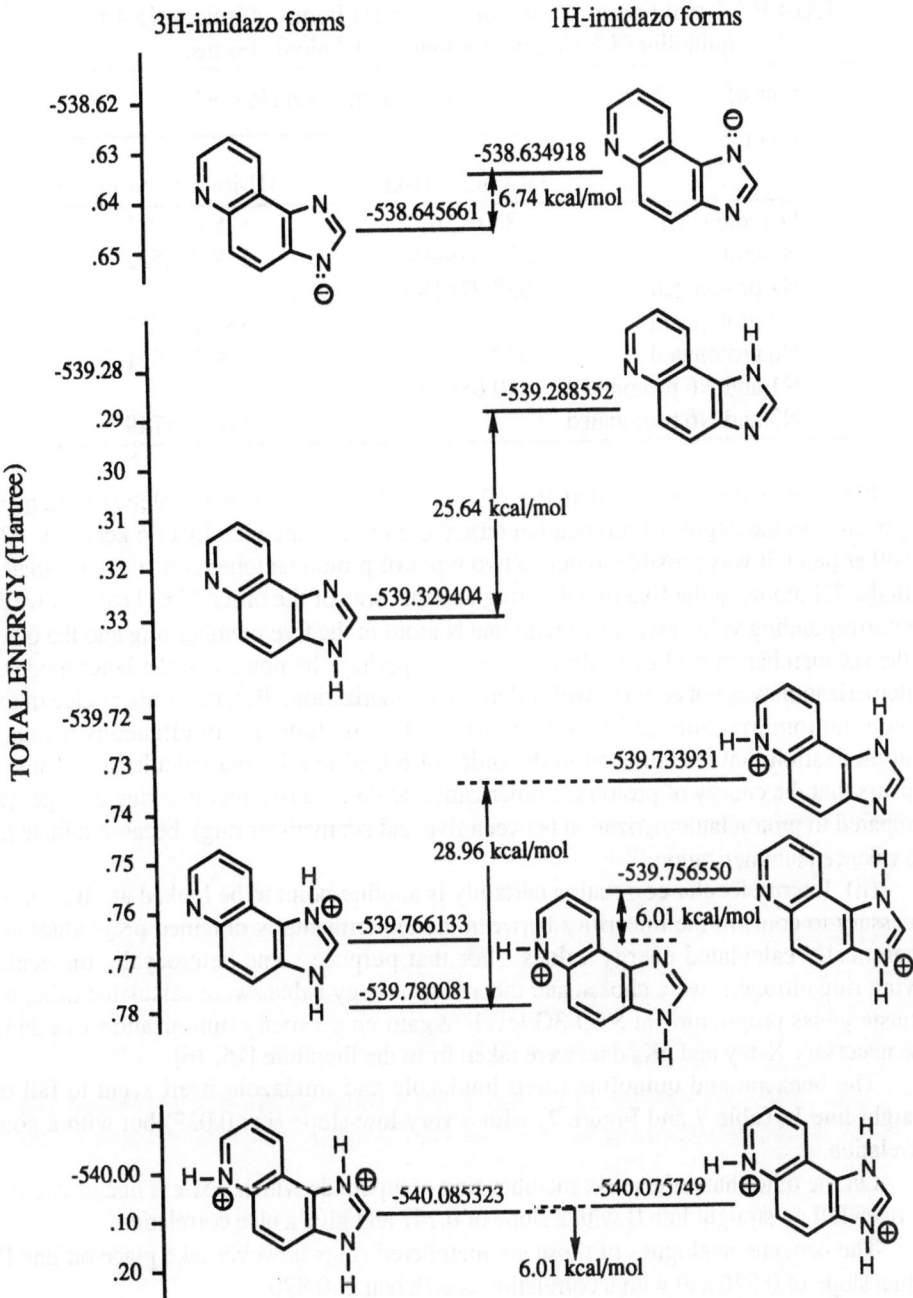

Figure 6. Energies and structure for 3H imidazo forms, 1H imidazo forms tautomers and their deprotonated and protonated forms.

TABLE 6. Total Energy Values of 3H and 1H Forms of Imidazo[3,4-f]-
quinolines for Neutral, Cationic and Anionic Forms.

type of process	total energies (in Hartree)	
	3H-imidazo form	1H-imidazo form
Deprotonated	-538.645661	-538.634918
Neutral	-539.329404	-539.288552
N1 protonated	-539.776133	-
N3 protonated	-	-539.756550
N6 protonated	-539.780081	-539.733931
N1 and N6 protonated	-540.085323	-
N3 and N6 protonated	-	-540.075749

It is interesting to note that the relative stability of the two valence tautomers irrespective to the degree of protonation turned out to be in the vicinity of 6 kcal/mol. On the other hand, it was possible to notice two types of proton tautomerism. One associated with the 2N atoms in the five member ring and that was of the order 27 ± 3 kcal/mol while the corresponding value associated with one N atom in the five member ring and the other in the six member ring 11 ± 3 kcal/mol. It should perhaps be noted that the latter mode of tautomerization was not coupled with valence tautomerization. But, the former value of the proton tautomerization (27 ± 3 kcal/mol) should include a simultaneous valence tautomerization that is estimated in the order of 6 kcal/mol by our calculation. Thus, it appears that the energy of proton tautomerization within the five member ring is larger (as compared to proton tautomerization between five and six member ring), because it includes the valence tautomerization.

(ii) Intermolecular correlation certainly is another point to be looked at. It is infact necessary to confirm the linearities between the experimentally obtained pK_a values and theoretically calculated energy values. For that purpose some heterocyclic molecules having ring nitrogens were chosen and theoretical energy values were calculated using the Monstergauss programme at STO-3G level. Again no geometry optimization was done. The necessary X-ray and pK_a data were taken from the literature [15, 16].

The benzene and quinoline fused imidazole and imidazole itself seem to fall on straight line I (Table 7 and Figure 7) with a very low slope (i.e. 0.039) but with a good correlation.

On the other hand those six member ring compounds which possess one or two *aza* nitrogen fall on straight line II with a slope of 0.247 and give a nice correlation.

The benzene analogues of those six membered rings however take place on line III with a slope of 0.546 and with a correlation coefficient of 0.990.

TABLE 7. Intermolecular Correlations Between pK_a and ΔE
Values of Some Heterocyclic Compounds.

compound	$-\Delta E = E_{prot.} - E_{neut.}$ kcal/mol	pK_a
imidazole	279.95	6.95
imidazo[4,5-f]quinoline	274.05	6.83
benzimidazole	242.65	5.53
pyridine	272.77	5.23
pyridazine	261.79	2.33
pyrimidine	257.06	1.30
pyrazine	253.96	0.60
cinnoline	275.24	2.70
quinazoline	272.68	1.50
quinoxaline	271.70	0.70

4. Conclusions and Future Work

To conclude it seems that it would be possible to have nice correlations between experimentally obtained pK_a values and theoretical calculated proton affinities both intramolecularly and intermolecularly.

A more detailed study on the correlation attempts for a great number of molecules including substituents and other heteroatoms will be done. Geometry optimizations of these compounds will be carried out using *semi-empirical* and/or *ab initio* programmes in the future.

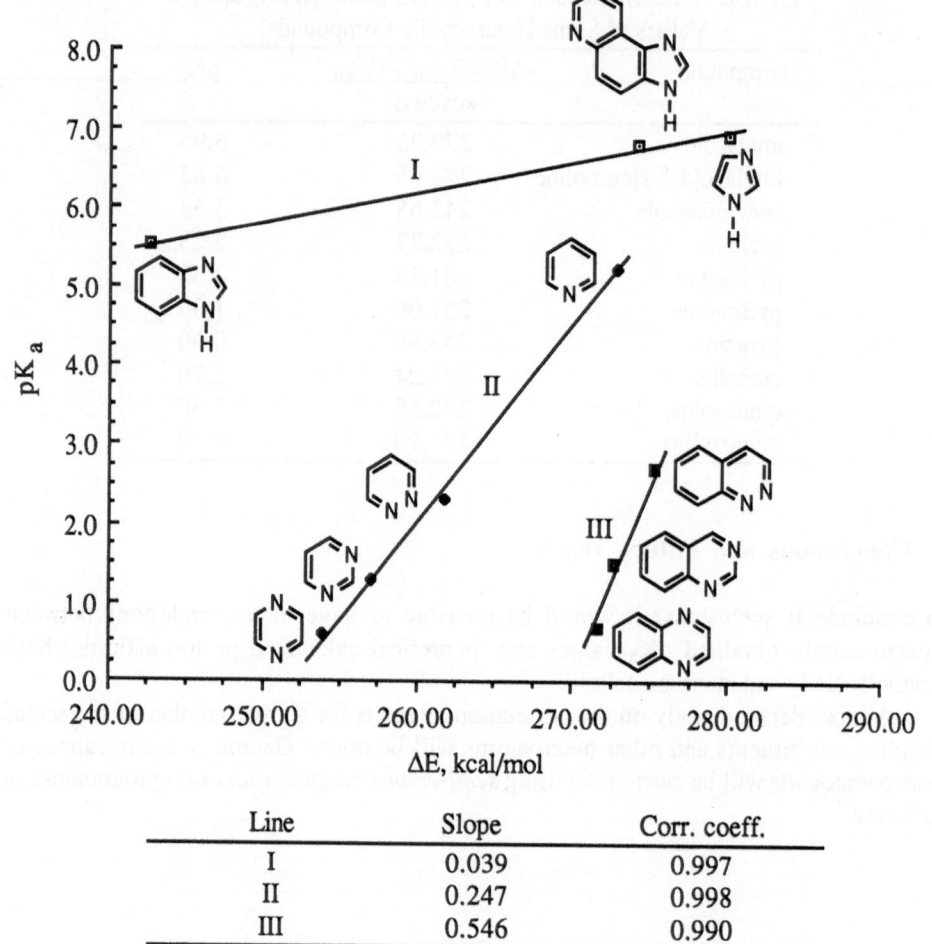

Line	Slope	Corr. coeff.
I	0.039	0.997
II	0.247	0.998
III	0.546	0.990

Figure 7. The graph of ΔE values vs. pK_a for some heterocyclic compounds.

5. Acknowledgement

The authors would like to thank Prof. I. G. Csizmadia from University of Toronto for his close cooperation and along with him Dr. M. R. Peterson from the same university for providing the Monstergauss Programme. The financial support of Technical and Scientific Council of Turkey is most greatfully appreciated. Finally, we are in debt to thank to Computer Center of Anadolu University (BAUM) for their time and concern.

6. References

[1] (a) Acar, F.; Badesha, S. S.; Flisch, W.; Gözogul, R.; Inel, O.; Jones, R. A.; Ögretir, C.; Rustidge, D. C. *Chim. Acta Turc.* **1981**, *1*, 225. (b) $K_{a\,taut} = [2]/[1] = 1/540$.

[2] Coulson, C. A.; *Valance*, Oxford Univ. Press, London and New York, 1979.

[3] Ridd, J. *Physical Methods in Heterocyclic Chemistry*, vol. I, Ed. A. R. Katritzky, Academic Press, New York and London, 1963, p.116.

[4] Brown, R. D.; Coller, B. A. W. *Aust. J. Chem.* **1959**, *12*, 152.

[5] Conant, J. B.; Wheland, G. W. *J. Am. Chem. Soc.* **1932**, *54*, 1212.

[6] Wheland, G. W. *J. Chem. Phys.* **1934**, *2*, 474.

[7] Mc Ewen, W. K.*J. Am. Chem. Soc.* **1936**, *58*, 1124.

[8] Morton, A. A. *Chem. Rev.* **1944**, *35*, 1.

[9] Gilman, H. *Organic Reactions*, vol.VIII, John Wiley and Sons, New York, 1954, p. 258.

[10] Shatenshtein, A. I. *Doklady Akad. Nauk. S. S. S. R.* **1950**, *60*, 1029.

[11] Streitwieser Jr., A. *Molecular Orbital Theory for Organic Chemists*, John Wiley and Sons, Inc. New York, 1961, Ch. 14.

[12] Longuet-Higgins, H. C. *J. Chem. Phys.* **1950**, *18*, 275.

[13] Ögretir, C.; Kaniskan, N. *unpublished work.*

[14] Hennig, H.; Tauchnitz, J. *J. Prakt. Chem.* **1970**, *6 (312)*, 1191.

[15] Wheatly, P. J. *Physical Methods in Heterocyclic Chemistry*, vol. V, Ed. A. R. Katritzky, Academic Press, New York and London, 1970.

[16] Alberth, A. *Physical Methods in Heterocyclic Chemistry*, vol. I, Ed. A. R. Katritzky, Academic Press, New York and London, 1962, Ch. 1.

MOLECULES WITH "VOLCANIC" GROUND HYPERSURFACES. STRUCTURE, STABILITY AND ENERGETICS.

Cleanthes A. NICOLAIDES and Petros VALTAZANOS
Theoretical and Physical Chemistry Institute
National Hellenic Research Foundation
48, Vas.Constantinou Ave., Athens 116 35
Greece

ABSTRACT. There exist molecular species whose ground state repulsive potential energy surfaces (PES) contain a local chemical minimum which may be stable upon symmetry breaking, thus trapping energy in vibrational levels lying above the dissociated products. Because of their special form, we have named these ground PES "volcanic". Volcanic ground PES owe their appearance to the geometry-dependent interaction of covalent-ionic structures and the concomitant (avoided) intersection between the repulsive ground and the attractive first excited singlet PES. This article contains our recent findings from MCSCF and CI calculations on the structure, stability and energetics of clusters and complexes which exhibit these features, such as $(H_2O)_2^*$, He_3^{++}, BeH_2^{++}, BH_2^{++}, $[Be_2H_8]^{++}$, $OBeH_2$ and $FBeH_2$. Except for $FBeH_2$, the other H_2 compounds constitute "nonclassical" hydrogen complexes, whereby the H_2 bond is weakened but not broken.

1. Extraordinary Ground Potential Energy Surfaces Caused by Ionic Structures

1.1. DIATOMIC MOLECULES

The vast majority of the known ground state potential energy surfaces (PES) either have a global minimum at the molecular geometry of stable equilibrium, or are repulsive and no molecule is formed. However, there are unusual situations- caused by the enhanced importance of ionic structures at geometries of avoided intersections-which lead to ground state PES that are repulsive asymptotically but have a higher-lying chemical minimum in the inner region. Such PES have a volcano-like form (see Figure 1) and thus, in analogy with

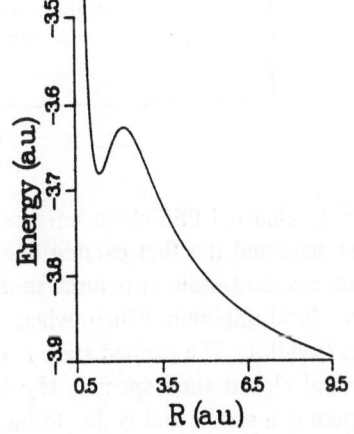

Figure 1. He_2^{++} ground state potential energy curve [1; 2; 3].

S. J. Formosinho et al. (eds.), Theoretical and Computational Models for Organic Chemistry, 355–373.

Kemble's [4] description of alpha particle decay, we have called them [1] "volcanic".

Until recently [5], volcanic ground PES were known only in doubly ionized diatomic molecules. The prototype of such diatomic dications, He_2^{++} $^1\Sigma_g^+$, was first computed by Pauling [6]who also provided the interpretation of its volcanic form in terms of a geometry dependent covalent-ionic structures mixing, i.e. in terms of $(He^+ + He^+) \leftrightarrow (He^{++} + He)$. Asymptotically, $E(He^+ + He^+)$ is below $E(He^{++} + He)$ by 29.8 eV. As R decreases, $E(R)$ $(He^+ + He^+)$ defines a repulsive curve given by $1/R$. On the other hand, the polarization of He by He^{++} yields an attractive curve with a minimum. The interaction of the two valence bond structures results in the volcanic form shown in Figure 2. In the MO language, the characteristic form is produced from the 2X2 CI of $1\sigma_g^2$ with $1\sigma_u^2$, with the coefficient of $1\sigma_u^2$ increasing outside the well [2].

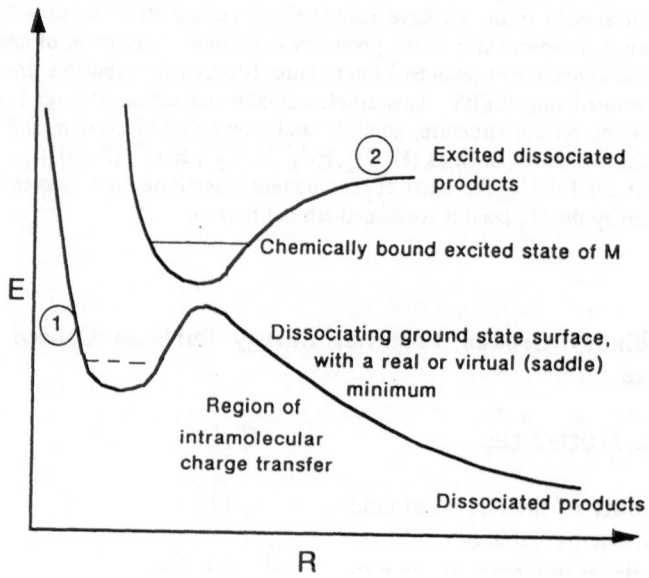

Figure 2. Unusual PES characteristics at an avoided intersection between a ground singlet state and the first excited one, along a fragmentation coordinate R which defines covalent-ionic structures mixing. The ground PES is repulsive but acquires a local minimum which, when all coordinates are considered, may be real or virtual (saddle). The excited state is chemically bound in most cases. In the case of neutral closed shell species, $H_2 + H_2$ or $H_2O + H_2O$ [7; 8; 9; 10], the ground PES minimum is a saddle and is due to intramolecular charge transfer involving the excited PES which breaks into a neutral ground state moiety plus the H_2^* $B^1\Sigma_u^+$ state. In the case of ionized species, such as $He^+ + He^+$, the excited curve dissociates into moieties with explicit charge exchange, (i.e. $He + He^{++}$).

1.2. POLYATOMIC MOLECULES

1.2.1. Saddles on repulsive PES of ground states of clusters of closed-shell molecules.

Research done at our institute in the years 83-84 [11; 7; 8; 9] revealed the existence of chemically bound excited clusters (CBEC) such as $(HeH_2)^*$, $(NeH_2)^*$, $(ArH_2)^*$, $(H_2)_n^*$, n=2-5, at geometries where there are avoided intersections between the repulsive ground state and the first excited state. Subsequent extensive calculations by different methods for HeH_2 [12] and for H_4 [13] verified these PES characteristics. The reason for their appearance is found in the strongly ionic character that the H_2 $B^1\Sigma_u^+$ state acquires at 4.0 a.u. and which gives rise to a localized intramolecular charge transfer in wavefunctions which dissociate into H_2 $B^1\Sigma_u^+$ plus the remaining moiety of the cluster [7; 8; 9]. Thus, the geometry-dependent ionic structure produces an attractive curve along one of the reaction coordinates which avoids explicit crossing with the repulsive covalent curve. The result is similar to that occuring in the charge transfer case of $(He^+ + He^+) \leftrightarrow (He^{++} + He)$. Its form is depicted in Figure 2 while Figure 3 shows a three-dimensional representation of the H_4 hypersurfaces at the geometry of avoided intersections. It can be seen that, again, a local minimum appears on the two-dimensional ground surface along the fragmentation coordinate R, $H_3^{"+"}...^R...H^{"-"}$. However, this time it corresponds to a saddle point. The ground state is completely repulsive. This fact, together with the avoided crossing with the bound excited state, allowed the explanation of the phenomenon of fluorescence quenching of H_2 $B^1\Sigma_u^+$ in the presence of He or H_2 [11; 7; 8; 14] in terms of the temporary formation of the CBEC which subsequently decays to the neutral closed shell ground states via nonadiabatic coupling with the ground PES [15].

The theory which allows the prediction of the approximate geometry at which the intramolecular charge transfer occurs and the consequent initialization of the computation of that part of the PES [7; 8; 9] has recently been applied to the case of the water dimer, $(H_2O)_2$. For this system, only ground state hypersurfaces had previously been computed for selected geometries. We found [10] the same PES features as those depicted in Figures 2 and 3. This is shown in Figure 4. Figure 5 shows the corresponding geometry.

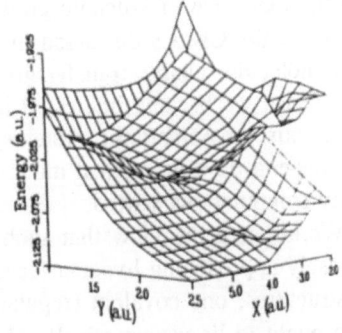

Figure 3. 3-D representation of H_4 hypersurfaces at the geometry of avoided intersections due to intramolecular charge transfer.

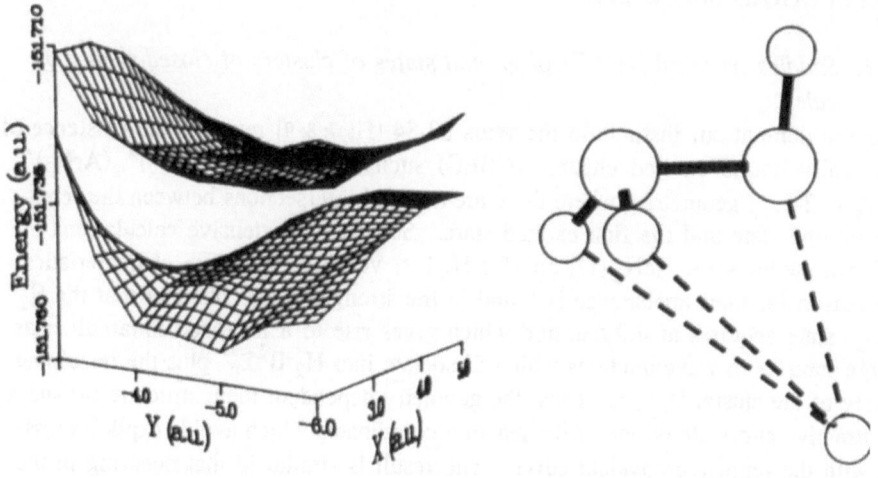

Figure 4. Ground and excited $^1A'$ surfaces of $(H_2O)_2$ obtained by the state averaged FORS-MCSCF method, using the MOLPRO [16] program package.

Figure 5. "Ball-and-stick" drawing of the $(H_2O)_2^*$ CBEC geometry, obtained by the state-averaged FORS-MCSCF method.

1.2.2. Volcanic Ground Hypersurfaces

Given the existence of volcanic ground states in certain diatomic dications and the fact that for the CBECs discussed in section 1.2.1 the avoided intersections caused by intramolecular charge transfer give rise only to saddles on the ground repulsive hypersufaces, we became interested in determining whether volcanic ground hypersurfaces exist in polyatomic systems. The search for such extraordinary surfaces should depend on a theoretical model which could lead to plausible predictions before any computations started.

We adopted the view that such a feature should again be a manifestation of two closely approaching hypersurfaces (perhaps intersecting), representing two diabatic structures, one covalent (repulsive) and one ionic (attractive). The attractive surface ought to lie asymptotically above the repulsive one. Whether a local minimum could be formed would depend on the character of both surfaces and on their interaction.

We chose [5] two small molecules, the He_3^{++} and BeH_2^{++}, and considered the interaction of the attractive (He_2^{++} $^1\Sigma_g^+$ + He) with the repulsive (He_2^+ $^2\Sigma_u^+$ + He$^+$) or (2He$^+$+He) surfaces or of (BeH^{++} $^2\Sigma^+$+H) with (BeH^+ $^1\Sigma^+$+H$^+$) or (Be$^+$+H$_2^+$). Such a consideration is legitimate since we know that the dications He_2^{++} and BeH^{++} are bound. (He_2^{++} + He) lies 11 eV above (He_2^+ + He$^+$) while (BeH^{++}+H) lies 5.8 eV above (BeH^++H$^+$).

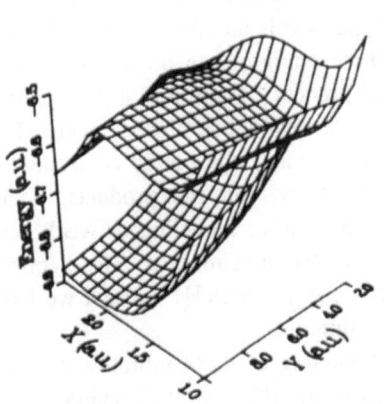

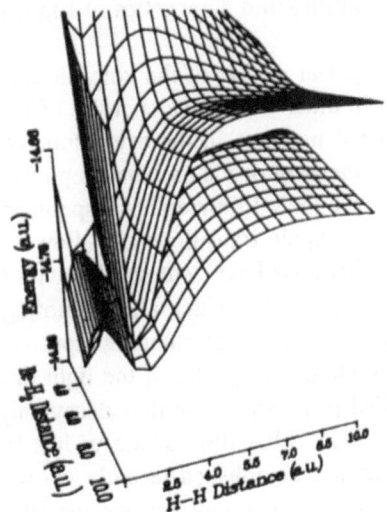

Figure 6. Three-dimensional representation of the C_{2v} potential energy surface of He_3^{++}. The upper one is the A_1 surface while the lower one is the B_2.

Figure 7. Three-dimensional representation of the C_{2v} potential energy surface of BeH_2^{++}. The two lowest A_1 surfaces are shown.

Multi-reference-single-and-double-excitation CI (MRCI), Full CI and MCSCF geometry optimization calculations indeed revealed the existence of volcanic ground states. These are shown in Figures 6 and 7. Computational information is given in section 3.1 However, whereas BeH_2^{++} has a true minimum in C_{2v} symmetry supporting vibrational levels, He_3^{++} does not, since the vibrational analysis showed that upon asymmetric stretching and consequent symmetry breaking, the C_{2v} minimum turns into a saddle and the system dissociates on the lower PES to $He_2^+ + He^+$.

Thus, in BeH_2^{++} we have determined the existence of the first volcanic ground hypersurface in a polyatomic molecule. The heuristic tool which we used for the anticipation of such a rarity and for the choice of suitable species involved the concept of covalent (repulsive)-ionic (attractive) diabatic states-mixing of ionized moieties. Considering our earlier work on the CBECs (section *1.2.1*) we might say that the result of such interactions, whether they occur in neutral closed-shell surface (e.g. HeH_2 or $(H_2)_2$) or in ionized species where charge transfer is seen explicitly $(BeH^{++}+H \leftrightarrow BeH^+ + H^+)$, is the unusual feature of a chemical well on the ground repulsive hypersurface which may be a saddle (e.g. $(H_2)_2$), a pseudominimum unstable under symmetry breaking (e.g. He_3^{++}), or a true minimum supporting vibrational levels (e.g. BeH_2^{++}).

2. Stability and Energetics of Light Diatomic Dications

The special feature of the volcanic ground state allows it to contain releasable energy. In order words, the energies of its rovibrational levels are above that of the dissociated products in the asymptotic region. These energies can be released either spontaneously or via an induced physicochemical process. Therefore, it is important to know their lifetimes. The spontaneous radiative channel is closed, even when electric dipole transitions are allowed by symmetry, since the bound rovibrational wavefunctions have zero overlap with those of the lower-lying free products. Thus, the decay modes which allow autofragmentation are either tunneling or weak coupling with a possible plunging repulsive state of a different symmetry. The electronic structure and spectra of the light dications He_2^{++}, Be_2^{++} or BeH^{++} which we have studied [1; 17] show that they decay only via tunneling.

The calculation of the widths (rates) of tunneling of these vibrational levels has been accomplished recently via semiclassical [1; 17; 18] as well as quantum [1; 19; 20] theory. The calculations for He_2^{++} employed extremely accurate PES [1; 2; 3], since tunneling widths turn out to be very sensitive to the shape of the potential [1]. This can be seen in the difference between our results on Be_2^{++} [17] and those of Bauschlicher and Rosi [18]. An important theoretical result from the work of refs 2, 19 and 20 is the fact that the quantum approach of Babb and Du [19] yields reliable results for widths of the order of 10^{-20} a.u., something which proved impossible in the case of numerical solution of the complex eigenvalue resonance equation satisfied by the decaying vibrational levels [1; 20].

Our results show that in the low-lying rovibrational levels, Be_2^{++} and BeH^{++} are dynamically completely stable, in spite of their small mass. On the other hand He_2^{++} in the (v=0, J=0, 2) levels lives for about 3.5 hrs. (For more discussion and results on tunneling see refs. 2, 18-20).

Having established that these high energy-density molecules are stable, energy generating (at the molecular level) physicochemical processes such as the following are possible [1; 17].

$$
\begin{array}{lll}
He_2^{++} + 27.6 \text{ Kcal} & \rightarrow 2He^+ + & 233.7 \quad \text{Kcal} \\
Be_2^{++} + 20.7 \text{ Kcal} & \rightarrow 2Be^+ + & 84.0 \quad \text{Kcal} \\
BeH^{++} + 19.7 \text{ Kcal} & \rightarrow Be^+ + H^+ + & 47.0 \quad \text{Kcal} \\
He_2^{++} + H_2 & \rightarrow 2HeH^+ + & 656.0 \quad \text{Kcal} \\
Be_2^{++} + H_2 & \rightarrow 2BeH^+ + & 206.0 \quad \text{Kcal} \\
BeH^{++} + H_2 & \rightarrow BeH^+ + H_2^+ + & 166.0 \quad \text{Kcal}
\end{array}
$$

(For reasons of comparison, $H_2 + \frac{1}{2}O_2 \rightarrow H_2O + 58$ Kcal)

The extremely high exothermicity and the small mass of the reaction products constitute favorable elements in the set of prerequisites for the decades-long quest for new and powerful propellants [21]. Whether such or similar species can be produced and kept in suitable molecular or solid matrices in relatively high densities is open to basic and technological research. In this regard, we have spent some of our research effort in two directions. The first concerns the effect on the molecular spectrum of an ionic crystal into which the molecule is embedded substitutionally [22; 23]. The second concerns the interactions from the immersion of light dications into a cluster of $(H_2)_n$ [5; 24]. The related results for the second direction are presented in section 3.

2.1. STABILITY AND VIBRATIONAL ANALYSIS

The calculation and graphical depiction of potential energy surfaces (PES) gives valuable information about the relative energies of various conformations of molecules. Nevertheless, definitive knowledge about stability and energy differences is acquired only after a vibrational analysis. Such an analysis shows whether a ground state can support vibrational levels and therefore whether the molecule exists in this state or not. For example, the potential energy curve of the dication BH^{++} has a local minimum which, however, cannot support any vibrational levels [17].

Another example is the case of the He_3^{++} (section *1.2.2*). Here, the PES has a definite minimum when symmetry constraints are imposed (see Figure 6). However, when the molecule is allowed to vibrate, thereby breaking its symmetry, the minimum turns into a saddle. The same result is obtained with the molecule $OLiH_2$ whose C_{2v} local minimum also turns into a saddle upon symmetry breaking (see section 4.2), as well as with $FLiH_2$ (section 4.3).

3. Nonclassical Hydrogen Complexes of Nontransition Metals

He_2^{++} has a very high electron affinity and no low-lying empty orbitals. Therefore, it presents serious difficulties when considering the problem of containment by liquid or solid environments. On the other hand, Be_2^{++} seemed to us as a convenient prototype for which systematic theoretical and computational analysis could be possible. The result on the BeH_2^{++} molecule (section *1.2.2.*) was the first step. In this section we turn to the analysis of its electronic structure and, based on the conclusions, on further results involving the stability of hydrogen complexes of Be^{++}, Be^+, B^{++} and Be_2^{++}.

Dunning-Hay double-zeta basis sets [25], augmented by one polarization function for each atom (p for hydrogen, d for everything else), were used for all calculations.

3.1. BeH$_2^{++}$

3.1.1. Features of the Hypersurfaces

Full-CI calculations using the MELDF system of programs [26] yielded the surfaces shown in Figure 7. The X-axis represents the distance from the Be atom to the H$_2$ moiety, while the Y-axis represents the distance between the two hydrogen atoms. Of the two surfaces, the one above is repulsive and has a deep valley at Y$\approx$2.0 a.u. A similar valley exists on the lower surface as well, with a clear minimum at X=3.0 a.u and Y=1.52 a.u. The Full CI energy at the minimum was determined to be -14.8563 a.u.

The two valleys exhibit an avoided crossing at X$\approx$10.5 a.u and Y$\approx$1.7 a.u.. The lower state energy at this point is -14.7750 a.u. This interaction between the two states yields a local minimum on the lower state along this valley. The dissociation limit is -14.8756 a.u., i.e. approximately 0.5 eV below the local minimum, and corresponds to (Be$^+$+H$_2^+$).

The lower surface is dominated by three configurations, $1a_1^2 2a_1^2$, $1a_1^2 3a_1^2$ and $1a_1^2 2a_1 3a_1$. The coefficients of the first range from 0.99 at the minimum to 0.4 at the valley crossing, to 0.66 at the (Be$^+$+H$_2^+$) limit, to 0.82 at the other avoided region and finally to 0.68 at the (Be$^+$+H$^+$+H) limit. Those of the second configuration range from near zero at the minimum, to 0.37 at both crossings and to 0.68 at the dissociation limits. Finally, those of the third range from near zero at the minimum, to 0.82 at the valley crossing to near zero at the (Be$^+$+H$_2^+$) limit, to 0.3 at the other crossing and to 0.21 at the (Be$^+$+H$^+$+H) limit.

The GAMESS program [27] was used to find the local minimum on the lower A$_1$ surface by MCSCF gradient-driven geometry optimization. The results are shown above. Subsequently, a vibrational analysis was performed on the minimum and this showed that symmetry breaking does not affect the character of the stationary point. I.e., this is a *true* minimum even in C$_1$ symmetry, having only real vibrational frequencies. These frequencies were found to be 995 cm^{-1} for the asymmetric stretch, 1248 cm^{-1} for the symmetric stretch and 3692 cm^{-1} for the bending mode (equivalent to the stretching mode of H$_2$). The harmonic zero point energy was found to lie 0.0135 a.u. above the well minimum.

3.1.2. The two-electron, three-center bond of BeH$_2^{++}$

The possible formation of BeH$_2^{++}$ was anticipated based on qualitative considerations of properties of ionic structures as explained in section *1.2.2*. Computation has confirmed the existence of bound BeH$_2^{++}$ in a local minimum. In this section we analyze the unusual bonding characteristics of this molecule.

The geometry is that of an isosceles triangle with a H-H distance of 1.520 a.u. (as compared with the free H$_2$ internuclear distance of 1.401 a.u.) and a Be-H distance of 3.069 a.u. (as compared with the free BeH^{++} internuclear distance [17] of 3.406 a.u.). The atomic Mulliken population charges are 1.573 for Be and 0.213 for each of the two hydrogens. The bending mode (see section *3.1.1.* above) has a fre-quency of 3692 cm^{-1} as compared to the stretching frequency of 4401 cm^{-1} for free H$_2$, showing the weakening but not the disappearance of the H$_2$ bond. Finally, a three center, two-electron bond is easily identifiable (shown in Figure 8a), having a natural orbital occupation of 1.9642. The corresponding π bond, which would nor-mally identify a dihydride (shown in Figure 8b) has a natural orbital occupation of only 0.0167.

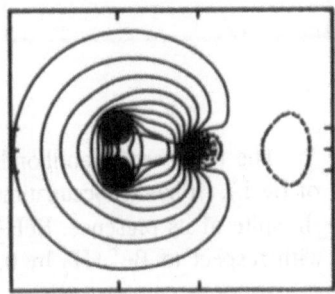 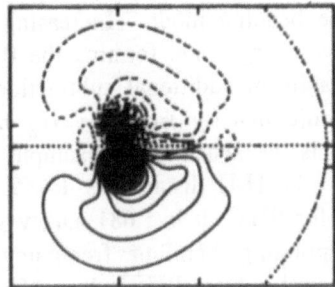

Figure 8a. The three-center two-elec-tron bonding orbital of BeH$_2^{++}$. It has an occupation of 1.964.

Figure 8b. The π bonding orbital of BeH$_2^{++}$. It has an occupation of only 0.017.

All of the above characteristics seem to indicate that BeH$_2^{++}$ is indeed a "nonclassi-cal hydrogen complex" [28; 29], Be$^{"++"}$...H$_2$, lying 2.2 eV below the energy of Be^{++} + H$_2$.

3.2. OTHER LIGHT METAL DIHYDROGEN AND POLYHYDROGEN COMPLEXES

In order to expand and generalize our understanding of bonding situations such as that of BeH$_2^{++}$, we examined the isoelectronic molecular ions LiH$_2^+$ and BH$_2^{+++}$ as well as ions with one more electron, i.e. BeH$_2^+$ and BH$_2^{++}$. In the first case, the dihy-drogen complex of the Li$^+$ ion is confirmed. Similar features as the ones presented for BeH$_2^{++}$ appear. Specifically, the vibrational frequencies are: Asymmetric stretch 531 cm^{-1}, symmetric stretch, 378 cm^{-1}, and bending 4420 cm^{-1}. The energy of Li$^{"+"}$...H$_2$ is 0.42 eV below that of Li$^+$+H$_2$. On the other hand, BH$_2^{+++}$ is unbound since, the large positive charge on the one hand breaks the H$_2$ bond and on the other does not allow the (H-B-H)$^{+++}$ formation.

The molecules of the second case, BeH_2^+ and BH_2^{++}, were chosen to test the effect of the σ antibonding orbital (Figure 9) which can now be occupied. Indeed, for BeH_2^+ its occupation number is 0.997 with a concomitant weakening but not cleavage of the overall bond. Thus, the features of the complex are enhanced as compared to those of BeH_2^{++}. BeH_2^+ lies only 0.25 eV below $Be^+ + H_2$ while its vibrational frequencies are 534 cm^{-1} for the symmetric stretch, 675 cm^{-1} for the asymmetric one and 4236 cm^{-1} for the bending mode. Increasing the positive charge, -i.e. forming the BH_2^{++} ion-results in additional subtraction of electronic density from the H_2 bond which is weakened. For example, for BeH_2^+ the H-H distance is 1.435 a.u. while for BH_2^{++} it is 1.681 a.u., with a corresponding bending frequency of 2948 cm^{-1}. Also, BH_2^{++} lies 3.96 eV below $B^{++} + H_2$. It should be noted that

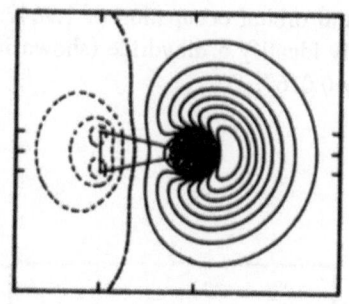

Figure 9. The three-center antibonding orbital of BeH_2^+. It has an occupation of 0.997. In spite of its presence, BeH_2^+ is stable with respect to $Be^+ + H_2$ by 0.25 eV.

the dissociation limits given here are intended to show the affinity of the various metal ions for H_2, and do not imply that lower ones do not exist (e.g. $B^+ + H_2^+$ for BH_2^{++}).

Can these concepts be utilized to predict metal ion polyhydrogen complexes? For example, what happens when BeH_2^+ reacts with a proton H^+, a H atom or a H_2 molecule ? Can it form complexes corresponding to BeH_3^{+++}, BeH_3^{++}, BeH_3^+ or BeH_4^{++} and in what geometry?

To answer this question we first considered the following. Classical hydrides, hydrogen complexes or mixed species are possible, depending on the overall charge distribution and the available MOs for bonding. For example, the BeH_3^{++} compound might be $[Be\cdots H_3]^{++}$ or $[H\text{-}Be\cdots H_2]^{++}$. In the first case the formation of the stable H_3^+ would suggest a complex $Be^+\cdots H_3^+$. However, electrostatic repulsion does not allow bond formation -as computations confirm. On the other hand, both BeH^{++} and BeH_2^{++} are bound in ionic structures with the positive charge on Be. Given the availability of empty metal ion-orbitals, this suggests that $H\text{-}Be^{"++"}\cdots H_2$ could indeed be bound in a geometry resembling the BeH^{++} dication [17] and the $Be^{++}\cdots H_2$ complex. The case of a $Be\cdots H_3^+$ conformation is also interesting. In this case, the p orbital of Be is already partially occupied due to the atomic zeroth order mixing $2s^2 \leftrightarrow 2p^2$ and it can interact with the H_3^+ doubly degenerate σ* orbital. A trigonal pyramid with Be at the apex and H_3^+ forming the equilateral base might be possible.

These hypotheses for tri- and tetrahydrogen compounds were put to test via MCSCF geometry optimization calculations. Inner 1s shells of metal atoms were kept frozen, while the rest of the orbitals were made available for occupation in all cases except BeH_4^{++} , where a smaller space was used, due to the number of electrons (and configurations) involved. The following results were obtained.

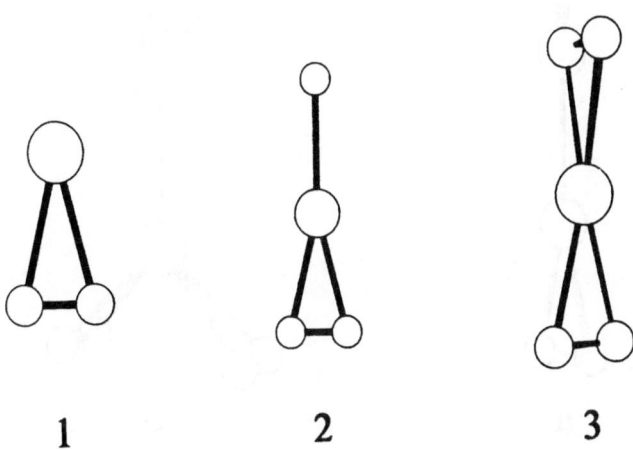

1 **2** **3**

Figure 10. The three structures of hydrogen complexes identified in this work. Structure 1 (C_{2v} symmetry) corresponds to BeH_2^{++}, BeH_2^+, LiH_2^+ and BH_2^{++}, structure 2 (C_{2v} symmetry) to BeH_3^{++} and structure 3 (D_{2d} symmetry) to BeH_4^{++}.

BeH_2^{++} , BeH_2^+, LiH_2^+ and BH_2^{++} were all found to have structure 1 (Figure 10) with a=1.520, 1.435, 1.410, 1.681 a.u and b=3.069, 3.543, 4.163, 2.924 a.u. respectively. BeH_3^{++} was found to have structure 2 (Figure 10) with a=1.503 a.u., b=3.041 a.u. and c=3.341 a.u. (almost exactly the parameters of BeH_2^{++} and BeH^{++} [17] respectively). BH_2^{+++}, BeH_3^{+++} and BeH_3^+ were found not to exist, while BeH_4^{++} has structure 3 (Figure 10) with a=1.500 and b=3.023 a.u. (practically two units of BeH_2^{++}). The H-H and Be-H symmetric and asymmetric stretching frequencies in BeH_3^{++} were found to be 3792, 1309 and 1050 cm^{-1} respectively, showing slight strengthening with respect to BeH_2^{++}, while the Be-H stretch with the third hydrogen atom was found to be 1037 cm^{-1}. Similarly, the shorter bond distances of BeH_4^{++} with respect to BeH_2^{++} and BeH_3^{++}, suggest slight strengthening of bonds, something born out by the vibrational analysis (e.g the H-H stretch is now 3820 cm^{-1}).

3.3. HYDROGEN COMPLEXES OF Be_2^{++} [24]

Given the results on the formation of Be^+ and Be^{++}-hydrogen complexes and the known characteristics of the Be_2^{++} bond, we considered it reasonable to anticipate the formation of stable $Be_2^{"++"}\cdots(H_2)_n$ complexes, at conformations predetermined to a good approximation by steric effects, MO considerations and the Be_2^{++} and $Be_2^{"++"}\cdots H_2$ geometries.

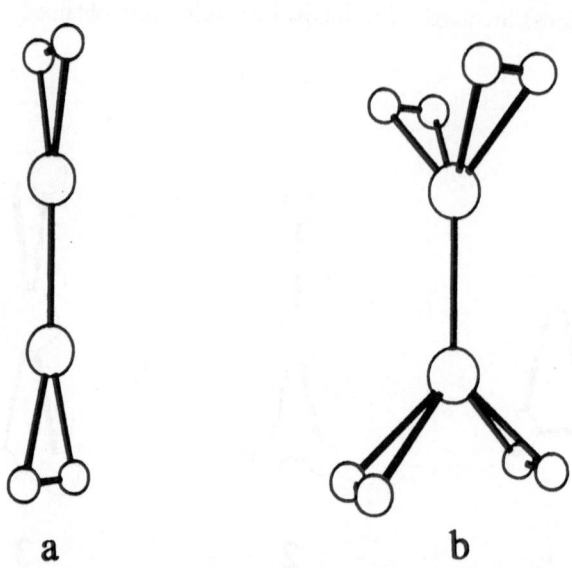

a   b

Figure 11. Structure of hydrogen complexes of $Be_2^{"++"}$. (a) Structure of $[Be_2H_4]^{++}$. For molecular parameters see table 2. The molecule has S_4 symmetry. (b) Structure of $[Be_2H_8]^{++}$. For molecular parameters see table 2. The dihedral angle between the two BeH_2 groups on the same side is $94.60°$. This molecule also has S_4 symmetry.

The GAMESS program [27] was used to locate the geometry minima and to perform the relevant vibrational analysis on them. A relatively small FORS-MCSCF [30] space was used to keep these calculations tractable. Larger single-and-double-excitation CI calculations in an extended space, using the MELDF system of programs [26] were subsequently performed at the previously located minima in order to determine more accurately the energetics and stabilization of the various species.

Figure 11a shows the geometry of $[Be_2H_4]^{++}$ while 11b shows that of $[Be_2H_8]^{++}$. Some of their most important geometrical parameters, vibrational frequencies and Mulliken charges, together with those of some Be^{++} hydrogen complexes (shown for comparison) are displayed in Table 2. Table 1 shows the stabilization of the same species relative to their separated components. Study of the two tables reveals that as hydrogens are added stabilization increases, with local minima eventually turning into absolute ones. Stabilization increase is accompanied by decrease of the Be charge and by shortening of the H-H bonds and, where appropriate, of the Be-Be bond. It is also accompanied by increase of the H-H stretching fre-

TABLE 1. Energy stabilization of Be^{++} and Be_2^{++} hydrogen complexes.

Species	Type of Minimum		Stabilization
$Be^{++}\cdots H_2$	Local	2.21 eV below $Be^{++} + H_2$;	0.53 eV above $Be^+ + H_2^+$
$H\cdots Be^{++}\cdots H_2$	Local	3.54 eV below $Be^{++} + H_2 + H$;	0.80 eV below $Be^+ + H_2^+ + H$; 0.95 eV above $Be^+ + H_2 + H^+$
$H_2\cdots Be^{++}\cdots H_2$	Absolute	4.27 eV below $Be^{++} + 2H_2$;	1.53 eV below $Be^+ + H_2^+ + H_2$
Be_2^{++}	Local		2.70 eV above $2Be^+$
$H_2\cdots Be_2^{++}\cdots H_2$	Local	2.45 eV below $Be_2^{++} + 2H_2$;	0.25 eV above $2Be^+ + 2H_2$
$(H_2)_2\cdots Be_2^{++}\cdots (H_2)_2$	Absolute	3.27 eV below $Be_2^{++} + 4H_2$;	0.57 eV below $2Be^+ + 4H_2$

TABLE 2. Properties of the Be^{++} and Be_2^{++} hydrogen complexes.

Species	Mulliken Charge		Internuclear Distances (a.u.)			Vib. Stretching Frequency (cm^{-1})	
	Be	H	Be-H	Be-Be	H-H	H-H	Be-Be
$Be^{++} \cdots H_2$	1.573	0.213	3.07		1.52	3692	
$H \cdots Be^{++} \cdots H_2$	1.231	0.214 (0.341)*	3.04 (3.34)*		1.50	3792	
$H_2 \cdots Be^{++} \cdots H_2$	1.128	0.218	3.02		1.50	3820	
Be_2^{++}				4.07			608
$H_2 \cdots Be_2^{++} \cdots H_2$	0.729	0.135	3.12	4.00	1.48	3880	649
$(H_2)_2 \cdots Be_2^{++} \cdots (H_2)_2$	0.603	0.099	3.30	3.97	1.46	3971	805

* Refers to the third hydrogen

quency and, again where appropriate, by increase of the Be-Be stretching frequency. All these observations are consistent with bond strengthening. The only parameter that shows bond weakening is the increase in the Be-H bond length going to $[Be_2H_4]^{++}$ and from that to $[Be_2H_8]^{++}$. This result, combined with the fact that the H-H bond in the same species is strengthened, suggests a weakening of the Be-H$_2$ bond in the Be_2^{++} complexes which increases as more such bonds are added.

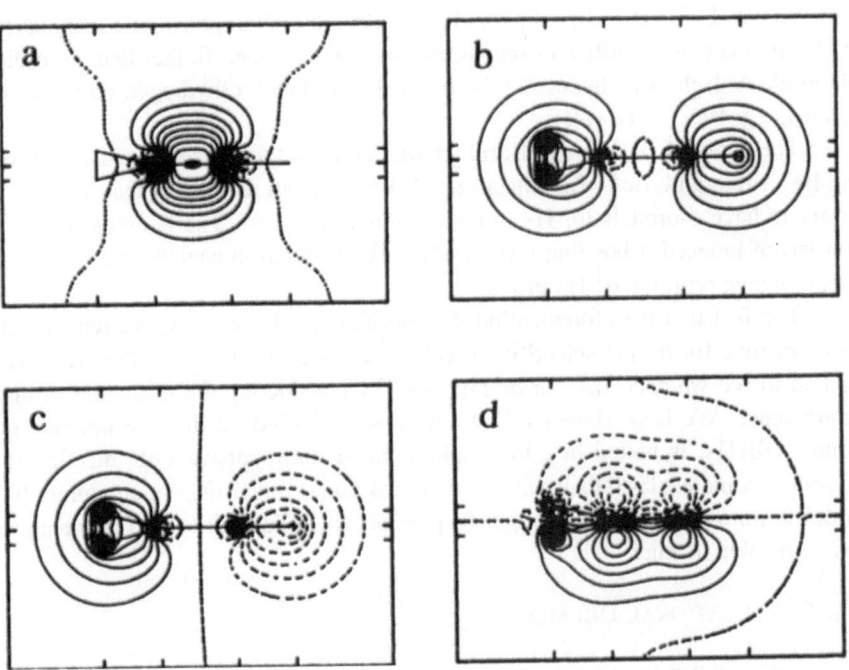

Figure 12. The bonding of $[Be_2H_4]^{++}$. (a) The Be-Be bond. Its occupation is 1.937. (b) The three-center bonds over the BeH$_2$ groups. Occupation is 1.978. (c) Same as in (b) only now there is a node across the Be-Be bond. Occupation is 1.978. (d) The back-bonding between the Be-Be π bond and the H$_2$ σ^* bond. Occupation is 0.026

The bonding picture of $[Be_2H_4]^{++}$ is revealed in Figures 12(a-d), where the four most important MCSCF natural orbitals are displayed. Figure 12a shows the Be-Be σ bond. Figures 12b and 12c both show pairs of three-center bonds over the BeH$_2$ groups. 12c has a node across the Be-Be bond. Should these two natural orbitals be localized by adding and subtracting them, they would give two separate two-electron three-center bonds on the two BeH$_2$ groups. All three of the orbitals shown in Figure 12(a-c) have occupations around 1.95. Finally, Figure 12d shows the orbital that provides the necessary for a non-classical hydrogen complex back-bonding between the Be-Be π bond and the H_2 σ^* orbital. The occupation of this orbital is 0.026.

4. Chemical Trapping of H_2 by BeO in a Volcanic Ground State

We have shown above that when H_2 interacts with nontransition metal atoms or diatoms such as Be^{++}, B^{++}, Be^+, Li^+ or Be_2^{++}, volcanic ground potential energy surfaces (PES) are generated while an exceptional bond takes place. Rather than an ordinary dihydride with the simultaneous scission of the H_2 bond, dihydrogen complexes are formed, e.g. $Be^{"++"}.....(H_2)$ [5].

The cases of BeH_2^{++} and BeH_2^+ as well as of polyhydrogen complexes involving Be^{++}, B^{++} and Be_2^{++} (i.e. metal-metal bond), point to the fact that it is not necessary to have d-orbitals for H_2 to bind without breaking. The bonding mechanism consists of induced σ-bonding involving the H_2 σ orbital in conjunction with weak π back-bonding with the σ^* H_2 orbital.

The fact that the aforementioned molecular species are ionized, renders them less attractive for further scientific or technological applications. Therefore, it is important to see whether this can be remedied by considering the formation of appropriate salts. We have shown [31] via accurate calculations that the neutral compound $OBeH_2$, is bound in a local minimum of the repulsive PES and that H_2 is present in a molecular (albeit with a weakened bond) rather than in an atomic (dihydride) form. It follows that it might be possible for solid BeO to trap H_2 chemically in a reversible manner.

4.1. COMPUTATIONAL DETAILS

Geometries of $OBeH_2$, H_2OBe, BeO, H_2 and H_2O were first optimized at the FORS-MCSCF level. Be and O core orbitals were kept frozen, as was an oxygen lone-pair orbital in all cases. The ground state energy of Be was also calculated at the same level of approximation. Single and double excitation (from a single SCF configuration) CI calculations were subsequently performed using the geometries and the optimized orbitals from the FORS-MCSCF optimization. A force-constant (vibrational analysis) calculation was also done on the located local minimum for $OBeH_2$. C_{2v} symmetry was used in all calculations except for the vibrational analysis.

Calculations were also performed for $OLiH_2$ and OBH_2, in order to determine whether similar minima existed for these molecules. The corresponding LiH_2^+ and BH_2^{++} bind hydrogen in molecular form, just like BeH_2^{++}, while their hypersurfaces also have the unusual volcano form.

4.2. RESULTS AND DISCUSSION

The geometry optimization calculations located a minimum for $OBeH_2$ in C_{2v} symmetry. In terms of absolute energy this lies 0.67 eV below the combined energies of

BeO and H_2. If, however, the first vibrational levels (zero-point energies) are compared instead, this difference becomes about 1.0 eV.

Force constant calculations in C_1 symmetry performed on this minimum confirmed that it holds up under symmetry breaking. The vibrational analysis yielded the following frequencies: 248 cm^{-1} for the (H_2Be)-O xz wag, 284 cm^{-1} for the (H_2Be)-O yz wag, 925 cm^{-1} for the BeH$_2$ symmetric stretch and 1357 cm^{-1} for the corresponding asymmetric stretch (compared to 995 and 1248 cm^{-1} for BeH$_2^{++}$ [5]), 1578 cm^{-1} for the BeO stretch (compared to an experimental value of 1487 cm^{-1} for free BeO [32]) and 4014 cm^{-1} for the H_2 stretch (compared to an experimental value of 4401 cm^{-1} for free H_2 [32] and a calculated value of 3692 cm^{-1} for the same stretch in BeH$_2^{++}$ [5]).

Analogous calculations on H_2O and Be showed their combined energies to lie about 1.0 eV below OBeH$_2$. Geometry optimization calculations on H_2OBe failed to locate a minimum.

Similar geometry optimization calculations on OLiH$_2$ indicated the presence of a minimum in C_{2v} symmetry. This minimum lies about 0.75 eV above the combined energies of LiO and H_2, and subsequent force constant calculations showed that on symmetry breaking ((H_2Li)-O yz wag) the minimum becomes a saddle point.

Finally, geometry optimization calculations on OBH$_2$, did not show the presence of a minimum, local or otherwise, in any symmetry.

4.3. FLUORINE COMPOUNDS

Calculations analogous to those above were performed on FBeH$_2$, FLiH$_2$ and FBH$_2$, in order to determine whether H_2 trapping similar to that of OBeH$_2$ occurs in this case also. Geometry optimization of FBH$_2$ failed to locate any minimum while that of FLiH$_2$ showed a minimum in C_{2v} symmetry which, however, turned into a saddle upon symmetry breaking.

FBeH$_2$, on the other hand, proved to have a true minimum even after symmetry-unconstrained vibrational analysis. This minimum, which is a local one, lies approximately 4.3 eV below the dissociation products (FH+H+Be) and approximately 1.4 eV above the dissociation products (FBe+H_2). More to the point, analysis of the bonding picture revealed it to be a classical hydride, unlike OBeH$_2$.

5. Conclusion

Unusual species can be formed due to the possibility of volcanic ground PES. A series of them have been computed, together with fundamental properties such as stability, structure and energetics. Future studies should aim at the very accurate mapping of such repulsive PES and at the understanding of the role of the local chemical minima on chemical dynamics.

372

References

[1] C. A. Nicolaides, *Chem. Phys. Lett.* **1989**, *161*, 547.

[2] A. Metropoulos, C. A. Nicolaides and R.J. Buenker, *Chem. Phys.* **1987**, *114*, 1.

[3] H. Yagisawa, H. Sato and T. Watanabe, *Phys. Rev. A* **1977**, *16*, 1352.

[4] E. C. Kemble, *"The Fundamental Physics of Quantum Mechanics"*, McGraw-Hill, N.Y. **1937**, section 31.

[5] P. Valtazanos and C. A. Nicolaides, *Chem. Phys. Lett.* **1990**, *172*, 254.

[6] L. Pauling, *J. Chem. Phys.* **1933**, *1*, 56.

[7] C. A. Nicolaides and A. Zdetsis, *J. Chem. Phys.* **1984**, *80*, 1900.

[8] C. A. Nicolaides, I. Petsalakis and G. Theodorakopoulos, *J. Chem. Phys.* **1984**, *81*, 748.

[9] C. A. Nicolaides, *J. Mol. Str. (Theochem)* **1989**, *202*, 285.

[10] C. A. Nicolaides and P. Valtazanos, *J. Mol. Str. (Theochem)*, in press.

[11] S. Farantos, G. Theodorakopoulos and C. A. Nicolaides, *Chem. Phys. Lett.* **1983**, *100*, 263.

[12] J. K. Perry and D. R. Yarkony, *J. Chem. Phys.* **1988**, *89*, 4945

[13] S. Yu Huang, Z. Sun and W. A. Lester, Jr., *"First Meeting on High Energy Density Materials"*, supported by the U.S. AFOSR, Rosslyn Va **1987**.

[14] C. D. Pibel, K. L. Carleton and C. B. Moore, *J. Chem. Phys.* **1990**, *93*, 323.

[15] I. D. Petsalakis, A. Metropoulos, G. Theodorakopoulos and C. A. Nicolaides, *Chem. Phys. Lett.* **1989**, *158*, 229.

[16] P. J. Knowles and H.-J. Werner, MULTI (**1984**); H.-J. Werner and P. J. Knowles, *J. Chem. Phys.* **1985**, *82*, 5053. P. J. Knowles and H.-J. Werner, *Chem. Phys. Lett.* **1985**, *115*, 259.

[17] C. A. Nicolaides, M. Chrysos and P. Valtazanos, *J. Phys. B* **1990**, *23*, 791.

[18] C. W. Bauschlicher and M. Rosi, *Chem. Phys. Lett.* **1989**, *159*, 485.

[19] J. F. Babb and M. L. Du, *Chem. Phys. Lett.* **1990**, *167*, 273.

[20] C. A. Nicolaides, H. J. Gotsis, M. Chrysos and Y. Komninos, *Chem. Phys. Lett.* **1990**, *168*, 570.

[21] W. B. Scott, in *Aviat. Week Space Technol.* (March 21, **1988**), p. 19.

[22] C. A. Nicolaides, P. Valtazanos and N. C. Bacalis, *Chem. Phys. Lett.* **1989**, *151*, 22. P. Valtazanos, N. C. Bacalis and C. A. Nicolaides, *Chem. Phys.***1990**, *144*, 363.

[23] C. A. Nicolaides and P. Valtazanos, *Chem. Phys. Lett.* **1990**, *173*, 195.

[24] C. A. Nicolaides and P. Valtazanos, *Chem. Phys. Lett.* **1990**, *174*, 489.

[25] T. H. Dunning and P. J. Hay, in *"Methods of Electronic Structure Theory"*, ed.

H. F. Schaeffer III (Plenum Press, N.Y. 1977), p. 1.

[26] The MELDF series of electronic structure programs was developed by L. E. McMurchie, S. T. Elbert, S. R. Langhoff and E. R. Davidson and extensively modified by D. Feller and D. C. Rawlings.

[27] M. W. Schmidt, J. A. Boatz, K. K. Baldridge, S. Koseki, M. S. Gordon, S. T. Elbert and B. Lam, *QCPE Bull.* 1987, 7, 115.

[28] G. J. Kubas, *Accounts Chem. Res.* 1988, *21*, 120; R. H. Crabtree, *Accounts Chem. Res.* 1990, *23*, 95.

[29] J. K. Burdett, J. R. Phillips, M. R. Pourian, M. Poliakoff, J. J. Turner and R. Upmacis, *Inorg. Chem.* 87, *26*, 3054. P. J. Hay, *J. Am. Chem. Soc.* 1987, *109*, 705.

[30] K. Ruedenberg, M. W. Schmidt, M. M. Gilbert and S. T. Elbert, *Chem. Phys.* 1982, *71*, 41,51,65.

[31] C. A. Nicolaides and P. Valtazanos, *Chem. Phys. Lett.*, to be published.

[32] K. P. Huber and G. Herzberg, *"Constants of Diatomic Molecules"*, (Van Nostrand Reinhold, N.Y. 1979).

MOLECULAR HYDROGEN AS A LIGAND IN TRANSITION METAL COMPLEXES

Feliu MASERAS, Miquel DURAN, Agusti LLEDOS and Juan BERTRAN
Departament de Química
Universitat Autònoma de Barcelona
08193 Bellaterra (Barcelona)
Spain

ABSTRACT. In the last few years, many experimental studies have shown the possibility that coordination can be made between a hydrogen molecule and a metallic complex with no breaking of the H-H bond. This discovery has given rise to new ideas on s-bond activation. In this paper, current theoretical explanations for the structure of the molecular hydrogen complexes are reviewed. Special interest is focused on two particular systems for which *ab initio* results are presented. First, analysis of the d^8 $ML_4(H_2)$ system shows the influence of the geometry of the ML_4 fragment on the coordination of H_2 (molecular hydrogen or dihydride). Second, a study on the d^6 $ML_4H(H_2)$ complex is presented, paying special attention to the interaction between the hydride and the molecular hydrogen ligands. Finnally, implications for the oxidative addition mechanism are discussed.

1. Introduction

Activation of small molecules by transition metal complexes is known to play a central role within the field of homogeneous catalysis [1]. Although much is known about the activation of multiple bonds like those of ethylene, carbon dioxide, dioxygen, or dinitrogen through coordination to a metallic center, this subject is still being actively worked upon. In contrast, the field of a σ-bond activation has lately been the subject of an explosion of new ideas. The reason for this is that, although ligands that act by donation of lone pairs, or π-bonding electrons, are common in transition metal chemistry, only recently has it become clear that σ-bonding pairs can also bind to metals. Classically, activation of σ bonds has been thought to proceed through an oxidative addition process that breaks one σ bond and creates two new bonds about the metallic center, the metal atom being oxidized formally by two units. However, the major discovery of recent years has been the possibility of stable intermolecular coordination of a σ bond to a metallic center, thus opening a potential new way for activation. In fact, the discovery of molecular hydrogen complexes by Kubas *et al* [2] has been considered to be "one of the most exciting results in inorganic chemistry in the 1980s". In a molecular hydrogen complex, hydrogen is bound as an intact molecule in an η^2 shape rather than in the usual dihydride form, as shown in scheme 1.

S. J. Formosinho et al. (eds.), Theoretical and Computational Models for Organic Chemistry, 375–396.
© 1991 *Kluwer Academic Publishers.*

Dihydride Complex   Molecular Hydrogen Complex

SCHEME 1

The unequivocal identification of this new class of compounds has supplied a challenge for the experimental techniques of detection. Infrared spectroscopy has been applied in some cases through identification of the H_2 vibration [3]. However, in many cases this technique cannot be used because the band is usually broad and can be masked. NMR spectroscopy has been a more fruitful technique [4], leading to studies of coupling between hydrogen and deuterium atoms, and of relaxation times. The use of relaxation time measurements as a technique of identification has, in fact, been proposed for the first time in molecular hydrogen complexes [4a]. This technique consists essentially of measuring the time that nuclear spins need to recover their equilibrium magnetization once they are perturbed. This relaxation time happens to be a function of the H - H distance. Generalization of the results obtained with this method, however, is still subject to discussion [4c]. Furthermore, the precise determination of the position of metal-bonded hydrogen atoms through X-ray diffraction is known to be problematic [2,5]. The only unequivocal method for identification of molecular hydrogen complexes is neutron diffraction [2,6,7], even though the requirement for big crystals makes the availability of results of this kind still scarce.

The first stable $\eta^2 H_2$ complex, characterized by Kubas in 1984, was $[W(CO)_3(PCy_3)_2(H_2)]^+$ [2]. Neutron diffraction has permitted a clear elucidation of its structure, which turns out to be octahedral with CO and H_2 occupying trans positions. A more remarkable feature, perhaps, is the value for the H - H distance (0.82 Å), significantly longer than that of free H_2 (0.74 Å), yet far shorter than that of dihydride. It is also noticeable that the hydrogen molecule is oriented in a direction parallel to the P - W - P axis. This suggests the presence of important electronic factors, since steric effects would orient H_2 parallel to the C - W - C axis [8]. Thermodynamic studies for this system show the binding energy between the hydrogen molecule and the metal to be at least 10.0 kcal/mol [9]. Thus, there is a real chemical bond between the hydrogen molecule and the metal. In solution, this particular complex presents an equilibrium mixture of the octahedral 6-coordinate $\eta^2 H_2$ form and a 7-coordinate dihydride derived through H - H bond cleavage. The molecular hydrogen isomer is more stable than the dihydride isomer by 1.2 kcal/mol, the energy barrier for the transformation from molecular hydrogen to dihydride being 10.1 kcal/mol.

This landmark discovery in 1984 has been followed by an explosion of publications on the subject [10]. So far, more than sixty stable molecular hydrogen complexes have been clearly characterized, the metals implied ranging from groups 6 to 10, with the metal atoms having any electron count (from d^0 to d^{10}). Furthermore, all these complexes are diamagnetic. In spite of this diversity, most metallic fragments tend to exhibit low basicity, have strong π-acceptor ligands and a positive charge. $d^6\ ML_5$ fragments, in particular, fulfill these requirements, making the class of octahedral d^6 $ML_5\ (H_2)$ complexes the most abundant among all complexes.

The universality of molecular hydrogen complexes has extended to dinuclear complexes [5b], bioinorganic compounds [11], and rare earth metals [6]. Apart from leading to synthesis of new compounds, the study of molecular hydrogen complexes has revived interest on compounds considered classically as polyhydrides. Many of these compounds have been reformulated to contain the dihydrogen ligand, with more reasonable coordination numbers and oxidation states than those predicted for polyhydrides. For instance, the $[Fe(PR_3)_3H_4$ complex that was considered to be a heptacoordinated Fe(IV) polyhydride [12] has recently been understood to contain a H_2 ligand and two hydrides with a much more common octahedral $d^6\ ML_6$ structure, and an oxidation state of (II) for the iron [4a].

Although many different synthetic routes have been described, the three major routes [10b] leading to H_2 complexes seem to be: (1) addition of H_2 to unsaturated 16-electron complexes; (2) photolytic displacement of CO; (3) protonation of metal hydrides.

These molecular hydrogen complexes are the first example of an unprecedented class of compounds in inorganic chemistry with intermolecular interaction between a σ bond and a metallic center. However, they can be understood using isolobal analogy [13], that establishes a bridge between organic and inorganic chemistry. Two fragments are called isolobal when the number, symmetry properties, approximate energy and shape of their frontier orbitals, and the number of electrons in them are similar. In general, a ML_n fragment can be considered isolobal with CH_3^+. The CH_5^+ species, which can be obtained by protonation of methane, actually consists of a H_2 subunit strongly bound to a CH_3^+ fragment [14].

CH_3^+ is a powerful Lewis acid, and so the bonding of an A - H fragment forming a $CH_3^+(A - H)^+$ molecule can better be described in terms of donation of the A - H σ-bonding electrons to the Lewis acid, A - H being the ligand. This weakens the A - H bond, but does not break it, because the resulting three-center molecular orbital is bonding over all three centers. When we consider a transition metal, d_π electrons are often available for back-bonding. This interaction can be described in terms of electron donation from the $M(d_\pi)$ orbital to the A - H σ^* orbital. This back-donation component, if strong enough, can break the A - H bond because the A - H σ^* orbital is being filled.

When A - H bond breaking takes place, an oxidative addition occurs involving conventional two-electron, two-center bonds.

The fact that these molecular complexes are formed more often with transition metals than with organic systems suggests that the back-donation component plays an important part. Other possible evidence for this is that Lewis acids without d_{π} electrons (e.g., BF_3) do not seem to bind H_2. In fact, the high proton acidity of the CH_5^+ system is consistent with an important donation from H_2 to CH_3^+ without a corresponding back-donation. Transition metals seem to decrease the proton acidity of the A - H bond to a much lesser degree.

Replacement of a hydrogen atom of the coordinated H_2 molecule by a CH_3 group would lead to a "molecular methane" complex. The intramolecular interaction between a C - H bond and a metal corresponds, in fact, to the well-known agostic interaction [15]. Moreover, the bridging hydrides in polynuclear complexes can be seen as M - H σ bonds coordinating a metallic center.

Ever since their discovery, the intimate relationship between molecular hydrogen complexes and homolytic activation of a σ bond through oxidative addition has been obvious and has allowed a deeper insight into this important process. Furthermore, new visions of heterolytic activation have recently emerged [10]: proton transfer from a previously coordinated σ bond to another ligand seems to play a role in heterolytic bond breaking reactions much more often than expected. Thus, σ complexes do not seem only to be possible intermediates in oxidative addition processes, but are also becoming compounds with a chemistry of their own (Scheme 2).

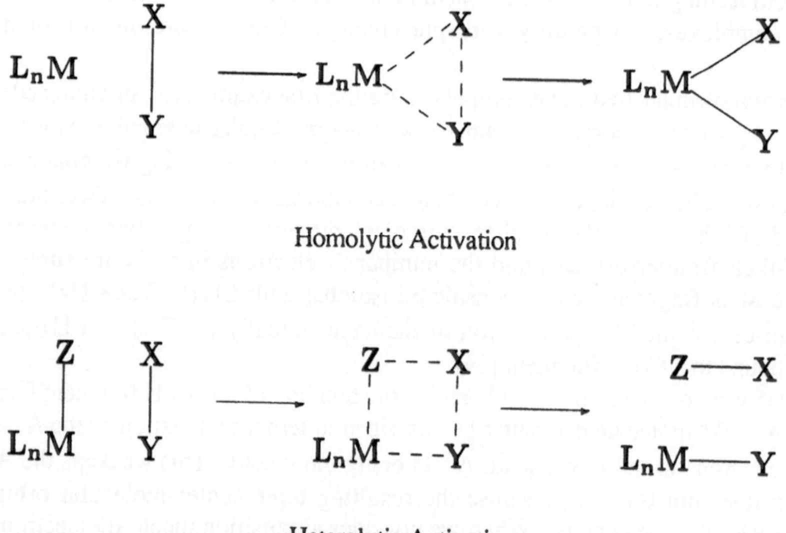

Homolytic Activation

Heterolytic Activation

SCHEME 2

In a new field such as this, which is the subject of a great deal of dynamic scientific activity, there is strong interaction between theory and experiment. In the following section we are going to present some of the theoretical results in this area. First, however, we will give a general view of the currently accepted explanation for the bonding process, obtained essentially from the Extended Hückel method. Second, a more accurate *ab initio* method will be applied to three specific problems, namely: influence of metallic fragment geometry on homolytic activation; interaction between molecular hydrogen and hydride ligands as a model for heterolytic activation, and extrapolation of the results obtained for H - H activation to the case of C - H activation.

2. Extended Hückel Results

The Extended Hückel method has been extensively and successfully applied to organometallic chemistry. This method, which is an extension, proposed by Hoffmann, of the classical Hückel method, provides a cheap and powerful tool for understanding the essential properties and reactivity of organometallic compounds using Molecular Orbital Theory. Following the pioneer work by Hoffmann, the bonding of molecular hydrogen to a metal atom has been shown to proceed fundamentally through two essential HOMO-LUMO interactions: donation of the H_2 σ electrons to an empty metal orbital, and back-donation from a filled metal orbital to the H_2 σ^* orbital.

The Extended Hückel method is especially fruitful when applied together with fragment analysis. The interaction between two fragments can then be reduced to the interactions between selected molecular orbitals of each fragment. The orbitals of the fragments making the leading contribution to the interaction can be determined through energetic and overlap considerations: a small energy difference and a big overlap lead to a strong interaction [16].

Application of this model to molecular hydrogen complexes has been shown to be quite appropriate. The complex is partitioned into two fragments: the hydrogen molecule and the metallic fragment. The hydrogen molecule has only two orbitals (σ and σ^*). Thus, the ability of the metal fragment to coordinate with hydrogen depends essentially on the characteristics of the orbitals of the former being suitable for interaction with the orbitals of the latter.

The three main interactions considered here are: (1) donation from the filled σ_{H_2} orbital to the metallic fragment, thus requiring an empty orbital of relatively low energy pointing in to the right direction; (2) back-donation from a filled orbital of the metallic fragment to the $\sigma^*_{H_2}$ orbital, thus requiring a d_π orbital with a relatively high energy, and (3) repulsive interaction between the σ_{H_2} and the occupied metallic orbitals. This scheme is complicated by the fact that an excess of the σ^* orbital population brings about the breaking of the H - H bond. A subtle compensation between these two effects, donation and back-donation, must, therefore, be established to allow for the formation of the molecular hydrogen complex *vs* the other competing processes, namely oxidation to dihydride and loss of free hydrogen.

The suitability of the d^6 ML_5 metallic fragment for the H_2 coordination has been assessed by this method [17], since these complexes have an empty d orbital pointing to the ligands that is able to accept σ–donation, and an occupied d orbital that is suitable for back-donation. Further, the preference for low basicity and for positive charge, which stabilizes the full set of d orbitals, is explained in terms of a double effect: first, the lowering of the metallic fragment LUMO favoring donation; second, the lowering of the occupied d_π orbital discouraging back-donation, thus avoiding the dihydride. Finally, and perhaps even more significantly, the effect of the different ligands in hydrogen coordination has been discussed. The usual presence of strong π-acceptor ligands like CO in trans with respect to the hydrogen molecule is rationalized as a factor lowering the energy of the d_π orbital and hybridizing it far away from the metal, thus reducing back-donation and impeding the oxidative addition of molecular hydrogen to yield dihydride.

The subtleties of the donation/back-donation equilibria are also analyzed in the effect of π-acceptors in cis complexes. In the d^6 $[W(PR_3)_2(CO)_3(H_2)]$ complex, with one CO ligand in a trans position and the other two CO molecules aligned with the Fe atom, the hydrogen molecule is oriented parallel to the P - Fe - P axis. In this case, the unfavourable back-donation by CO ligands is avoided by the rotation of the H_2 molecule, hence providing an interpretation of the rotational barrier [17].

Apart from the d^6 $ML_5(H_2)$ case, other more extreme cases have been analyzed, d^{10} $ML_3(H_2)$ complexes, in particular [18]. In this case, all metal d orbitals are occupied and donation from the σ_{H_2} orbital does not seem favorable. Further, there is a four electron repulsion interaction between two occupied orbitals. For this kind of complex, the presence of a very strong π-acceptor like NO^+ is required to withdraw electron density from the metal. The tetrahedral arrangement of the ligands around the metal atom is shown to favor the back-donation to the $\sigma^*_{H_2}$ orbital without significantly altering the three-orbital four-electron interaction where the σ_{H_2} orbital is involved.

Burdett *et al* [19] have built a general scheme for this kind of complex, starting from isolobal considerations on polyhydrogen systems, which pays special attention to the possible existence of H_3 units coordinated to metallic complexes. The importance of the electron count (the H_3 fragment being either H_3^- or H_3^+) on the possible structure for this ligand (i.e., open or closed) has also been analyzed.

Despite the success obtained through the use of the Extended Hücked method, its limitations are obvious. The consideration of the electrons as independent and non-interacting, although helpful for the analysis, can lead to incorrect results. Moreover, the total inability of this method to predict reasonable bond distances precludes its application in the comparison of absolute energies, or for the study of reactivity. The need for a more accurate description in some problems of interest has led to the use of the more elaborate *ab initio* method.

3. *Ab Initio* Results

The first *ab initio* calculation of a molecular hydrogen complex was performed by Hay [20] and addressed the analysis of the bonding in the Kubas' complex $[W(CO)_2(PR_3)_2(H_2)]$. This work, which was carried out at the RHF-SCF level, modelled the phosphored ligands as PH_3, and considered a rigid octahedral structure for the metallic fragment. The inner orbitals of tungsten were represented by pseudopotentials, whilst a double -ζ basis set was used for all atoms, together with an additional set of polarization functions on the atoms of the molecular hydrogen. Using this model, optimization of molecular hydrogen and the dihydride complex was performed. The results obtained agreed with previous experiments and confirmed the qualitative predictions of the Extended Hückel method. The main results turned out to be an H - H distance of 0.796 Å and a binding energy of -16.8 kcal/mol for the molecular hydrogen complex, together with an energetic difference of 10.0 kcal/mol between the molecular hydrogen complex and the dihydride complex.

Another interesting *ab initio* study of molecular hydrogen complexes was recently presented by Pacchioni [21]. This study was addressed to the $[Cr(CO)_5(H_2)]$ and $[Cr(CO)_4(H_2)_2]$ complexes, with special interest in the mechanism of exchange of hydrogen atoms in the second complex. This work used pseudopotentials along with a double-ζ basis set for the metal, a double-ζ plus polarization basis set for hydrogens, and a minimal basis set for carbons and oxygens. A rigid structure was adopted for the metallic fragment, and independent one-dimensional optimizations of the position of the hydrogens were performed. The most interesting results turned out to be the nature of the possible intermediates for the hydrogen exchange reaction, and the prediction of the possible existence of a trihydrogen unit.

Our group has carried out some *ab initio* calculations on two subjects in particular, namely, the influence of the ML_4 geometry on hydrogen coordination [22], and molecular hydrogen complexes with a hydride ligand [23].

3.1 INFLUENCE OF THE ML_4 FRAGMENT GEOMETRY IN THE HYDROGEN COORDINATION

Of the molecular hydrogen complexes that were first synthesized, none were the d^8 $ML_4(H_2)$ type. This is not at all surprising because it is well known that d^8 ML_4 systems yield oxidative addition with formation of a dihydride. Application by Bianchini *et al* of the novel tripodal phosphine ligand to the rhodium atom led to the synthesis of the first compound in this class, the $[Rh(P(CH_2CH_2PPh_2)_3(H_2)]^+$ complex [24].

This system also has the interesting property of exhibiting two isomeric forms of the metallic fragment, one with C_{2v} symmetry and another with C_{3v} symmetry. As shown in Scheme 3, each isomer is associated to a different H_2 coordination, namely dihydride and molecular hydrogen. The need to devise a general scheme which is able to explain these experimental data and which can be extended to related complexes led us to perform calculations on this system.

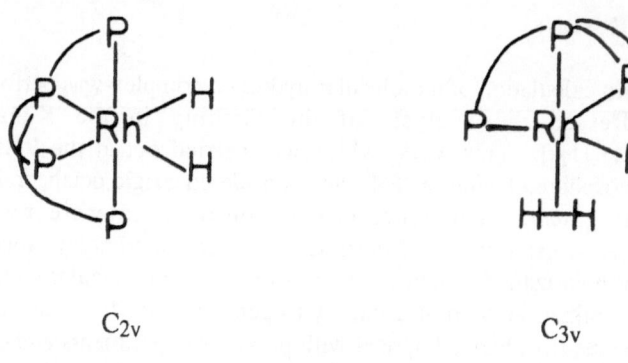

C_{2v} C_{3v}

SCHEME 3

The two different isomers were modelled through rigid $[Rh(PH_3)_4]^+$ fragments. Calculations at the RHF-SCF level were made through use of pseudopotentials for the rhodium and phosphorus atoms. A double-ζ basis set was used for Rh and active hydrogen atoms, whereas a minimal basis set was employed for phosphines. The relative positions of the hydrogen atoms was completely optimized.

Optimization of the system with the different fixed metallic fragment structures led to two clearly different minima, which are presented in Figure 1. For the C_{2v} structure a dihydride was obtained (Figure 1a), whereas for the C_{3v} structure a molecular hydrogen was obtained (Figure 1b), in good agreement with experimental data. As expected, the optimized H - H distance of 0.80 Å for the molecular complex is appreciably longer than that of the free H_2 molecule. As regards the energetics, dihydride is by far the most stable complex, because it is 30.9 kcal/mol more stable than the molecular hydrogen complex, 41.5 kcal/mol more stable than the isolated fragments C_{2v} $[Rh(PH_3)_4]^+ + H_2$ and 51.6 kcal/mol more stable than C_{3v} $[Rh(PH_3)_4]^+ + H_2$.

This behaviour was explained by using molecular orbital analysis, which was restricted to the orbitals of the $[Rh(PH_3)_4]^+$ fragments for the sake of clarity. In what follows, the axis orientation will be that depicted in Figure 1. The first problem to be solved was the choice of the molecular orbitals which are meaningful to this particular analysis. A first criterion was to consider only the orbitals having a large contribution of rhodium d orbitals. In both structures, they happen to be the four highest energy occupied orbitals and the lowest unoccupied orbital. This is a perfectly satisfactory result, since these complexes are formally d^8. Two of these five orbitals were eliminated, namely; those having a node on the incoming plane of the hydrogen molecule, i.e., the molecular orbitals that are fundamentally d_{yz} and d_{xy}, which turn out to be filled. Thus, three orbitals were considered for each isomeric form, one occupied and two unoccupied. Energies for this selected set of molecular orbitals are presented in Table 1.

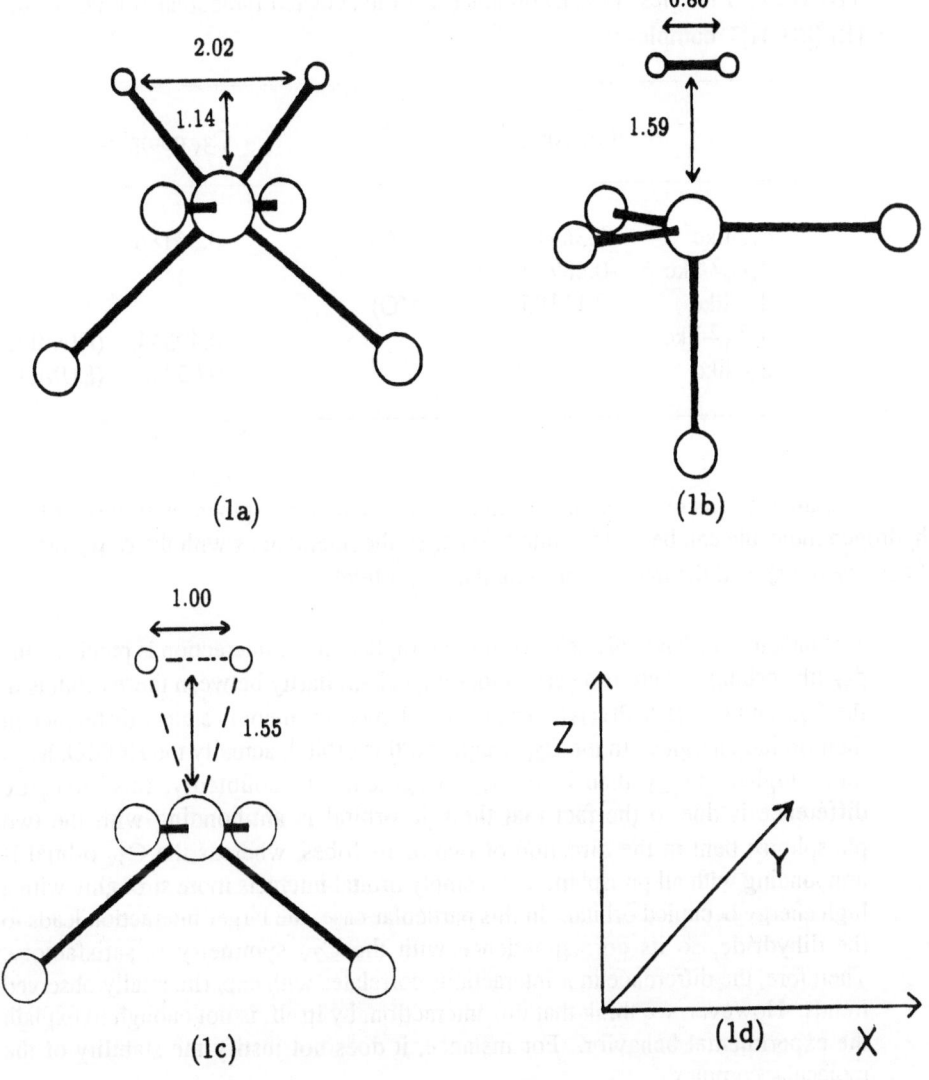

Figure 1. Optimized structures of the dihydride complex C_{2v} $[Rh(PH_3)_4H_2]^+$ (1a), the molecular hydrogen complex $[Rh(PH_3)_4H_2]^+$ (1b), and the transition state for the direct oxidative addition of H_2 to the square-planar complex (1c). In 1d, the axis orientation that is used in the discussion of the text is shown.

TABLE 1: Energies (in a.u.) obtained for the selected molecular orbitals of the $[Rh(PH_3)_4]^+$ complexes.

	C_{2v} complex		C_{3v} complex	
d_{xz}-like	-0.36919	(HOMO)	-0.55186	
$d_{x^2-z^2}$-like	-0.48767			
d_{y^2}-like	-0.11185	(LUMO)		
$d_{x^2-y^2}$-like			-0.40544	(HOMO)
d_{z^2}-like			-0.05990	(LUMO)

The interactions of these three orbitals of the metallic fragment with those of the hydrogen molecule can be divided into two parts: the interactions with the $\sigma^*_{H_2}$ orbital (π-interactions), and the interactions with the σ_{H_2} orbital.:

1. π-interaction. The only selected orbital implied in a π interaction is precisely the d_{xz}-like orbital. There is a certain topological similarity between these orbitals in the C_{2v} and C_{3v} $[Rh(PH_3)_4]^+$ fragments. However, there is a clear difference in their orbital energies. In the C_{2v} fragment, this orbital, actually the HOMO, has a much higher energy than in the C_{3v} fragment. Undoubtedly, this energetic difference is due to the fact that the C_{2v} orbital is antibonding with the two phosphines bent in the direction of two of its lobes, whereas the C_{3v} orbital is nonbonding with all phosphines. An empty orbital interacts more stronghly with a high energy occupied orbital. In this particular case, the larger interaction leads to the dihydride, so its correspondence with the C_{2v} symmetry is satisfactory. Therefore, the difference in π interactions correlates with experimentally observed trends. However, we think that this interaction, by itself, is not enough to explain the experimental behavior. For instance, it does not justify the stability of the molecular complex.

2. σ interactions. It is not possible to compare directly each implied orbital of the C_{2v} fragment with those of the C_{3v} symmetry because they correspond to different hybridizations. The two molecular orbitals that remain to be analyzed must be d_{x2-y2}- and d_{z2}-like. It must be noted that in the C_{3v} complex both orbitals can be recognized having this shape, but in the C_{2v} system they are recombined to give a pair of orbitals that could be identified as being essentially d_{x2-y2}- and d_{z2}-like. We think that the difference in electronic occupation of each complex symmetry is the key factor in explaining their differing behavior towards the hydrogen molecule. Let us, therefore, study the resulting situation for each complex:

(a) C_{2v} $[Rh(PH_3)_4]^+$. The LUMO here, which should receive electrons from the hydrogen molecule, orientates its larger lobes perpendicularly. Furthermore, there is a relatively high-energy occupied orbital, the $d_{x^2-z^2}$-like one, orientated towards the vacant site of the coordination sphere. This last orbital will cause a strong repulsion with the H_2 σ^* orbital.

(b) C_{3v} $[Rh(PH_3)_4]^+$. The most external lobe of the LUMO is orientated towards the vacant site, while the $d_{x^2-y^2}$-like orbital, actually the HOMO, concentrates most of its electron density in another region of the molecule, so repulsions are minimized.

These results showed clearly that the geometrical arrangement of the ligands around the metallic atom has a decisive influence on the reactivity of the complex. In other words, apart from electronic and steric effects, there is a more subtle way in which they can entirely alter their reactivity. In this particular case, a change in their geometry led to completely different species. Other complexes with strongly restricted geometry have also been shown to contain coordinated molecular hydrogen, probably owing to the impossibility of becoming dihydrides.

In a supplementary set of calculations performed on this system without any geometrical restrictions, d^8 ML_4 complexes behaved according to predictions. The interaction of the stable square-planar complex with free hydrogen led to only one minimum: the octahedral dihydride. Dihydride was estimated to be 10.1 kcal/mol less stable than the reactants, showing the importance of distortion from the square-planar complex caused by the tripodal phosphine in the system of Bianchini et al. The transition state for the homolytic activation process was located, its energy lying 33.4 kcal/mol higher than the reactants, which in turn means an energy 23.3 kcal/mol above that of the dihydride complex. The most interesting feature of its geometry, which is presented in Figure 1c, is that the H - H distance is quite close to that of the C_{3v} molecular hydrogen complex. In fact, the H - H distance for this transition state is 1.00 Å, much closer to the the molecular hydrogen complex distance than to the 2.02 Å calculated for the dihydride complex. In contrast, the geometry of the metallic fragment is similar to that of the C_{2v} dihydride. The intimate relationship between the molecular hydrogen complexes and homolytic activation processes was again demonstrated.

Intervention of molecular hydrogen complexes in hydrogenation processes can be discussed in relation to these results. The classically known process, e.g. Wilkinson catalysis, proceeds through the initial formation of a dihydride. Direct intervention of a molecular hydrogen complex in catalic hydrogenation processes has been proposed for $[OsH_3(H_2)(PR_3)_3]^+$ [25]. However, in this case the molecular hydrogen complex acts merely as a ready source for the unsaturate hydride transient species $[OsH_3(PR_3)_3]^+$. This transient species reacts through π-cordination of ethylene followed by attack of a hydride on the double bond.

A genuine direct intervention of a molecular hydrogen complex in a catalytic hydrogenation process was suggested by Bianchini *et al* from a comparison of the different behavior of the complexes $[M(PP_3)]^+$ (M=Co, Rh, Ir; PP_3=tripodal phosphine) in the presence of a mixture of hydrogen and an alkene such as dimethyl maleate [26]. When the metal is iridium, H_2 is oxidatively added to give a classical dihydride that is unreactive *vs* the olefin. The metal-hydrogen bonds are too strong to permit olefin insertion. In the case of rhodium, the molecular hydrogen complex is formed, and the extent of the backdonation from the metal to the σ_{H_2} orbital is small enough to prevent direct formation of the dihydride complex, but big enough to allow easy activation of the H - H bond. Thus, when the metal is rhodium, there is an efficient hydrogenation of olefin under very mild conditions. When the metal is cobalt, the molecular hydrogen complex is also formed, but the H - H bond is not activated enough to allow for the transference of hydrogen atoms to the alkene.

Moreover, direct reaction of H_2 ligands in catalysis has been proved. Recently, hydrogen coordinated as a molecular hydrogen to a metal has been proved to be able to hydrogenate acetone to give isopropyl alcohol [27]. In order to understand these new kinds of processes, a study on heterolytic activation of hydrogen complexes is presented in the next section.

3.2 MOLECULAR HYDROGEN COMPLEXES WITH A HYDRIDE LIGAND

As mentioned above, protonation of hydrides is one of the major routes to molecular hydrogen complexes. If a dihydride is protonated, the product will contain three hydrogen atoms coordinated in two different forms: one molecular hydrogen ligand and one hydride ligand.

This kind of compound is worthy of special attention on the part of theoretical chemists for several reasons. On the one hand, their identification is complicated, because of the problems associated with the experimental localization of hydrogen atoms bonded to metallic atoms. On the other hand, experimental observation of the interchange reaction between the molecular H_2 and the hydrogen hydride shows the power of these compounds as models for the heterolytic activation reaction.

Several molecular hydrogen-hydride compounds have been considered in recent literature. An especially common type is constituted by octahedral iron complexes with four PR_3 groups, one molecular hydrogen and one hydride as ligands. One of these complexes, trans-$[Fe(R_2PCH_2CH_2PR_2)_2H(H_2)]^+$, is among the best known molecular hydrogen complexes. For this complex, experimental data from neutron diffraction is available [6], thermodynamics on the acidity of the molecular hydrogen coordinated to the metal [28], and inelastic neutron scattering on the rotation barrier of the hydrogen molecule [29]. For cis isomers, with the molecular hydrogen and hydride ligand adjacent to each other, there is less experimental data available. Neverhteless, the existence of the cis isomer in equilibrium with the trans isomer has been postulated from NMR studies of the $[Fe(P(OEt)_3)_4H(H_2)]^+$ complex [30]. Likewise, the $[Fe(P(CH_2CH_2PR_3)_3)H_3]^+$ species, containing the tripodal phosphine ligand which sterically prevents the trans disposition between the ligands, seems to be a molecular hydrogen-hydride complex [31].

The interest of polyhydrides and the copious amount of unexplained data on complexes of the type [Fe(PR$_3$)$_4$H(H$_2$)] led us to carry out theoretical calculations on these compounds. In particular, calculations were made on the four systems presented in scheme 4. In all complexes, the PR$_3$ ligands were modelled by PH$_3$ groups.

All calculations were made with an all-electron molecular-orbital *ab initio* method at the RHF-SCF level. The basis set used was of double-ζ quality for the metal, whereas a double-ζ basis set with a polarization set for the active hydrogens was used. Finally, a minimal basis set was employed for phosphines. The internal structure of phosphines and the Fe - P distances were frozen, but the remaining geometrical parameters, including all angles and dihedral angles about the metal atom, were optimized. The optimization of the geometrical structure of the metallic fragment was necessary because of the confusing existing experimental data for the cis isomer. Binding energies of the molecular hydrogen ligand to the metallic fragment were calculated as the difference between the energy of the optimized ML$_5$(H$_2$) system and those of the isolated systems H$_2$ and ML$_5$, the geometry of the last fragment being frozen to that in the related ML$_5$(H$_2$) complex.

Complex 1, with five phosphines and a molecular hydrogen as ligands, was introduced only for comparative purposes. Its geometry is depicted in Figure 2a and its features are discussed below.

The optimized geometry of complex 2 (Figure 2b), the trans isomer, allows the validity of the computational method to be tested through comparison with experimental data for the [Fe(P(CH$_2$CH$_2$PR$_3$)$_3$)H$_3$]$^+$ complex [6]. The result is quite satisfactory, as the Fe - H$_1$ distance is very well reproduced (1.631 Å, *vs* the experimental value of 1.62 Å), and the phosphines are bent in the same direction. Other parameters are not so

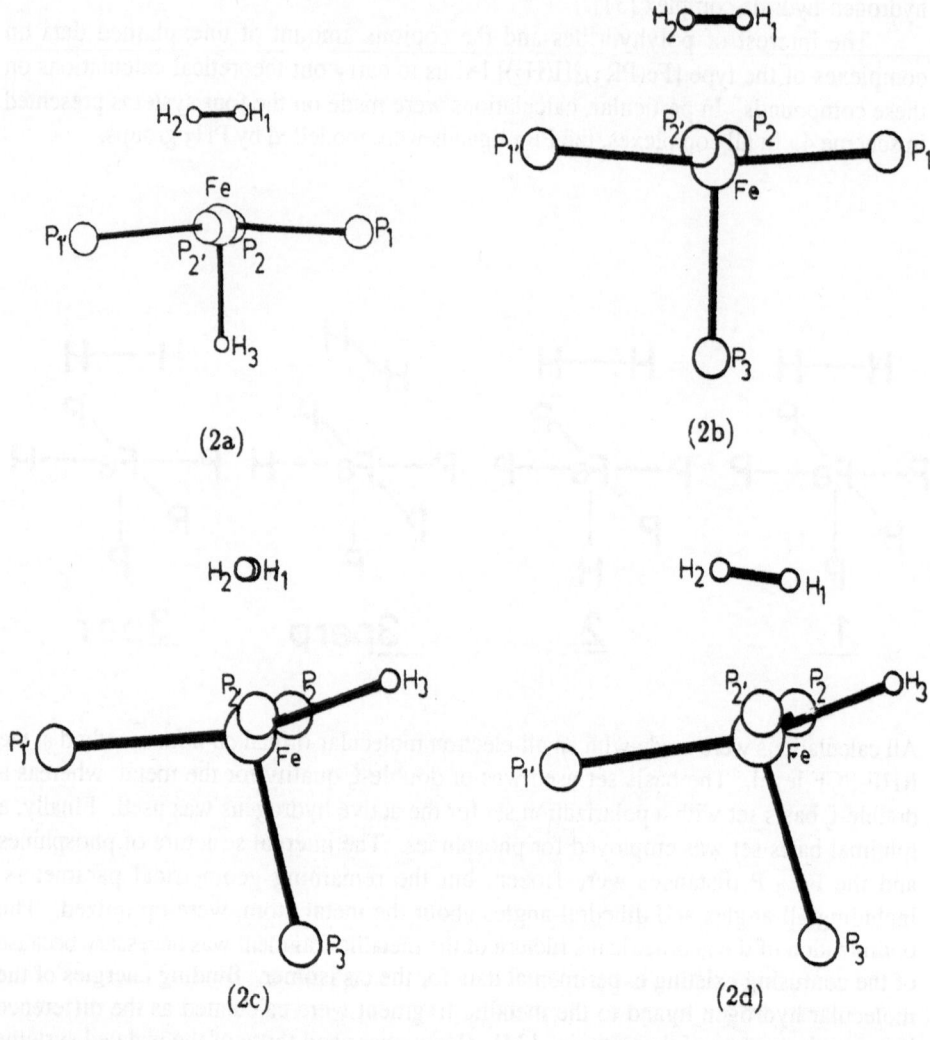

Figure 2. Optimized geometries for the complexes $[Fe(PH_3)_5(H_2)^{2+}]$ (2a), trans-$[Fe(PH_3)_4H(H_2)^{2+}]$ (2b), cis-$[Fe(PH_3)_4H(H_2)^{2+}]$-perp. (2c) and cis-$[Fe(PH_3)_4H(H_2)^{2+}]$-par. (2d).

well reproduced (e.g., the H - H bond length, which is 0.82 Å experimentally, but is computed to 0.770 Å), yet it seems that the overall result of the comparison is fairly good.

Comparison of the optimized geometries for $\underline{1}$ and $\underline{2}$ permits understanding of the different trans effects of phosphine and hydride ligands. Two remarkable differences appear: the bending direction of phosphines is opposite in the two complexes; and the bonding between the metal and hydrogen molecule seems to be stronger in $\underline{2}$ than in $\underline{1}$, as can be observed both in the Fe - H_1 and H_1 - H_2 distances.

The effect of bending of the cis ligands (in this case phosphines) on the strength of the bond with ligands bearing π-acceptor properties has already been discussed by other authors from Extended Hückel calculations [32]. In short, such a deformation produces a hybridization of the orbital of the metal atom responsible for the back-donation to the σ^* orbital of the hydrogen molecule; this fact favors this process from both the energetic and overlap points of view. This theory is consistent with the fact that the phosphines lying in the plane defined by the metal and the hydrogen molecule are bent away to a greater degree than are those lying outside that plane. Nevertheless, one may ask why, in complex $\underline{1}$, phosphines bend towards the hydrogen molecule if this leads to a weakening of the bonding between the metallic center and the hydrogen molecule. This behavior is attributed to the fact that for complex $\underline{1}$ the interaction with the hydrogen molecule is so weak that the metallic fragment retains some memory of the d^6 ML_5 $[Fe(PH_3)_5PR_3)_5]^{2+}$ complex from which it originates: a square pyramidal structure having the basal phosphines bent towards the vacant site.

The different strength of the bonding with the hydrogen molecule is at first sight surprising. The hydride, when situated in the trans position with respect to the hydrogen molecule, causes a stronger bond than phosphine does. Actually, this is just the opposite of what one would expect from the traditional trans effect, since the hydride is a σ-donor ligand which is obviously stronger than phosphine. This strange trans effect of hydride favoring molecular hydrogen coordination to a metallic center has already been mentioned by other authors [10b]. Although no definitive explanation for this behavior has been found, we think that it could be attributed to an effect of hydride (in fact, carrying a negative charge) increasing the energy of the metal orbitals responsible for backdonation, thus favoring the π interaction.

As shown in scheme 4, two different orientations of the hydrogen molecule were considered for the cis isomer, one with the H - H bond perpendicular to the P - Fe - H axis, and another orientation with the bond parallel to that axis. This was done because complex $\underline{3}$ is likely to be more sensitive to the orientation of the hydrogen molecule. In fact, the difference in the position of the hydrogen ligand in the two structures, shown in Figure 2, is significant.

The most remarkable aspect of the bonding in the hydrogen molecule in the $\underline{3}$-perp. isomer (Figure 2c) is its great similarity to that found in complex $\underline{1}$: Distances between the Fe_1, H_1, and H_2 atoms are coincident to within 0.01 Å, their value being 1.70 for the iron-hydrogen bond and 0.74 for the hydrogen-hydrogen bond. An immediate consequence is that the ligands in an octahedral complex affecting the coordination of a hydrogen molecule are those in trans position, or those in cis positions which are orientated in such a way that they lie parallel to the H - H bond; however, ligands in cis orientated perpendicular to the H - H bond do not affect the coordination of the hydrogen molecule.

In complex 3-par., (Figure 2d) the bonding between the hydrogen molecule and the metal is much stronger than in complex 3-perp., as demonstrated by the longer H - H_2 distance. Perhaps the most striking aspect of its geometry is the asymmetry in the bonding between both hydrogens and the metal, the hydrogen molecule being bent towards the hydride. This kind of asymmetry, which otherwise is not found in any other optimized complex, is also present to a lesser extent in the experimental structure of the only molecular hydrogen complex with a hydride in a cis configuration known to date [7]. This behavior demonstrates the existence of an attraction between the two ligands which are cis to each other: the hydride and the hydrogen molecule. This attractive effect leads to a difference of 7.0 kcal/mol between the 3-par. and 3-perp isomers.

In order to gain further insight into the peculiarities of the bonding between the iron atom and the hydrogen molecule in this configuration, we calculated the total electron density in the plane containing the metal and the hydrogen molecule. The most interesting stationary points of the electron density function are the so-called bond critical points, which have only one positive eigenvalue of the second derivative of such function [33]. The bond critical points indicate the presence of a covalent chemical bond. In Figure 3, where a contour plot of the electron density for complex 3-par. is presented, no bond critical point is detected between the hydrogen molecule and the hydride ligand. The interaction between the hydride and the hydrogen molecule, therefore, cannot be described in terms of the incipient formation of a covalent bond.

Another reason that could account for the strong interaction between the two ligands in cis, the hydride and the molecular hydrogen, is an electrostatic effect. This is feasible if one considers the negative charge supported by the hydride. Although Mulliken populations are not the best method to analyze calculations on organometallic complexes, inspection of atomic charges deduced from them is, in this case, quite illuminating. In complex 3-par., H_3 (the hydride) supports a charge of -0.40 au, H_1 supports +0.10 au, and H_2 supports -0.13 au. Thus, the difference in charge between H_1 and H_2, which is minimal in other cases, increases in complex 3-par. to 0.23 au. This fact can be explained by the close proximity of the hydride ligands; this polarizes the hydrogen molecule, which in turn is coordinated to the metal. The hydride ligand, in fact, has special characteristics which make it suitable for this kind of interaction: (a) it is very close to the metal, and is thus close to other ligands; and (b), there is no directionality in its spheric 1s orbital.

Previous discussion in this section has been based on structural results, assuming they correlate well with the implied energetics. This hypothesis is fully supported by the calculated binding energies, which are presented in Table 2. This is the obvious interpretation for the fact that the bond is markedly stronger in complexes 2- and 3-par. than in the other two systems.

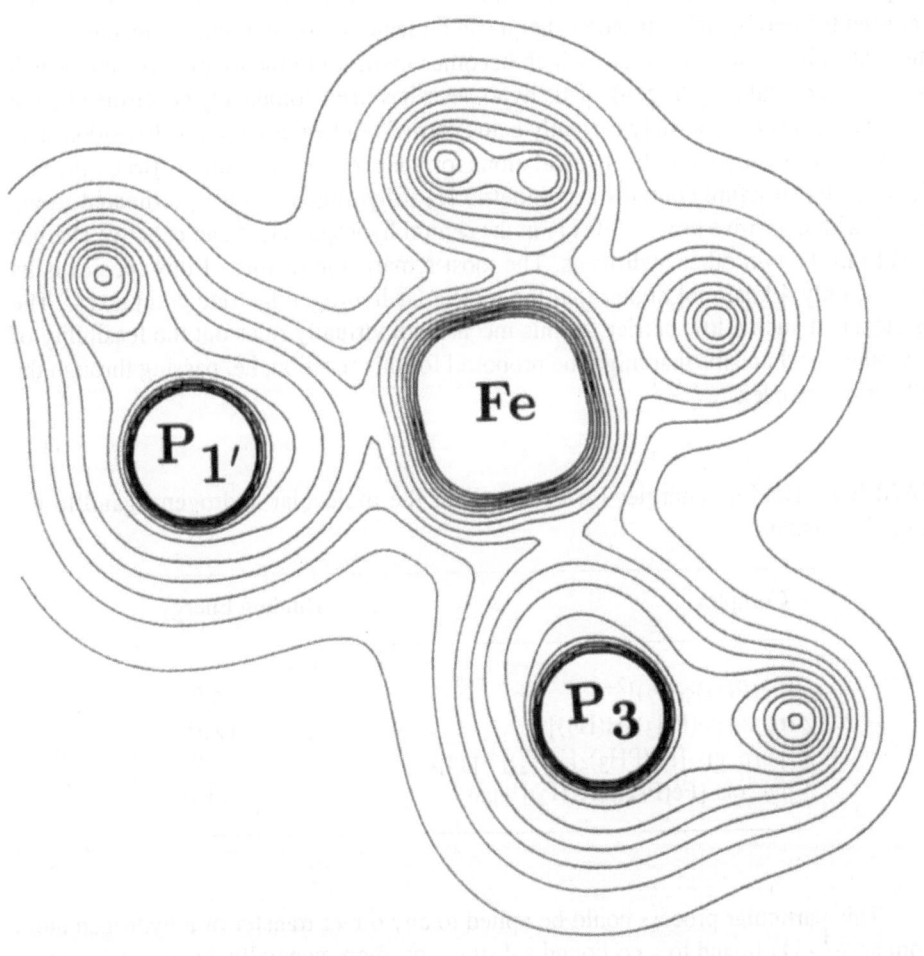

Figure 3. Electron isodensity contour map of the optimized system <u>3</u>-par. The lines plotted correspond to the values of 0.01, 0.03, 0.07, 0.09, 0.12, 0.16, 0.20, 0.24, 0.28, 0.34, and 0.40 a.u.

The existence of a hydrogen atom with a positive charge suggests a relationship with the proton acidity of molecular hydrogen compounds. Analysis of the structure of the 3-par. complex can provide a first clue to the mechanism of the exchange reaction between the two ligands. It seems that it can be seen as a proton transfer reaction from the molecular hydrogen ligand (which becomes hydride) to the hydride ligand (which becomes molecular hydrogen). This hypothesis has been completely confirmed by the preliminary results of a study currently being carried out by our group. A transition state has been located by theoretical calculations for the exchange reaction between the two ligands. Its structure contains a symmetric H_3 trihydrogen ligand coordinated to the metal, with a positive charge (+0.11) in the central hydrogen atom and negative charges (-0.28) in the two other hydrogens. The most remarkable feature of this species is its energy, only 4 kcal/mol higher that the molecular-hydrogen hydride complexes. The existence of such a low barrier for this mechanism virtually rules out the feasibility of any other mechanisms that might be proposed for this reaction, i.e. passing through the trihydride complex.

TABLE 2: Binding energies (in kcal/mol) of the molecular hydrogen ligand to the metallic fragment.

Complex	Binding Energy
1 $[Fe(PH_3)_5(H_2)]^{2+}$	-4.9
2 trans-$[Fe(PH_3)_4H(H_2)]^+$	-12.0
3 -perp. cis-$[Fe(PH_3)_4H(H_2)]^+$-perp	-3.9
3-par. cis-$[Fe(PH_3)_4H(H_2)]^+$-par	-11.0

This particular process could be aplied to any direct transfer of a hydrogen atom from an η^2 - H_2 ligand to a co-bound substrate, or, more generally, heterolytic σ-bond activation. The existence and structure of molecular hydrogen complexes suggests that homolytic and heterolytic activation may be just two aspects of the same reaction. In the heterolytic process, a bound H_2 ligand transfers a proton to an electron pair of a base of the medium, or to a bonding electron pair of the complex, while in the homolytic process, the proton is effectively transferred to a lone pair of the metal.

The H_2/hydride equilibrium gives an unprecedented opportunity to study the most fundamental bond-breaking/forming process at a metal site. It appears that the H_2 ligand has a chemistry of its own and that complete H - H rupture at a metal center may not be a necessary initial step in catalytic hydrogenation, since deprotonation of H_2 ligands has been demonstrated. This new perspective of σ-bond activation leads to a new vision of organometallic chemistry, the outcome of which must await further study.

4. Generalization to Other Systems

The possibilities opened in the field of σ-bond activation by the study of molecular hydrogen complexes appear to be only the very tip of the iceberg. Some conclusions of this study can be extrapolated to the activation of other bonds. At present, special interest is being focused on the C - H σ bond, whose selective activation is an obvious goal of any catalytic or synthetic process involving organic compounds.

Now, one must wonder if there is any limit to the ability of metals to bond in a stable way to other σ bonds, including those in non-reactive molecules like alkanes. In fact, evidence for an intermediate methane complex has been found at low temperature in the reductive elimination of methane from a cationic rhenium methyl hydride [34]. The *ab initio* theoretical study of the intermolecular process of oxidative addition of a methane C - H bond has led to the location of transition states where the bond is partially broken [35]. The same results have been fond for intramolecular oxidative additions which are related to agostic interactions. In fact, agostic interaction itself is a kind of non-oxidative coordination [15,36]. For unsaturated substrates like ethylene, the activation of a C - H bond seems to follow an intermolecular path, without any previous coordination of the double bond. A feasible explanation consists here of the fact that metal orbitals suitable for ethylene coordination are the same as those which are responsible for oxidative addition, thus making the processes competitive [37].

In the cases described above, the partial breaking of C - H bond is observed only in transition states. The present research, for theoreticians and experimentalists alike, involves searching for the existence of a "molecular C-H" complex. Nevertheless, there are some differences between the C - H and H - H bonds that should be mentioned. In agostic interactions, which are actually C - H chelations, available crystalographic data look quite different from those available for molecular hydrogen complexes [38]. For instance, coordination is not strictly side-on, but the C - H bond is usually skewed in such a way that the H atom is closer to the metal. It is also significant that the lengthening of the C - H bond, on coordination, is very slight. Since backdonation from the metallic d_π orbital is expected to be the main factor affecting the C - H bond lengthening, the explanation may be that interaction with the σ^*_{C-H} is more difficult than with σ^*_{H-H}. Reasons for this behavior seem to be the greater hindrance of alkanes and the accessibility of only one lobe of the σ^*_{C-H} orbital [39]. Two electron three-center bonds seem to be formed more readily when two of the three centers are H. Thus, methane is probably intrinsically a poorer ligand than hydrogen.

Theoretical calculations comparing the reactivity of H - H and C - H bonds towards organometallic complexes confirm these different reactivities. For instance, Goddard *et al* have found barriers for oxidative additions of H_2, CH_4 and CH_3CH_3 to palladium of 5.1, 30.5 and 38.6 kcal/mol, respectively [35b]. If the energetic relationship between transition states is preserved for molecular complexes, a "molecular C - H" complex must be much weaker than a "molecular H - H" complex.

This notwithstanding, the differences between the two systems should not conceal their similarities. Just as H_2 complexes tend to behave as proton acids, agostic C - H systems tend to be acidic. Further, Siegbahn *et al* [35] have shown how the transition state for the methane elimination from a methyl hydride nickel complex has the same charge distribution as that already presented in this paper for the exchange of molecular hydrogen and hydride. Hence, the study of molecular hydrogen complexes will undoubtedly provide deeper understanding of the important C - H activation reaction.

It is quite clear that molecular hydrogen complexes open a new field in experimental and theoretical organometallic chemistry. Only future exploration will show if this new, promising land is merely an island or if it is, on the contrary, an entire continent, open to organometallic chemistry.

5. References

[1] Collman, J. P.; Hegedus, L. S.; Norton, J. R.; Finke, R. G. *Principles and Applications of Organotransition Metal Chemistry* ; University Science Books: Mill Valley, California, 1987.

[2] Kubas, G. J.; Tyan, R. R.; Swanson, B. I.; Vergamini, P. J.; Wasserman, H. J. *J. Am. Chem. Soc.* **1984**, *106*, 451-452.

[3] (a) Sweany, R. L. *J. Am. Chem. Soc.* **1985**, *107*, 2374-2379. (b) Gadd, G. E.; Upmacid, R. K.; Poliakoff, M.; Turner. J. J. *J. Am. Chem. Soc.* **1986**, *108*, 2547-2552.

[4] (a) Hamilton, D. G.; Crabtree, R. H. *J. Am. Chem. Soc.* **1988**, *110*, 4126-4133. (b) Arliguie, T.; Border, C.; Chaudret, B.; Devillers, J.; Poilblanc, R. *Organometallics* **1989**, *8*, 1308-1314. (c) Cotton, F. A.; Luck, R. L. *J. Am. Chem. Soc.* **1989**, *111*, 5757-5761.

[5] (a) Morris, R. H.; Sawyer, J. F.; Shiralian, M. ; Subkowski, J. D. *J. Am. Chem. Soc.* **1985**, *107*, 5581-5582. (b) Hampton, C. ; Cullen, W. R.; James, B. R.; Charland, J.-P. *J. Am. Chem. Soc.* **1988**, *110*, 6918-6919.

[6] Ricci, J. S. ; Koetzle, T. F.; Bautista, M. T.; Hofstede, T. M.; Morris, R. H.; Sawyer, J. F. *J. Am. Chem. Soc.* **1989**, *111*, 8823-8827.

[7] Van der Sluys, L. S.; Eckert, J.; Eisenstein, O.; Hall, J. H.; Huffman, J. C.; Jackson, S. A.; Koetzle, T. F.; Kubas, G. J.; Vergamini, P. J.; Caulton, K. G. *J. Am. Chem. Soc.* **1990**, *112*, 4831-4841.

[8] Eckert, J.; Kubas, G. J.; Hall, J. H.; Hay, P. J.; Boyle, C. M. *J. Am. Chem. Soc.* **1990**, *112*, 2324-2332.

[9] (a) Gonzalez, A. A.; Zhang, K.; Nolan, S. P.; Lopez de la Vega, R.; Mukerjee, S. L.; Hoff, C. D. *Organometallics* **1988**, *7*, 2429-2435. (b) Gonzalez, A. A.; Hoff, C. D. *Inorg. Chem.* **1989**, *28*, 4295-4297.

[10] (a) Kubas, G. J., *Acc. Chem. Res.* **1988**, *21*, 120-128. (b) Kubas, G. J. *Comments Inorg. Chem.* **1988**, *7*, 17-40. (c) Crabtree, R. H.; Hamilton, D. G. *Adv. Organomet. Chem.* **1988**, *28*, 299-338. (d) Crabtree, R. H. *Acc. Chem. Res.* **1990**, *23*, 95-101. (e) Ginzburg, A. G.; Bagatur'yants, A. A. *Organometallic Chemistry in the USSR*, **1989**, *2*, 111-126.

[11] Collman, J. P.; Wagenknecht, P. S.; Hembre, R. T. Lewis, N. S. *J. Am. Chem. Soc.* **1990**, *112*, 1294-1295.

[12] Aresta, M.; Giannoccaro, P.; Rossi, M.; Sacco, A. *Inorg. Chim. Acta* **1971**, *5*, 115-118.

[13] Hoffman, R. *Angew. Chem. Int. Ed. Eng.*, **1982**, *21*, 711-724.

[14] Olah, G. A. *Curr. Top. Chem.* **1979**, *80*, 19.

[15] Brookhart, M. ;. Green, M. L. H. *J. Organomet. Chem.* **1983**, *250*, 395-408.

[16] Albright, T. A.; Burdett, J. K.; Whangbo, M.-H. *Orbital Interactions in Chemistry*; Wiley: New York, 1985.

[17] Jean, Y.; Eisenstein, O.; Volatron, F.; Maouche, B.; Sefta, F. *J. Am. Chem. Soc.* **1986**, *108*, 6587-6592.

[18] Volatron, F.; Jean, Y.; Lledós, A. *Nouv. Jour. Chim.* **1987**, *11*, 651-656.

[19] (a) Burdett, J. K.; Pourian, M. R. *Organometallics* **1987**, *6*, 1684-1691. (b) Burdett, J. K.; Pourian, M. R. *Inorg. Chem.* **1988**, *27*, 4445-4450.

[20] Hay, P. J. *J. Am. Chem. Soc.* **1987**, *109*, 705-710.

[21] Pacchioni, G. *J. Am. Chem. Soc.* **1990**, *112*, 80-85.

[22] Maseras, F.; Duran, M.; Lledós, A.; Bertrán, J. *Inorg. Chem.* **1989**, *28*, 2984-2988.

[23] Maseras, F. Duran, M.; Lledós, A.; Bertrán, J. *J. Am. Chem. Soc.*, submitted.

[24] (a) Bianchini, C.; Mealli, C.; Peruzzini, M.; Zanobini, F. *J. Am. Chem. Soc.* **1987**, *109*, 5548-5549. (b) Bianchini, C.; Masi, D.; Meli, A.; Peruzzini, M.; Zanobini, F. *J. Am. Chem. Soc.* **1988**, *110*, 6411-6423.

[25] Lundquist, E. G.; Folting, K.; Streib, W. E.; Huffman, J. C.; Eisenstein, O.; Caulton, K. G. *J. Am. Chem. Soc.* **1990**, *112*, 855-863.

[26] Bianchini, C.; Mealli, C.; Meli, A.; Peruzzini, M.; Zanobini, F. *J. Am. Chem. Soc.* **1988**, *110*, 8725-8726.

[27] Harman, W. D.; Taube, H. *J. Am. Chem. Soc.* **1990**, *112*, 2261-2263.

[28] Baker, M. V.; Field, L. D.; Young, D. J. *J. Chem. Soc., Chem. Commun.* **1988**, 546-548.

[29] Eckert, J.; Blank, H.; Bautista, M. T.; Morris, R. H. *Inorg. Chem.* **1990**, *29*, 747-750.

[30] Albertin, G.; Antoniutti, S.; Bordignon, E. *J. Am. Chem. Soc.* **1989**, *111*, 2072-2077.

[31] Bianchini, C.; Peruzzini, M.; Zanobini, F. *J. Organomet. Chem.* **1988**, *354*, C19-C22.

[32] Lundquist, E. G.; Folting, K.; Streib, W. E.; Huffman, J. C.; Eisenstein, O.; Caulton, K. G. *J. Am. Chem. Soc.* **1990**, *112*, 855-863.

[33] Bader, R. W. F.; Anderson, S. G.; Duke, A. J. *J. Am. Chem. Soc.* **1979**, *101*, 1389-1395.

[34] Gould, G. L.; Heinekey, D. M. *J. Am. Chem. Soc.* **1989**, *111*, 5502-5504.

[35] (a) Blomberg, M. R. A.; Brandemark, U.; Siegbahn, P. E. M. *J. Am. Chem. Soc.* **1983**, *105*, 5557-5563. (b) Low, J. J.; Goddard, W. A., III *J. Am. Chem. Soc.* **1984**, *106*, 821. (c) Low, J. J.; Goddard, W. A., III *Organometallics* **1986**, *5*,

609-622. (d) Ziegler, T.; Tschinke, V.; Fan, L., Becke, A. D. *J. Am. Chem. Soc.* **1989**, *111*, 9177-9185. (e) Koga, N.; Morokuma, K. *J. Phys. Chem.* **1990**, *94*, 5454-5462.

[36] (a) Eisenstein, O.; Jean, Y. *J. Am. Chem. Soc.* **1985**, *107*, 1177-1186. (b) Koga, N.; Obara, S.; Kitaura, K.; Morokuma, K. *J. Am. Chem. Soc.* **1985**, *107*, 7109-7116.

[37] Morokuma, K., personal communication.

[38] Dawoodi, Z.; Green M. L. H.; Mtetwa, V. S. B.; Prout, K. *J. Chem. Soc., Chem. Commun.* **1982**, 1410.1411.

[39] Saillard, J.-Y.; Hoffmann, R. *J. Am. Chem. Soc.* **1984**, *106*, 2006-2026.

MOLECULAR ORBITAL STUDIES OF REDUCTIVE ELIMINATION REACTIONS

M. J. CALHORDA
Centro de Química Estrutural
Complexo I, I. S. T.
Av. Rovisco Pais
1096 Lisboa Codex
Portugal

ABSTRACT. Reductive elimination reactions are very important in the synthesis of organic compounds and have been extensively studied. In this work, two alternative pathways for the formation of carbon-carbon bonds between saturated and unsaturated carbon atoms are considered, starting from a model square planar complex $Pd(PH_3)_2(CH_3)(CH=CH_2)$. In the first, which will be called the concerted mechanism, the formation of the new C-C bond is simultaneous with the breaking of the two metal-C bonds. When the organic product, CH_3-$CH=CH_2$ is formed it is no longer coordinated to the metal. This kind of mechanism has been the object of several theoretical studies involving elimination of saturated fragments. On the other hand, in the second pathway, which will be called the migration mechanism, the methyl group is allowed to migrate to the vinylic α carbon to form a distorted η^1-olefin, which in a second step will separate from the $Pd(PH_3)_2$ fragment. Extended Hückel molecular orbital calculations suggest that this last pathway is energetically favored over the first one, specially when the P-Pd-P angle is allowed to change during the reaction. These results are in accordance with experimental data which show the easiest elimination when an unsaturated fragment is present.

1. Introduction

Reductive elimination reactions play a major role in organic synthesis and in homogeneous catalysis by allowing the formation of a bond between two residues bound to a metal fragment and their later separation as a new organic species [1]. Such reactions can proceed from several kinds of compounds, ranging from square planar complexes, to three coordinated species where the fourth ligand was lost prior to reaction, octahedral complexes where a previous oxidative addition induces reductive elimination, five coordinated derivatives, etc. Among all these, 16 electron square planar complexes, generally containing two phosphine ligands (or, sometimes, a bidentate phosphine) are rather important species, as elimination can proceed directly from them or indirectly from some of their derivatives.

Palladium is the metal involved in most reductive elimination reactions, although they can also occur with nickel, platinum and others. New H-H, C-H, and C-C bonds may be formed, the last being specially significant for synthetic purposes [2]. Many studies, both experimental [3] and theoretical [4] have been made in order to understand the mechanism of reductive elimination reactions and to account for the effect of different factors, such as the specific metal involved, type of fragments to be eliminated, role of other ligands.

397

S. J. Formosinho et al. (eds.), Theoretical and Computational Models for Organic Chemistry, 397–410.
© 1991 *Kluwer Academic Publishers.*

The relative ease of formation of H-H, C-H, and C-C bonds, for instance, has been addressed theoretically [4e, f]. C-C bonds are more difficult to form than C-H bonds, and these than H-H bonds, the main reason being that the methyl groups have to reorient during the process. Such difficulty does not arise, however, in H-H bond formation.

A common feature assumed in all these studies is that the mechanism of reductive elimination is concerted, the new bond being formed as the two old bonds are broken. A closer look at a wider variety of reactions involving several kinds of carbon fragments reveals different reactivity patterns. For instance, reactions are normally much faster when one of these fragments is attached to the metal through an unsaturated carbon atom [2]. This suggests that another pathway becomes available for the reaction.

In the following study, the formation of a carbon-carbon bond between a vinylic and a methylic carbon will be addressed and two different pathways will be considered. One of them requires the presence of an unsaturated carbon atom coordinated to the metal.

2. The Model System

The calculations are of the extended Hückel type [5] with modified H_{ii}"s [6]. The basis set for palladium consisted of 5s, 5p, and 4d orbitals. The s and p orbitals were described by single Slater type wave functions and the d orbitals were taken as contracted linear combinations of two Slater type wave functions. Only 3s and 3p orbitals were used for phosphorus.

The geometry of the model complexes $(PH_3)_2PdRR'$ was square planar. All the coordination angles around palladium were 90°. The following distances (Å) were considered: Pd-P 2.30, P-H 1.42, Pd-C(vinyl) 2.05, Pd-C(methyl) 2.00, C-H 1.09, C=C 1.33. The parameters used in the calculations were, for palladium: 5s (H_{ii} -7.32 eV, ζ 2.19), 5p (H_{ii} -3.75 eV, ζ 2.152), 4d (H_{ii} -12.02 eV, ζ_1 5.893, ζ_2 2.613, C_1 0.5264, C_2 0.6373). For the other elements the standard parameters were used.

Square planar complexes are an important starting point for reductive elimination reactions. We shall take as a model a 16 electron, square planar palladium complex containing two phosphine ligands (PH_3) and two organic fragments, R and R'. The geometry is essentially the same as the one used in a previous study of reductive elimination by Tatsumi and coworkers [4a].

The reaction to be studied is the following

$$(PH_3)_2PdRR' \rightarrow (PH_3)_2Pd + R-R'$$

where R = CH_3 (Me) and R'= $CH=CH_2$ (vinyl), and a comparison will be made with the case when R=R'=CH_3.

$Pd(PH_3)_2$ is a d^8 ML_2 fragment [7]. The empty d orbital is xy (the coordinate system was chosen with the x and y axes passing through the ligands), which has y mixed in, so that it is hybridized away from the two phosphines. A consequence of this mixing is that the orbital becomes less antibonding and its energy drops. It is the LUMO of this metallic fragment. The second LUMO is a sp hybrid orbital, which also points away from the phosphines, towards incoming ligands. Here, it is mixing of the s orbital into x which makes the orbital less antibonding, thus more stable. These two empty orbitals are shown in Figure 1, in two different representations.

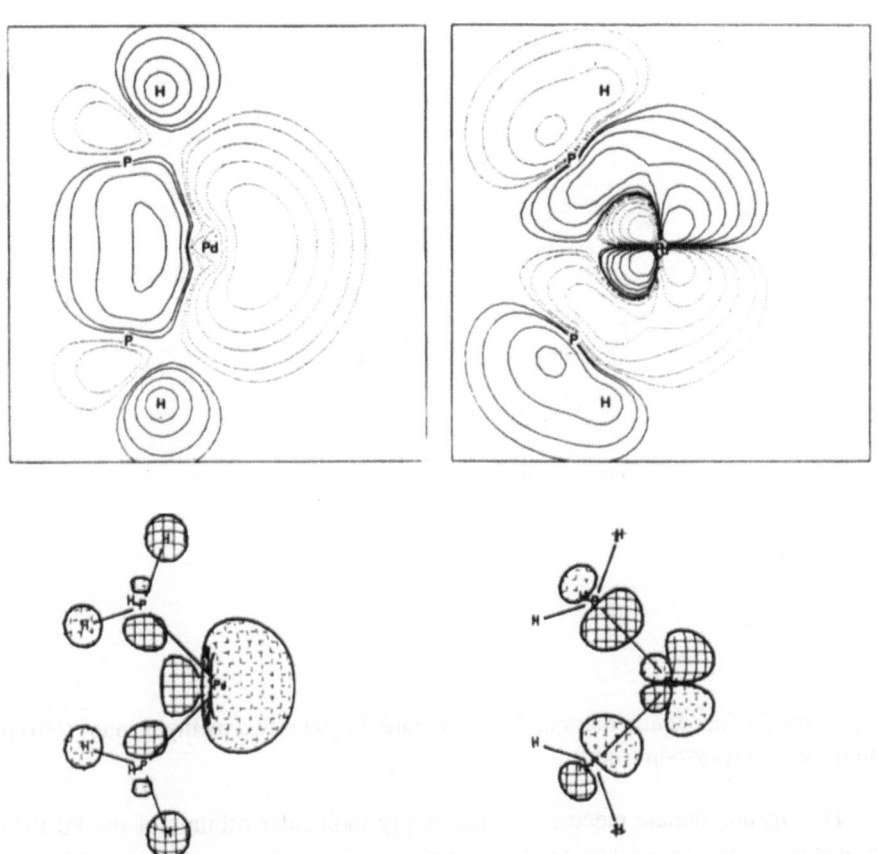

Figure 1 - The two LUMOs of the Pd(PH$_3$)$_2$ fragment: wave function contour on the xy plane (values for ψ are 0.2, 0.1, 0.055, 0.025, 0.01, 0.005; solid lines indicate positive ψ and dotted lines negative ψ) shown on the top; perspective view (outline contour is 0.06) shown on the bottom [8].

They are extremely adequate, both in energy and composition, for interaction with other ligands. They are the acceptor ligands of the Pd(PH$_3$)$_2$ fragment. The four highest occupied molecular orbitals are the remaining d orbitals, z^2, x^2-y^2, xz, and yz, respectively. In Figure 2, the interaction diagrams between this metallic fragment and two methyl groups (left) or a methyl and a vinyl (right) are represented.

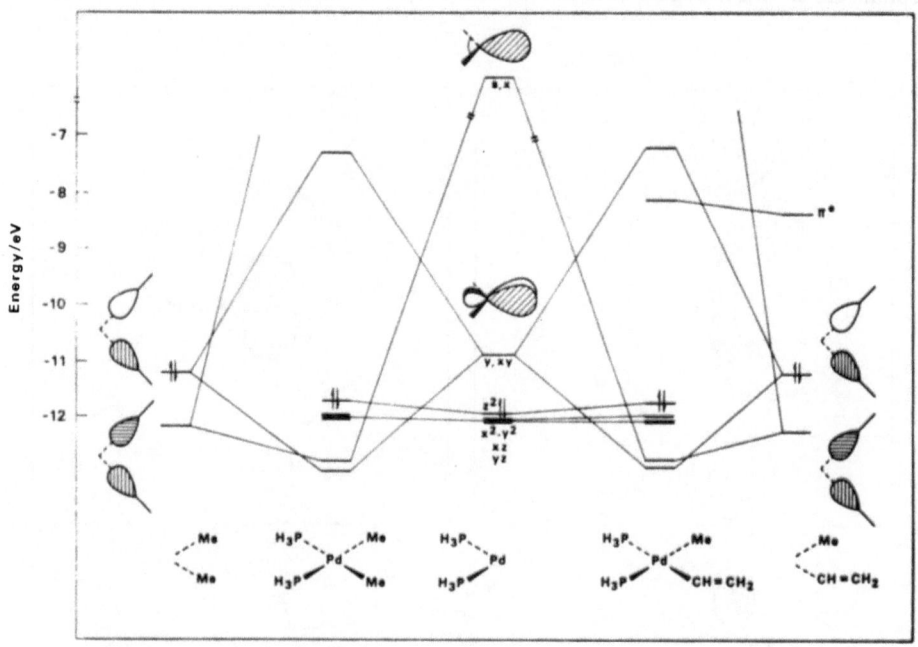

Figure 2 - Interaction diagram between $Pd(PH_3)_2$ and two methyl groups (left) or a methyl and a vinyl group (right).

The ligands donate electrons to the empty molecular orbitals of the $Pd(PH_3)_2$ fragment using the carbon lone pairs. Their symmetric combination interacts with the sp hybrid and the antisymmetric combination with xy, y. Although their energies are different (the vinyl group has an sp^2 carbon) the type of interaction is essentially the same for the two complexes. In the vinylic complex there is, however, a low energy empty orbital, π^*, due to the presence of the carbon-carbon double bonds.

Another question to be addressed refers to the position of the vinyl group relative to the P-P-Pd-C-C plane. It can rotate around the Pd-C bond and the change in energy for that movement is shown in Figure 3.

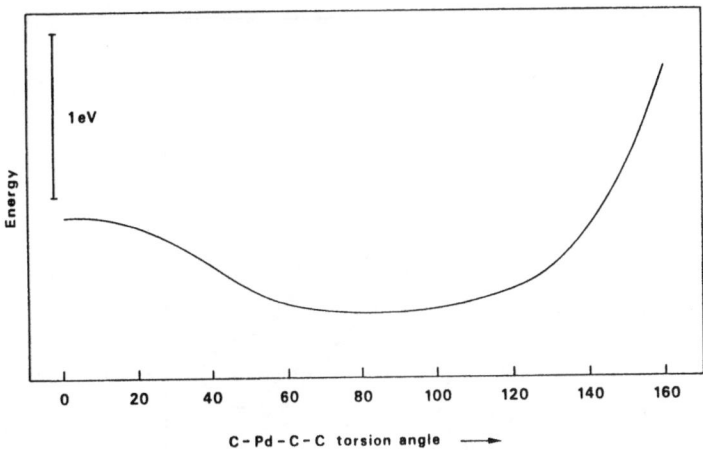

Figure 3 - Change in energy of a square planar $Pd(PH_3)_2(Me)(vinyl)$ complex as a function of the C-Pd-C-C torsion angle.

A torsion angle of 0° means that the double bond lies in the plane of the complex. The potential energy surface is very flat for an interval around 90° (60° - 110°), when the plane of the double bond is perpendicular to the plane of the complex, with a minimum at 82°. This result is in agreement with experimental data, as for most complexes containing coordinated vinyl derivatives that angle lies between 70 and 90°, the number of known structures decreasing with the torsion angle [9]. The reason for this behavior is steric, but it can be traced in the molecular orbital diagram. If the interaction between the metallic fragment $Pd(PH_3)_2Me$ and $CH=CH_2$ is studied, it is possible to see that it is very similar for the two limiting geometries (torsion angles 0 and 90°, respectively), one pair of electrons being donated by the vinylic carbon to the acceptor orbital of the Pd fragment. One of the orbitals is markedly different, however: the one shown in Figure 4, for the in plane (0° torsion angle) conformation.

It accounts for the energy difference between the two species, which is about 0.6 eV. There is a strong repulsion between the methylic carbon and one of the hydrogen atoms of the vinyl (reflected in a negative overlap population), as can be seen in the wave function.

All the calculations, including the previous interaction diagram of Figure 2, were then done for a 90° torsion angle, that is, for the conformation having the plane of the double bond perpendicular to the plane of the complex.

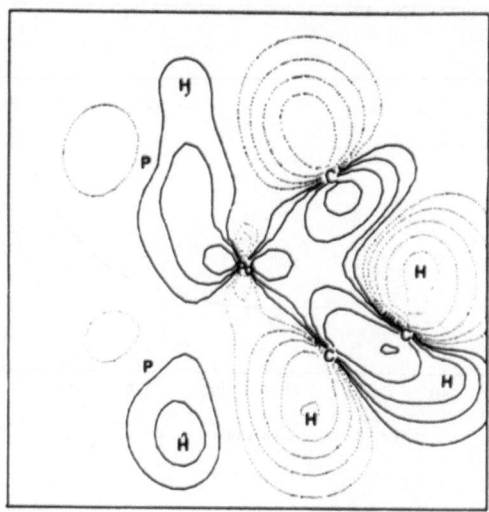

Figure 4 - A molecular orbital of Pd(PH$_3$)$_2$(Me)(vinyl) in the in plane conformation showing the repulsion between a vinyl hydrogen and the methyl carbon (wavefunction contours as in Figure 1).

3. The Reaction. Concerted Mechanism.

This mechanism has been studied in detail for reductive elimination of ethane from (PH$_3$)$_2$Pd(Me)$_2$ [4a]. We are going to consider the same pathway and compare the result with that obtained when one of the methyl groups is replaced by a vinyl. The reaction is depicted in the following scheme

and these changes take place:
- The new C-C bond starts to form, the two atoms approaching till they are only 1.68 Å away, which is getting closer to the normal C-C bond length (or, in a different way, the C-Pd-C angle goes from 90 to 30°).
- The Pd-C bonds break, the distance increasing from 2.05 Å to 3.25 Å.
- The methyl groups reorient. In the beginning, each carbon lone pair was pointing towards the palladium atom, while in the final geometry they point towards each other.

- The P-Pd-P angle opens from 90° to 150° and the $Pd(PH_3)_2$ d^{10} fragment which results from the reaction becomes much more stable.

All these changes take place simultaneously and in the end ethane (or propene) is formed along with unsaturated $Pd(PH_3)_2$. It is possible to follow changes in the total energy and the overlap populations during the reaction. These are shown in Figure 5, both for R = Me (right) and R = vinyl (left).

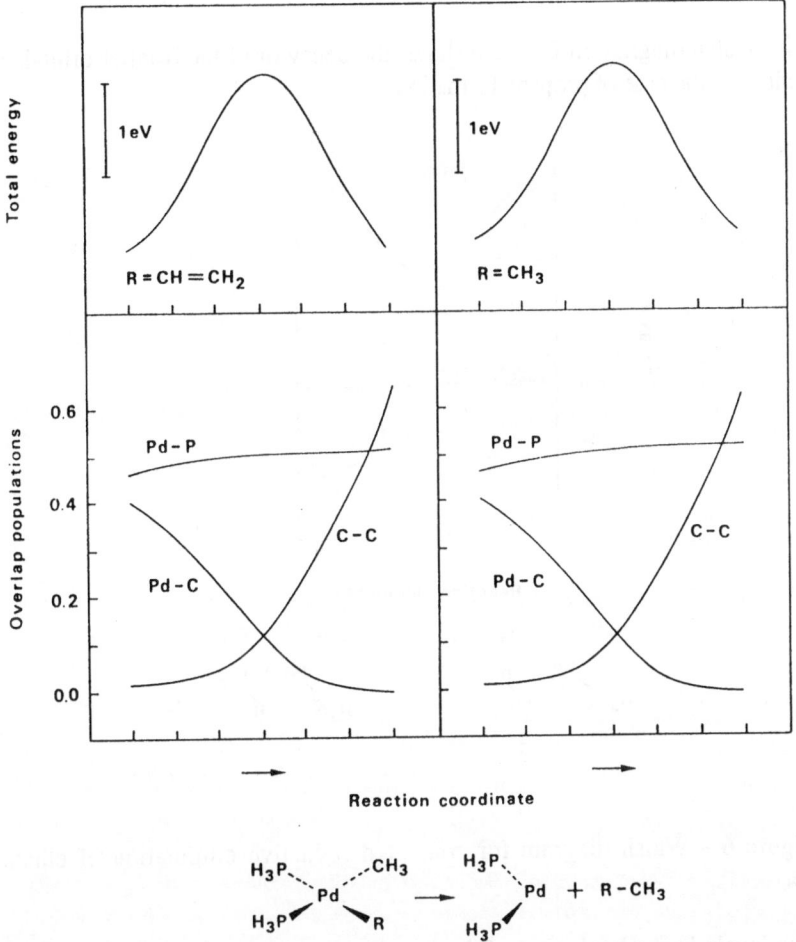

Figure 5 - Change in total energy and overlap population for concerted reductive elimination of ethane and propene.

The activation energies are practically the same (ca. 1.9 eV), although the metal-carbon(vinyl) bond is stronger than the metal-carbon(methyl) bond. This is given in the calculations by a larger overlap population (0.47 compared to 0.41) and agrees with an experimental determination of bond strengths for the system $(\eta^5\text{-}C_5Me_5)(PMe_3)IrX_2$

(71 kcal mol[-1] for X=vinyl, 50 kcal mol[-1] for X=alkyl) [10]. The overlap populations change smoothly along the reaction coordinate. The Pd-C overlap population decreases slowly as this bond is broken, while the C-C overlap population increases due to the formation of this bond. Notice that although there are two different Pd-C bonds in Pd(PH$_3$)$_2$Me(vinyl), only the Pd-C(Me) bond is shown in Figure 5. The variation of overlap population for the other follows the same trend. On the other hand, the Pd-P bond becomes slightly stronger during the process, due to the opening of the P-Pd-P angle.

The Walsh diagram in Figure 6 shows the behavior of the frontier orbitals during the reaction for the case of propene formation.

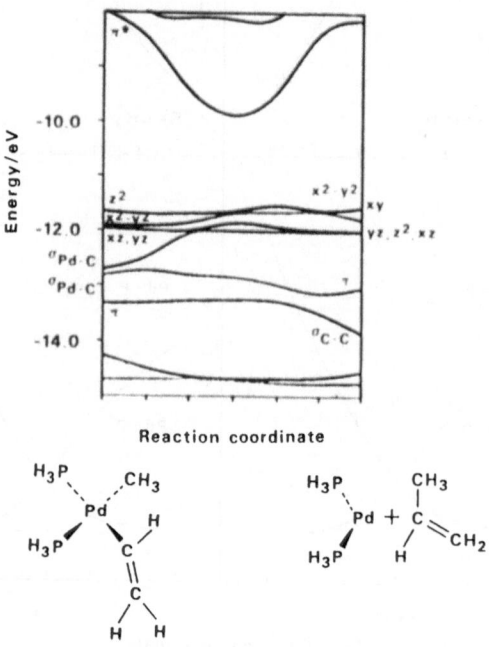

Figure 6 - Walsh diagram for concerted reductive elimination of ethane and propene.

The levels in the beginning of the reaction are the same as shown earlier in the molecular orbital diagram (Figure 2). The two bonding molecular orbitals correspond to the two Pd-C bonds. Immediately higher in energy are four of the d orbitals and finally π* and the antibonding xy. There is no symmetry element both in the initial complex Pd(PH$_3$)$_2$Me(vinyl) in its perpendicular (or out of the plane) conformation, although there would be a mirror plane for the in plane conformation, and along the reaction. All levels have the same symmetry, so they cannot cross. It is possible to see, however, that the level which is more destabilized and thus responsible for the high energy barrier is the higher energy σ Pd-C bond. On the other hand, the lowering of energy during the last steps of the reaction parallels the behavior of the level which becomes the σ C-C bond.

Mixing between levels makes a further analysis very difficult, but the qualitative trends are analogous to those described by Tatsumi and coworkers for the reaction of the more symmetrical $Pd(PH_3)_2Me_2$ [4a].

This concerted mechanism affords comparable results for the two systems studied, suggesting that the double bond plays no major role. If this mechanism were operative, reductive elimination from these palladium complexes should proceed in the same way both in the presence and absence of unsaturated fragments.

4. The Reaction. Migration.

It is possible to imagine other mechanisms which lead to the same products as the concerted one. One possible pathway was suggested by the conversion of $(\eta^5\text{-}C_5Me_5)(PMe_3)IrH(C_2H_3)$ in $(\eta^5\text{-}C_5Me_5)(PMe_3)Ir(C_2H_4)$. It was proposed that this reaction proceeds through hydride migration from the metal to the vinylic carbon [11].

The following mechanism will then consist of two successive steps, the first of which is a migration of the methyl group from the metal to the vinylic carbon, as shown in the scheme.

During the migration of the methyl group to the vinylic carbon, the bond between this carbon atom and palladium is not broken. Only Pd-C(Me) breaks, as the new C-C bond is formed. There is, however, a rearrangement, so that the Pd-C(vinyl) bond moves to the x axis from its initial position 45° away. In the first set of calculations, the P-Pd-P angle was kept constant, while in the second it was allowed to open till 120°. In this situation, all the angles around palladium will be 120° at the end of the migration. The propene molecule is coordinated to the metal in a η^1 fashion and is distorted. The carbon atom bound to palladium is tetrahedral, not planar. Some reorientation of the vinyl group is needed to achieve this coordination.

The second step of the reaction includes the relaxation of the olefin to a planar skeleton and its movement away from the metal (the final Pd-C distance considered is 3.25 Å, the same as assumed in the first mechanism). This is a downhill process. It is the first step that controls the reaction and, for this reason, most of what follows deals with it.

The change in total energy and overlap population of the relevant bonds is given in Figure 7 (left), for the calculations with the P-Pd-P angle increasing from 90 to 120º.

The energy barrier becomes significantly smaller than for the concerted pathway when relaxation of the $Pd(PH_3)_2$ fragment is allowed. The overlap population curves show the progressive formation of the C-C bond and the disappearance of the Pd-C(Me) bond (this overlap population drops to zero). On the other hand, we can see that some Pd-C(vinyl) bonding is kept, although it becomes weaker as expected. Also, this carbon atom forms four σ bonds and the vinylic C-C π bond has to be at least partly broken. This is observed in the overlap population, but not shown in Figure 7.

As referred before, the molecule has no symmetry elements, so that all levels represented in the Walsh diagram of Figure 7 (right) can mix. The levels at the final step (right side of the diagram) reflect these changes. The highest occupied molecular orbitals are the five d orbitals, followed by the remaining Pd-C bond and the new C-C bond.

This methyl to vinyl migration is also formally similar to the well known migration of methyl groups to an adjacent carbonyl, when both are coordinated to the same metal atom. Although this reaction has been widely studied [12], for both octahedral and square planar complexes, it is interesting to compare it to methyl migration to a vinyl, all the factors being the same.

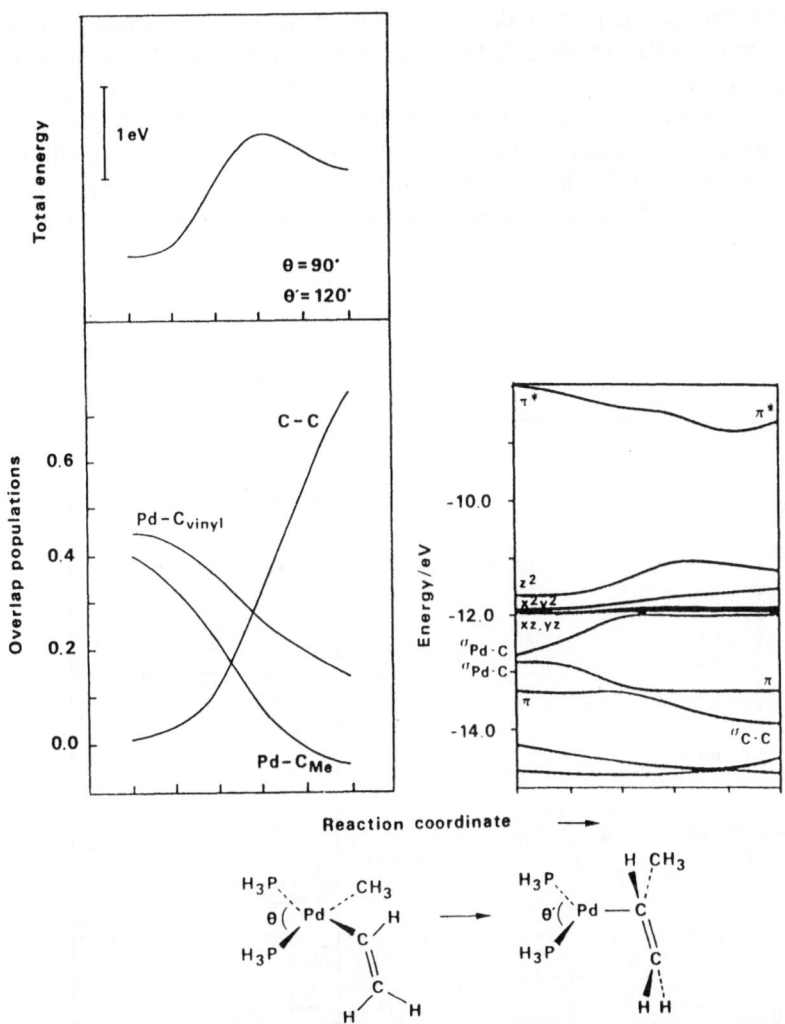

Figure 7 - Change in total energy and overlap population (left) and Walsh diagram (right) for migration of the methyl group to the vinylic carbon in Pd(PH3)2Me(vinyl).

We use, then, a very similar model, square planar Pd(PH3)2Me(CO)+, where CO has replaced the vinyl group. The complex has a symmetry plane, xy, which is kept through the proposed pathway, as shown in the following scheme.

As in the previous migration, the new C-C bond starts to form when the methyl group moves towards the carbonyl. At the same time, the P-Pd-P angle widens and the CO group approaches the x axis (the angle between the Pd-CO bond and x changes from 45° to 0° at the end of the reaction). Important differences are, on one side, that the CO group needs much less reorientation than the one required for the vinyl, and on the other hand that the new C-C bond will lie on the xy plane. The changes in overlap populations along the reaction coordinate are shown in Figure 8 (left) and the total energy and the Walsh diagram in Figure 8 (right).

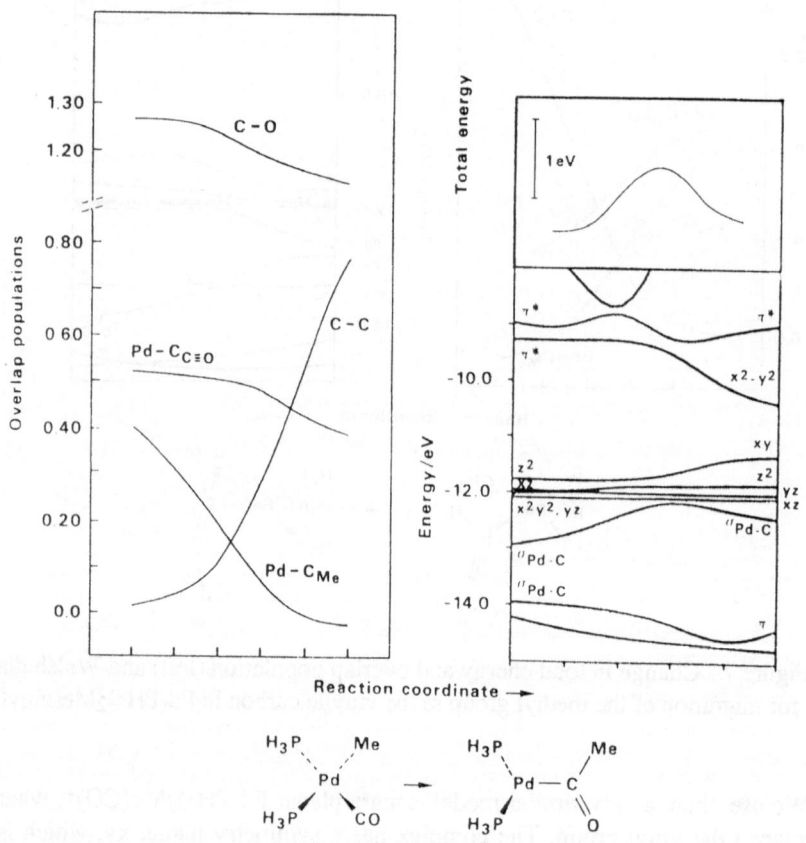

Figure 8 - Changes in total energy and overlap population (left) and Walsh diagram (right) for migration of the methyl group to the vinylic carbon in $Pd(PH_3)_2Me(CO)^+$.

The energy barrier is lower than for the methyl to vinyl migration in parallel with the easier motion needed to accomplish the reaction. The Pd-CO bond remains relatively strong throughout the reaction, while Pd-C(Me) breaks and C-C is being formed. The decrease in C-O overlap population reflects the lower multiplicity of that bond as the carbon atom becomes involved in another bond. The Walsh diagram is similar to the previous one and the small barrier is due to breaking of the Pd-C bond. In spite of the C_S symmetry being kept along the reaction, most of the energy levels still mix and the result is thus not clearer than before.

The geometry of the resulting acyl complex has not been optimized. A sudy of conformational preferences of acyl complexes has been published recently [13].

5. Conclusions.

The above results suggest that a migration mechanism should be favored over the concerted mechanism when at least one of the eliminating fragment contains an unsaturated carbon atom coordinated to the metal. The absolute values found for the energy barriers should not be taken very seriously, as extended Hückel calculations are not the ideal kind to deal with this type of study, where bonds are being formed and broken simultaneously. For instance, a computed ab initio barrier for concerted reductive elimination of ethane from $Pd(PH_3)_2Me_2$ [4g] is only ~10 kcal mol^{-1}, much lower than the one calculated here. On the other hand, the sharp decrease of the activation energy obtained when choosing a different mechanism must reflect a real trend and is in accordance with experimental findings. The only available comparable ab initio studies are those of methyl to CO migration in $Pd(PH_3)HMe(CO)$ and the proposed transition state [12c, d] is certainly very similar to the one we proposed both for methyl to CO and to vinyl migration. The calculated barriers for these two processes (smaller one for migration to carbon) are compatible with the smaller need of reorganization in this situation.

6. Acknowledgements.

M.J.C. thanks Dr. J.M.Brown from Oxford University, UK, for the information and discussion about his experimental work on reductive elimination, Prof. R.Hoffmann from Cornell University, USA, for his interest in this problem, Dr. C.Mealli and Dr. D.M.Proserpio for having made available their CACAO program, and NATO for a fellowship.

410

7. References.

[1] Collman, J.P.; Hegedus, L.; Norton, J.R.; Finke, R.G. "Principles of Organotransition Metal Chemistry", University Science Books, Mill Valley, 1987.

[2] Brown, J.M.; Cooley, N.A. *Chem Rev.* **1988**, *88*, 103.

[3] (a) Brown,J.M.; Cooley,N.A.; Price,D.W. *J.Chem.Soc., Chem.Comm.* **1989**,458. (b) Stang, P.J.; Kowalski, M.H. *J.Am.Chem.Soc.* **1989**, *111*, 3356. (c) Gillie, A.; Stille, J.K. *J.Am.Chem.Soc.* **1980**, *102*, 4933. (d) Loar, M.K.; Stille, J.K. *J.Am.Chem.Soc.* **1981**, *103*, 4174. (e) Moravskii, A.; Stille, J.K. *J.Am.Chem.Soc.* **1981**, *103*, 4182. (f) Young, S.J.; Kellenberger, B.; Reibenspies, J.H.; Himmel, S.E.; Manning, M.; Anderson, O.P.; Stille, J.K. *J.Am.Chem.Soc.* **1988**, *110*, 5744. (g) Byers, P.K.; Canty, A.J.; Crespo, M.; Puddephatt, R.J.; Scott, J.D. *Organometallics* **1988**, *7*, 1363.

[4] (a)Tatsumi, K.; Hoffmann, R.; Yamamoto, A.; Stille, J.K. *Bull. Chem.Soc.Japan* **1981**, *54*, 1857. (b) Tatsumi, K.; Nakamura, A.; Komiya, S.; Yamamoto, A.; Yamamoto, T. *J.Am.Chem.Soc.* **1984**, *106*, 8181. (c) Hoffmann, R. in "IUPAC Frontiers of Chemistry", K. J. Laidler ed., Pergamon Press, Oxford, 1982. (d) Low, J.L.; Goddard, W.A., III *J.Am.Chem.Soc.* **1984**, *106*, 6924. (e) Low, J.L.; Goddard, W.A.,III *J.Am.Chem.Soc.* **1984**, *106*, 8321. (f) Low, J.L.; Goddard, W.A., III *Organometallics* **1986**, *5*, 609. (g) Low, J.L.; Goddard, W.A., III *J.Am.Chem.Soc.* **1986**, *108*, 6115. (h) Obara, S.; Kitaura, K.; Morokuma, K. *J.Am.Chem.Soc.* **1984**, *106*, 7482. (i) Komiya, S.; Albright, T.A.; Hoffmann, R.; Kochi, J.K *J.Am.Chem.Soc.* **1976**, *98*, 7255.

[5] (a) Hoffmann, R. *J.Chem.Phys.* **1963**, *39*, 11397. (b) Hoffmann, R.; Lipscomb, W.N. *J.Chem.Phys.* **1962**, *36*, 2179.

[6] Ammeter, J.; Burgi, H.-B.; Thibeault, J.C.; Hoffmann, R. *J.Am.Chem.Soc.* **1978**, *100*, 36.

[7] Albright, T.A.; Burdett, J.K.; Whangbo, M.H. "Orbital Interactions in Chemistry", John Wiley & Sons, NY, 1985.

[8] Mealli, C.; Proserpio, D.M. *J.Chem.Ed.* **1990**, *67*, 399.

[9] Calhorda, M.J.; Brown, J.M.; Cooley, N.A. submitted for publication.

[10] Stoutland,P.O.;Bergman, R.G.;Nolan,S.P.;Hoff,C.D. *Polyhedron* **1988**, *7*, 1429.

[11] Silvestre, J.; Calhorda, M.J.; Hoffmann, R.; Stoutland, P.O.; Bergman, R.G. *Organometallics* **1986**, *5*, 1841.

[12] (a) Berke, H.; Hoffmann, R. *J.Am.Chem.Soc.* **1978**, 100, 7224. (b) Sakaki, S.; Kitaura, K.; Morokuma, K.; Ohkubo, K. *J.Am.Chem.Soc.* 1983, 105, 2280. (c) Koga, N.; Morokuma, K. *J.Am.Chem.Soc.* **1985**, *107*, 7230. (d) Koga, N.; Morokuma, K. *J.Am.Chem.Soc.* **1986**, *108*, 6136. (e) Ziegler, T.; Versluis, L.; Tschinke, V. *J.Am.Chem.Soc.* **1986**, *108*, 612. (f) Dedieu, A.; Sakaki, S.; Strich, A.; Siegbahn, P.E.M. *Chem. Phys.Lett.* **1987**, *133*, 317. (g) Shusterman, A.J.; Tamir, I.; Pross, A. *J.Organometal.Chem.* **1988**, *340*, 203.

[13] Blackburn, B.K.; Davies, S.G.; Sutton, K.H.; Whittaker, H. *Chem.Soc.Rev.* **1988**, *17*, 147.

LABORATORY PROJECTS IN COMPUTATIONAL ORGANIC CHEMISTRY

G.A. ARTECA [1], A. BOTTONI [2], M. DURAN [3], R. FAUSTO [4],
M.R. PETERSON [5], H.D. THOMAS [6]

[1]*Department of Chemistry, University of Saskatchewan, Saskatoon, Saskatchewan, Canada S7N 0W0*
[2]*Dipartimento di Chimica, Università di Bologna, Via Selmi 2, 40126 Bologna, Italy*
[3]*Departament de Química, Universitat Autònoma de Barcelona, 08193 Bellaterra (Barcelona), Catalonia, Spain*
[4]*Departamento de Química, Universidade de Coimbra, 3000 Coimbra, Portugal*
[5]*Department of Chemistry, University of Toronto, 80 St. George St., Toronto, Ontario, Canada M5S 1A1*
[6]*School of Chemical Sciences, Department of Chemistry, University of Georgia, Athens, Georgia 30602, U.S.A.*

ABSTRACT. This chapter outlines the computational projects proposed to the students of the NATO Advanced Study Institute as a complement to the topics covered by the lecturers. A general description of methods and strategies is given, followed by the complete index of problems. Some selected examples of chemical relevance are exposed in greater detail.

1. Introduction

In this chapter we outline the computational projects related to the preceding lectures. The projects deal primarily with *ab initio* methods involving the geometry optimization of stable species, the optimization of transition structures, and the treatment of correlation effects. Moreover, some problems cover other areas, such as basis set effects, the use of open-shell computational methods, the optimization of structures containing rings, and illustrations of some anomalies and hazards of *ab initio* molecular orbital calculations (using the GAUSSIAN 88 and 90 programs). In addition, conformational studies with molecular mechanics are discussed involving the use of the MM2 program. Since some of the preceding lectures dealt with the variational transition state theory and chemical kinetics, we have also included applications involving the POLYRATE program for the computation of reaction rate constants. Finally, we present examples of topological methods of molecular shape analysis (program GSHAPE), using electron isodensity contours as models for the

S. J. Formosinho et al. (eds.), Theoretical and Computational Models for Organic Chemistry, 411–428.
© 1991 *Kluwer Academic Publishers.*

3D molecular envelope surface.

The computational facilities at the NATO ASI were composed of a cluster of two Digital Equipment Corporation Vax Station 3100's under VMS, together with one Convex C120 superminicomputer under Convex/05 (Unix). Two terminals were attached to the Vax cluster, fourteen to the Convex, and three printers were available to obtain hardcopy results.

The GAUSSIAN 90 series of programs was implemented on the Vax cluster, while the GAUSSIAN 88 and the MM2 programs run on the Convex. The POLYRATE program was linked to GAUSSIAN 88 on the Convex, while the GSHAPE program was implemented on both computers. This disposition of programs allowed an optimum use of the available computational resources.

2. General Description of Methods and Strategies

One of the major uses of molecular orbital theory has been, and continues to be, the determination of the geometries and relative energies of stable (or meta-stable) species. These species correspond to the reactant, intermediates, and products of chemical reactions. These structures are local minima of the potential energy surface.

The location of critical points requires addressing a problem of unconstrained optimization, for the solution of which there exist a number of efficient algorithms. In this chapter a number of simple optimization projects are proposed to those students with little or no experience with geometry optimizations. As well, we have included some more challenging projects for the experienced students.

When dealing with chemical reaction paths, a required step is the location of transition structures on the potential energy surface. The location of transition structures of equal or lower symmetry than either reactants or products is far more difficult than finding minima. The transition structures correspond to saddle points of signature index $\lambda=1$ (*i.e.*, they are maxima in one and only one direction on the potential energy hypersurface). A number of methods for finding transition states are proposed as projects. Some of these methods are developed specifically to solve the chemical saddle point problem.

We have included some calculations involving problems which frequently pose difficulties, such as the analysis of anionic species. Such calculations generally require the use of diffuse functions in the basis set to assure obtaining bound occupied orbitals.

Another area of concern has been the use of methods beyond the single-determinant wavefunction level of theory (Hartree-Fock). In this context, some computational projects deal with the inclusion of electron correlation at different levels. Applications of Møller-Plesset perturbation theory to various orders (MP2, MP3, and MP4), as well as single and double excitation configuration interaction (CISD) and quadratic configuration interaction (QCI) are presented for some small systems.

Other projects introduce the multiconfigurational SCF method (MCSCF), in particular, the complete active space SCF (CASSCF) method. The MCSCF method represents a natural extension of the Hartree-Fock procedure for the description of bond breaking and bond formation. It consists of optimizing both the coefficients of the molecular orbitals on which the configurations are obtained and the mixing of several configurations. One of the major aims of these projects is to make students more familiar with the MCSCF strategies for studying chemically relevant problems. These include investigating a reaction surface, locating critical points of chemical interest, and finding their optimum geometries using gradient techniques. To solve the various MCSCF exercises, the CASSCF option, recently available in GAUSSIAN 90 should be used. Analytical derivatives of the CASSCF wavefunction are included in this program. As a rule, most of the problems proposed for the tutorial sessions can be addressed with the GAUSSIAN 88 and 90 series of programs [1].

Given the limited time available at the NATO ASI, it has been encouraged that many of the projects be treated with small basis sets (for example STO-3G [2] and STO-3G* [3]) and at only exploratory theoretical levels. However, these projects should illustrate the techniques available with modern computers and computer programs, and also the principles for computing molecular properties (energy, geometry, and others). For large systems (more than 6 non-hydrogen atoms), calculations are usually performed with the split-valence 3-21G [4] or 3-21G* [4c, 5], or even 6-31G* [6] basis sets, which give much more reliable results than the minimal basis STO-3G, at only about two times the computational effort (for 3-21G). Such optimizations are often followed by single point calculations with the 6-31G* [6] basis set at the MP2 or MP4 level for more reliable energy differences. Smaller systems are usually optimized at the 6-31G* SCF level, or even with the 6-311G*(*) [7] at the MP2 level. Single points are then computed using the MP4 level of theory. Other projects require even more sophisticated basis sets. For example, negative ions generally require diffuse functions (3-21G+(+), [8], 4-31G+(+) [9], 6-311G+(+) [10]) for proper descriptions of the electron distribution. All of these basis sets are available in GAUSSIAN 88 and GAUSSIAN 90.

An alternative method of studying the structures of molecules is the molecular mechanics empirical force-field method. This method requires considerably less computational time. Molecular mechanics force fields have shown considerable versatility in being able to offer insights to a large variety of chemical problems. The method has been proven to provide reliable answers by its ability to reproduce a large amount of experimental data with errors that are equal to or less than the experimental ones. The main drawback to using a molecular mechanics method is that the chemical system under study cannot involve the making or breaking of covalent bonds or other electronic effects.

Two computer programs were available during the computational laboratory sessions for all of the participants to become better acquainted with the computational power and versatility of the molecular mechanics force-field method. The MINP program is an interactive program that permits the user to create three-dimensional structures in a format that is then used as an input file for the MM2 program. Neither the MINP nor the MM2 program requires a terminal with graphics capabilities. Several additional exercises were designed for the participants to understand the types of chemical systems that can be studied with the molecular mechanics force-field method.

The methods discussed above provide a description of molecular properties based on the knowledge of (at least a part of) the potential energy surface. Similarly, a theoretical study of chemical reactions can be formulated on the basis of this piece of information. In principle, one should be able to predict reaction rate constants from the trajectories on the potential energy surface. Upper bounds to these rate constants can be computed within the framework of the conventional transition state theory. A better description can be obtained by using the improved Canonical Variational Transition State theory [11], which takes into consideration the recrossing of trajectories across a surface dividing reactants from products. Within this theory, the location of the dividing surface (a generalized transition structure) over the minimum energy path is optimized variationally. This procedure locates the transition states at the highest free energy points on the minimum energy path for each temperature, thus properly taking into consideration entropic and zero-point energy effects, in the form of generalized normal modes.[12]

The practical implementation of this methodology has also been included among the topics covered in this computational laboratory chapter. A number of projects involving the use of the program POLYRATE [13] for the calculation of rate constants are proposed as problems. The program POLYRATE has an interface which opens the possibility of using *ab initio* potential energy surfaces, previously generated with the GAUSSIAN series of programs.

Finally, in this chapter we covered another aspect of molecular theory which is not usually addressed, i.e., the case of the characterization of molecular shape. The knowledge of equilibrium geometries and the total energy of chemical species does not provide much information on some of its other properties, such as size and three-dimensional shape. The description of these properties is actually outside the realm of quantum mechanics, since molecules have no boundary. However, it is possible to give an approximate description to these classical notions by introducing appropriate models of molecular surfaces. Constant electron density contour surfaces provide a reasonable family of surfaces which represent the approximate three-dimensional molecular "body" [14]. Simpler surfaces, such as fused-sphere (van der Waals) surfaces can also provide a reasonable description of a molecular surface [15].

For a number of practical applications, it is necessary to characterize the shape features of molecular surfaces. This is of importance in problems of molecular shape recognition, in correlations between molecular structure and biological (or biochemical) activity, and other problems related to computer-assisted rational drug design. Accordingly, we have proposed some laboratory projects for the NATO ASI dealing with the characterization of molecular shape. The model chosen to represent the 3D boundary of a molecule is the continuum of constant electron density surfaces associated with a given nuclear configuration. The projects deal with the characterization of these surfaces using the tools of molecular topology [14], implemented in the program GSHAPE [16]. This program uses the electron density computed by a previous *ab initio* calculation using the GAUSSIAN programs, to provide a characterization in terms of topological invariants and the so-called shape matrix [17]. The description is algebraic, thus making it possible to assess in a very compact fashion the degrees of shape similarity between the surfaces of related molecules.

3. Problem Index

The laboratory projects proposed to the students have been the following:

I.   *Ab initio quantum mechanics*
I.1   Geometry optimization of methanol.
I.2   The effect of d-orbitals on $HOOH$ and H_2SO geometries.
I.3   Open-shell optimizations - Formaldehyde.
I.4   Open-shell optimizations - Ethylene.
I.5   Three-member rings (Cyclopropane and Oxirane).
I.6   Four-member rings (Dioxycyclobutane).
I.7   Rearrangements: HCN $\leftrightarrow$ HNC.
I.8   Addition reactions: $SiH_2 + H_2 \rightarrow SiH_4$.

I.9   Planar methane.

I.10   Singlet and triplet oxygen.

I.11   Singlet and triplet methylene.

I.12   Convergence of the MP methods.

I.13   Convergence of the UMP methods.

I.14   Electron correlation effects on ethylene hydrogenation.

I.15   Electron correlation effects on the nuclear geometry.

I.16   Isomers of negative molecular ions.

I.17   MCSCF active space.

I.18   MCSCF critical points.

I.19   MCSCF critical point optimization.

II.   *Molecular Mechanics*

II.1   Construction of the MM2 input deck and interpretation of the output.

II.2   Use of MM2 replacement options.

II.3   Anti and gauche conformers of 1-propanol.

II.4   Conformational study of 2-bromocyclohexanone and ethyleneglycol.

II.5   Structural studies of naphtalene and indole.

II.6   Thermodynamical studies of Friedel-Crafts alkylations.

III.   *Variational transition state theory*

III.1   Reaction paths on the potential energy surface for the CH_5 system: study of the $CH_3 + H_2 \rightarrow CH_4 + H$ reaction with the POLYRATE program.

III.2   Computation of rate constants for the $HCN \leftrightarrow HNC$ isomerization reaction, using an *ab initio* (GAUSSIAN 88) potential surface with POLYRATE.

IV.   *Three-dimensional topological molecular shape characterization*

IV.1   Molecular shape characterization of an isodensity contour surface of water using the program GSHAPE.

IV.2   Construction of the electronic density - reference curvature map $((a,b)$-parameter map) for water.

4. Problem Examples

I.1   METHANOL GEOMETRY OPTIMIZATION AND CALCULATION OF VIBRATIONAL FREQUENCIES

Methanol is one of the simplest asymmetric molecules capable of hindered internal rotation and has been the subject of numerous theoretical and experimental studies [18-21]. It is now well established that its minimum energy conformation and the transition structure for the internal rotation correspond to the staggered and eclipsed conformations, respectively, both of them belonging to the C_S symmetry point group (Fig.1).

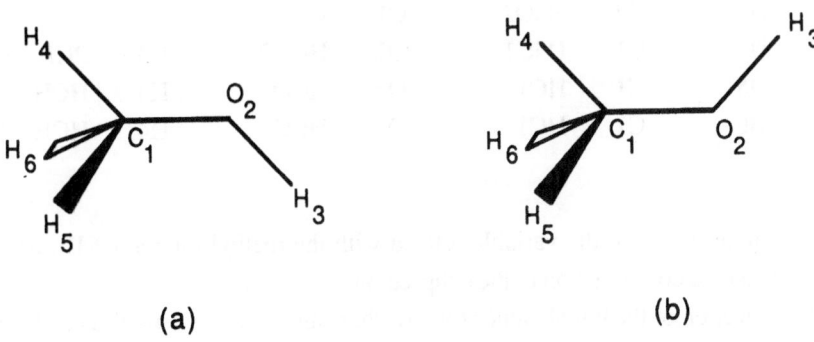

(a)   (b)

Figure 1- Staggered (a) and eclipsed (b) conformations of methanol.

Although simple, this molecule presents some interesting conformational properties and is a good example to illustrate both (i) *ab initio* molecular geometry optimization of minima and transition structures (which may be located by symmetry analysis), and (ii) *ab initio* calculations of vibrational frequencies.

The proposed computational project involves the optimization of the staggered and eclipsed conformations of methanol, using the STO-3G minimum basis set, followed by the calculation of the vibrational frequencies, in order to confirm the nature of the critical point associated with each conformation. Prior to each step of calculation, the students should be encouraged to make educated guesses on the relative stability and nature of the studied conformations.

Many of the problems encountered in a geometry optimization result from an incorrect or poor choice of the coordinate system. Symmetry and an appropriate number of coordinates must be used, and the coupling between "stiff" and "flexible" coordinates should be avoided to improve the performance of the computation.

Internal coordinates for the GAUSSIAN system of programs are defined by using the symbolic Z-matrix notation [1]. The Z-matrix specification must take into account the decoupling of the flexible methyl rotation mode from the harder HCH bend modes. Thus, while there are different possibilities, the following Z-matrix constitutes an appropriate choice:

```
C1
O2      C1    C1O2
H3      O2    H3O2    C1    HOC
H4      C1    H4C1    O2    H4CO    H3    PHI   0
H5      C1    HC1     O2    HCO     H4    HCH  +1
H6      C1    HC1     O2    HCO     H4    HCH  -1
```

Assigning values to the variable related with the methyl torsion (PHI), one passes from the staggered conformation to the eclipsed one.

The selection of the initial values for stretching and bending variables can be made in consonance with the values recommended in ref. 22. However, it is frequent (and convenient) in conformational analysis to adopt a previously optimized conformation as an initial estimate for the geometries of other conformations. In the case of methanol, using the optimized geometry of the staggered conformer as initial geometry for the eclipsed conformation, the number of optimization cycles reduces to 5, with the consequent decrease in the computational time.

The results of the calculations are summarized in Table 1, where the experimental R_s (average) geometry [20] (obtained from the analysis of the microwave spectra of several isotopically substituted molecules) is also presented for comparison. It is evident from this table that the *ab initio* equilibrium geometrical parameters (R_e) agree with experiment within the experimental errors, except for C-O and C-O-H coordinates. The less reasonable agreement obtained for the coordinates involving the oxygen atom could be anticipated considering the small dimension of the basis set, as well as the lack of polarization functions. These latter functions are usually required to obtain reasonably accurate values for coordinates involving strong electronegative atoms. It should be pointed out that the R_e and R_s structures are not directly comparable structures, as the experimental one is affected by vibrational contributions. Moreover, in the experimental work [20] it was assumed local C_{3v} methyl symmetry.

TABLE 1 - Experimental and calculated (STO-3G) molecular geometries and relative energies for the staggered and eclipsed conformations of methanol.[a]

Parameter	Exp. [20,23]	Ab initio	
	staggered	staggered	eclipsed
C-H4	109.36 (32)	109.14	109.19
C-H5,6	109.36 (32)	109.51	109.42
O-H	94.51 (34)	99.11	98.91
C-O	142.46 (24)	143.30	143.88
H4-C-O	110.30 (70)	107.63	113.05
H5,6-C-O	110.30 (70)	112.37	109.98
H5-C-H6	108.63 (70)	108.20	107.88
H4-C-H5,6	108.63 (70)	108.12	107.91
COH	108.53 (48)	103.85	104.65
ΔE	4.48	0.00	8.42

[a]Energies in kJ/mol; bond lengths in pm; angles in degrees. The numbers in parentheses are experimental errors in units of the last digit.

Looking at the calculated relative energies of the conformations (cf. Table 1), and using simple stereochemical considerations, it could easily be predicted that the staggered form corresponds to the minimum energy conformation, while the eclipsed conformation should correspond to a transition structure. However, by taking into account only the calculated relative energies, the nature of the critical points corresponding to these conformations on the PES could not be determined. The characterization of a critical point on the PES can only be made by analyzing the local Hessian. Thus, the calculation of the vibrational frequencies constitutes an adequate way for critical point characterization. Transition structures must have one imaginary frequency which is associated with the normal mode that transforms the reactants to the products, or, as in the present case, with the normal mode associated with the internal rotation. By contrast, a minimum possesses 3N-6 (3N-5 for linear molecules) nonzero positive frequencies (where N is the number of nuclei). The results obtained for methanol are presented in Table 2. The presence of one imaginary frequency for the eclipsed conformer confirms that it is a transition structure.

The comparison between calculated frequencies and the experimental values [24], also presented in Table 2, reveals the usual overestimation of the SCF *ab initio* values.

420

Despite this disagreement with experiment, *ab initio* computation of frequencies can be very useful in vibrational spectroscopy. Moreover, accurate methods for systematically scaling the theoretical force constants to reproduce the experimental frequencies have been developed [25].

TABLE 2. Experimental and calculated vibrational frequencies for methanol (minimum energy conformation).[a]

Description	Symmetry	Exp. [24]	Calculated
OH stretch	A'	3681	4228
CH$_3$ asymmetric stretch	A'	3000	3716
CH$_3$ asymmetric stretch	A"	2960	3679
CH$_3$ symmetric stretch	A'	2844	3516
CH$_3$ asymmetric bend	A'	1477	1831
CH$_3$ asymmetric bend	A"	1477	1814
CH$_3$ symmetric bend	A'	1455	1769
COH bend	A'	1345	1723
CH$_3$ rock	A'	1060	1308
CH$_3$ rock	A"	---	1209
C-O stretch	A'	1033	1317
C-O torsion	A"	270	398

[a]Frequencies in cm^{-1}. Experimental values correspond to gas-phase frequencies in the infrared spectrum. The calculated frequencies for the corresponding normal modes in the eclipsed conformation are 397i, 1258, 1307, 1320, 1698, 1772, 1814, 1830, 3522, 3691, 3706, and 4250 cm^{-1}, respectively.

I.12-15 MØLLER-PLESSET METHODS

It is well known that the major deficiency of the Hartree-Fock model is its incapacity to account for the correlation effect associated with the motions of electrons of opposite spin. In principle, this contribution can be computed using a full configuration interaction (CI) method, where the wavefunction corresponds to a variationally optimized combination of all possible electronic configurations. However, the application of this method to molecules of chemical interest can involve a number of configurations which rapidly increases with the molecular size, leading to serious computational problems.

The perturbation approach originally proposed by Møller and Plesset [26] (MP method) provides a simpler and less time-consuming scheme for computing the electron correlation effect. Within this scheme, the full CI Hamiltonian is treated as a perturbed Hamiltonian and the energy and the wavefunction are expanded in power series following the Rayleigh-Schrödinger perturbation theory.

The perturbation series can be truncated to various orders and one indicates the accuracy of MP methods applied within the Restricted Hartee-Fock (RHF) scheme by referring to the highest-order term allowed in the energy expansion. Thus a truncation to second-order corresponds to an MP2 approach, to third-order to an MP3 approach and so forth [27]. MP theory may also be used in the spin-Unrestricted Hartree-Fock (UHF) model. In this case, second- and third-order approximations of MP theory are indicated as UMP2 and UMP3.

The aim of the projects I.12 to I.15 is to examine the convergence of MP perturbation series for very simple molecules (Projects I.12 and I.13) and to analyze the effect that the introduction of correlation energy has on the description of simple organic reactions (Project I.14) and on geometry optimizations (Project I.15).

As an illustration, we will discuss in detail project I.12. In this proposed problem, the students examine the convergence of the MP series in the case of the molecule H_2O in three different C_{2v} geometries characterized by the following values of the geometrical parameters:

(1) $\phi(HOH)=\phi_e(HOH)$, $R(OH)=R_e$.

(2) $R(OH)=1.5\ R_e$.

(3) $R(OH)=2.0\ R_e$.

Where $\phi_e(HOH)$ and R_e are the HOH bond angle and the OH bond length at the equilibrium geometry ($R_e=0.97821$Å, $\phi_e(HOH)=107.6°$). Furthermore, the MP results must be compared with the values provided by other methods like CISD (CI including all single and double substitution), QCI (quadratic CI) and CCD (coupled-cluster using double substitutions). It is suggested to perform computations at different geometries since the weight of the dominant configuration in a full CI computation changes significantly with the change of the OH bond length. Accordingly, these computations could show the effect of the different configuration mixing on the convergence of the MP series. To this purpose the students perform, for the three above geometries and using a 6-21G basis set, the following computations: RHF, RHF/MP2, RHF/MP3, RHF/MP4, CISD, QCI, and CCD. All these methods are available in the Gaussian 90 series of programs [1]. In order to interpret the results more easily, it is suggested to plot in a diagram the energy values obtained at the various MPn computational levels, for each geometry, versus n, where n corresponds to the order of the MP series (n=1 for the Hartree-Fock SCF approximation).

I.17-21 THE MULTICONFIGURATION SCF (MCSCF) METHOD

Reactivity problems, which usually describe the rupture and the formation of chemical bonds, are the typical examples of the breaking down of the Hartree-Fock approximation. To deal with this type of problems, where one has degeneracy or near-degeneracy of several configurations in the region of incipient bonding, a Multiconfiguration SCF (MCSCF) [28, 29] approach is required. This method uses a limited CI expansion for the wavefunction that includes the set of near-degenerate configurations. The procedure consists of optimizing both the CI coefficients as well as the molecular orbital coefficients from which the various configurations are obtained. In this method it is convenient to divide the total orbital space in three subspaces:

(i) the subspace of core-orbitals which remain doubly occupied during the computation;

(ii) the subspace of valence or active orbitals which are characterized by a variable occupancy;

(iii) the subspace of virtual-orbitals which have zero occupancy.

At this school several computational exercises concerning to that application of MCSCF method to organic reactivity problems have been proposed to the students. The MCSCF code is implemented in the Gaussian 90 series of programs [1] and thus automatically available using the appropriate keyword and options. The aim of the various exercises is to make people more familiar with the fundamental concepts needed to perform MCSCF computations and with the strategy necessary to investigate a reaction potential energy surface (PES) at this computational level.

As mentioned before, the investigation of a PES involves the location of the important critical points (intermediates, transition structures), the geometry optimization of the molecular structures corresponding to these critical points, and their characterization by means of the eigenvalues of the Hessian matrix.

Let us consider Project I.17 as an illustrative example of these ideas. The aim of this project is to make the students familiar with the concept of "active space" (from which the various configurations of the CI expansion are generated) and with the criteria which should be followed to choose an appropriate set of "active orbitals". Usually the choice of the active space is made in such a way that it includes all the orbitals which are involved in the bond-making and bond-breaking. From a practical point of view it is convenient to perform, using a minimal basis set (STO-3G basis [30] available in Gaussian 90), an MCSCF computation on reactant molecules at infinite separation. In this type of computation, it is useful to employ the MO's obtained from one of the semiempirical methods available in Gaussian 90 (Extended Huckel, CNDO, INDO) as starting orbitals for the MCSCF procedure. Since at infinite separation the orbitals are still localized on the reactant molecules, it becomes easier in this way to recognize the orbitals which are appropriate for the valence space [29]. To illustrate this point, the following examples of prototype organic reactions have been proposed to the students in the first project:

(a) the [2+2] cycloaddition reaction between two ethylene molecules to give cyclobutane [31].

(b) the 1,3-dipolar cycloaddition reaction which involves fulminic acid and acetylene and gives the product isoxazole [29].

(c) the [4+2] cycloaddition reaction between 1,3-butadiene and ethylene (Diels-Alder reaction) [32].

The above three examples can be described using an active space of four orbitals with four electrons coupled in such a way to have an overall singlet spin. In the first case (ethylene+ethylene), for instance, the active space must be able to describe correctly the rupture of two π ethylene bonds and the formation of the two new σ bonds involved in the cyclization. Thus, the orbitals which must be considered are the two π (HOMO) and the two π^* (LUMO) orbitals of the two isolated ethylenes.

The second case is slightly more complicated. The fulminic acid can be thought of as having two allyl-like π systems of three π orbitals: the first one is an in-plane π orbital system and is associated with the incipient σ bonds of the cyclic product; the second one is an out-of-plane π system which takes part in the formation of the final delocalized π orbitals. In a similar way the acetylene has two sets of orthogonal ethylene-type π orbitals, one in the plane and one out of the plane of the two approaching molecules. To describe with equivalent accuracy the reactants, the products and the transient species involved in the bond breaking and forming, the minimal active space which is required consists of four in-plane π orbitals, i.e., the HOMO and LUMO of each isolated molecule. The third case is now evident. Among the four π orbitals of butadiene, we must consider the HOMO and the LUMO, together with the HOMO and LUMO of ethylene in order to form an appropriate active space suitable to describe the formation of the cyclic structure.

For each of the three above reactions the students have to consider some of the possible approaches between the reactant molecules. For each approach, in order to recognize the orbitals of the valence space, the students should perform a semiempirical computation on the molecules at large separation using standard values for the various geometrical parameters. After having defined the active space, the semiempirical orbitals can be used as starting orbitals for an MCSCF computation at large separation. The MCSCF orbitals obtained in this way are now a good starting point for MCSCF computations near the equilibrium configurations and for geometry optimizations.

Regarding other reactions, for the cases (a) and (c) mentioned before the students must compute, using the MCSCF gradient method [33] implemented in Gaussian 90, the optimum geometries of the possible transition structures and intermediates associated with the various approaches between the reactant molecules. As a final step, in order to characterize the optimized structures, one should compute the matrix of energy-second derivatives (Hessian matrix) by finite differences with respect to the various geometrical parameters and find the corresponding eigenvalues and eigenvectors.

II.4 A STUDY OF ETHYLENE GLYCOL USING MOLECULAR MECHANICS

In conformational studies, one would like to know the possible conformations that a molecule may have and the relative energy differences between the different conformations. Generally speaking, the conformation with the lowest steric energy will be the conformation with the atoms that are farthest apart, while maintaining reasonable bond lengths and valence angles. An example is n-butane with the anti conformation that is at the global minimum. By analogy, one might naïvely expect the all-trans anti conformation of ethylene glycol to be also the global minimum conformation. One would expect the conformation with the O-C-C-O dihedral angle between 0° to 60° (the gauche conformer) to be stable due to the stabilizing influence of hydrogen bonding. We shall examine several aspects of the conformational behavior of ethylene glycol using Allinger's MM2 force field [34,35].

The easiest conformation to generate is the all-trans form. Once the trans conformer has been obtained, one may use the dihedral angle driver option in the MM2 program to create the gauche conformer. (Use of the dihedral angle driver option is studied in previous exercises, II.1 and II.2.) When the structures of both conformations have been optimized, the difference in the final steric energies indicates the gauche conformer to be 1.512 kcal/mole lower in energy than the all-trans conformer. In the gauche conformer one hydroxyl group should be directed towards the other oxygen atom to create the hydrogen bond. (Failure to create the hydrogen bond caused many participants to believe the trans conformer to be more stable.) An *ab initio* study of ethylene glycol using the 4-21G basis set located ten stable conformations [36]. The O-C-C-O dihedral angle for the gauche conformer is calculated to be 54.9° by the MM2 program and is 57.3° by the *ab initio* method [36].

One must be careful about which conformations are examined, otherwise one may obtain wrong answers. Like most computer programs, the MM2 molecular mechanics program should not be used as a "black box". In order to aid the user of the MM2 program, the output lists all bond lengths, valence and dihedral angles, and the distances between all atoms which are not covalently bonded. By examining the output from the MM2 program, one can find where the largest amount of strain is present in the structure. Many participants who studied compounds of interest to their own research falsely believed they had obtained an optimized structure when the program was successfully executed. An examination of the final output from the MM2 program revealed several valence angles that had an unacceptably high amount of steric energy. One must examine the output to ensure that the steric energy is equitably distributed throughout the structure and that the optimized structure has chemically reasonable bond lengths, valence and dihedral angles.

The population of each conformer in equilibrium at a given temperature can be found by use of the Boltzmann distribution. Ideally, one needs to know the Gibbs free energy, which is not equivalent to the final steric energy calculated by the MM2 program. Without knowing ΔH and ΔS, several approximations can be made. Since $\Delta(PV)=0$, $\Delta H=\Delta E$ is a good approximation in the gas phase. In order to get a reasonable value for ΔS, we need to know the number of conformations of the structure and the symmetry number of each conformation. The symmetry number of a molecule, σ, represents the number of indistinguishable orientations of the molecule. In other words, it is the number of different ways that a molecule can be rotated into a different configuration that is indistinguishable from the original configuration.

The gauche conformer exists in two conformations of identical steric energy, since the O-C-C-O dihedral angle may be either $55°$ or $-55°$. Consequently, there is an entropy contribution equal to R ln 2, where R is the ideal gas constant. The trans conformer also has a symmetry number $\sigma=2$, giving the same entropy contribution R ln 2. The ΔG at T=298 K for this equilibrium is given by: $\Delta G=\Delta E-2RT$ ln $2=(1.512-0.822)$ kcal/mole=0.690 kcal/mole .

Another option in the MM2 program is to calculate the heat of formation of ethylene glycol. The algorithm used by the MM2 program is known as the "bond increment method" and is described in refs. [37-39]. It is given by : $H_f = S_E$ + bond enthalpy + PFC, where S_E is the final steric energy from the MM2 program, and PFC is the partition function contribution. Using the values from the output one finds the calculated heat of formation to be -92.93 kcal/mole, which is in agreement with the experimental value of -92.6±1.5 kcal/mole [40]. One word of caution concerning this option: calculation the heat of formation is meaningful only for ground state geometries and is useless for transition structures, as well as being useless for structures that have not been fully optimized.

Acknowledgments

We would like to thank Profs. H.B. Schlegel (Department of Chemistry, Wayne State University, Detroit, USA) and B.C. Garrett (Pacific Northwest Laboratories,

Molecular Science Research Center, Richland, USA) for suggesting some of the problems included in this chapter.

References

[1]   Frisch M.J.; Binkley J.S.; Schlegel H.B.; Raghavachari K.; Melius C.F.; Martin R.L.; Stewart J.J.P.; Bobrowicz F.W.; Rohlfing C.M.; Kahn L.R.; DeFrees D.J.; Seeger R.; Whiteside R.A.; Fox D.J.; Fluder E.M.; Pople J.A. GAUSSIAN 90, Gaussian Inc., Pittsburgh, 1990.

[2]   (a) Clementi E.; Raimondi D.L. *J. Chem. Phys.* **1963**, *38*, 2686. (b) Hehre W.J.; Stewart R.F.; Pople J.A. *J. Chem. Phys.* **1969**, *51*, 2657. (c) Hehre W.J.; Ditchfield R.; Stewart R.F.; Pople J.A. *J. Chem. Phys.* **1970**, *52*, 2657. (d) Pietro W.J.; Levi B.A.; Hehre W.J.; Stewart R.F. *Inorg. Chem.* **1980**, *19*, 2225. (e) Pietro W.J.; Hehre W.J. *J. Comp. Chem.* **1983**, *4*, 241. (f) Clementi E.; Raimondi D.L. *J. Chem. Phys.* **1963**, *38*, 2686. (g) Pietro W.J.; Blurock E.S.; Hout R.F.; Hehre W.J.; DeFrees D.J.; Stewart R.F. *Inorg.Chem.* **1981**, *20*, 3650.

[3]   Collins J.B.; Schleyer P.v.R.; Binkley J.S.; Pople J.A. *J. Chem. Phys.* **1976**, *64*, 5142.

[4]   (a) Binkley J.S.; Pople J.A.; Hehre W.J. *J. Am. Chem. Soc.* **1980**, *102*, 939. (b) Gordon M.S.; Binkley J.S.; Pople J.A.; Pietro W.J.; Hehre W.J. *J. Am. Chem. Soc.* **1982**, *104*, 2797. (c) Dobbs K.D.; Hehre W.J. *J. Comp. Chem.* **1986**, *7*, 359.

[5]   (a) Binkley J.S.; Whiteside R.A.; Krishnan R.; Seeger R.; DeFrees D.J.; Schlegel H.B.; Topiol S.; Kahn L.R.; Pople J.A. GAUSSIAN 80, Carnegie-Mellon Publishing Unit, Pittsburgh, 1980. (b) Pietro W.J.; Francl M.M.; Hehre W.J.; DeFrees D.J.; Pople J.A.; Binkley J.S. *J. Am. Chem. Soc.* **1982**, *104*, 5039.

[6]   (a) Ditchfield R.; Hehre W.J.; Pople J.A. *J. Chem. Phys.* **1971**, *54*, 724. (b) Hehre W.J.; Ditchfield R.; Pople J.A. *J. Chem. Phys.* **1972**, *56*, 2257. (c) Hehre W.J.; Pople J.A. *J. Chem. Phys.* **1972**, *56*, 4233. (d) Hehre W.J.; Lathan W.A. *J. Chem. Phys.* **1972**, *56*, 5255. (e) Hariharan P.C.; Pople J.A. *Theor. Chim. Acta* **1973**, *28*, 213. (f) Dill J.D.; Pople J.A. *J. Chem. Phys.* **1975**, *28*, 213. (g) Binkley, J.S.; Pople J.A. *J. Chem. Phys.* **1977**, *66*, 879. (h) Francl M.M.; Pietro W.J.; Hehre W.J.; Binkley J.S.; Gordon M.S.; DeFrees D.J.; Pople J.A. *J. Chem. Phys.* **1982**, *77*, 3654.

[7]   Krishnan R.; Binkley J.S.; Seeger R.; Pople J.A. *J. Chem. Phys.* **1980**, *72*, 650.

[8]   (a) Clark T.; Chandrasekhar J.; Spitznagel G.W.; Schleyer P.v.R. *J. Comp. Chem.* **1983**, *4*, 294. (b) G.W. Spitznagel; Clark T.; Schleyer P.v.R.; Hehre W.J.; *J. Comp. Chem.* **1987**, *8*, 1109.

[9]   Chandrasekhar J.; Andrade J.G.; Schleyer P.v.R. *J. Am. Chem. Soc.* **1981**, *103*, 5609.

[10]   Frisch M.J.; Pople J.A.; Binkley J.S. *J. Chem. Phys.* **1984**, *80*, 3265.

[11]   Truhlar D.G.; Isaacson A.D.; Garrett B.C., *Theory of chemical reaction dynamics*, vol. IV, Baer M. (ed.), CRC Press, Boca Ratón, 1985.

[12]   Miller W.H.; Handy N.C.; Adams J.E. *J. Chem. Phys.* **1980**, *72*, 99.

[13]   Isaacson, A.D.; Truhlar, D.G.; Rai, S.N.; Steckler, R.; Hancock, G.C.; Garrett, B.C.; Redmon, M.J. *Comp. Phys. Comm.* **1987**, *47*, 91.

[14]   (a) Mezey P.G. *Int. J. Quantum Chem., Quantum Biol. Symp.* **1986**, *14*, 133. (b) Mezey P.G. *J. Comp. Chem.* **1987**, *8*, 462.

[15]   Arteca G.A.; Mezey P.G. *J. Comp. Chem.* **1988**, *7*, 554.

[16]   Walker P.D.; Arteca G.A.; Mezey P.G. *J. Comp. Chem.* **1990**, *9*, in press.

[17]   Mezey P.G. *J. Math. Chem.* **1988**, *2*, 299.

[18]   Basharov M.A,;Volkenshtein M.V.; Golovanov I.B.; Sobolev V.M *Zhur. Strukt. Khim.* **1984**, *25*, 36.

[19]   Morokuma K.; Umeyama H. *Chem. Phys. Letters* **1977**, *49* , 333.

[20]   Lees R.M.; Baker J.G. *J. Chem. Phys.* **1968**, *48* , 5299.

[21]   Teixeira-Dias J.J.C.; Fausto R. *J. Mol. Struct.* **1986**, *144* , 199.

[22]   Peterson M.R.; Csizmadia I.G. *J. Mol. Struct. (Theochem)* **1985**, *123*, 399.

[23]   Ivash E.V.; Dennison D.M. *J. Chem. Phys.* **1953**, *21*, 1804.

[24]   Shimanouchi T., *Tables of molecular vibrational frequencies*, Vol.1, NSRDS-NBS 6, U.S. Govt. Printing Office, Washington, 1967.

[25]   Fogarasi G.; Pulay P., *Ab initio calculation of force fields and vibrational spectra*, in*Vibrational Spectra and Structure*, Vol.14, Durig J.R. (ed.), Elsevier, Amsterdam, 1984 (and references therein).

[26]   Møller C.; Plesset M.S. *Phys. Rev.* **1934**, *46*, 618.

[27]   (a) Pople J.A.; Binkley J.S.; Seeger R. *Int. J. Quantum Chem.Symp.* **1976**, *10*, 1. (b) Pople J.A.; Seeger R.; Krishan R. *Int. J. Quantum Chem.Symp.* **1977**,*11*, 149

[28]   Hegarty D.; Robb M.A. *Mol. Phys.* **1979**, *38* , 1795.

[29]   Bernardi F.; Bottoni A.; McDouall J.J.W.; Robb M.A.; Schlegel H.B. *Faraday Symp. Chem. Soc.* **1984**, *19*, 137.

[30]   Hehre W.J.; Stewart R.F.; Pople J.A. *J. Chem. Phys.* **1969**, *51*, 2657.

[31]   Bernardi F.; Bottoni A.; Robb M.A.; Schlegel H.B.; Tonachini G. *J. Am. Chem. Soc.* **1985**, *107*, 2260.

428

[32]   (a) Bernardi F.; Bottoni A.; Robb M.A.; Field M.J.; Hillier I.H.; Guest M.F.   J. *Chem. Soc., Chem. Commun.* **1985**, 1052. (b)   Bernardi F.; Bottoni A.; Venturini A.,; Robb M.A.; Field M.J.; Hillier I.H.; Guest M.F *J. Am. Chem. Soc.* **1988**, *110*, 3050.

[33]   Schlegel H.B.; Robb M.A.   *Chem. Phys. Letters* **1982**, *93*, 43.

[34]   Allinger N.L.   *J. Am. Chem. Soc.* **1977**, *99*, 8127.

[35]   Allinger N.L.; Chang S.H.M.; Glaser D.H.; Honig H.   *Isr. J. Chem.* **1980**, *20*, 51.

[36]   Van Alsenoy C.; Van Den Enden L.; Schäfer L.   *J. Mol. Struct. (Theochem)* **1984**, *108* , 121.

[37]   Allinger N.L.; Hirsch J.A.; Miller M.A.,;Tyminski I.J.; Van Catledge F.A.   *J. Am. Chem. Soc.* **1968**, *90*, 1199.

[38]   Pedley J.B.; Naylor R.D.; Kirby S.P., *Thermochemical Data of Organic Compounds*, Second Edition, Chapman and Hall, London, 1986.

[39]   Cox J.D.; Pilcher G., *Thermochemistry of Organic and Organometallic Compounds*, Academic Press, London, 1970.

[40]   Gardner P.J.; Hussain K.S.   *J. Chem. Thermodynamics* **1972**, *4*, 819, as quoted in ref. 38.

Subject Index

434